Lecture Notes in Physics

Springer

Berlin
Heidelberg
New York
Barcelona
Hong Kong
London
Milan
Paris
Singapore
Tokyo

The Editorial Policy for Proceedings

The series Lecture Notes in Physics reports new developments in physical research and teaching – quickly, informally, and at a high level. The proceedings to be considered for publication in this series should be limited to only a few areas of research, and these should be closely related to each other. The contributions should be of a high standard and should avoid lengthy redraftings of papers already published or about to be published elsewhere. As a whole, the proceedings should aim for a balanced presentation of the theme of the conference including a description of the techniques used and enough motivation for a broad readership. It should not be assumed that the published proceedings must reflect the conference in its entirety. (A listing or abstracts of papers presented at the meeting but not included in the proceedings could be added as an appendix.)

When applying for publication in the series Lecture Notes in Physics the volume's editor(s) should submit sufficient material to enable the series editors and their referees to make a fairly accurate evaluation (e.g. a complete list of speakers and titles of papers to be presented and abstracts). If, based on this information, the proceedings are (tentatively) accepted, the volume's editor(s), whose name(s) will appear on the title pages, should select the papers suitable for publication and have them refereed (as for a journal) when appropriate. As a rule discussions will not be accepted. The series editors and Springer-Verlag will normally not interfere with the detailed editing except in fairly obvious cases or on technical matters.

Final acceptance is expressed by the series editor in charge, in consultation with Springer-Verlag only after receiving the complete manuscript. It might help to send a copy of the authors' manuscripts in advance to the editor in charge to discuss possible revisions with him. As a general rule, the series editor will confirm his tentative acceptance if the final manuscript corresponds to the original concept discussed, if the quality of the contribution meets the requirements of the series, and if the final size of the manuscript does not greatly exceed the number of pages originally agreed upon. The manuscript should be forwarded to Springer-Verlag shortly after the meeting. In cases of extreme delay (more than six months after the conference) the series editors will check once more the timeliness of the papers. Therefore, the volume's editor(s) should establish strict deadlines, or collect the articles during the conference and have them revised on the spot. If a delay is unavoidable, one should encourage the authors to update their contributions if appropriate. The editors of proceedings are strongly advised to inform contributors about these points at an early stage.

The final manuscript should contain a table of contents and an informative introduction accessible also to readers not particularly familiar with the topic of the conference. The contributions should be in English. The volume's editor(s) should check the contributions for the correct use of language. At Springer-Verlag only the prefaces will be checked by a copy-editor for language and style. Grave linguistic or technical shortcomings may lead to the rejection of contributions by the series editors. A conference report should not exceed a total of 500 pages. Keeping the size within this bound should be achieved by a stricter selection of articles and not by imposing an upper limit to the length of the individual papers. Editors receive jointly 30 complimentary copies of their book. They are entitled to purchase further copies of their book at a reduced rate. As a rule no reprints of individual contributions can be supplied. No royalty is paid on Lecture Notes in Physics volumes. Commitment to publish is made by letter of interest rather than by signing a formal contract. Springer-Verlag secures the copyright for each volume.

The Production Process

The books are hardbound, and the publisher will select quality paper appropriate to the needs of the author(s). Publication time is about ten weeks. More than twenty years of experience guarantee authors the best possible service. To reach the goal of rapid publication at a low price the technique of photographic reproduction from a camera-ready manuscript was chosen. This process shifts the main responsibility for the technical quality considerably from the publisher to the authors. We therefore urge all authors and editors of proceedings to observe very carefully the essentials for the preparation of camera-ready manuscripts, which we will supply on request. This applies especially to the quality of figures and halftones submitted for publication. In addition, it might be useful to look at some of the volumes already published. As a special service, we offer free of charge LaTeX and TeX macro packages to format the text according to Springer-Verlag's quality requirements. We strongly recommend that you make use of this offer, since the result will be a book of considerably improved technical quality. To avoid mistakes and time-consuming correspondence during the production period the conference editors should request special instructions from the publisher well before the beginning of the conference. Manuscripts not meeting the technical standard of the series will have to be returned for improvement.

For further information please contact Springer-Verlag, Physics Editorial Department II, Tiergartenstrasse 17, D-69121 Heidelberg, Germany

Friedrich W. Hehl   Claus Kiefer
Ralph J.K. Metzler (Eds.)

Black Holes:
Theory and Observation

Proceedings of the 179th W.E. Heraeus Seminar
Held at Bad Honnef, Germany, 18–22 August 1997

 Springer

Editors

Friedrich W. Hehl
Ralph J.K. Metzler
Institut für Theoretische Physik
Universität zu Köln
D-50923 Köln, Germany

Claus Kiefer
Fakultät für Physik
Universität Freiburg
Hermann-Herder-Strasse 3
D-79104 Freiburg, Germany

Library of Congress Cataloging-in-Publication Data.

Die Deutsche Bibliothek - CIP-Einheitsaufnahme

Black holes: theory and observation / 179. WE-Heraeus-Seminar,
held at Physikzentrum Bad Honnef, Germany, 18 - 22 August 1997 /
F. Hehl ... (ed.). - Berlin ; Heidelberg ; New York ; Barcelona ; Hong
Kong ; London ; Milan ; Paris ; Singapore ; Tokyo : Springer, 1998
  (Lecture notes in physics ; Vol. 514)
  ISBN 3-540-65158-6

ISSN 0075-8450
ISBN 3-540-65158-6 Springer-Verlag Berlin Heidelberg New York

Typesetting: Camera-ready by the authors/editors
Cover design: *design & production*, Heidelberg

SPIN: 10644238   55/3144 - 5 4 3 2 1 0 – Printed on acid-free paper

It is surprising how soon after the publication of Newton's Principia in 1687 somewhat realistic models of the universe with spherical 'galaxies' were developed, as is documented in Thomas Wright's Plate XXXI of his book on "An Original Theory or New Hypothesis of the Universe" (1750)[1]:

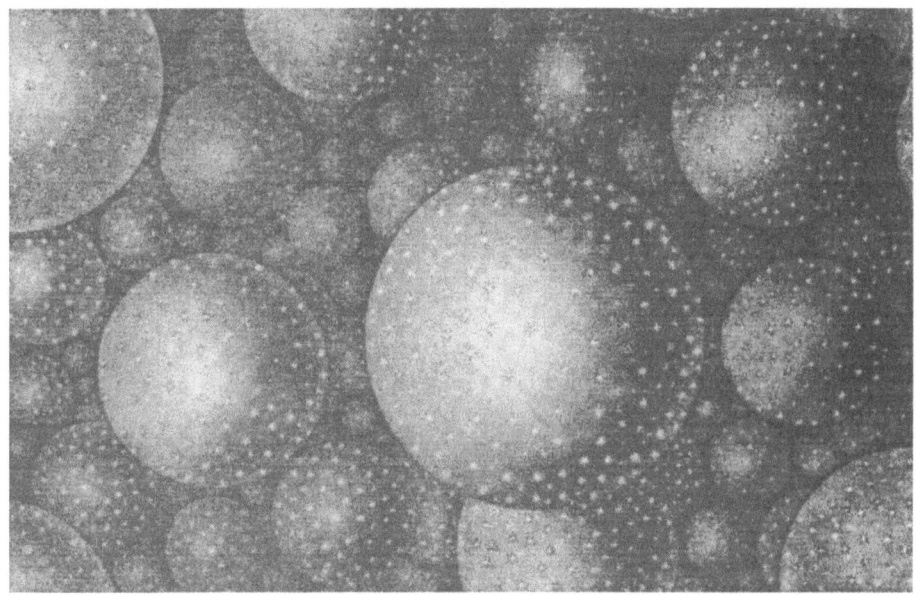

And already in 1784, John Michell, in a letter to Henry Cavendish[2], gave the first description of a black hole:

> *... if the semi-diameter of a sphære of the same density with the sun were to exceed that of the sun in the proportion of 500 to 1, a body falling from an infinite height towards it, would have acquired at its surface a greater velocity than that of light, and consequently, supposing light to be attracted by the same force in proportion to its vis inertiæ, with other bodies, all light emitted from such a body would be made to return towards it, by its own proper gravity...*

It is amusing to note that the black hole in the center of our galaxy (see Chapter 3 by A. Eckart & R. Genzel) has the same linear dimension, in orders of magnitude, as the black star model of Michell. The Michell hypothesis was soon followed by a similar one of Pierre-Simon de Laplace, see Chapter 1 by J.-P. Luminet.

---

[1] Facsimile reprint, with an introduction by Michael A. Hoskin. Macdonald, London and American Elsevier Inc., New York 1971.

[2] Phil. Trans. Roy. Soc. London, **74** (1784) Part I, page 16.

Preface

Black holes are among the most fascinating objects in Nature. Being originally only of interest to theoretical physicists, they are now – thirty years after the name "black holes" was introduced – also an important target in observational astronomy.

The purpose of the 179th WE-Heraeus Seminar, which we organized in Bad Honnef in August 1997, was to give an overview of our present knowledge on black holes. It was our aim to cover all aspects of black hole physics: classical and quantum aspects, astrophysical, observational, and numerical aspects. From this wide range of topics it becomes clear that black hole physics constitutes an interdisciplinary subject par excellence. We hope that the consideration of all these aspects in one book will give a useful addition to other, more specialized, volumes.

We thank all speakers for their work in preparing and holding their lectures, leading to a most successful meeting. Most of the lectures can be found in the present volume, and we are grateful for the willingness of the speakers to write up their talks. We also include some articles which are not based on lectures given at the seminar, but which constitute a useful supplement to them. In addition, the reader will also find short statements which were made at the panel discussion on "The definite proofs of the existence of black holes". The general consensus from this discussion was that the existence of black holes has now practically been proved with certainty.

Last but not least we want to thank the WE-Heraeus Foundation for the generous support without which the seminar in this form could not have taken place.

Cologne and Freiburg, September 1998

Friedrich W. Hehl
Claus Kiefer
Ralph J.K. Metzler

Table of Contents

Part IV: Beyond Classical General Relativity

Part V: Thermodynamics

List of Contributors

Sandip K. Chakrabarti
S.N. Bose National Centre
for Basic Sciences
Dept. of Theoretical Astrophysics
Calcutta 700091, India

Werner Collmar
Max-Planck-Institut für
Extraterrestrische Physik
Postfach 16 03
D-85748 Garching, Germany

Andreas Eckart
Max-Planck-Institut für
Extraterrestrische Physik
Postfach 16 03
D-85748 Garching, Germany

Alberto A. García
Centro de Investigacion y de
Estudios Avanzadros del I.P.N.
Departamento de Física
Apartado Postal 14-740
México D.F. 07000, Mexico

Reinhard Genzel
Max-Planck-Institut für
Extraterrestrische Physik
Postfach 16 03
D-85748 Garching, Germany

Domenico Giulini
Institut für Theoretische Physik
Universität Zürich
Winterthurer Str. 190
CH-8057 Zürich, Switzerland

Friedrich W. Hehl
Institut für Theoretische Physik
Universität zu Köln
D-50923 Köln, Germany

Markus Heusler
Institut für Theoretische Physik
Universität Zürich
Winterthurerstraße 190
CH-8057 Zürich, Switzerland

Gerard 't Hooft
Instituut voor Theoretische Fysica
Universiteit Utrecht
Princetonplein 5
3584 CC Utrecht, The Netherlands

Werner Israel
Department of Physics
and Astronomy
University of Victoria
P.O. Box 3055
Victoria, B.C.
V8W 3P6 Canada

Claus Kiefer
Fakultät für Physik
Universität Freiburg
Hermann-Herder-Str. 3
D-79104 Freiburg, Germany

Andrew R. Liddle
Astronomy Centre
University of Sussex
Falmer, Brighton BN1 9QJ
United Kingdom

Jean-Pierre Luminet
Observatoire de Paris-Meudon
Département d'Astrophysique
Relativiste et de Cosmologie
CNRS UPR-176
F-92195 Meudon Cedex, France

Alfredo Macías
Departamento de Física
Universidad Autonoma
Metropolitana
Av. Michocan La Purisima S/N
Apartado Postal 55-534
México D.F. 09340, Mexico

Bahram Mashhoon
Department of Physics
and Astronomy
University of Missouri-Columbia
Columbia, Missouri 65211, USA

Gernot Neugebauer
Theoretisch-Physikalisches Institut
Friedrich-Schiller-Universität Jena
Max-Wien-Platz 1
D-07743 Jena, Germany

Darío Núñez
Instituto de Ciencias Nucleares
Universidad Nacional Autónoma
de México
Apartado Postal 70–543
México D.F. 04510, Mexico

Yuri N. Obukhov
Department of Theoretical Physics
Moscow State University
117234 Moscow, Russia

Hernando Quevedo
Instituto de Ciencias Nucleares
Universidad Nacional Autónoma
de México
Apartado Postal 70–543
México D.F. 04510, Mexico

Cristopher S. Reynolds
JILA
University of Colorado
Boulder, CO 80309-0440, USA

Volker Schönfelder
Max-Planck-Institut für
Extraterrestrische Physik
Postfach 16 03
D-85748 Garching, Germany

Franz E. Schunck
Astronomy Centre
University of Sussex
Falmer, Brighton BN1 9QJ
United Kingdom

Edward Seidel
Max-Planck-Institut für Gravitation
Albert-Einstein-Institut
Schlaatzweg 1
D-14473 Potsdam, Germany

Roland Speith
Institut für Astronomie und Astrophysik
Abteilung Theoretische Astrophysik
Universität Tübingen
Auf der Morgenstelle 10
D-72076 Tübingen, Germany

Norbert Straumann
Institut für Theoretische Physik
Universität Zürich
Winterthurer Str. 190
CH-8057 Zürich, Switzerland

Daniel Sudarsky
Instituto de Ciencias Nucleares
Universidad Nacional Autónoma
de México
Apartado Postal 70–543
México D.F. 04510, Mexico

Jörn Wilms
Institut für Astronomie und Astrophysik
Abteilung Astronomie
Universität Tübingen
Waldhäuser Str. 64
D-72076 Tübingen, Germany

Andreas Wipf
Theoretisch-Physikalisches Institut
Friedrich-Schiller-Universität Jena
Max-Wien-Platz 1
D-07743 Jena, Germany

Part I

Overview

Black Holes: A General Introduction

Jean-Pierre Luminet

Observatoire de Paris-Meudon, Département d'Astrophysique Relativiste et de Cosmologie, CNRS UPR-176, F-92195 Meudon Cedex, France

Abstract. Our understanding of space and time is probed to its depths by black holes. These objects, which appear as a natural consequence of general relativity, provide a powerful analytical tool able to examine macroscopic and microscopic properties of the universe. This introductory article presents in a pictorial way the basic concepts of black hole's theory, as well as a description of the astronomical sites where black holes are suspected to lie, namely binary X–ray sources and galactic nuclei.

1 The Black Hole Mystery

Let me begin with an old Persian story. Once upon a time, the butterflies organized a summer school devoted to the great mystery of the flame. Many discussed about models but nobody could convincingly explain the puzzle. Then a bold butterfly enlisted as a volunteer to get a real experience with the flame. He flew off to the closest castle, passed in front of a window and saw the light of a candle. He went back, very excited, and told what he had seen. But the wise butterfly who was the chair of the conference said that they had no more information than before. Next, a second butterfly flew off to the castle, crossed the window and touched the flame with his wings. He hardly came back and told his story; the wise chairbutterfly said "your explanation is no more satisfactory". Then a third butterfly went to the castle, hit the candle and burned himself into the flame. The wise butterfly, who had observed the action, said to the others: "Well, our friend has learned everything about the flame. But only him can know, and that's all".

As you can guess, this story can easily be transposed from butterflies to scientists confronted with the mystery of black holes. Some astronomers, equipped with powerful instruments such as orbiting telescopes, make very distant and indirect observations on black holes; like the first butterfly, they acknowledge the real existence of black holes but they gain very little information on their real nature. Next, theoretical physicists try to penetrate more deeply into the black hole mystery by using tools such as general relativity, quantum mechanics and higher mathematics; like the second butterfly, they get a little bit more information, but not so much. The equivalent of the third butterfly would be an astronaut plunging directly into a black hole, but eventually he will not be able to go back and tell his story. Nevertheless, by using numerical calculations, such as those performed at the Observatoire de Meudon which I will show you later, outsiders can get some idea of what happens inside a black hole.

2   Physics of Black Holes

2.1   Light imprisoned

Let us begin to play like the second butterfly, and explore the black hole from the point of view of theoretical physics. An elementary definition of a black hole is a region of space-time in which the gravitational potential, GM/R, exceeds the square of the speed of light, c^2. Such a statement has the merit to be independent of the details of gravitational theories. It can be used in the framework of Newtonian theory. It also provides a more popular definition of a black hole, according to which any astronomical body whose escape velocity exceeds the speed of light must be a black hole. Indeed, such a reasoning was done two centuries ago by John Michell and Pierre-Simon de Laplace. In the *Philosophical Transactions of the Royal Society* (1784), John Michell pointed out that "if the semi diameter of a sphere of the same density with the sun were to exceed that of the sun in the proportion of 500 to 1, (...) all light emitted from such a body would be made to return towards it", and independently, in 1796, Laplace wrote in his *Exposition du Système du Monde*: "Un astre lumineux de même densité que la terre et dont le diamètre serait deux cents cinquante fois plus grand que celui du soleil, ne laisserait, en vertu de son attraction, parvenir aucun de ses rayons jusqu'à nous ; il est donc possible que les plus grands corps lumineux de l'univers soient, par cela même, invisibles". Since the density imagined at this time was that of ordinary matter, the size and the mass of the associated "invisible body" were huge - around 10^7 solar masses, corresponding to what is called today a "supermassive" black hole. Nevertheless, from the numerical figures first proposed by Michell and Laplace, one can recognize the well-known basic formula giving the critical radius of a body of mass M:

$$R_S = \frac{2GM}{c^2} \approx 3\frac{M}{M_\odot}\,\text{km}\,, \tag{1}$$

where $M_\odot$ is the solar mass. Any spherical body of mass M confined within the critical radius R_S must be a black hole.

These original speculations were quickly forgotten, mainly due to the development of the wave theory of light, within the framework of which no calculation of the action of gravitation on light propagation was performed. The advent of general relativity, a fully relativistic theory of gravity in which light is submitted to gravity, gave rise to new speculations and much deeper insight into black holes.

To pictorially describe black holes in space-time, I shall use light cones. Let me recall what a light cone is. In figure 1, a luminous flash is emitted at a given point of space. The wavefront is a sphere expanding at a velocity of about $c = 300\ 000\,\text{km/s}$, shown in a) at three successive instants. The light cone representation in b) tells the complete story of the wavefront in a single space-time diagramspacetime!diagram. As one space dimension is removed, the spheres become circles. The expanding circles of light generate a cone originating at the emission point. If, in this diagram, we choose the unit of length as 300 000 km and the unit of time as 1 second, all the light rays travel at 45°.

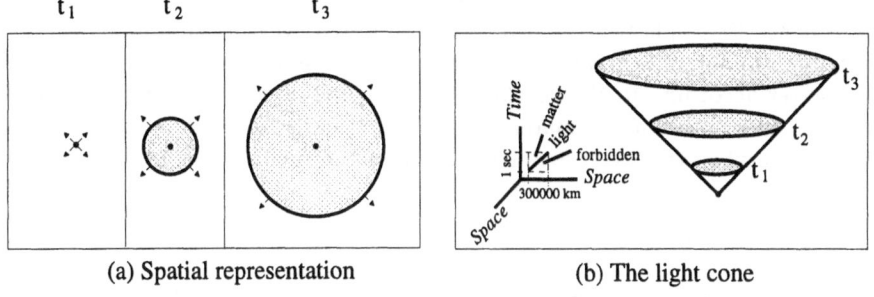

(a) Spatial representation   (b) The light cone

Fig. 1 The light cone.

The light cone allows us to depict the causal structure of any space-time. Take for instance the Minkowski flat space-time used in Special Relativity (figure 2). At any event E of space-time, light rays generate two cones (shaded zone). The rays emitted from E span the future light cone, those received in E span the past light cone. Physical particles cannot travel faster than light: their trajectories remain confined within the light cones. No light ray or particle which passes

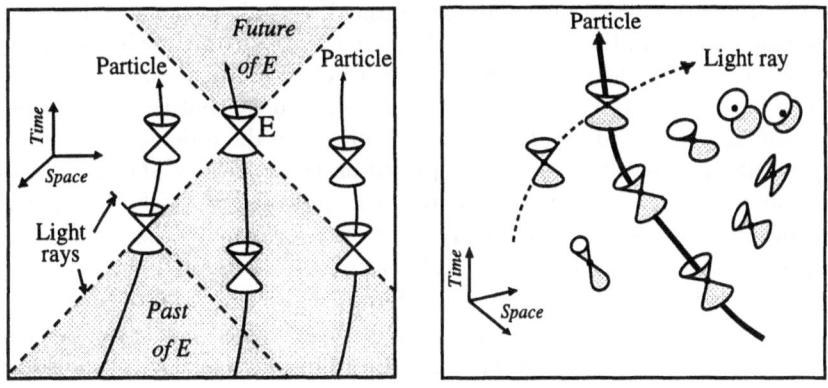

Figs. 2 and 3 The space-time continuum of Special Relativity and the soft space-time of General Relativity.

through E is able to penetrate the unshaded zone. The invariance of the speed of light in vacuum is reflected by the fact that all the cones have the same slope. This is because the space-time continuum of Special Relativity, free from gravitating matter, is flat and rigid. As soon as gravity is present, space-time is curved and Special Relativity leaves room to General Relativity. Since the Equivalence Principle states the influence of gravity on all types of energy, the light cones follow the curvature of space-time (figure 3). They bend and deform themselves according to the curvature. Special Relativity remains locally valid however: the worldlines of material particles remain confined within the light cones, even when the latter are strongly tilted and distorted by gravity.

2.2   Spherical collapse

Let us now examine the causal structure of space-time around a gravitationally collapsing star - a process which is believed to lead to black hole formation. Figure 4 shows the complete history of the collapse of a spherical star, from its initial contraction until the formation of a black hole and a singularity. Two space dimensions are measured horizontally, and time is on the vertical axis, measured upwards. The centre of the star is at $r = 0$. The curvature of space-time is visualized by means of the light cones generated by the trajectories of light rays. Far away from the central gravitational field, the curvature is so weak that the light cones remain straight. Near the gravitational field, the cones are distorted and tilted inwards by the curvature. On the critical surface of radius $r = 2M$, the cones are tipped over at 45° and one of their generators becomes vertical, so that the allowed directions of propagation of particles and electromagnetic waves are oriented towards the interior of this surface. This is the *event horizon*, the boundary of the black hole (grey region). Beyond this, the stellar matter continues to collapse into a singularity of zero volume and infinite density at $r = 0$. Once a black hole has formed, and after all the stellar matter has disappeared into the singularity, the geometry of space-time itself continues to collapse towards the singularity, as shown by the light cones.

The emission of the light rays at E_1, E_2, E_3 and E_4 and their reception by a distant astronomer at $R_1, R_2, R_3, \ldots$ well illustrate the difference between the *proper time*, as measured by a clock placed on the surface of the star, and the *apparent time*, measured by an independent and distant clock. The (proper) time interval between the four emission events are equal. The corresponding reception intervals become longer and longer. At the limit, light ray emitted from E_4, just when the event horizon is forming, takes an infinite time to reach the distant astronomer. This phenomenon of "frozen time" is just an illustration of the extreme elasticity of time predicted by Einstein's relativity, according to which time runs differently for two observers with a relative acceleration - or, from the Equivalence Principle, in different gravitational potentials. A striking consequence is that any outer astronomer will *never* be able to see the formation of a black hole. The figure 5 shows a picturesque illustration of frozen time. A spaceship has the mission of exploring the interior of a black hole – preferably a

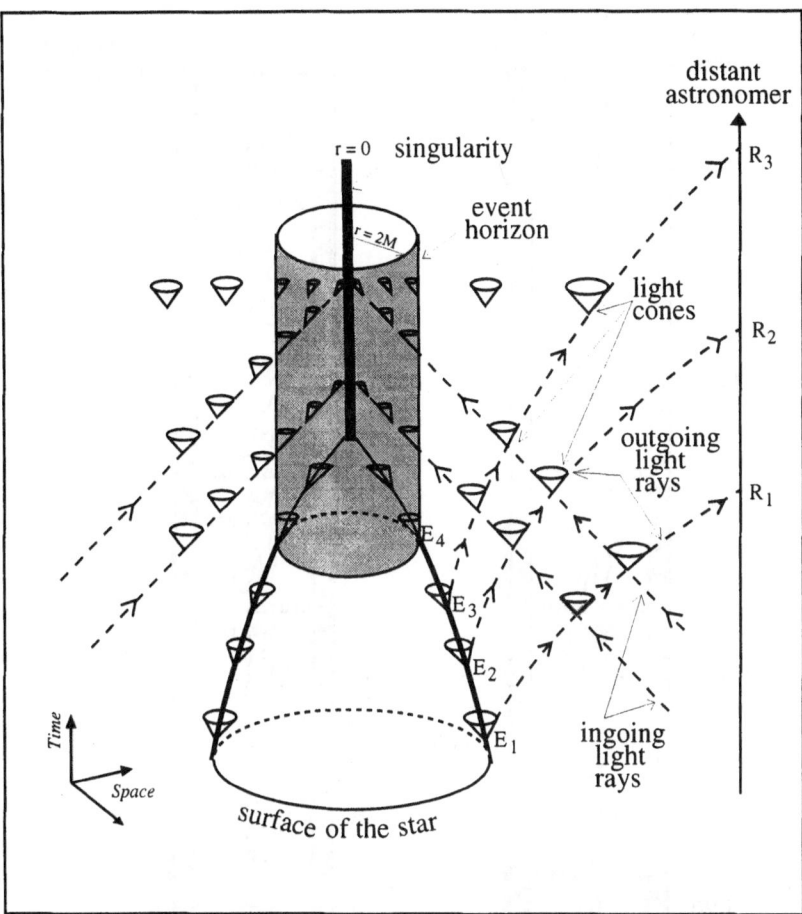

Fig. 4 A space-time diagram showing the formation of a black hole by gravitational collapse.

big one, so that it is not destroyed too quickly by the tidal forces. On board the ship, the commander sends a solemn salute to mankind, just at the moment when the ship crosses the horizon. His gesture is transmitted to distant spectators via television. The film on the left shows the scene on board the spaceship in proper time, that is, as measured by the ship's clock as the ship falls into the black hole. The astronaut's salute is decomposed into instants at proper time intervals of 0.2 second. Crossing of the event horizon (black holes have not a *hard* surface) is not accompanied by any particular event. The film on the right shows the scene received by distant spectators via television. It is also decomposed into intervals of apparent time of 0.2 second. At the beginning of his gesture, the salute is slightly slower than the real salute, but initially the delay is too small to be

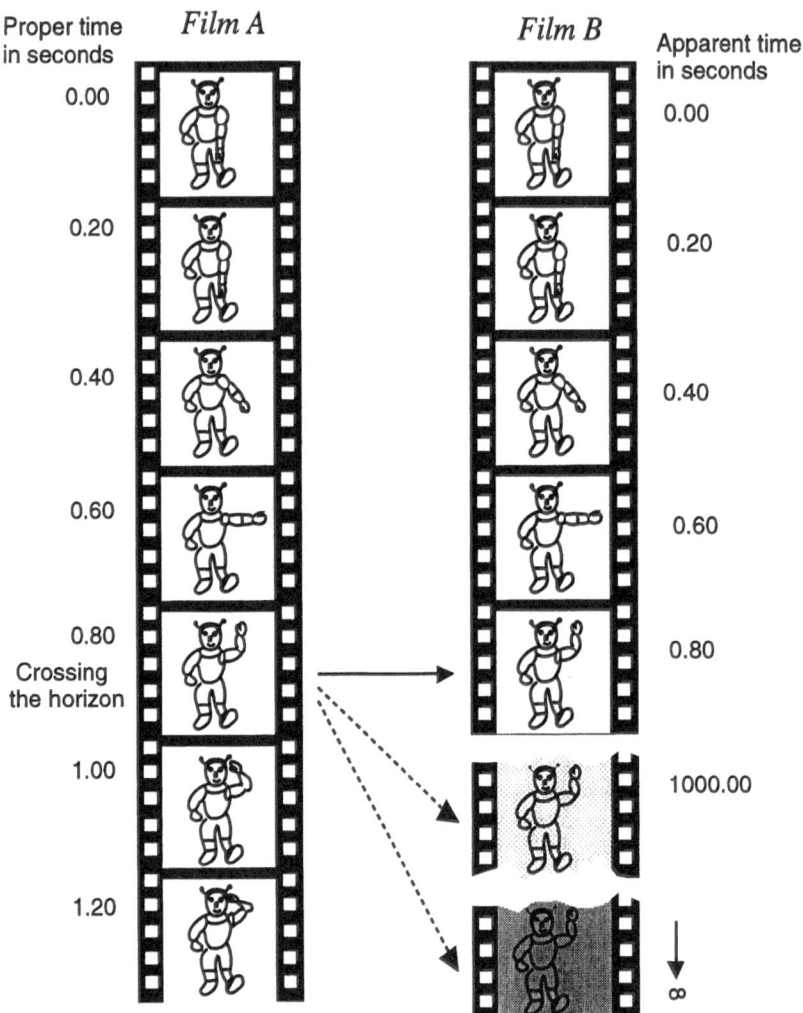

Fig. 5 The astronaut's salute.

noticed, so the films are practically identical. It is only very close to the horizon that apparent time starts suddenly to freeze; the film on the right then shows the astronaut eternally frozen in the middle of his salute, imperceptibly reaching the limiting position where he crossed the horizon. Besides this effect, the shift in the frequencies in the gravitational field (the so-called Einstein effect) causes the images to weaken, and they soon become invisible.

All these effects follow rather straightforwardly from certain equations. In General Relativity, the vacuum space-time around a spherically symmetric body

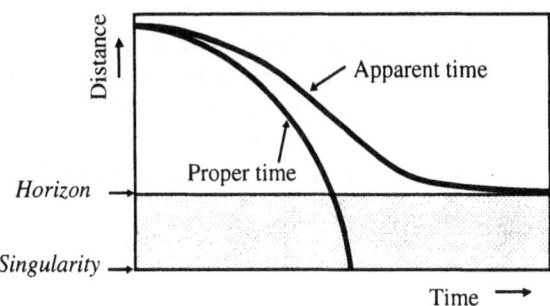

Fig. 6 The two times of a black hole.

is described by the Schwarzschild metric:

$$ds^2 = -\left(1 - \frac{2M}{r}\right) dt^2 + \left(1 - \frac{2M}{r}\right)^{-1} dr^2 + r^2 d\Omega^2, \qquad (2)$$

where $d\Omega^2 = d\theta^2 + \sin^2\theta \, d\phi^2$ is the metric of a unit 2-sphere, and we have set the gravitational constant G and the speed of light c equal to unity. The solution describes the external gravitational field generated by any spherical mass, whatever its radius (Birkhoff's theorem, 1923).

When the radius is greater than $2M$, there exists "interior solutions" depending on the equation of state of the stellar matter, which are non-singular at $r = 0$ and that match the exterior solution. However, as soon as the body is collapsed under its critical radius $2M$, the Schwarzschild metric is the unique solution for the gravitational field generated by a spherical black hole. The event horizon, a sphere of radius $r = 2M$, is a coordinate singularity which can be removed by a suitable coordinate transformation (see below). There is a true gravitational singularity at $r = 0$ (in the sense that some curvature components diverge) that cannot be removed by any coordinate transformation. Indeed the singularity does not belong to the space-time manifold itself. Inside the event horizon, the radial coordinate r becomes timelike. Hence every particle that crosses the event horizon is unavoidably catched by the central singularity. For radial free-fall along a trajectory with $r \to 0$, the proper time (as measured by a comoving clock) is given by

$$\tau = \tau_0 - \frac{4M}{3} \left(\frac{r}{2M}\right)^{3/2} \qquad (3)$$

and is well-behaved at the event horizon. The apparent time (as measured by a distant observer) is given by

$$t = \tau - 4M \left(\frac{r}{2M}\right)^{1/2} + 2M \ln \frac{\sqrt{r/2M} + 1}{\sqrt{r/2M} - 1}, \qquad (4)$$

and diverges to infinity as $r \to 2M$, see figure 6.

The Schwarzschild coordinates, which cover only $2M \leq r < \infty, -\infty < t < +\infty$, are not well adapted to the analysis of the causal structure of space-time near the horizon, because the light cones, given by $dr = \pm(1 - \frac{2M}{r})\,dt$, are not defined on the event horizon. We better use the so-called Eddington-Finkelstein coordinates — indeed discovered by Lemaître in 1933 but they remained unnoticed. Introducing the "ingoing" coordinate

$$v = t + r + 2M \ln\left(\frac{r}{2M} - 1\right), \tag{5}$$

the Schwarzschild metric becomes

$$ds^2 = -(1 - \frac{2M}{r})\,dv^2 + 2dv\,dr + r^2 d\Omega^2. \tag{6}$$

Now the light cones are perfectly well behaved. The ingoing light rays are given by

$$dv = 0, \tag{7}$$

the outgoing light rays by

$$dv = \frac{2dr}{1 - \frac{2M}{r}}. \tag{8}$$

The metric can be analytically continued to all $r > 0$ and is no more singular at $r = 2M$. Indeed, in figure 4 such a coordinate system was already used.

2.3   Non-spherical collapse

A black hole may well form from an asymmetric gravitational collapse. However the deformations of the event horizon are quickly dissipated as gravitational radiation; the event horizon vibrates according to the so–called "quasi-normal modes" and the black hole settles down into a final axisymmetric equilibrium configuration.

The deepest physical property of black holes is that asymptotic equilibrium solutions depend only on three parameters: the mass, the electric charge and the angular momentum. All the details of the infalling matter other than mass, electric charge and angular momentum are washed out. The proof followed from efforts for over 15 years by half a dozen of theoreticians, but it was originally suggested as a conjecture by John Wheeler, who used the picturesque formulation "a black hole has no hair". Markus Heusler's lectures in Chap. 7 will develop this so–called "uniqueness theorem".

As a consequence, there exists only 4 exact solutions of Einstein's equations describing black hole solutions with or without charge and angular momentum:

- The Schwarzschild solution (1916) has only mass M; it is static, spherically symmetric.

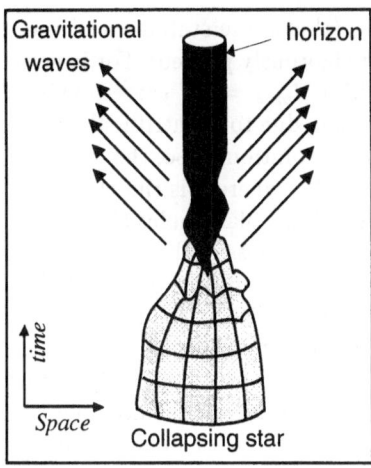

Fig. 7 Gravitational collapse of a star.

- The Reissner–Nordström!solution (1918), static, spherically symmetric, depends on mass M and electric charge Q.

- The Kerr solution (1963), stationary, axisymmetric, depends on mass M and angular momentum J.

- The Kerr–Newman solution (1965), stationary and axisymmetric, depends on all three parameters M, J, Q.

The 3-parameter Kerr-Newman family is the most general solution, corresponding to the final state of black hole equilibrium. In Boyer-Lindquist coordinates, the Kerr-Newman metric is given by

$$ds^2 = -(1 - \frac{2Mr}{\Sigma}) \, dt^2 - 4Mra\frac{\sin^2\theta}{\Sigma} \, dt \, d\phi$$
$$+(r^2 + a^2 + \frac{2Mra^2\sin^2\theta}{\Sigma}) \sin^2\theta d\phi^2 + \frac{\Sigma}{\Delta} \, dr^2 + \Sigma \, d\theta^2 \,, \qquad (9)$$

where $\Delta \equiv r^2 - 2Mr + a^2 + Q^2$, $\Sigma \equiv r^2 + a^2 \cos^2\theta$, and $a \equiv J/M$ is the angular momentum per unit mass. The event horizon is located at distance $r_+ = M + \sqrt{M^2 - Q^2 - a^2}$.

From this formula we can see, however, that the black hole parameters cannot be arbitrary. Electric charge and angular momentum cannot exceed values corresponding to the disappearance of the event horizon. The following constraint must be satisfied: $a^2 + Q^2 \leq M^2$.

When the condition is violated, the event horizon disappears and the solution describes a naked singularity instead of a black hole. Such odd things should not exist in the real universe (this is the statement of the so–called Cosmic Censorship Conjecture, not yet rigorously proved). For instance, for uncharged rotating configuration, the condition $J_{max} = M^2$ corresponds to the vanishing of surface gravity on the event horizon, due to "centrifugal forces"; the corresponding solution is called extremal Kerr solution. Also, the maximal allowable electric charge is $Q_{max} = M \approx 10^{40} e\, M/M_\odot$, where e is the electron charge, but it is to be noticed that in realistic situations, black holes should not be significantly charged. This is due to the extreme weakness of the gravitational interaction compared to the electromagnetic interaction. Suppose a black hole forms with initial positive charge Q_i of order M. Under realistic conditions, the black hole is not isolated in empty space but is surrounded by charged particles of the interstellar medium, e.g. protons and electrons. The black hole will predominantly attract electrons with charge $-e$ and repel protons with charge $+e$ by its electromagnetic field, and predominantly attract protons of mass m_p by its gravitational field. The repulsive electrostatic force on protons is larger than the gravitational pull by a factor of $eQ/m_p M \approx e/m_p \approx 10^{18}$. Therefore, the black hole will neutralize itself almost instantaneously. As a consequence, the Kerr solution, obtained in equation (9) by putting $Q = 0$, can be used for any astrophysical purpose involving black holes. It is also a good approximation to the metric of a (not collapsed) rotating star at large distance, but it has not been matched to any known solution that could represent the interior of a star.

The Kerr metric in Boyer-Lindquist coordinates has singularities on the axis of symmetry $\theta = 0$ – obviously a coordinate singularity – and for $\Delta = 0$. One can write $\Delta = (r - r_+)(r - r_-)$ with $r_+ = M + \sqrt{M^2 - a^2}$. The distance r_+ defines the outer event horizon (the surface of the rotating black hole), whereas r_- defines the inner event horizon. Like in Schwarzschild metric (where r_+ and r_- coincide at the value $2M$), the singularities at $r = r_+$, $r = r_-$ are coordinate singularities which can be removed by a suitable transformation analogous to the ingoing Eddington-Finkelstein coordinates for Schwarzschild space–time. For full mathematical developments of Kerr black holes, see Chandrasekhar (1992) and O'Neill (1995).

2.4   The black hole maelstrom

There is a deep analogy between a rotating black hole and the familiar phenomenon of a vortex - such as a giant maelstrom produced by sea currents. If we cut a light cone at fixed time (a horizontal plane in figure 8), the resulting spatial section is a "navigation ellipse" which determines the limits of the permitted trajectories. If the cone tips over sufficiently in the gravitational field, the navigation ellipse detaches itself from the point of emission. The permitted trajectories are confined within the angle formed by the tangents of the circle, and it is impossible to go backwards.

This projection technique is useful to depict the causal structure of spacetime around a rotating black hole (figure 9). The gravitational well caused by a rotating black hole resembles a cosmic maelstrom. A spaceship travelling in

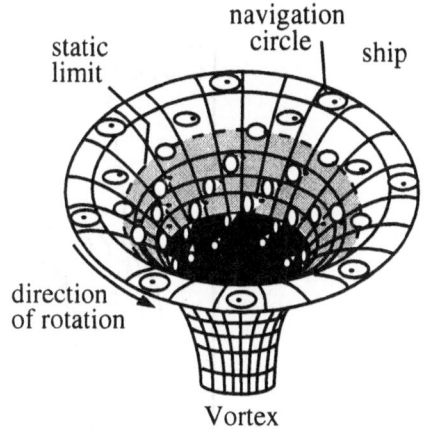

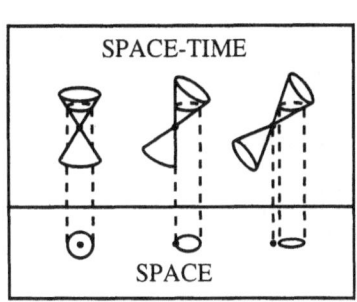

Figs. 8 and 9 Navigation circles in the black hole maelstrom.

the vicinity is sucked towards the centre of the vortex like a boat. In the region outside the so-called *static limit* (clear), it can navigate to wherever it wants. In the zone (in grey) comprised between the static limit and the event horizon, it is forced to rotate in the same direction as the black hole; its ability to navigate freely is decreased as it is sucked inwards, but it can still escape by travelling in an outwards spiral. The dark zone represents the region inside the event horizon: any ship which ventured there would be unable to escape even if it was travelling at the speed of light. A fair illustration is the Edgar Poe's short story: *A descent into the maelstrom* (1840).

The static limit is a hypersurface of revolution, given by the equation $r = M + \sqrt{M^2 - a^2 cos^2\theta}$. As we can see from figure 10, it intersects the event horizon at its poles $\theta = 0, \pi$ but it lies outside the horizon for other values of θ. The region between the static limit and the event horizon is called ergosphere. There, all stationary observers must orbit the black hole with positive angular velocity. The ergosphere contains orbits with negative energy. Such a property has lead to the idea of energy extraction from a rotating black hole. Roger Penrose (1969) suggested the following mechanism: A distant experimentalist fires a projectile in the direction of the ergosphere along a suitable trajectory (figure 10). When it arrives the projectile splits into two pieces: one of them is captured by the black hole along a retrograde orbit, while the other flies out of the ergosphere and is recovered by the experimentalist. Penrose has demonstrated that the experimentalist could direct the projectile in such a way that the returning piece has a greater energy than that of the initial projectile. This is possible if the fragment captured by the black hole is travelling in a suitable retrograde orbit (that is orbiting in the opposite sense to the rotation of the black hole), so that when it penetrates the black hole it slightly reduces the hole's angular momentum. The net result is that the black hole loses some of its rotational

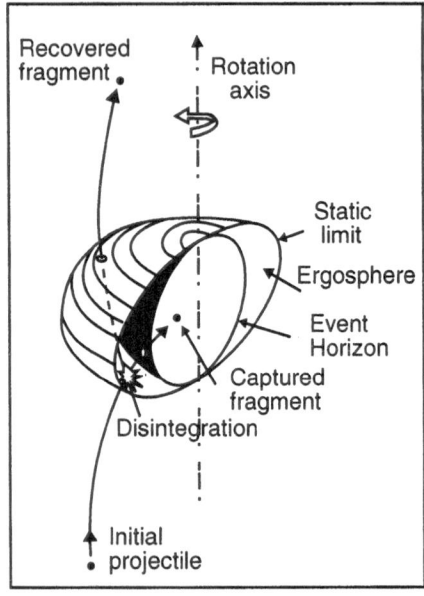

Fig. 10 Cross-section of a rotating black hole.

energy and the difference is carried away by the escaping fragment.

The amount of energy that can theoretically be extracted from a black hole has been calculated by Christodolou and Ruffini (1971). The total mass-energy of a black hole is

$$M^2 = \frac{J^2}{4M_{ir}^2} + \left(\frac{Q^2}{4M_{ir}} + M_{ir}\right)^2, \tag{10}$$

where $M_{ir} \equiv \frac{1}{2}\sqrt{\left(M + \sqrt{M^2 - Q^2 - a^2}\right)^2 + a^2}$. The first term corresponds to the rotational energy, the second one to the Coulomb energy, the third one to an "irreducible" energy. The rotational energy and the Coulomb energy are extractable by physical means such as the Penrose process, the superradiance (analogous to stimulated emission in atomic physics) or electrodynamical processes (see Norbert Straumann's lectures in Chap. 6 for details), while the irreducible part cannot be lowered by classical (e.g. non quantum) processes. The maximum extractable energy is as high as 29 per cent for rotational energy and 50 per cent for Coulomb energy. It is much more efficient that, for instance, nuclear energy release (0.7 per cent for hydrogen burning).

2.5   Black hole thermodynamics

It is interesting to mention that the irreducible mass is related to the area A of the event horizon by $M_{ir} = \sqrt{A/16\pi}$. Therefore the area of an event horizon cannot decrease in time by any classical process. This was first noticed by Stephen Hawking, who drew the striking analogy with ordinary thermodynamics, in which the entropy of a system never decreases in time. Such a property has motivated a great deal of theoretical efforts in the 1970's to better understand the laws of black hole dynamics – i.e. the laws giving the infinitesimal variations of mass, area and other black hole quantities when a black hole interacts with the external universe – and to push the analogy with thermodynamical laws. For the development of black hole thermodynamics, see G. Neugebauer's and W. Israel's lectures in Chaps. 16 to 18. Let me just recall that black hole mechanics is governed by four laws which mimic classical thermodynamics:

– *Zeroth law.*

  In thermodynamics: all parts of a system at thermodynamical equilibrium have equal temperature T.

  In black hole mechanics: all parts of the event horizon of a black hole at equilibrium have the same surface gravity g. The surface gravity is given by Smarr's formula $M = gA/4\pi + 2\Omega_H J + \Phi_H Q$, where Ω_H is the angular velocity at the horizon and Φ_H is the co-rotating electric potential on the horizon. This is a quite remarkable property when one compares to ordinary astronomical bodies, for which the surface gravity depends on the latitude. Whatever a black hole is flattened by centrifugal forces, the surface gravity is the same at every point.

– *First Law.*

  In thermodynamics: the infinitesimal variation of the internal energy U of a system with temperature T at pressure P is related to the variation of entropy dS and the variation of pressure dP by $dU = TdS - PdV$.

  In black hole dynamics: the infinitesimal variations of the mass M, the charge Q and the angular momentum J of a perturbed stationary black hole are related by $dM = \frac{g}{8\pi}dA + \Omega_H dJ + \Phi_H dQ$.

– *Second Law.*

  In thermodynamics, entropy can never decrease: $dS \geq 0$.

  In black hole dynamics, the area of event horizon can never decrease: $dA \geq 0$.

  This law implies, for instance, that the area of a black hole resulting from the coalescence of two parent black holes is greater than the sum of areas of the two parent black holes (see figure 11). It also implies that *black holes cannot bifurcate*, namely a single black hole can never split in two parts.

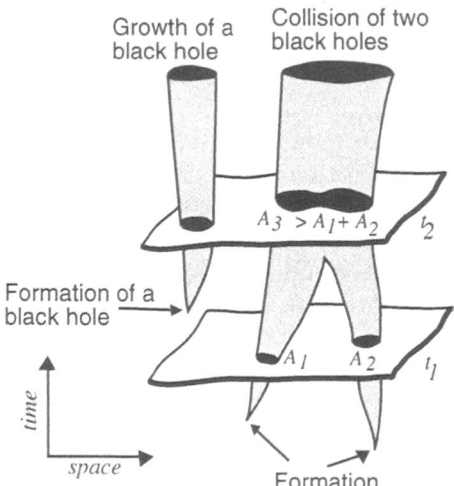

Fig. 11 The irreversible growth of black holes.

– *Third law.*

In thermodynamics, it reflects the inaccessibility of the absolute zero of temperature, namely it is impossible to reduce the temperature of a system to zero by a finite number of processes.

In black hole mechanics, it is impossible to reduce the surface gravity to zero by a finite number of operations. For Kerr black holes, we have seen that zero surface gravity corresponds to the "extremal" solution $J = M^2$.

It is clear that the area of the event horizon plays formally the role of an entropy, while the surface gravity plays the role of a temperature. However, as first pointed out by Bekenstein, if black holes had a real temperature like thermodynamical systems, they would radiate energy, contrarily to their basic definition. The puzzle was solved by Hawking when he discovered the evaporation of mini-black holes by quantum processes.

2.6   The quantum black hole

The details of Hawking radiation and the - not yet solved - theoretical difficulties linked to its interpretation are discussed by other lecturers (by Gerard 't Hooft in Chap. 21, Andreas Wipf in Chap. 19, and Claus Kiefer in Chap. 20). Therefore I shall only present the basic idea in a naive pictorial way (figure 12). The black hole's gravitational field is described by (classical) general relativity, while the surrounding vacuum space–time is described by quantum field theory. The quantum evaporation process is analogous to pair production in a strong

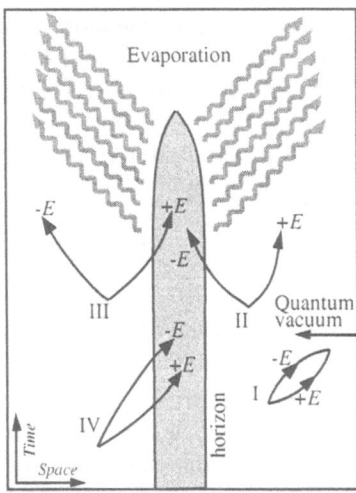

Fig. 12 The quantum evaporation of a mini-black hole by polarization of the vacuum.

magnetic field due to vacuum polarization. In the Fermi sea populated by virtual pairs of particles-antiparticles which create and annihilate themselves, the four various possible processes are depicted schematically in figure 12.

Some virtual pairs emerging from the quantum vacuum just annihilate outside the horizon (process I). Some pairs produced in the vicinity of the black hole disappear completely in the event horizon (process IV). Some pairs are split, one particle (or antiparticle) escaping the black hole, the other one being captured (processes II and III). The calculations show that the process II is dominant, due to the (classical) gravitational potential which polarizes the quantum vacuum. As a consequence, a black hole radiates particles with a thermal spectrum characterized by a blackbody temperature precisely given by the formula suggested by the thermodynamical analogy:

$$T = \hbar \frac{g}{2\pi} = 10^{-7} \frac{M_\odot}{M} \text{ Kelvin}, \tag{11}$$

where $\hbar$ is Planck's constant. We immediately see that the temperature is completely negligible for any astrophysical black hole with mass comparable or greater to the solar mass. But for mini-black holes with masses 10^{15} g (the typical mass of an asteroid), the Hawking temperature is 10^{12} K. Since the black hole radiates away, it loses energy and evaporates on a timescale approximately given by

$$t_E \approx 10^{10} \, years \left(\frac{M}{10^{15} \, grams} \right)^3 \tag{12}$$

Thus, mini-black holes whose mass is smaller than that of an asteroid (and size less than $10^{-13} cm$) evaporate on a timescale shorter than the age of the universe. Some of them should evaporate now and give rise to a huge burst of high energy radiation. Nothing similar has ever been observed (γ-ray bursts are explained quite differently). Such an observational constraint thus limits the density of mini-black holes to be less than about $100/(lightyear)^3$.

The black hole entropy is given by

$$S = \frac{k_B}{\hbar} \frac{A}{4} \tag{13}$$

(where k_B is Boltzmann's constant), a formula which numerically gives $S \approx 10^{77} k_B (\frac{M}{M_\odot})^2$ for a Schwarzschild black hole. Since the entropy of a non–collapsed star like the Sun is approximately $10^{58} k_B$, we recover the deep meaning of the "no hair" theorem, according to which black holes are huge entropy reservoirs. By Hawking radiation, the irreducible mass, or equivalently the event horizon area of a black hole decreases, in violation of the Second Law of black hole mechanics. The latter has to be generalized to include the entropy of matter in exterior space–time. Then, the total entropy of the radiating black hole is $S = S_{BH} + S_{ext}$ and, since the Hawking radiation is thermal, S_{ext} increases, so that eventually S is always a non-decreasing function of time.

To conclude briefly the subject, even if mini-black holes are exceedingly rare, or even if they do not exist at all in the real universe because the big bang could not have produced such fluctuations, they represent a major theoretical advance towards a better understanding of the link between gravity and quantum theory.

2.7   Space-time mappings

Various mathematical techniques allow the geometer to properly visualize the complex space-time structure generated by black holes.

Embedding diagram – The space-time generated by a spherical mass M has the Schwarzschild metric:

$$ds^2 = -\left(1 - \frac{2M(r)}{r}\right) dt^2 + \left(1 - \frac{2M(r)}{r}\right)^{-1} dr^2 + r^2 d\Omega^2 \tag{14}$$

where $M(r)$ is the mass comprised within the radius r. Since the geometry is static and spherically symmetric, we do not lose much information in considering only equatorial slices $\theta = \pi/2$ and time slices $t = constant$. We get then a curved 2–geometry with metric $(1 - \frac{2M(r)}{r})^{-1} dr^2 + r^2 d\phi^2$. Such a surface can be visualized by embedding it in Euclidean 3–space $ds^2 = dz^2 + dr^2 + r^2 d\phi^2$. For a non–collapsed star with radius R, the outer solution $z(r) = \sqrt{8M(r - 2M)}$, for $r \geq R \geq 2M$, is asymptotically flat and matches exactly the non–singular inner solution $z(r) = \sqrt{8M(r)(r - 2M(r))}$ for $0 \leq r \leq R$ (figure 13). For a black hole, the embedding is defined only for $r \geq 2M$. The corresponding surface is the Flamm paraboloid $z(r) = \sqrt{8M(r - 2M)}$. Such an asymptotically flat surface exhibits two sheets separated by the "Schwarzschild throat" of radius $2M$. The two sheets can be either considered as two different asymptotically flat "parallel" universes (whatever the physical meaning of such a statement may be) in which a black hole in the upper sheet is connected to a time–reversed "white hole" in the lower sheet (figure 14), or as a single asymptotically flat space–time containing a pair of black/white holes connected by a so–called "wormhole" (figure 15). The freedom comes from the topological indeterminacy of general relativity, which allows us to identify some asymptotically distant points of space–time without changing the metric.

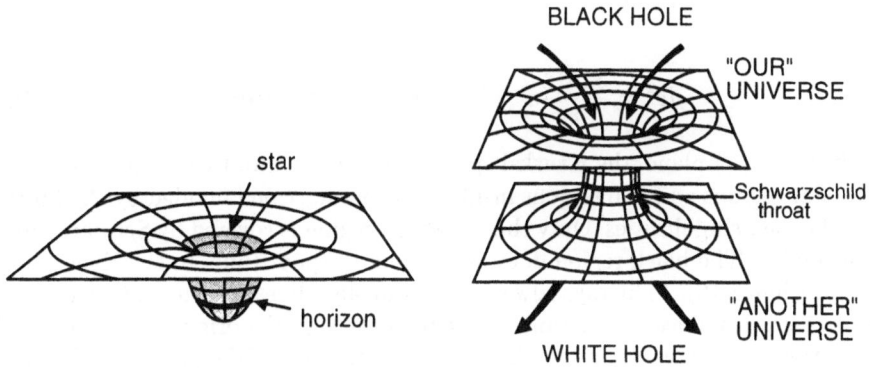

Figs. 13 and 14 Embedding of a non-collapsed spherical star and of Schwarzschild space-time.

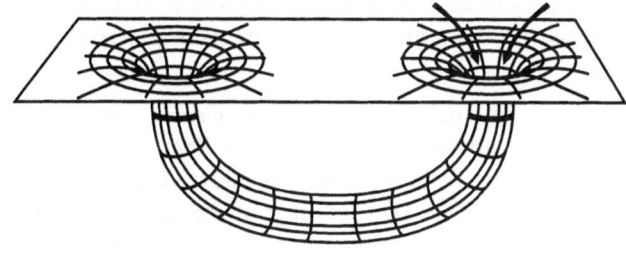

Fig. 15 A wormhole in space-time.

However, the embedding technique does not give access to the regions of space-time inside the event horizon.

Kruskal diagram – To explore inner space-time we use the maximal analytical extension of the Schwarzschild metric. This is achieved by means of a coordinate transformation discovered by Kruskal:

$$u^2 - v^2 = (\frac{r}{2M} - 1)e^{r/2M}$$

$$\frac{v}{u} = \left\{ \begin{array}{c} \coth \frac{t}{4M} \\ 1 \\ \tanh \frac{t}{4M} \end{array} \right\} \quad \text{for} \quad r \left\{ \begin{array}{c} < 2M \\ = 2M \\ > 2M \end{array} \right. . \tag{15}$$

The metric then becomes

$$ds^2 = \frac{32M^3}{r} e^{-r/2M}(-dv^2 + du^2) + r^2 d\Omega^2 . \tag{16}$$

In the (v, u) plane, the Kruskal space-time divides into two outer asymptotically flat regions and two regions inside the event horizon bounded by the future and the past singularities. Only the unshaded region is covered by Schwarzschild coordinates. The black region does not belong to space-time. In the Kruskal diagram (figure 16), light rays always travel at 45°, lines of constant distance r are hyperbolas, lines of constant time t pass through the origin. The interior of the future event horizon is the black hole, the interior of the past event horizon is the white hole. However it is clear that the wormhole cannot be crossed by timelike trajectories: no trajectory can pass from one exterior universe to the other one without encountering the $r = 0$ singularity.

Moreover, the Kruskal extension is a mathematical idealization of a spherical black hole since it implicitly assumes that the black hole exists forever. However in the physical universe, a black hole is not inscribed in the initial conditions of the universe, it may form only from gravitational collapse. In such a case, one gets a "truncated" Kruskal diagram (figure 17), which indicates that only the future event horizon and the future singularity occur in a single asymptotically flat space–time. Such a situation offers no perspective to space–time travelers!

Penrose–Carter diagrams – The Penrose–Carter diagrams use conformal transformations of coordinates such that $g_{\alpha\beta} \rightarrow \Omega^2 g_{\alpha\beta}$ which put spacelike and timelike infinities at finite distance, and thus allow to depict the full space–time into square boxes. The Penrose–Carter diagram for the Schwarzschild black hole does not bring much more information than the Kruskal one, but it turns out to be the best available tool to reveal the complex structure of a rotating black hole. Figure 18 shows the "many–fingered" universe of a Kerr black hole; it suggests that some timelike trajectories (B, C) may well cross the outer and inner event horizons and pass from an asymptotically flat external universe to another one

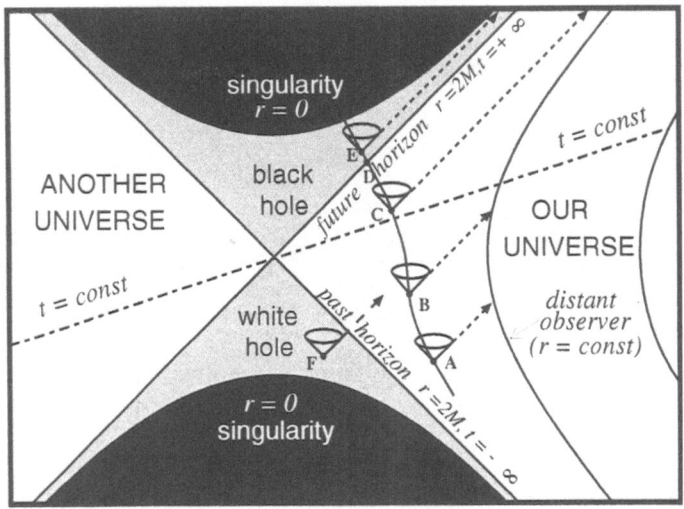

Fig. 16 Exploring a spherical black hole using Kruskal's map. (u-axis $\rightarrow$, v-axis $\uparrow$)

without encountering a singularity. This is due to the fact that the singularity S is timelike rather than spacelike. Also, the shape of the singularity is a ring within the equatorial plane, so that some trajectories (A) can pass through the ring and reach an asymptotically flat space–time inside the black hole where gravity is repulsive. However, the analysis of perturbations of such idealized Kerr space–times suggests that they are unstable and therefore not physically plausible. Nevertheless the study of the internal structure of black holes is a fascinating subject which is more deeply investigated by Werner Israel's lecture in Chap. 18.

3 Astrophysics of Black Holes

The fact that General Relativity does predict the existence of black holes and that General Relativity is a reliable theory of gravitation does not necessarily prove the existence of black holes, because General Relativity does not describe the astrophysical processes by which a black hole may form.

Thus, the astronomical credibility of black holes crucially depends on a good understanding of gravitational collapse of stars and stellar clusters.

In this section we first examine briefly the astrophysical conditions for black hole formation, next we describe the astronomical sites where black hole candidates at various mass scales lurk.

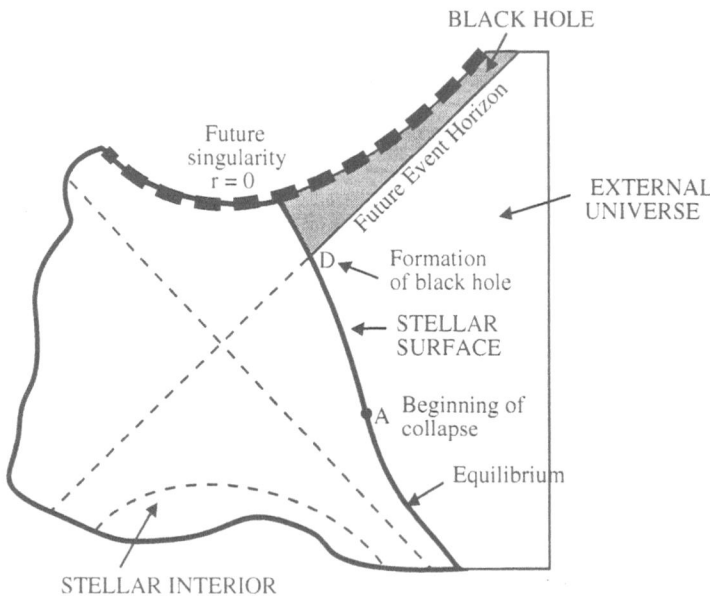

Fig. 17 Truncated Kruskal map representing the collapse of a star into a black hole.

3.1   Formation of stellar black holes

The basic process of stellar evolution is gravitational contraction at a rate controlled by luminosity. The key parameter is the initial mass. According to its value, stars evolve through various stages of nuclear burning and finish their lives as white dwarfs, neutron stars or black holes. Any stellar remnant (cold equilibrium configuration) more massive than about $3M_\odot$ cannot be supported by degeneracy pressure and is doomed to collapse to a black hole.

The figure 19 shows the stellar paths in a density-mass diagram according to the most recent observational and theoretical data. Below $8M_\odot$ stars produce white dwarfs, between 8 and $45M_\odot$ they produce neutron stars; black holes are formed only when the initial mass exceeds $45M_\odot$ (we note on the diagram that stars with initial mass between 20 and $40M_\odot$ suffer important mass losses at the stage of helium burning). Taking account of the stellar initial mass function, one concludes that approximately 1 supernova over 100 generates a black hole rather than a neutron star. Another possibility to form a stellar mass black hole is accretion of gas onto a neutron star in a binary system until when the mass of the neutron star reaches the maximum allowable value; then, gravitational collapse occurs and a low mass black hole forms. Taking into account these various processes, a typical galaxy like the Milky Way should harbour $10^7 - 10^8$ stellar black holes.

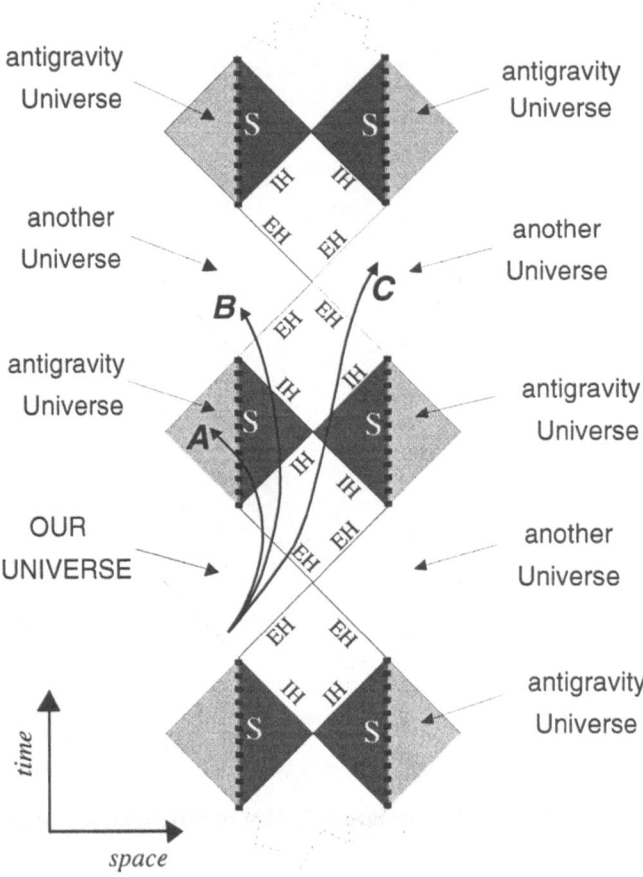

Fig. 18 Penrose map of a rotating black hole. EH denotes the outer and IH the inner event horizon.

3.2   Formation of giant black holes

A massive black hole can form by continuous growth of a "seed" stellar mass black hole, by gravitational collapse of a large star cluster or by collapse of a large density fluctuation in the early universe (see next subsection). A well nourished stellar mass black hole can grow to a supermassive black hole in less than a Hubble time. Such a process requests large amounts of matter (gas and stars) in the neighborhood, a situation than can be expected in some galactic nuclei.

A dense cluster of ordinary stars, such that the velocity dispersion $v_c \leq v_*$,

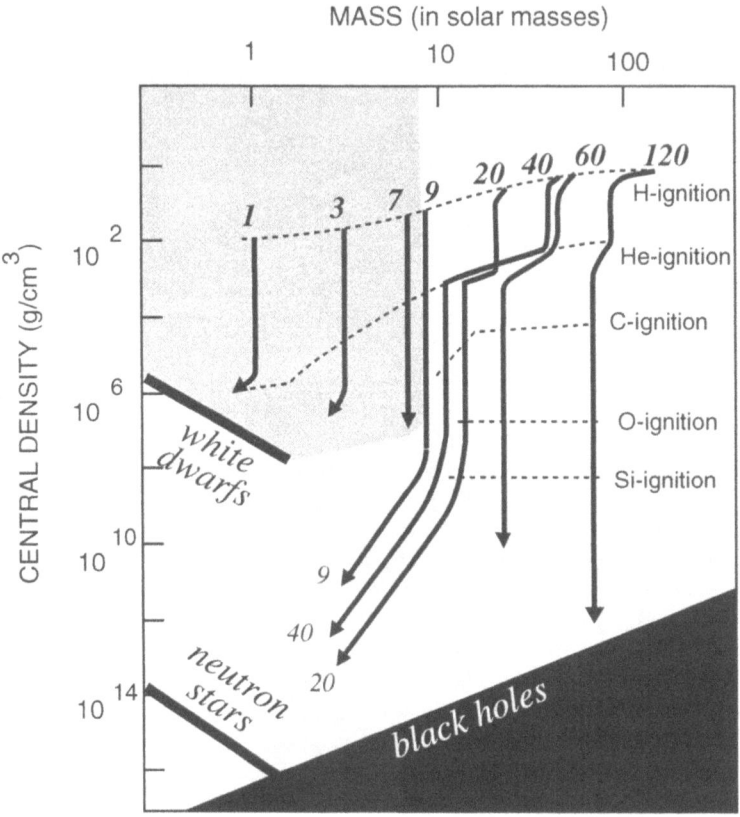

Fig. 19 The density-mass diagram of astronomical objects.

where $v_* \approx 600\,km/s$ is the typical escape velocity for main sequence stars, first evolves through individual stellar burning; supernovae explosions lead to the formation of compact remnants, e.g. neutron stars and stellar mass black holes. A cluster of compact stars becomes relativistically unstable at sufficiently high central gravitational redshift $1 + z_c = (1 - 2M)^{-1/2} \geq 1.5$ (Zeldovich and Podurets, 1965). Numerical simulations (e.g. Shapiro and Teukolsky, 1987, Bisnovatyi-Kogan, 1988) confirm this scenario. Starting with $\approx 10^7 - 2.10^8$ compact stars of $1 - 10M_\odot$ within a cluster radius $r \leq 0.01 - 0.1\,pc$ and velocity dispersion $800 - 2000\,km/s$, the evolution proceeds through three stages:

– secular core collapse via the gravothermal catastrophe (long timescale)

– short epoch dominated by compact star collisions and coalescences, leading to the formation of black holes with mass $M \approx 90M_\odot$

– relativistic instability leading to a massive black hole surrounded by a halo of stars.

3.3   Formation of mini-black holes

Zeldovich in 1967 and Hawking in 1971 pointed out that it was in principle possible to create a black hole with small mass (e.g. below the Chandrasekhar limit) by applying a sufficiently strong external pressure. Such conditions could have been achieved only in the very early universe. Gravitational forces may locally halt the cosmic expansion of a clump of matter and reverse it into collapse if the self-gravitational potential energy of the clump exceeds the internal energy:

$$\frac{GM^2}{R} \approx G\rho^2 R^5 \geq pR^3 \tag{17}$$

During the radiation era, $p \approx \rho c^2$, so the condition (17) is equivalent to $GM/c^2 \geq R$, where R is the size of the fluctuation. Then a primordial black hole of mass M forms. Due to the relation between density and time $G\rho \approx t^{-2}$ in an Einstein-De Sitter model of the early universe, the maximum mass of a collapsing fluctuation is related to the cosmic time by $M(grams) \approx 10^{38} t \, (seconds)$. Thus at Planck time $t \approx 10^{-43}$s, only mini-black holes may form with the Planck mass $\approx 10^{-5}$g, at time $t \approx 10^{-4}$s, black holes may form with $\approx 1 M_\odot$, at the time of nucleosynthesis $t \approx 100$ s supermassive black holes with $10^7 M_\odot$ may form. The observational status of primordial black holes is poor and unclear. On one hand, mini-black holes with mass $\leq 10^{15}$g could be detected by a burst of γ–radiation corresponding to the last stage of quantum evaporation in less than a Hubble time. Nothing similar having been observed, this puts severe upper limits on the actual average density of mini-black holes. On the other hand, the fact that most galactic nuclei seem to harbour massive black holes (see below) and that supermassive black holes are suspected to feed quasars at very high redshift, favour the hypothesis of the rapid formation of primordial massive black holes in the early universe.

3.4   Black Hole candidates in binary X-ray sources

Light cannot escape (classical) black holes but one can hope to detect them indirectly by observing the electromagnetic energy released during accretion processes.

Accretion of gas onto a compact star (neutron star or black hole in a binary system) releases energy in the X-ray domain, see S. Chakrabarti's lecture in this volume for the details. Search for stellar mass black holes thus consists in locating rapidly variable binary X-ray sources which are neither periodic (the corresponding X-ray pulsars are interpreted as rotating neutron stars) nor recurrent (the corresponding X-ray bursters are interpreted as thermonuclear explosions on a neutron star's hard surface). In spectroscopic binaries, the Doppler curve of the spectrum of the primary (visible) star provides the orbital period P of the binary and the maximum velocity v_* of the primary projected along the line-of-sight. Kepler's laws gives the following mass function which relates observed quantities to unknown masses:

$$\frac{Pv_*^3}{2\pi G} = \frac{(M_c \sin i)^3}{(M_* + M_c)^2},$$

(18)

where M_c and M_* are the masses of the compact star and of the optical primary, i the orbital inclination angle. A crucial fact is that M_c cannot be less than the value of the mass function (the limit would correspond to a zero-mass companion viewed at maximum inclination angle). Therefore the best black hole candidates are obtained when the observed mass function exceeds $3M_\odot$ – since, according to the theory, a neutron star more massive than this limit is unstable and will collapse to form a black hole. Otherwise, additional information is necessary to deduce M_c: the spectral type of the primary gives approximately M_*, the presence or absence of X-ray eclipses gives bounds to $\sin i$. Hence M_c is obtained within some error bar. Black hole candidates are retained only when the lower limit exceeds $3M_\odot$. At present day, about ten binary X-ray sources provide good black hole candidates. They can be divided into two families: the high–mass X–ray binaries (HMXB), where the companion star is of high mass, and the low–mass X–ray binaries (LMXB) where the companion is typically below a solar mass. The latter are also called "X–ray transients" because they flare up to high luminosities. Their mass properties are summarized in the table 1 below.

Table 1 Stellar mass black hole candidates

	mass function	$M_c/M_\odot$	$M_*/M_\odot$
HMXB			
Cygnus X-1	0.25	11–21	24–42
LMC X-3	2.3	5.6 –7.8	20
LMC X-1	0.14	≥ 4	4–8
LMXB (X–ray transients)			
V 404 Cyg	6.07	10–15	≈ 0.6
A 0620-00	2.91	5–17	0.2–0.7
GS 1124-68 (Nova Musc)	3.01	4.2–6.5	0.5–0.8
GS 2000+25 (Nova Vul 88)	5.01	6–14	≈ 0.7
GRO J 1655-40	3.24	4.5 – 6.5	≈ 1.2
H 1705-25 (Nova Oph 77)	4.65	5–9	≈ 0.4
J 04224+32	1.21	6–14	$\approx 0.3 - 0.6$

Other galactic X-ray sources are suspected to be black holes on spectroscopic or other grounds, see Chakrabarti's lectures in this volume for developments. For instance, some people argue that gamma-ray emission (above $100\,keV$) emitted from the inner edge of the accretion disc would attest the presence of a black hole rather than a neutron star, because the high-energy radiation is scattered back by the neutron star's hard surface and cools down the inner disc. If this is true, then many "gamma–ray novae" in which no measurement of mass can be done (due to the absence of optical counterpart or other limitations) are also good black hole candidates. This is specially the case for Nova Aquila 1992 and 1 E 17407-2942, two galactic sources which also exhibit radio jets. Such "microquasars" involving both accretion and ejection of matter provide an interesting link between high energy phenomena at the stellar and galactic scales.

3.5   Evidence for massive black holes in galactic nuclei

After the original speculations of Michell and Laplace, the idea of giant black holes was reintroduced in the 1960's to explain the large amounts of energy released by *active galactic nuclei* (AGNs). This generic term covers a large family of galaxies including quasars, radiogalaxies, Seyfert galaxies, blazars and so on, for the classification see W. Collmar's and V. Schönfelder's lecture in this volume. The basic process is accretion of gas onto a massive black hole. The maximum luminosity for a source of mass M, called the Eddington luminosity, is obtained by balance between gravitational attraction and radiation pressure repulsion acting on a given element of gas. It is given by

$$L \approx 10^{39} \, \frac{M}{10^8 M_\odot} \mathrm{W} \tag{19}$$

The observed luminosities of AGNs range from $10^{37} - 10^{41}$ W, where the higher values apply to the most powerful quasars. Then the corresponding masses range from $10^6 - 10^{10} M_\odot$.

Due to constant improvements of observational techniques, it turned out in the 1990's that most of the galactic nuclei (active or not) harbour large mass concentrations. Today the detection of such masses is one of the major goals of extragalactic astronomy. The most convincing method of detection consists in the dynamical analysis of surrounding matter: gas or stars near the invisible central mass have large dispersion velocities, which can be measured by spectroscopy. It is now likely that giant black holes lurk in almost all galactic nuclei, the energy output being governed by the available amounts of gaseous fuel. The best candidates are summarized in Table 2.

For instance, our Galactic Centre is observed in radio, infrared, X–ray and gamma–ray wavelengths (other wavelengths are absorbed by dust clouds of the galactic disc). A unusual radiosource has long been observed at the dynamical centre, which can be interpreted as low–level accretion onto a moderately massive black hole. However, a definite proof is not yet reached because gas motions are hard to interpret. Recently Eckart and Genzel (1996) obtained a full three-dimensional map of the stellar velocities within the central 0.1 pc of our Galaxy.

The values and distribution of stellar velocities are convincingly consistent with the hypothesis of a $2.5 \times 10^6 M_\odot$ black hole.

The nucleus of the giant elliptical M87 in the nearby Virgo cluster has also a long story as a supermassive black hole candidate. Several independent observations are consistent with a $1 - 3 \times 10^9 M_\odot$ black hole accreting in a slow, inefficient mode. A disc of gas is orbiting in a plane perpendicular to a spectacular jet; recent spectroscopic observations of the Hubble Space Telescope show redshifted and blueshifted components of the disc, which can be interpreted by Doppler effect as parts of the disc on each side of the hole are receding and approaching from us.

The spiral galaxy NGC 4258 (M 106) is by far the best massive black hole candidate. Gas motions near the centre has been precisely mapped with the 1.3 cm maser emission line of H_2O. The velocities are measured with accuracy of 1 km/s. Their spatial distribution reveals a disc with rotational velocities following an exact Kepler's law around a massive compact object. Also the inner edge of the disc, orbiting at 1080 km/s, cannot comprise a stable stellar cluster with the inferred mass of $3.6 \times 10^7 M_\odot$.

Table 2 Massive black hole candidates

dynamics	host galaxy	galaxy type	$M_h/M_\odot$
maser	M 106	barred	4×10^7
gas	M 87	elliptical	3×10^9
gas	M 84	elliptical	3×10^8
gas	NGC 4261	elliptical	5×10^8
stars	M 31	spiral	$3 - 10 \times 10^7$
stars	M 32	elliptical	3×10^6
stars	M 104	(barred?) spiral	$5 - 10 \times 10^8$
stars	NGC 3115	lenticular	$7 - 20 \times 10^8$
stars	NGC 3377	elliptical	8×10^7
stars	NGC 3379	elliptical	5×10^7
stars	NGC 4486B	elliptical	5×10^8
stars	Milky Way	spiral	2.5×10^6

The black hole in our galaxy and the massive black holes suspected in nearby ordinary galaxies would be small scale versions of the cataclysmic phenomena occurring in AGNs. But AGNs are too far away to offer a spectral resolution good enough for dynamical measurements. Indeed, estimates of luminosities of AGNs and theoretical arguments involving the efficiency of energy release in strong gravitational fields invariably suggest that central dark masses are comprised between $10^7 - 10^9 M_\odot$. Variability of the flux on short timescales also indicates

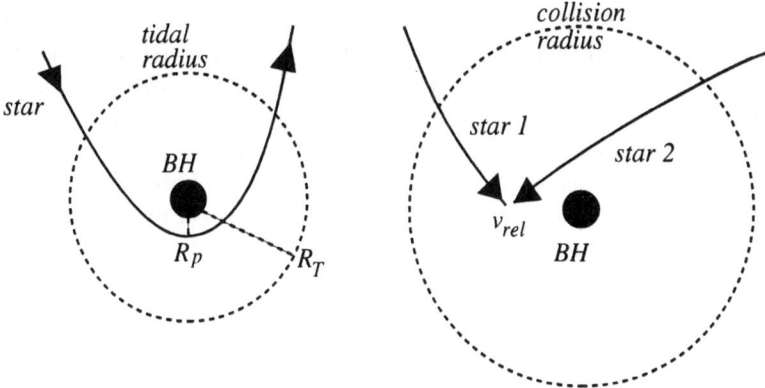

Fig. 20 Tidal and collision radius.

that the emitting region has a small size; many AGNs exhibit large luminosity fluctuations over timescales as short as one hour, which indicate that the emitting region is smaller than one light-hour. Such large masses in such small volumes cannot be explained by star clusters, so that accreting massive black holes remain the only plausible explanation.

3.6 Stellar disruption

The accretion of gas at rate dM/dt and typical efficiency $\epsilon \approx 0.1$ produces a luminosity

$$L \approx 10^{39} \left(\frac{\epsilon}{0.1}\right) \frac{dM/dt}{1 M_\odot/year} \, \mathrm{W} \tag{20}$$

By comparing the luminosity of the accretion model with the observed luminosities in AGNs, we conclude that the gas accretion rate must lie in the range $10^{-2} - 10^2 \, M_\odot/year$. One is thus led to the question about the various gas production mechanisms able to fuel a giant black hole. An efficient process is mass loss from stars passing near the hole. Current models of galactic nuclei involve a massive black hole surrounded by a dense large cloud of stars. Diffusion of orbits makes some stars to penetrate deeply within the gravitational potential of the black hole along eccentric orbits. Disruption of stars can occur either by tidal forces or by high-velocity interstellar collisions (figure 20). The collision radius $R_{coll} \approx 7 \times 10^{18} \frac{M}{10^8 M_\odot}$ cm for a solar–type star is the distance within which the free–fall velocity of stars becomes greater than the escape velocity at the star's surface v_* (typically 500 km/s for ordinary stars); if two stars collide inside R_{coll} they will be partially or totally disrupted.

Also stars penetrating the critical tidal radius $R_T \approx 6 \times 10^{13} \left(\frac{M}{10^8 M_\odot}\right)^{1/3}$ cm for solar–type stars will be ultimately disrupted, about 50 per cent of the released gas will remain bound to the black hole. In some sense, the tidal encounter of a star with a black hole can be considered as a collision of the star with itself ...

In the collision process the factor $\beta = v_{rel}/v_*$ plays a role analogous to the penetration factor $\beta = R_T/R_p$ in the tidal case (where R_p is the periastron distance). As soon as $\beta \geq 1$ the stars are disrupted, and when $\beta \geq 5$ the stars are strongly compressed during the encounter. Thus, in both cases, β appears as a *crushing factor*, whose magnitude dictates the degree of maximal compression and heating of the star.

The first modelisation of the tidal disruption of stars by a big black hole was done in the 1980's by myself and collaborators (see Luminet and Carter, 1986 and references therein). We have discovered that a star deeply plunging in the tidal radius without crossing the event horizon is squeezed by huge tidal forces and compressed into a short-lived, ultra-hot "pancake" configuration. Figure 21 shows schematically the progressive deformation of the star (the size of the star has been considerably over-emphasized for clarity). The left diagram represents the deformation of the star in its orbital plane (seen from above), the right one shows the deformation in the perpendicular direction. From a to d the tidal forces are weak and the star remains practically spherical. At e the star penetrates the tidal limit. It becomes cigar-shaped. From e to g a "mangle" effect due to tidal forces becomes increasingly important and the star is flattened in its orbital plane to the shape of a curved "pancake". The star rebounds, and as it leaves the tidal radius, it starts to expand, becoming more cigar-shaped again. A little further along its orbit the star eventually breaks up into fragments.

If the star chances to penetrate deeply (say with $\beta \geq 10$), its central temperature increases to a billion degrees in a tenth of a second. The thermonuclear chain reactions are considerably enhanced. During this brief period of heating, elements like helium, nitrogen and oxygen are instantaneously transformed into heavier ones by rapid proton or alpha–captures. A thermonuclear explosion takes place in the stellar pancake, resulting in a kind of "accidental supernova". The consequences of such an explosion are far reaching. About 50 per cent of the stellar debris is blown away from the black hole at high velocity (propelled by thermonuclear energy release), as a hot cloud able of carrying away any other clouds it might collide with. The rest of the debris falls rapidly towards the hole, producing a burst of radiation. Like supernovae, the stellar pancakes are also crucibles in which heavy elements are produced and then scattered throughout the galaxy. Thus, observation of high–velocity clouds and enrichment of the interstellar medium by specific isotopes in the vicinity of galactic nuclei would constitute an observational signature of the presence of big black holes.

Explosive or not, the tidal disruption process would induce a burst of luminosity in the host galactic nucleus on a timescale of a few months (the time required for the debris to be digested). To describe the evolution of the star we developed a simplified "affine model" in which we assumed that the layers of constant density keep an ellipsoidal form. Many astrophysicists were skeptical about the predictions of the model until when full hydrodynamical calculations were performed all around the world, using 3D Smooth Particle Hydrodynami-

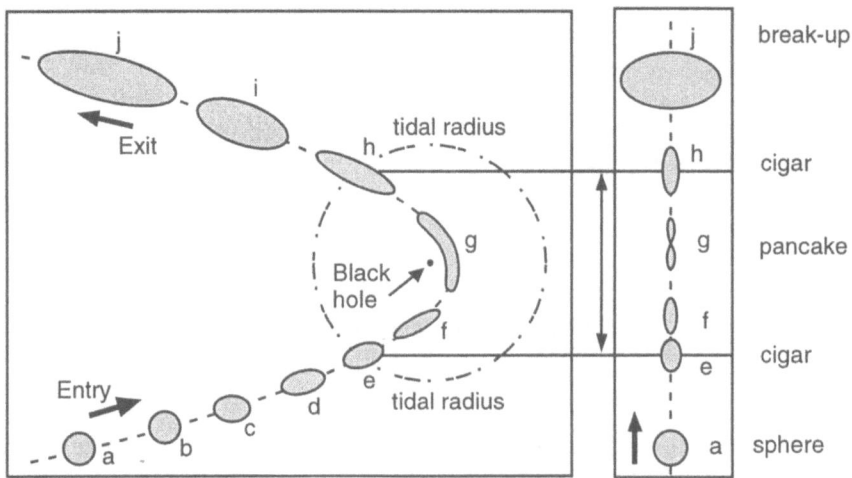

Fig. 21 The disruption of a star by tidal forces of a black hole.

cal codes (Laguna and Miller, 1993, Khlokov, Novikov and Pethick, 1993, Frolov et al., 1994) or spectral methods (Marck et al., 1997). The main features and quantitative predictions of the affine star model were confirmed, even if shock waves may decrease a little bit the maximum pancake density.

The nucleus of the elliptical galaxy NGC 4552 has increased its ultra-violet luminosity up to $10^6 L_\odot$ between 1991 and 1993 (Renzini et al, 1993). The timescale was consistent with a tidal disruption process, however the luminosity was $\approx 10^{-4}$ lower than expected, suggesting only a partial disruption of the star.

4   A Journey Into a Black Hole

Imagine a black hole surrounded by a bright disc (Figure 22). The system is observed from a great distance at an angle of 10° above the plane of the disc. The light rays are received on a photographic plate (rather a bolometer in order to capture all wavelengths). Because of the curvature of space–time in the neighborhood of the black hole, the image of the system is very different from the ellipses which would be observed if an ordinary celestial body (like the planet Saturn) replaced the black hole. The light emitted from the upper side of the disc forms a direct image and is considerably distorted, so that it is completely visible. There is no hidden part. The lower side of the disc is also visible as an indirect image, caused by highly curved light rays.

The first computer images of the appearance of a black hole surrounded by an accretion disc were obtained by myself (Luminet, 1978). More sophisticated calculations were performed by Marck (1993) in Schwarzschild and Kerr space–times. A realistic image, e.g. taking account of the space–time curvature, of the

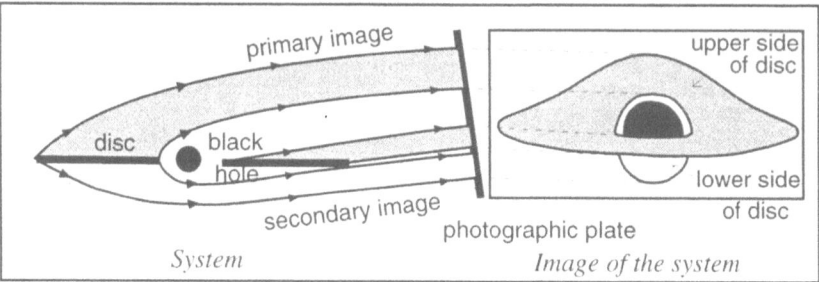

Fig. 22 Optical distortions near a black hole.

blue- and redshift effects, of the physical properties of the disc and so on, can be precisely calculated at any point of space–time – including inside the event horizon. A movie showing the distortions observed along any timelike trajectory around a black hole was produced (Delesalle, Lachièze-Rey and Luminet, 1993). The figure 23 is a snapshot taken along a parabolic plunging trajectory. During such a "thought journey" the vision of the third butterfly becomes accessible, all external spectators can admire the fantastic landscape generated by the black hole.

For a long time considered by astronomers as a mere theoretical specula-tion, black holes are now widely accepted as the basic explanation for X–ray massive binaries and galactic nuclei. Allowing for the elaboration of the most likely models, black holes also respond to the principle of simplicity, according to which among equally plausible models, the model involving the least num-ber of hypotheses must be preferred. However, for such a wide acceptance to be settled down, the basic picture of a black hole had to be drastically changed. The conjunction of theoretical and observational investigations allowed for such a metamorphosis of the black hole image, passing from the primeval image of a naked black hole perfectly passive and invisible, to the more sophisticated im-age of a thermodynamical engine well-fed in gas and stars, which turns out to be the key of the most luminous phenomena in the universe. Then the modern astronomer won over to such a duality between light and darkness may adopt the verse of the french poet Léon Dierx: "Il est des gouffres noirs dont les bords sont charmants".

Bibliography

[1] Begelman, M., Rees, M. (1996): Gravity's Fatal Attraction: Black Holes in the Universe (New York: Scientific American Library)

[2] Chandrasekhar, S. (1992): The Mathematical Theory of Black Holes, (Ox-ford: Oxford University Press)

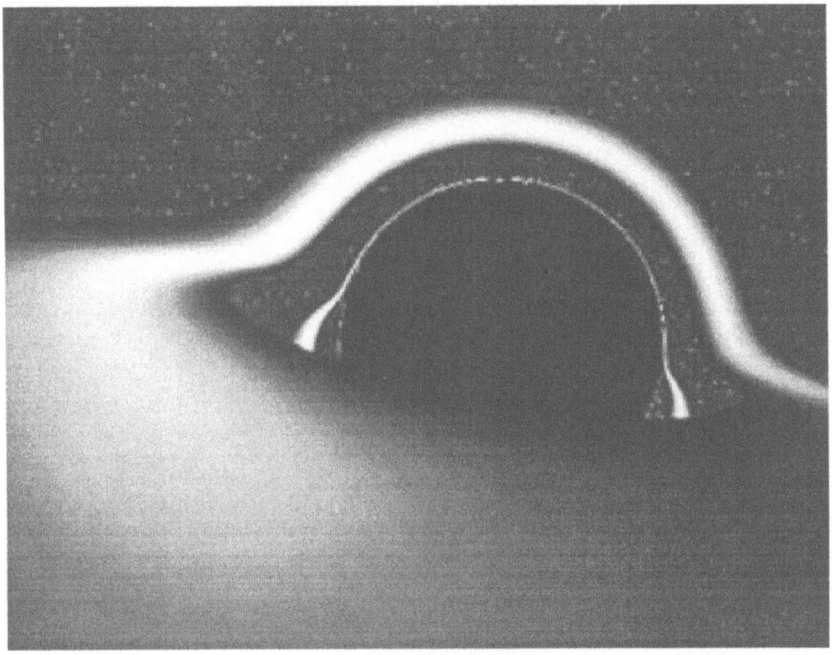

Fig. 23 The appearance of a distant black hole surrounded by an accretion disc. (©J.-A.Marck/Sygma)

[3] Delesalle, L., Lachièze-Rey, M., Luminet J.-P. (1994): Infinitely Curved, video VHS 52 mn, Arte/CNRS Audiovisuel

[4] DeWitt, C., DeWitt, B.S. (Eds.) (1973): Black Holes (Les Houches School, Gordon and Breach: New York)

[5] Hawking, S.W., Israel, W. (Eds.) (1989): 300 Years of Gravitation (New York, Cambridge: Cambridge University Press), pp. 199–446

[6] Luminet, J.-P. (1979): *Astron. Astrophys.* **75**, 228

[7] Luminet, J.-P., Carter, B. (1986): *Astrophys. J. Suppl.* **61**, 219

[8] Luminet, J.-P. (1992): Black Holes (New York, Cambridge: Cambridge University Press). German translation: *Schwarze Löcher* (Vieweg, 1997)

[9] Marck, J.-A. (1993): *Class. Quantum Grav.* **13**, 393

[10] Misner, C.W., Thorne, K.S., Wheeler, J.A. (1973): Gravitation (San Francisco: Freeman)

[11] O'Neill, B. (1995): The Geometry of Kerr Black Holes (Wellesley: Peters)

[12] Wheeler, J.A. (1990): A Journey into Gravity and Spacetime. (New York, Cambridge: Cambridge University Press)

Part II

Observations, Astrophysics

Evidence for Massive Black Holes in the Nuclei of Active Galaxies from Gamma-Ray Observations

Werner Collmar and Volker Schönfelder

Max-Planck-Institut für Extraterrestrische Physik, Postfach 16 03, D-85748 Garching, Germany

Abstract. According to our present knowledge, Active Galactic Nuclei (AGNs) consist of massive black holes which accrete matter from their environments. By this process the gravitational energy is converted into electromagnetic radiation. Before the launch of the Compton Gamma-Ray Observatory (CGRO) only four AGNs had been reported to emit detectable gamma-rays. Now, after roughly six years in orbit, the different CGRO experiments have discovered about 90 AGNs in the energy range between 50 keV and 20 GeV. Accordingly, CGRO finally opened the field of extragalactic gamma-ray astronomy. Two main conclusions can be drawn from the CGRO-results: i) Seyfert galaxies are hard X-ray sources rather than gamma-ray sources. They emit thermal radiation which cuts off at energies above a few hundreds of keV (~200 keV). ii) Radio-loud quasars and BL Lac-objects — so called blazars — however, can be strong gamma-ray emitters up to energies of at least several GeV. They show a highly variable non-thermal emission component. The assumption of isotropic emission together with the observed variability would lead to a high radiation density in a small emission region. At least for some sources, the derived compactness is in conflict with the assumption of isotropy and therefore relativistic beaming of the gamma emission is strongly suggested. The observed broad band νF_ν-spectra show a maximum at gamma-rays. However, taking into account the relativistic beaming, the gamma-ray energy output is a factor of up to 10^4 smaller than suggested by the observed intensity spectra.

We discuss these CGRO results in the context of the gamma-ray emission processes, AGN emission models, and finally collect the evidence for massive black holes in the center of active galaxies.

1 What Are 'Active' Galaxies in Comparison to 'Normal' Galaxies

The observable universe consists of billions of galaxies, many of which containing 100 billion stars or more, like our 'Milky Way', for example. On a morphological basis these galaxies are grouped into three main categories: spirals, ellipticals, and irregulars.

In 1943, the US astronomer Carl Seyfert noticed a distinct class of galaxies with stellar appearing cores and broad nuclear emission lines. Galaxies, showing these two characteristica, were subsequently named 'Seyfert' galaxies and mark today a sub-class of the so-called 'Active Galaxies'. Active galaxies comprise names like 'quasars' (quasi-stellar radio sources), 'QSOs' (quasi-stellar objects), 'Markarian galaxies', 'radio galaxies', and 'blazars' for example. The names of the different sub-classes of active galaxies developed historically according to the

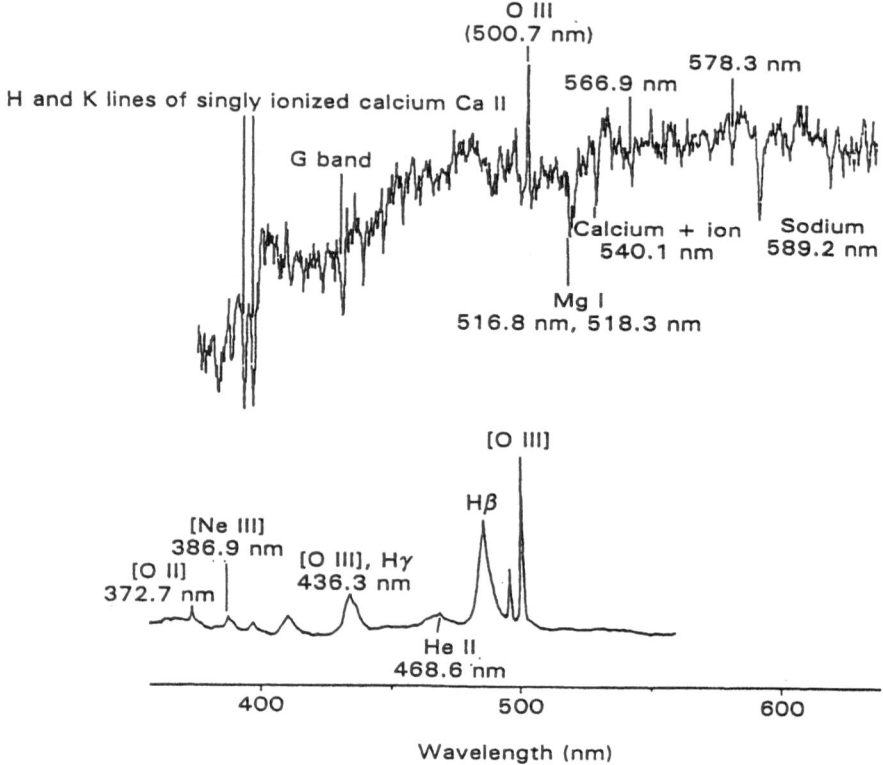

Fig. 1 Comparison of optical spectra from a 'normal' (spiral) galaxy above and an 'active' (Seyfert) galaxy below. The spectrum of the spiral galaxy shows mainly absorption lines due to stars, while the spectrum of the Seyfert galaxy is dominated by strong and broad emission lines. This figure is from Robson (1997).

name of the reporting astronomer (e.g. 'Seyferts', 'Markarians'), morphological characteristica (QSOs), or from other properties.

There are distinct observational differences between ordinary or 'normal' galaxies and 'active' galaxies. In contrast to the images of 'normal' galaxies, which are basically an assembly of stars, the images of active galaxies show bright nuclei. Their optical spectra typically show broadened emission lines instead of absorption lines (Fig. 1). Their overall spectra show luminosity maxima in IR, UV, X- or even γ-rays, and are often dominated by non-thermal emission, while normal galaxies radiate most of their energy in the optical band simply being the sum of the starlight. Another major difference between 'normal' and 'active' galaxies is the variability. While normal galaxies always look the same — maybe a supernova is brightening occasionally for a few months — the emission of active galaxies changes significantly on short time scales (Fig. 2) down

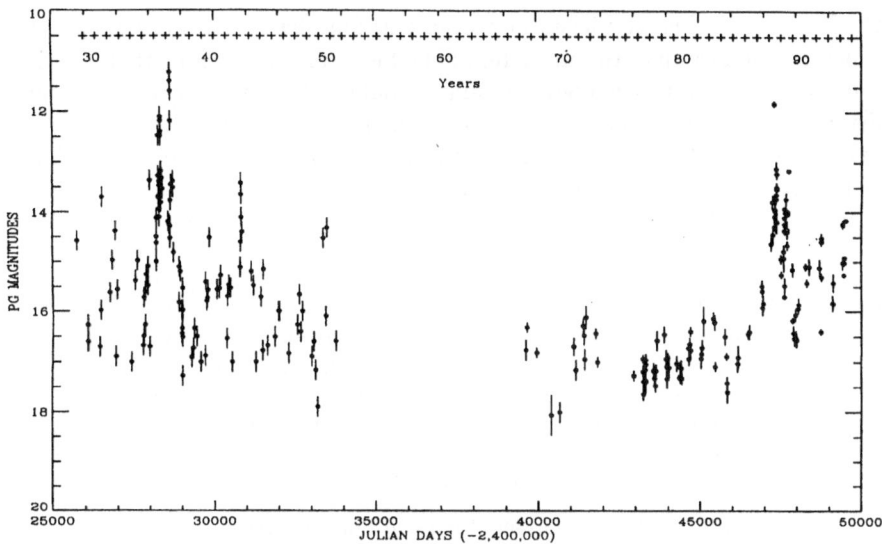

Fig. 2 Time history of the quasar 3C 279 in the optical band from 1930 to 1994. Variations up to $\sim$7 magnitudines (a factor of $\sim$630) are observed. The 'activity' of the source is obvious. The figure is from Hartman et al. (1996).

to days or even below, which — in our mind — gave rise to the term 'active'. This short-term variability points to small sizes of the emission regions which are consistent with the bright nuclei. In the specific case of quasars or QSOs the nucleus overshines the rest of the galaxy by far resulting in a star-like image. On the basis of light-travel arguments, the emission region radius is estimated to

$$r < c\Delta t/(1+z) , \qquad (1)$$

with Δt the observed time variability, z the redshift of the source, and c the speed of light. Some active galaxies show jets like the famous quasar 3C 273 for example.

2   A Unification Scheme for AGN

Roughly 3% of all galaxies are classified as being 'active'. Although there is no overall agreement amongst astronomers about a precise definition of active galaxies, one common criterium is the generation of large luminosities in small core regions. In many cases the active and bright core overshines the remaining galaxy by far.

As described above there exist a large number of classes and sub-classes, which developed historically and sometimes overlap in their parameters, and which are confusing physicists and even astrophysicists not working in AGN research. For example, the most luminous Seyferts are brighter than the low-luminous quasars. If such a Seyfert galaxy would be 'put' far out to cosmological distances, only the bright core would remain visible, and, subsequently, it would be classified as a QSO ('quasi-stellar object'). So, in a first step of unification, it is suggested that some Seyferts are simply 'nearby quasars'.

Table 1 This table (from Urry and Padovani 1995) shows the simplified classification of AGN according to the two parameters 'radio-loudness' and 'optical emission line properties'. The main types of AGN are sorted into the table. The abbreviations have the following meaning: NLRG, BLRG: narrow, broad line radio galaxies, SSRQ, FSRQ: steep, flat spectrum radio quasars, FR I, II: Fanaroff-Riley Type 1,2 radio galaxies, and BL Lac: BL Lacertae object. For a more detailed description of the different types see e.g. Urry and Padovani (1995) or Robson (1997).

Radio Loudness	Optical Emission Line properties			
	Type 2 (Narrow Line)	Type 1 (Broad Line)	Type 0 (unusual)	
radio-quiet (85-90%)	Seyfert-2	Seyfert-1 QSO		? ↑
radio-loud (10-15%)	NLRG (FR I, FR II)	BLRG SSRQ, FSRQ	Blazars (BL Lac, FSRQ)	↓ ?
	decreasing jet angle to line of sight →			

Observationally, AGN are classified on the basis of three parameters: the strength of their radio emission, the emission line properties in the optical and UV range, and the luminosity. However, the question was asked, whether there is some common underlying physics or physical structure involved. One of the first steps towards unification concerned the difference between the two sub-classes of Seyfert galaxies simply called 'Type 1' and 'Type 2'. While the 'Type 1' Seyferts show broad - ($\Delta v > 2000$ km/s) and narrow emission lines ($\Delta v < 2000$ km/s), the 'Type 2' sources show only narrow emission lines. A key observation by Antonucci and Miller (1985) revealed weak broad lines in polarized light from the Seyfert-2 galaxy NGC 1068. Since scattered light is polarized, this observation was interpreted that NGC 1068 also generates broad emission lines, which however, are somehow hidden from direct observation. From such observational facts a simplified classification scheme evolved recently which is shown in Table 1 ([31]). The different classes of AGN are ordered according to two parameters only: their 'optical emission line properties' and their 'radio loudness'.

With respect to their line properties, AGN are subdivided into so called 'Type-2' sources showing only narrow emission lines, 'Type-1' sources showing also broad emission lines, and 'Type-0' sources showing an unusual behaviour, e.g. weak or even absence of any lines. AGN are considered to be radio-loud if the ratio of the radio flux at 5 GHz to the optical B-band ('blue') flux is larger than 10 ($f_{5GHz}/f_B \geq 10$) ([15]). According to this definition roughly 10% to 15% of all AGN are radio-loud.

The different main AGN types are sorted into this two-dimensional scheme accordingly (Table 1). It is believed that the horizontal axis reflects an orientation effect, namely the decreasing angle to the line of sight towards the jet of an active galaxy (see Fig. 3). There is no commonly accepted interpretation why some AGN are radio-loud and others not. Whether this has to do with the host galaxy — radio-loud sources are mainly found in elliptical galaxies, while radio-quiet ones in spirals — or with the spin of the putative central BH or something else remains unclear to date. This question is currently one of the major unresolved problems in AGN research.

Fig. 3 shows the 'standard model' for AGN, which explains the different types as simply being an orientational effect. Despite the fact that many details are not yet understood, the scenario — which is described in the following — is commonly accepted because it explains the major observational facts reasonably well and consistently. The central object is thought to be a supermassive Black Hole (BH) with masses of the order of $\sim 10^6$ to $\sim 10^{10}$ M$_\odot$. The Schwarzschild radius ($R_s = 2GM_{BH}/c^2$) of an 10^8 M$_\odot$ BH is $\sim 3 \cdot 10^8$ km, which is approximately 2 Astronomical Units (AU) or $\sim 10^{-5}$ parsecs (pc). The BH is surrounded by an accretion disc consisting of ionized material reaching out to several hundreds of Schwarzschild radii. This very center of an active galaxy is surrounded by an extended molecular torus with an inner diameter of ~ 1.5 pc and an outer diameter of the order of ~ 30 pc. Within the molecular torus and near the center of the active galaxy fast moving (v≥ 2000 km/s) gas clouds (dark spots in Fig. 3) exist which are ionized by the accretion disk radiation and which emit the observed broad emission lines. These fast moving clouds are located within ~ 2-$20 \cdot 10^{16}$ cm and are marking the so called 'broad-line region' (BLR). Further out such clouds (grey spots in Fig. 3) move slower (v≤ 2000 km/s) and therefore give rise to the observed narrow emission lines. This so-called narrow-line region (NLR) ranges roughly from 10^{18} cm to 10^{20} cm. In addition to these clouds a hot electron corona (dark dots in Fig. 3) populates the inner region which can scatter some continuum and broad line emission as was observed in NGC 1068 ([1]). In a radio-loud AGN a strong jet of relativistic leptons emanates perpendicular to the plane of the accretion disc. The generation of such jets is still not understood, however it is believed that strong magnetic fields play a fundamental role. Radio jets have been observed on scales from 10^{17} cm to $\sim 10^{24}$ cm, which is significantly larger than the largest galaxies.

The 'unification-by-orientation' scenario assumes such a general structure for all AGN. Depending on the spatial orientation with respect to us we observe the

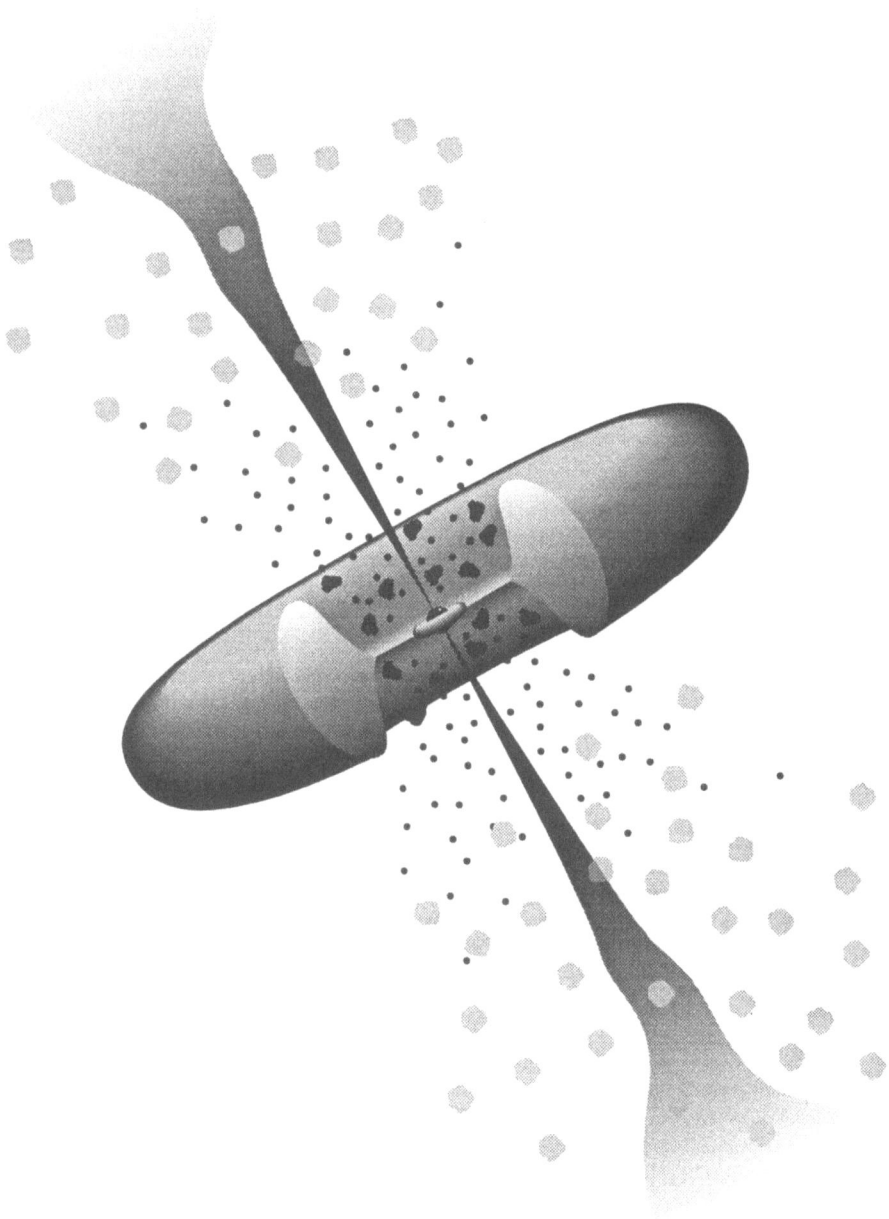

Fig. 3 The 'standard model' for radio-loud AGN. For an explanation see the text. The plot is from Urry and Padovani (1995).

different types of AGN. If we look towards the 'central engine' along the jet axis ($<10°$), i.e. basically directly into the jet, we observe a blazar. A regular quasar or a Seyfert-1 galaxy is observed if we look at an offset angle of the order of $30°$, at which both the narrow- and broad-line region are visible. At larger angular offsets the broad-line region will be hidden by this extended molecular torus giving rise to Seyfert-2 galaxies. A typical radio galaxy, showing two strong opposite jets, is observed at angles approximately perpendicular to the jet axis.

This scenario is very intriguing and is widely accepted among astronomers. However, there are still several unresolved questions like 'Where are the radio-quiet and narrow-line (Type 2) QSOs?', and 'What causes the difference between radio-loud and radio-quiet AGN?'. In addition, this scenario neglects evolution, which probably is another important parameter for active galaxies.

3   Gamma-ray Emission of AGN

The γ-ray domain of the electromagnetic spectrum is located at photon energies above ~500 keV, still containing the 511 keV electron-positron annihilation line. Due to absorption in the earth atmosphere, γ-ray observations in the MeV to GeV range have to be carried out from space. Currently this is mainly done by the Compton Gamma-Ray Observatory (CGRO), which was launched on April 5, 1991 (Fig. 4). CGRO is a NASA satellite, with contributional experiments from Europe, mainly Germany. CGRO carries four different experiments, which in total cover an energy range from 20 keV to ~30 GeV and which are supplementary in energy range. Their main characteristica are given in Table 2.

Table 2 The four scientific instruments aboard CGRO

Instrument	Energy Range [MeV]	Field-of-View
OSSE (Oriented Scintillation Spectroscopy Exp.)	0.05 - 10	3.8° x 11°
COMPTEL (Compton Telescope)	0.75 - 30	~1 sr
EGRET (Energetic Gamma-Ray Telescope Exp.)	20 - 30000	~0.5 sr
BATSE (Burst and Transient Source Exp.)	0.02 - 10	all-sky

Before CGRO only four AGN had been reported to emit detectable γ-rays. The radio galaxy Centaurus A ([2]) and the Seyfert galaxies NGC 4151 and MCG-8-11-11 (Perotti et al. 1981a, 1981b) up to energies of ~20 MeV. The quasar 3C 273 was the fourth one, which was only detected at the high-energy

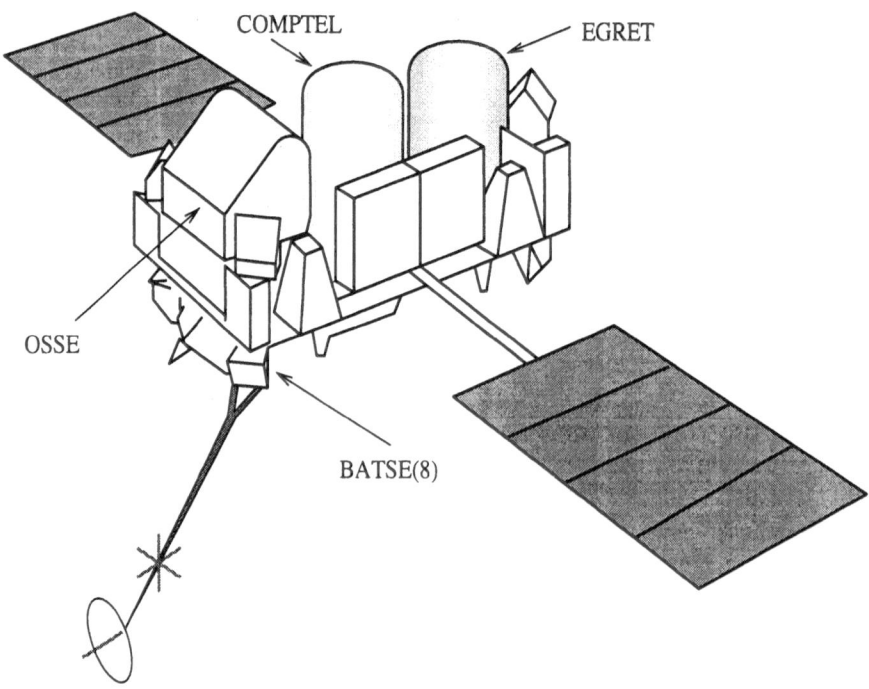

Fig. 4 Schematic view of the Compton Gamma-ray Observatory. The four different experiments are indicated. BATSE, in contrast to the others, is designed to be an all-sky monitor with the prime goal to observe γ-ray bursts. Therefore it consists of 8 detectors, which are located at the edges of the spacecraft and are pointing to different directions.

end above 50 MeV by the Cos-B satellite ([30]). In view of the improved sensitivities of the CGRO instruments with respect to previous experiments, it was expected prior to the launch that CGRO would detect several quasars, Seyfert, and radio galaxies. In addition, there was the hope for new and unexpected discoveries which is usually the case when improved instruments come into operation.

3.1   Blazars

Because 3C 273 was a prime candidate for γ-ray emission, shortly after launch the CGRO was pointed towards this quasar. Indeed, a strong γ-ray signal was immediately observed by the EGRET experiment. However, this signal was not at the 'right' location. It was offset by $\sim 8^{o}$ from the position of 3C 273, thereby

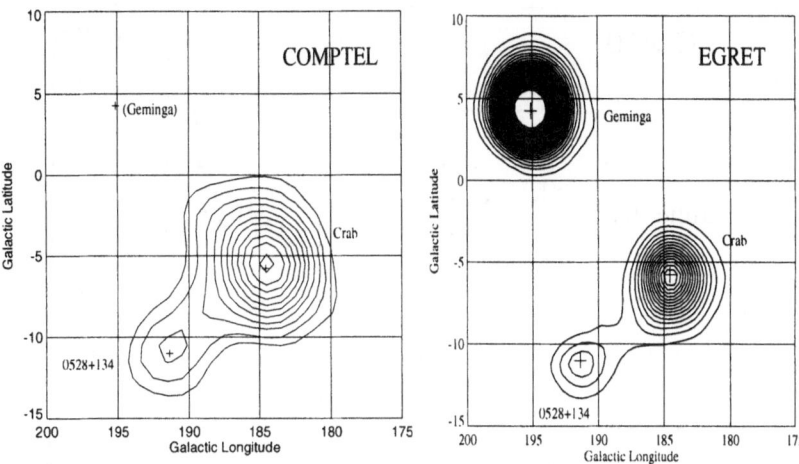

Fig. 5 Comparison of COMPTEL (10-30 MeV) and EGRET (>100 MeV) significance maps at the galactic anticenter. Three significant sources are seen by EGRET: the previously known galactic pulsars Crab and Geminga, and the newly discovered γ-ray blazar PKS 0528+134. The absence of Geminga in the COMPTEL maps is due to its hard spectrum, which causes it to fall below the COMPTEL detection threshold despite its prominence in the EGRET data. This figure is from Hartman et al. (1997).

causing some confusion within the EGRET and satellite operations teams. After the satellite pointing and the EGRET data analysis was confirmed to be correct, the strong γ-ray source was identified with the blazar-type quasar 3C 279. 3C 279, known to be a radio source and an optically violently variable quasar, was the most 'dramatic' object inside the EGRET location error box. Further observations of other sky regions often showed a fairly bright γ-ray source in the EGRET field-of-view and their error boxes always contained a blazar-type AGN, supporting the identification of 3C 279. So, after some time of operation, the conclusion that blazars can — at least occasionally — be strong emitters of γ-ray radiation became firm. Blazars are radio-loud quasars or BL Lac objects, which show a flat radio spectrum, strong and rapid variability in both optical and radio bands, and strong optical polarization. In addition, many blazars show superluminal motion of radio 'knots' resolved by very long baseline interferometry (VLBI).

After 5 years of operation, roughly 70 blazars have been detected by EGRET at energies above 100 MeV ([14]). Eight of them are observed also at low-energy γ-rays by COMPTEL ([6]), and also eight by the OSSE experiment ([20]) mainly at hard X-ray energies (∼50 keV to ∼500 keV). Figure 5 shows simultaneous COMPTEL and EGRET maps of the galactic anticenter region. The evidence for the newly discovered γ-ray blazar PKS 0528+134 is obvious in both maps.

The γ-ray fluxes are observed to be highly variable with variability time

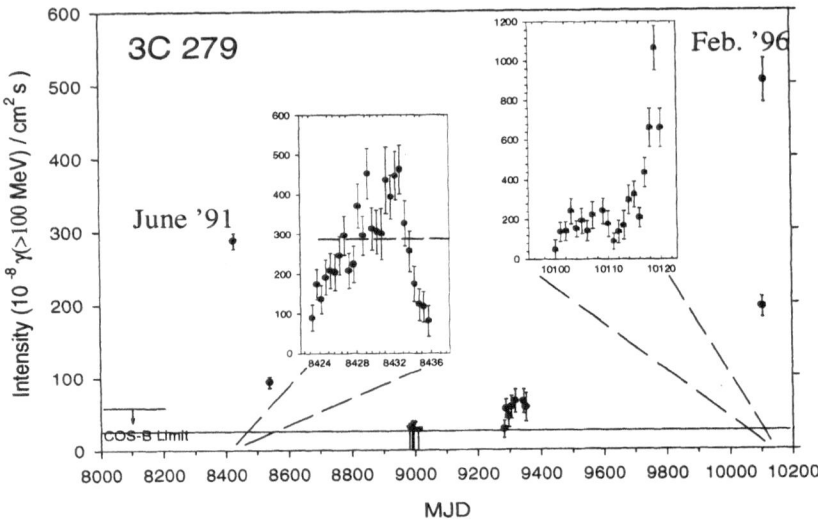

Fig. 6 Time history of fluxes, including flaring events, of 3C 279. For details see the text. The figure is from Hartman et al. (1997).

scales, e.g. doubling of the flux, of years down to less than one day. In some sources strong flux variations, up to a factor of 80, have been observed. A prominent example, the long-term γ-ray light curve of 3C 279 as observed by EGRET at energies above 100 MeV is shown in Figure 6. Short-term variability ($\sim$days) is clearly seen in the insets, which resolve the two major γ-ray flares observed from this blazar.

The energy spectra measured by the different CGRO instruments are consistent with power-law shapes. However, if they are combined a spectral turnover at MeV-energies becomes evident (Fig. 7) for several blazars. This indicates that the MeV-range is a spectral transition region for blazars (see also Figs. 8, 9). Integrating the observed spectra and assuming isotropic emission of the distant sources results in source luminosities of $\sim 10^{48}$ erg/s on average with a maximum of $\sim 5 \cdot 10^{49}$ erg/s for the γ-ray band above 1 MeV.

3.2   Seyfert - and Radio Galaxies

Seyfert galaxies, Type 1 as well as Type 2, have not fulfilled the expectations about prominent γ-ray emission. It was found by the OSSE experiment (e.g. [13]) that the spectra of Seyfert galaxies cut off at energies around $\sim$100 to $\sim$200 keV showing the signature of a thermal spectrum (see Fig. 10). No Seyfert galaxy has yet been detected by COMPTEL and EGRET.

Cen A, the closest active galaxy ($\sim$4 Mpc), is the only radio galaxy detected so far at hard X-rays and γ-rays. Cen A is a peculiar object with a Seyfert-2 type

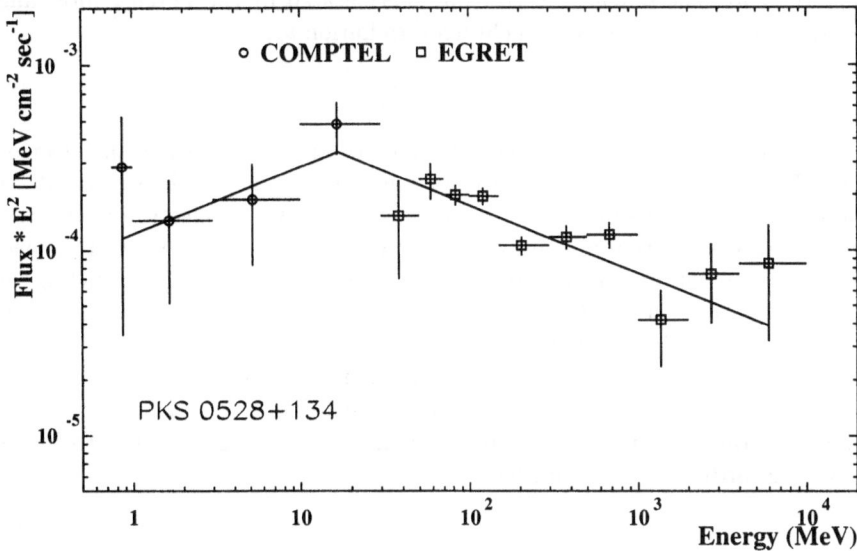

Fig. 7 Simultaneous EGRET and COMPTEL spectrum of the blazar PKS 0528+134 for a γ-ray flare. The solid line represents the best-fitting 'broken' power-law function. The spectral turnover at MeV-energies is evident. The figure is from Collmar et al. (1997).

nucleus with strong optical extinction. Cen A was re-detected by COMPTEL ([29]) and probably also by EGRET ([22]). However, source confusion is not completely ruled out in the case of the EGRET detection. Apart from Cen A no other radio galaxy has yet been detected by the CGRO experiments.

Because Seyferts are not γ-ray emitters and Cen A is a single and peculiar case we shall concentrate on the blazars for the rest of the paper.

4   Broad-band Properties of AGN

4.1   Gamma-Ray Production Processes

For the generation of γ-ray photons mainly non-thermal processes are responsible. Non-thermal means that the primary particles have a non-Maxwellian energy distribution, which is usually considered to be a power-law shape. Thermal emission is only of minor importance for the γ-ray emission, because in astrophysical environments the necessary plasma temperatures are only reached in exceptional cases. There are several processes to generate γ-ray photons (e.g. [27]). Here we shall concentrate on the ones important for the γ-ray blazars.

Synchrotron radiation is emitted when relativistic electrons interact with a magnetic field. The moving electrons are accelerated by the Lorentz force and radiate. The frequency of the synchrotron radiation ν_s,

$$\nu_s \propto B \cdot E_e^2, \tag{2}$$

is proportional to the magnetic field strength B, and the square of the electron energy E_e. For example, 10^{13} eV electrons moving in a magnetic field of 10^{-4}T would generate γ-ray synchrotron photons of $\sim$1 MeV. An electron cloud with a power-law distribution in energy will generate a power-law synchrotron spectrum. For a particle number spectrum of the form $N(E) \propto E^{-s}$ and a synchrotron photon number spectrum $N(E) \propto E^{-\alpha}$ the simple relation $\alpha=(s+1)/2$ between the two power-law indices holds.

The inverse Compton process describes the interaction between a low-energy photon and a high-energy electron. In this scattering process energy from the electron to the photon is transferred and the photons are scattered to higher energies according to the equation

$$\nu_{IC} \approx E_e^2 \cdot \nu_0, \tag{3}$$

where ν_{IC} is the frequency of the inverse Compton photon, ν_0 the frequency of the soft photon, and E_e the energy of the electron. For example, a MeV γ-ray is generated by the interaction of a UV-photon ($\sim 10^{15}$ Hz) with a 0.5 GeV electron. Like in the synchrotron process also in the IC-process a power-law shaped electron distribution in energy generates a power-law shaped IC-spectrum with the same simple relation between the spectral indices.

The pair-production process is important in regions of high photon densities. This process describes the interaction of a high-energy γ-ray with another photon or the field of an atomic nucleus. Schematically this can be written as

$$\gamma + \gamma \rightarrow e^- + e^+, \quad \gamma + nucleus \rightarrow e^- + e^+ . \tag{4}$$

Because of simultaneous energy and momentum conservation this conversion cannot occur for a single γ-ray photon. In the case of photon-photon interaction the following condition is valid:

$$E_1 \cdot E_2 \geq \frac{2(m_e c^2)^2}{(1 - cos\alpha)}, \tag{5}$$

with $E_{1,2}$ the photon energies, m_e the electron rest mass, c the speed of light, and α the interaction angle between the photons. The inverse process

$$e^- + e^+ \rightarrow \gamma + \gamma \tag{6}$$

is also possible. The 511-keV annihilation line has been observed from the inner part of our galaxy proving that the pair-production process (generation of positrons) occurs in real astrophysical environments.

4.2   Multifrequency Observations of Gamma-Blazars

The CGRO observations of γ-ray blazars have shown that these sources can
be enormously luminous at γ-ray energies and at the same time show short-
term variability. Simultaneous multiwavelength spectra, and especially multiwavelength monitoring have the potential of significantly increasing our knowledge about the physics of these spectacular sources. In simultaneous spectra
from radio to γ-ray energies the luminosity of the objects in different bands
can be directly compared providing the possibility to identify the major emission components and processes. Especially valuable would be the multifrequency
monitoring because clues about correlated/uncorrelated time variability of different frequency bands are provided thereby revealing informations about their
connections and possibly about their emission sites. For example, the information whether an X-ray or optical flare occurs simultaneously, earlier, later, or
is even absent during a γ-ray flare would be very constraining for the blazar
modelling. To proceed in this direction several multifrequency campaigns have
been organized around CGRO blazar observations. While the individual observations are easy to carry out, simultaneous observations of a particular source
are difficult to achieve. Because telescope time is precious, especially on space
experiments, severe pointing constraints exist, and scientific review panels have
to be convinced, only a few examples of such campaigns were carried out with
good coverage from the radio to the γ-ray band.

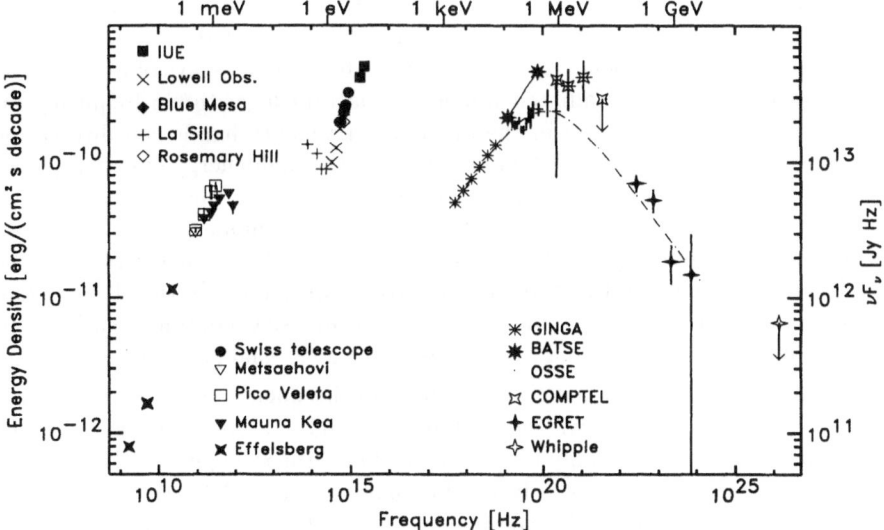

Fig. 8 Quasi-simultaneous broad-band spectrum of the blazar 3C 273 (Lichti et al.
1995).

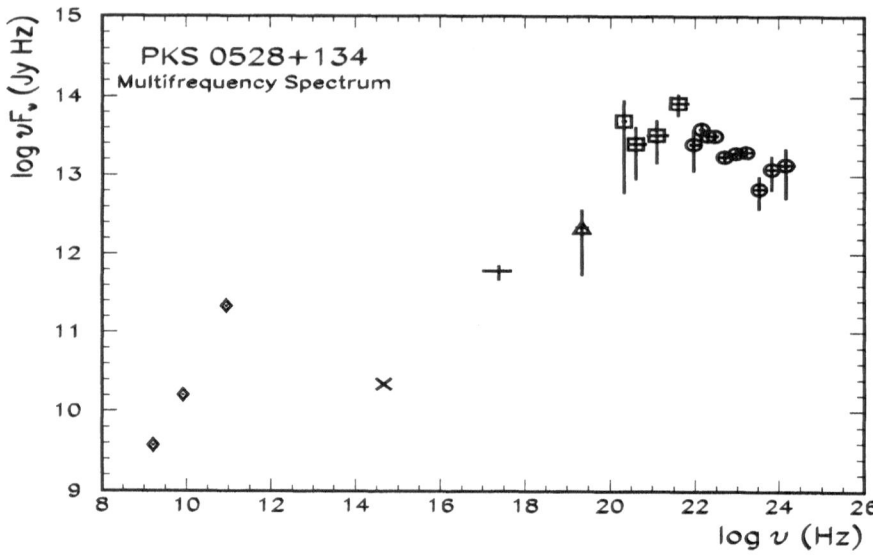

Fig. 9 Broad-band spectrum of PKS 0528+134 during a γ-ray high state (Collmar et al. 1997).

To date, probably the best result was derived for the blazar 3C 273, which is one of the best studied sources because it is the closest (z = 0.158) blazar-type quasar and is bright at all wavelength bands. Figure 8 ([16]) shows a quasi-simultaneous multifrequency spectrum from radio to TeV γ-rays of 3C 273. The ordinate represents the radiated power per natural logarithmic frequency interval, so the energy release in the different wavelength bands can directly be compared. First of all the spectrum shows that the high-energy emission (X- to γ-rays) is a significant part of the bolometric luminosity. The spectrum shows (probably) four maxima indicating different emission components and processes, which are interpreted as synchrotron emission from relativistic electrons in the radio- and far-IR band, thermal emission from a dust torus in the IR and from an accretion disk in the UV ('blue bump'), and inverse Compton radiation generated by relativistic electrons and soft photons at X- and gamma-rays.

Figure 9 ([7]) shows a non-simultaneous multifrequency spectrum of the blazar PKS 0528+134, which with a redshift of z=2.06 ([12]) is a very far one. This spectrum shows probably only two maxima. One in the radio/infrared region and one in the γ-ray band, which clearly dominates the overall power output. The spectrum is interpreted to be completely of non-thermal origin with only two visible emission mechanisms at work: synchrotron emission from radio to optical and inverse Compton emission from X-rays to γ-rays (e.g. [5]). The source is too far away to show the thermal signatures from the dust torus and the accretion disk.

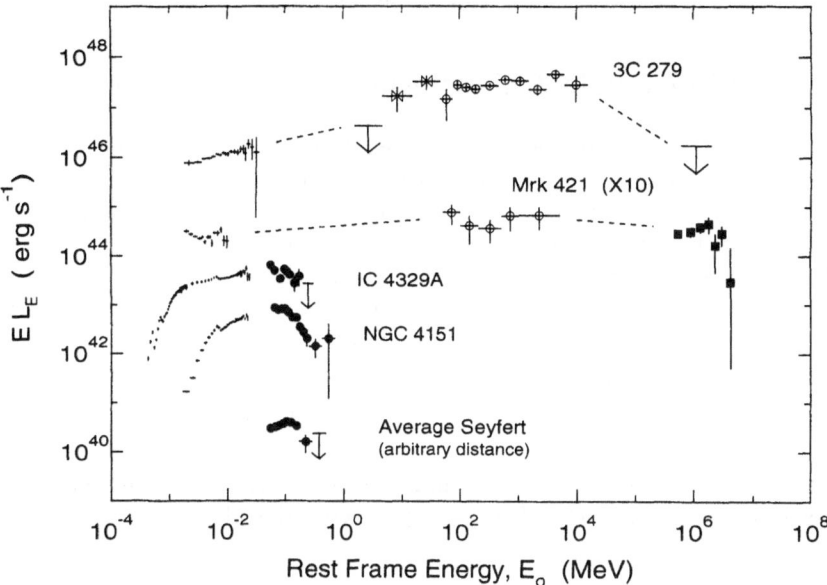

Fig. 10 Multiwavelength power spectra for the 'average' Seyfert galaxy, the bright Seyferts NGC 4151 and IC 4329A, the BL Lac-object Markarian 421, and the blazar 3C 279. The figure clearly shows, that in contrast to both blazars, the high-energy Seyfert spectra cut off around 100 keV, indicating thermal instead of a non-thermal spectra. The figure is from Dermer and Gehrels (1995).

Figure 10 (from [9]) compares the high-energy luminosity spectra (assuming isotropic emission) of different sources and source classes as function of their rest frame energy. This figure allows to compare the amounts of radiated energy. The different spectral behaviour of blazars and Seyferts becomes clearly obvious from this diagram. It is obvious that the flat spectrum radio quasars (3C 279 as a bright example is shown) are by far the most luminous sources at these energies. The BL Lac objects (e.g. Mkn 421) can also be γ-ray emitters, but are usually less luminous than the quasars. However, in contrast to the quasars three BL Lac objects have been detected at TeV γ-rays ([32]). Only the closest Seyfert galaxies are observable at all (z<0.06). They are no γ-ray emitters because thermal spectra are observed which are assumed to originate from a hot thermal plasma (kT$\sim$30 keV) and cut off at energies around $\sim$100 to $\sim$200 keV (e.g. NGC 4151 and IC 2943A).

Correlated monitoring observations have shown that detectable γ-ray emission in blazars is generally present in coincidence with enhanced radio activity. Sometimes correlated activity between γ-rays and other wavelength bands (e.g. X-rays, optical) has been observed. However, since the emission of these sources is always variable, these activities cannot be correlated unambiguously.

5   Model-independent Conclusions from Observations

Several observational facts for AGN in general and γ-ray blazars in particular have been summarized in the previous sections. In this section we want to apply some basic physical concepts to derive conclusions on physical parameters of active galaxies like their source of energy and their central masses.

The compactness of the central source is defined by the L/R-ratio, with L being its luminosity and R its radius. For a normal galaxy like our 'Milky Way' for example we derive an order of magnitude value of

$$\frac{L}{R} = \frac{10^{44} \text{ erg/s}}{15 \text{ kpc}} = 2 \cdot 10^{21} \frac{\text{erg/s}}{\text{cm}} labelcoll : sect4 - 1 \tag{7}$$

However, for γ-ray blazars with luminosities of more than $\sim 10^{49}$ erg/s and time variability of less than a day, a compactness value of

$$\frac{L}{R} = \frac{10^{49} \text{ erg/s}}{10^{15} \text{ cm}} = 10^{34} \frac{\text{erg/s}}{\text{cm}} \tag{8}$$

is derived. The difference in compactness of more than 10 orders of magnitude clearly indicates the different source natures.

The luminosity of a normal galaxy is roughly the sum of the luminosity of its individual stars, and so is generated by nuclear fusion. In active galaxies there is considerably more energy generated in much smaller volumes, even in a time-variable fashion. Only gravitation can provide more energy than nuclear fusion. If one assumes that a galaxy is a sphere of radius R, then for a given mass M the energy release from fusion and gravitation is equal, if

$$\frac{3}{5} \frac{GM^2}{R} = 0.007 Mc^2, \tag{9}$$

with G the gravitational constant. Therefore the radius R_{crit}, for equal energy release is

$$R_{crit} = 43 \cdot \frac{2GM}{c^2} = 43 \cdot R_S. \tag{10}$$

If R_{crit} is less than 43 Schwarzschild radii (R_S), energy generation by gravitation is more efficient than by nuclear fusion. Let us assume that an active galaxy would be powered by fusion, then its maximal compactness is

$$\frac{L}{R} = \frac{E}{R_{crit} \cdot T_{life}} = \frac{0.007 Mc^2}{43 \cdot \frac{2GM}{c^2} \cdot T_{life}} = 5 \cdot 10^{29} \frac{\text{erg/s}}{\text{cm}} \tag{11}$$

for a lifetime (T_{life}) of 10^8 years. This result shows that fusion can at most provide a compactness of $10^{30} \frac{\text{erg/s}}{\text{cm}}$, which is in conflict to observations of AGN.

Even this value is only an upper limit because nuclear fusion provides thermal energy which would have to be converted into non-thermal radiation energy as observed in blazars. In addition, the observed short-term variability in blazars is hard to understand if fusion is powering active galaxies. These arguments lead to the conclusion that active galaxies are powered by gravitational energy of a central massive object.

The AGN luminosity is generated by the release of gravitational energy of the infalling matter in the gravitational field of a central massive object. The mass of this object can be estimated by the physics of the accretion process. The so-called Eddington limit describes the balance between the gravitational force and the radiation pressure on the accreting material. If the radiation pressure dominates, the accretion stops. For spherical accretion in the Thomson regime the Eddington limit for the luminosity of a source as function of the central mass is given by

$$L_{Edd} = 1.3 \cdot 10^{38} \frac{M}{M_\odot} \frac{erg}{s}. \tag{12}$$

If a source is radiating at its Eddington limit with a luminosity of 10^{46} erg/s, its mass is of the order of $10^8 \, M_\odot$. Therefore nuclei of active galaxies should have masses in the range between $10^6 \, M_\odot$ and $10^{10} \, M_\odot$. If the source is radiating below the Eddington limit, even larger masses would result.

The Eddington limit of the accretion process provides also clues on the build-up times and the lifetimes of AGN. The released gravitational energy has to be converted into the observed radiation energy:

$$L = \epsilon \cdot \dot{m}_{acc} c^2, \tag{13}$$

with L being the source luminosity, ϵ the efficiency for converting gravitational energy into radiation, $\dot{m}_{acc}$ the mass accretion rate, and c the speed of light. This equation has two consequences. Firstly, massive central objects of $10^8 \, M_\odot$ cannot be build up quickly, because at high accretion rates the corresponding strong outpushing radiation stops the accretion process. Secondly, the 'active' lifetimes of AGN are limited. After the source has accreted all material from its surroundings the activity stops. An ϵ of 0.1, which is the favoured number to date, leads to build-up timescales of the order of 10^8 to 10^9 years and lifetimes of the order of several times 10^8 years for typical accretion rates of ~ 1 to $\sim 2 \, M_\odot$ per year. The powering of AGN by gravitational energy implies that bright active galaxies could not exist very early in the universe (unless they are born very massive) and cannot shine forever. Their brightness ultimately has to stop, when they simply have 'eaten up' their host galaxies. Because they are not radiating forever, many 'normal' galaxies might contain 'quiet' massive centers, which are not 'fed' anymore.

Blazars emit polarized non-thermal radio emission: synchrotron radiation. This fact proves the presence of magnetic fields and relativistic electrons as well as an operating acceleration mechanism for the electrons. For large compactness,

$\frac{L}{R} \geq 10^{30} \frac{\text{erg/s}}{\text{cm}}$, the relativistic electrons cannot escape without interacting with their own synchrotron photons by the inverse Compton process, i.e. boosting the self-generated synchrotron photons to higher energies like X- and γ-ray energies. So high-energy emission is a natural consequence of high compactness and the presence of relativistic electrons.

The large γ-ray luminosities of blazars together with the observed short-term variability implies a highly compact emission region of the γ-ray radiation as well. For photons above the threshold energy $m_e c^2 = 511$ keV, the optical depth for pair production of a source of size R is given by $\tau_{\gamma\gamma} = n_\gamma \sigma_p R$, with n_γ the photon density and σ_p the cross section for pair production. A source will be optically thick for pair production if (e.g. [24])

$$\tau_{\gamma\gamma} = \left(\frac{L_\gamma}{4\pi R^2 m_e c^3} \right) \sigma_e R = \frac{\sigma_e}{4\pi m_e c^3} \cdot \frac{L_\gamma}{R} = 2 \cdot 10^{-30} \frac{\text{cm s}}{\text{erg}} \cdot \frac{L_\gamma}{R} \geq 1, \quad (14)$$

where σ_e is the Thomson cross section, L_γ the γ-ray luminosity, and R the source radius. For isotropic luminosities of 10^{48} erg/s and radii of $2.6 \cdot 10^{15}$ cm (1 light day), as observed in the γ-ray blazar 3C 279 for example, an optical depth for pair production of

$$\tau_{\gamma\gamma} = 770 \quad (15)$$

is derived, which simply means that γ-rays cannot escape from such a source region without generating electron/positron pairs. Nevertheless, these γ-rays are observed! So, they have to come from an optically thin region. To resolve this contradiction, beamed emission from a relativistic jet is considered instead of central isotropic emission which was assumed in the calculations above. A jet origin of the observed γ-rays is consistent with the facts that for many blazars superluminal motion has been observed which is indicative for jets with a small offset angle from our line-of-sight, and with the redshift distribution of these sources, which shows that the source distance is not the critical parameter for their detection because γ-ray blazars are observed far out into the universe up to redshifts of z=2.3 ([11]). A relativistic jet origin of the γ-rays would imply beamed emission. For beamed emission the angle between the photons is nearby zero, therefore absorption by pair production is shifted to extremely high γ-ray energies (see equ. 5). The beamed emission has another effect: the observed γ-ray luminosity overestimates the internal generated luminosity by a factor of up to $\sim 10^4$ due to relativistic Doppler boosting of the emitted photons and the solid angle effect. Therefore internally the optical depth for pair production is well below 1, which is in accord with the observation of these γ-rays. A luminosity reduction of up to 10^4 does not change the conclusion reached above on the issue whether nuclear fusion or gravitational energy powers active galaxies, because 1) the given upper limit for compactness does not contain an efficiency factor for conversion of thermal energy to non-thermal energy and 2) the observed short-term variability in blazars is not conceivable to be powered by fusion.

The Center of an AGN

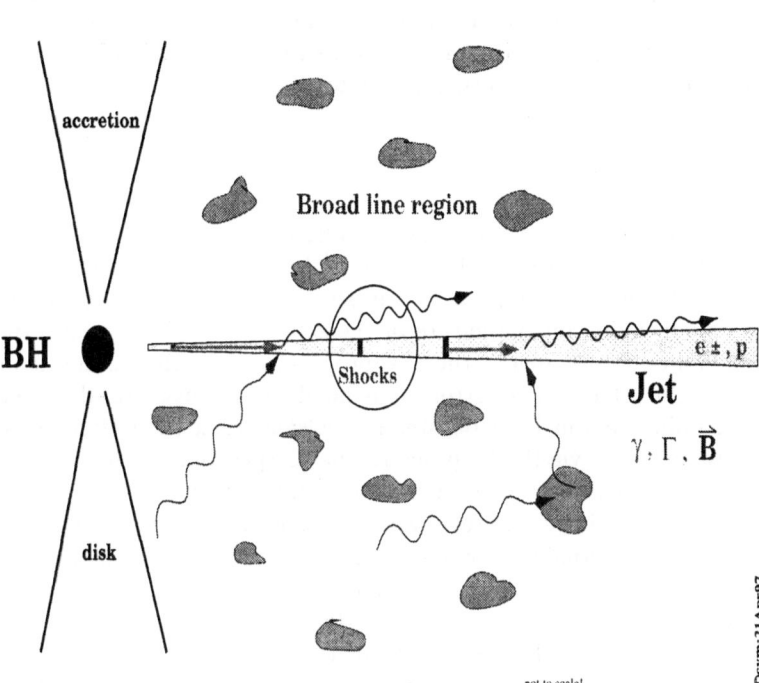

Fig. 11 Sketch (not scaled) of the inner part of an active galaxy. The γ-ray emission is thought to be generated in the inner jet region, a few hundreds of Schwarzschild radii apart from the central supermassive BH, where blobs of relativistic particles are injected and accelerated in the jet. The figure is from v. Montigny (1997).

6   Models for the Gamma-ray Emission of Blazars

Since the CGRO discovery of γ-ray blazars, the origin of their γ-ray emission has been widely discussed. The models have been developed within the framework of the standard picture of radio-loud AGN (Fig. 3). Isotropical core emission has to be excluded. The compactness of the emission region inferred from the observed γ-ray fluxes and time variability would lead to severe absorption of the γ-ray radiation by pair production. The favoured scenario to explain the blazar radio-through-optical continua is that we are viewing nearly along the axis of a relativistically outflowing plasma jet which has been ejected from an accreting supermassive black hole. This broadband radiation is thought to be produced by non-thermal electron synchrotron radiation in outflowing plasma blobs. Figure 11 sketches the central part of an AGN, where according to models the γ-ray emission is generated. The high-energy blazar continuum emission appears to constitute a distinct second component in the broadband spectral energy distribution of blazars. Two classes of models have been proposed to explain the blazar γ-radiation, where either leptons or hadrons are the primary accelerated particles, which then radiate directly or through the production of secondary particles which in turn emit photons.

6.1   Leptonic Models

In leptonic models, the γ-ray emission of blazars is produced by non-thermal relativistic leptons (electrons and positrons) which scatter soft photons to γ-ray energies via the inverse-Compton (IC) process. These leptonic models come in two flavors depending on the nature of the soft photons. The synchrotron-self Compton models (SSC) assume the SSC process to be dominant. In this process (e.g. [19, 4]) the relativistic electrons moving along the magnetized jet generate synchrotron photons which are boosted by the same relativistic electron population to γ-ray energies via the IC-process. The IC-spectrum follows to first order the shape of the synchrotron spectrum (just shifted to higher energies), which explains the observed spectral turnover at MeV-energies in the γ-ray spectra.

The so-called external Compton scattering (ECS) models consider a different origin for the soft photons. They assume the soft X-ray and UV-photons which are radiated from the accretion disk directly into the jet (e.g. [8]) or scattered into the jet by the broad line region clouds ([28]), to be the soft target photons for the relativistic jet electrons. The ECS-models also can reproduce the broadband non-thermal spectral shape of the γ-ray blazars. These models explain the spectral bending at MeV-energies by the so called incomplete Compton cooling of the electrons. When a blob of relativistic electrons is injected into the jet, a power-law shaped IC-spectrum is generated with low - and high-energy cutoffs corresponding to the low - and high-energy cutoffs in the electron spectrum. Because the high-energy photons cool first, the high-energy cutoff in the IC-spectrum moves towards lower energies with time. The electron cooling by the IC-process stops when the blob has moved out into regions where the photon field becomes too thin to maintain this process. Integrating these spectra over time and over many outmoving blobs generates the spectral turnover at MeV energies observed by CGRO in time-averaged (days to weeks) γ-ray spectra ([8]).

6.2   Hadronic Models

Models have also been proposed in which accelerated hadronic particles (mainly protons) carry the bulk of the energy (e.g. [17]). Because protons do not suffer severe radiation losses, they can be accelerated up to energies of 10^{20} eV ([18]), reaching the thresholds for photo-pion production, and the threshold for pion production in inelastic proton-matter collisions ([3]). In these processes the protons transfer energy into photons, pairs, and neutrinos via pion production. The photons and pairs are reprocessed and initiate a cascade through inverse Compton and synchrotron processes to form a power-law photon spectrum in the end. Electrons and positrons generated via charged pion decay will be accompanied by energetic neutrino production. The detection of a strong neutrino flux from blazar jets would definitively identify hadrons as the primary radiating particles.

7 Summary and Conclusions

Recent γ-ray observations of AGN by CGRO have revealed a wealth of new informations on the γ-ray properties of AGN. Unexpectedly, blazar-type AGN have been found to be occasionally strong emitters of γ-ray radiation. On the other hand, Seyfert-type AGN have not fulfilled the expectations. They are found to be strong emitters of hard X-ray radiation but no γ-ray emitting sources. These results fit well into the standard model of AGN and therefore provide strong support for this model, which assumes a supermassive BH as the central engine of active galaxies. The γ-ray blazars are sources in which a relativistic jet is more or less directly pointed towards us. The γ-ray luminosity produced in the jet is relativistically boosted and therefore becomes visible for the CGRO γ-ray telescopes.

Taking relativistic beaming into account, the internally generated γ-ray luminosities are estimated to be of the order of 10^{44} to 10^{45} erg/s (roughly the overall luminosity of our whole Milky Way), which is switched on and off on timescales of days. Only the release of gravitational energy in the field of a massive central object, which is the most efficient process known in nature, can provide the necessary energetics.

With isotropic luminosities up to 10^{44} erg/s ([9]) mass estimates via the Eddington limit provide central masses of the order 10^6 to $10^7\,M_\odot$ for Seyferts. For blazars 10^8 to $10^{10}\,M_\odot$ have been estimated from γ-ray observations (e.g. [20, 7]). However, since the emission in blazars is probably beamed, the derived blazar masses have to be taken with caution.

According to models explaining the γ-ray emission of blazars, the γ-ray observations probe the physics of the inner jet region. They assume the injection of γ-ray radiating blobs at distances of several hundreds Schwarzschild radii ($100 < R_S < 300$) from the central engine. For a central mass of $10^8 M_\odot$, 500 R_S are $\sim 1.5 \cdot 10^{11}$ km, ~ 1000 AU or ~ 25 times the size of the solar system. If these models are correct, only a BH is viable to be the central massive object. No other object like a massive star cluster, for example, is comprehensible of being stable in such a 'massive' environment.

The γ-ray blazars are found to be jet sources. The large energies involved in the γ-ray emission ultimately provided by the even larger kinetic energy of the AGN jet, which in turn is in the end supplied by the central engine, is - according to our current understanding of physics - only conceivable to be due to gravitational energy release in the vicinity of a massive BH.

In the line of these arguments, the recent CGRO γ-ray observations of AGN provide clear evidence for massive BH in the nuclei of active galaxies.

Bibliography

[1] Antonucci, R.R.J., Miller, J.S. (1995): ApJ **297**, 621

[2] Ballmoos, P., Diehl, R., Schönfelder, V. (1987): ApJ **312**, 134

[3] Bednarek, W. (1993): ApJ **402**, L29

[4] Bloom, S.D., Marscher, A.P. (1993): *AIP Conference Proceedings 280*, (St. Louis), 578

[5] Böttcher, M., Collmar, W. (1998): A&A, submitted

[6] Collmar, W. (1996): *Workshop on Gamma-Ray Emitting*, eds. J.G. Kirk, M. Camenzind, C. v. Montigny, S. Wagner (Heidelberg), 9

[7] Collmar, W., Bennett, K., Bloemen, H., et al. (1997): A&A, in press

[8] Dermer, C.D., Schlickeiser, R., (1993): ApJ **416**, 458

[9] Dermer, C.D., Gehrels, N. (1995): ApJ **447**, 103

[10] Hartman, R.C., Webb, J.R., Marscher, A.P. et al. (1996): ApJ **461**, 698

[11] Hartman, R.C., Collmar, W., v. Montigny, C., Dermer, C.D. (1995): in *Proc. of the 4th Compton Symposium*, eds. J. Kurfess, C. Dermer, in press

[12] Hunter, S.D., Bertsch, D.L., Hartman, R.C., et al. (1993): ApJ **409**, 134

[13] Johnson, W.N., Zdziarski, A.A., Madejski, G.M., et al. (1997): in *Proc. of the 4th Compton Symposium*, eds. J. Kurfess, C. Dermer, in press

[14] Kanbach, G. (1996): *Workshop on Gamma-Ray Emitting*, eds. J.G. Kirk, M. Camenzind, C. v. Montigny, S. Wagner (Heidelberg), 1

[15] Kellermann, K.I., Sramek, R., Schmidt, M., et al. (1989): AJ **98**, 1195

[16] Lichti, G.G., Balonek, T., Courvoisier, T.J.-L., et al. (1995): A&A **298**, 711

[17] Mannheim, K., Biermann, P.L. (1992): A&A **253**, L21

[18] Mannheim, K. (1992): A&A **269**, 67

[19] Maraschi, L., Ghisellini, G., Celotti, A. (1992): ApJ **397**, L5

[20] McNaron-Brown, K., Johnson, W.N., Jung, G.V., et al. (1995): ApJ **451**, 692

[21] v. Montigny, C. (1997): Talk at the '4th Compton Symposium', Williamsburg, 1997

[22] Nolan, P., Bertsch, D.L., Chiang, J., et al. (1996): ApJ **159**, 100

[23] Perotti, F., Della Ventura, A., Villa, G., et al. (1981a): ApJ **247**, L63

[24] Peterson, B. M. (1997): *An Introduction to active galactic nuclei* (University Press, Cambridge)

[25] Robson, I. (1997): *Active Galactic Nuclei* (Wiley, Chichester, New York, Brisbane, Toronto, Singapure)

[26] Perotti, F., Della Ventura, A., Villa, G., et al. (1981b): Nature **292**, 133

[27] Schönfelder, V. (1995): *Physik in unserer Zeit* (Weinheim), 6/95, 262

[28] Sikora, M., Begelman, M.C., Mitchel, C., et al. (1994): ApJ **421**, 153

[29] Steinle, H., Bennett, K., Bloemen, H., et al. (1998): A&A, accepted

[30] Swanenburg, B.N., Hermsen, W., Bennett, K., et al. (1978): Nature **275**, 298

[31] Urry, M., Padovani, P. (1995): PASP**107**, 803

[32] Weekes, T.C., Aharonian, F., Fegan, D.J., Kifune, T. (1997): in *Proc. of the 4th Compton Symposium*, eds. J. Kurfess, C. Dermer, in press

First Conclusive Evidence for a Massive Black Hole in the Center of the Milky Way

Andreas Eckart and Reinhard Genzel

Max-Planck Institut für extraterrestrische Physik, Postfach 16 03, D-85748 Garching, Germany

Abstract. In the last few years near-infrared imaging and spectroscopy at a high angular resolution has made it possible to determine stellar velocities down to separations of less than five light days from the compact radio source SgrA* that is located in the constellation Sagittarius at the dynamic center of the Milky Way. These measurements make a convincing case for the presence of a compact, central dark mass of 2.6×10^6 solar masses. Via simple physical considerations one can show that this dark mass cannot consist of a stable cluster of stars, stellar remnants, substellar condensations or even a degenerate gas of elementary particles. Energy equipartition requires that at least 10^5 out of the 2.6×10^6 solar masses must be associated with the source SgrA* itself. This mass is very likely enclosed within less than 8 light minutes which corresponds to 15 Schwarzschild radii of a million solar mass black hole. Accepting these arguments one must conclude that a massive black hole is located at the core of the Milky Way.

1 Introduction

Do massive black holes exist at the centers of galaxies? The answer to this question is of considerable importance for the understanding of 'active galactic nuclei' (AGNs, quasars, e.g.) and their evolution in the early Universe. In the centers of these objects luminosities of up to 10^{14} $L_\odot$ (one solar luminosity $L_\odot$ corresponds to 4×10^{26} Watt) are produced within a light year or less. Highly collimated jets of relativistic electrons and rapidly varying X- and γ-ray emission provide strong but indirect evidence that AGNs cannot be powered by stars but by the conversion of gravitational energy to radiation in accretion flows onto massive black holes. For a direct proof of the 'black hole' paradigm it is necessary, however, to determine the characteristic mass concentration and to show the existence of an event horizon. Probably the most unambiguous method for carrying out such a proof is the determination of the form of the gravitational potential from the velocity field of stars and gas orbiting the hole candidate. Using this technique, ground-based and Hubble Space Telescope observations of the Doppler shifts of spectral lines from gas and stars have shown indeed that many (and perhaps most) nearby galaxies have massive dark mass concentrations in their nuclei (e.g. Kormendy and Richstone 1995). With the exception of detailed radio Very-Long-Baseline-Interferometry (VLBI) observations of the galaxy NGC 4258 (Myoshi et al. 1995), none of these measurements have sufficiently high resolution so far for proving that the central dark mass must be a black hole and could not be, for instance, a dense compact cluster of stellar remnants. In contrast, the nucleus of the Milky Way (distance ~8 kpc corresponding to 26100 light years) is a thousand times closer than the nearest AGN and one hundred thousand times closer than the nearest quasar. Thus it is a unique laboratory for testing the black hole paradigm. The Galactic Center, however, is hidden by dust so that observations in the visible are impractical. With the advent of sensitive infrared detectors, high resolution images and imaging spectrometers, it has recently become possible to study the stellar velocity field at unprecedented resolution and to provide the best evidence so far for a massive black hole at the nucleus of a galaxy.

2   Initial evidence for a mass concentration in the Galactic Center

The first indications for a central mass concentration in the Milky Way emerged in the late seventies from spectroscopic observations of a mid-infrared fine structure line of Ne^+ (Wollman et al. 1977, Lacy et al. 1979). These measurements showed unusually large Doppler shifts (± 250 km/s) of ionized gas clouds in the central parsec towards the maximum stellar density. As radio interferometric observations led to the discovery of a compact, non-thermal radio source SgrA* in the same region (Balick and Brown 1974), a plausible interpretation – in analogy to quasars – was that the large gas velocities indicate orbital motions in the vicinity of a million solar mass black hole, coincident with SgrA* (Lynden-Bell and Rees 1971, Lacy, Hollenbach and Townes 1982). Further infrared (and radio) spectroscopic data, taken by various groups in the eighties, strengthened the gas dynamic evidence for this central mass concentration (e.g. Serabyn and Lacy 1985, Genzel and Townes 1987) but were not considered as compelling by many researchers in the field. In addition to gravitational forces, gas may be affected by magnetic fields, radiation pressure, stellar winds and friction with other gas components – all known to be present in the Galactic Center – thus making the interpretation uncertain. Beginning in the late eighties, several groups began measuring radial velocities of late type red giants and supergiants (e.g. Rieke and Rieke 1988, Sellgren et al. 1990, Krabbe et al. 1995, Haller et al. 1996). These measurements confirm and strengthen the evidence for the presence of a 1 to 3×10^6 $M_\odot$ central mass that cannot be accounted for solely by the stellar cluster that is sampled by the near-infrared light.

3   Status as of 1997

In the last few years, it has become possible to measure the stellar velocity field down to scales as small as 5 light days. Thus one is able to place decisive constraints on the nature of the central mass concentration. This significant progress was made possible on the one hand due to the discovery (by Forrest et al. 1987, Allen et al. 1990 and Krabbe et al. 1991) of a ompact cluster of hot, luminous emission line stars (the so called 'HeI stars'). Apart from being interesting in their own right (these stars must have formed in the last few million years and now power the central parsec), they provide radial velocity measurements to a scale of $\sim 1''$ (0.04 pc). The other and – as it turns out – most important development has been the first measurement of stellar proper motions.

Figure 1 shows a $0.15''$ resolution, $2\,\mu m$ image obtained with the MPE SHARP camera on the 3.5 m New Technology Telescope (NTT) of the European Southern Observatory (ESO). The excellent resolution, image quality and high dynamic range of these images (the ratio between weakest and strongest sources is $\sim 10^{-4}$) are the result of combining 'speckle imaging' techniques (coadding many short exposure images) with non-linear deconvolution techniques that remove the very substantial image artifacts in speckle imaging. These near-infrared images show close to 10^3 stars in the central parsec, concentrated and centered on or very near the compact radio source SgrA*. Using a novel field imaging spectrometer, 3D, Genzel et al. (1996) have been able to determine radial velocities for about 220 of these stars (see also Krabbe et al. 1995, Haller et al. 1996). Combining about 60 independent high resolution images between 1992 and 1997, Eckart and Genzel (1996, 1997 and unpublished) and Genzel et al. (1997) derived (relative) proper motions for about 70 stars. Figure 2 (bottom) shows two examples of the data obtained. In the upper section of Figure 2, the derived proper motion vectors (without error bars) are plotted for a number of stars on the high resolution image, assuming a Sun–Galactic Center distance of 8 kpc.

Of special interest is, naturally, the immediate vicinity of SgrA* (right upper inset in Figure 2) where one finds a $\sim 1''$ diameter concentration of faint stars. Several of these stars in this so-called SgrA* cluster show proper motions in excess of 1000 km/s (stars further out have velocities of only a few hundred km/s), the fastest one (S1: v$\sim$1400 km/s) also being the closest one ($\sim 0.13''$) to SgrA*. This finding is exactly what one would expect if SgrA* were coincident with a large compact mass. Because of the obvious importance of these large motions and the substantial technical difficulties in deriving reliable proper motions of faint stars in such a crowded environment, a confirmation of these results was critical. This has now happened. A group from the University of California, Los Angeles, has used the 10 m Keck telescope on Mauna Kea, Hawaii, and carried out yet higher resolution $2\,\mu m$ imagery of the central few arcseconds. Combining data from three epochs, 1995, 1996 and 1997, this group fully confirms the very high velocities of the SgrA* cluster stars (Ghez et al. 1997). It is also important to ascertain that the observed positional changes indeed present orbital motions in the central gravitational field and that the stars are actually located in the Galactic Center. One can easily show that the positional changes cannot be caused, for instance, by variability in a double or multiple stellar system or due to a central gravitational lens (Eckart and Genzel 1996, 1997). More random variability of unrelated stars can also cause apparent motions but the consistency of the data within a given epoch and the continuous and steady positional

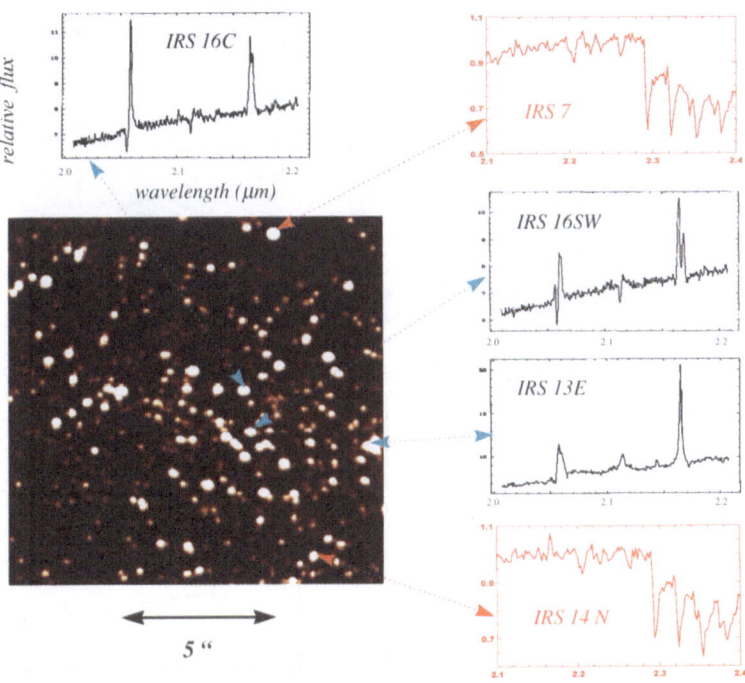

Fig. 1 False color near-infreared (2 μm) image of the central 10″ (0.39 pc) of the Galaxy, obtained with the MPE SHARP camera in 1994 on the ESO NTT (Eckart et al. 1995), along with a selection of stellar spectra obtained with the MPE 3D spectrometer (Genzel et al. 1996 and unpublished). The brightest stars are overexposed and are about 200 times brighter than the faintest stars. Stars with emission lines (from HeI and HI) are luminous and massive blue supergiants. Stars with absorption bands (due to CO, NaI and CaI) are cool red giants and supergiants. From these spectra radial velocities are obtained for individual stars.

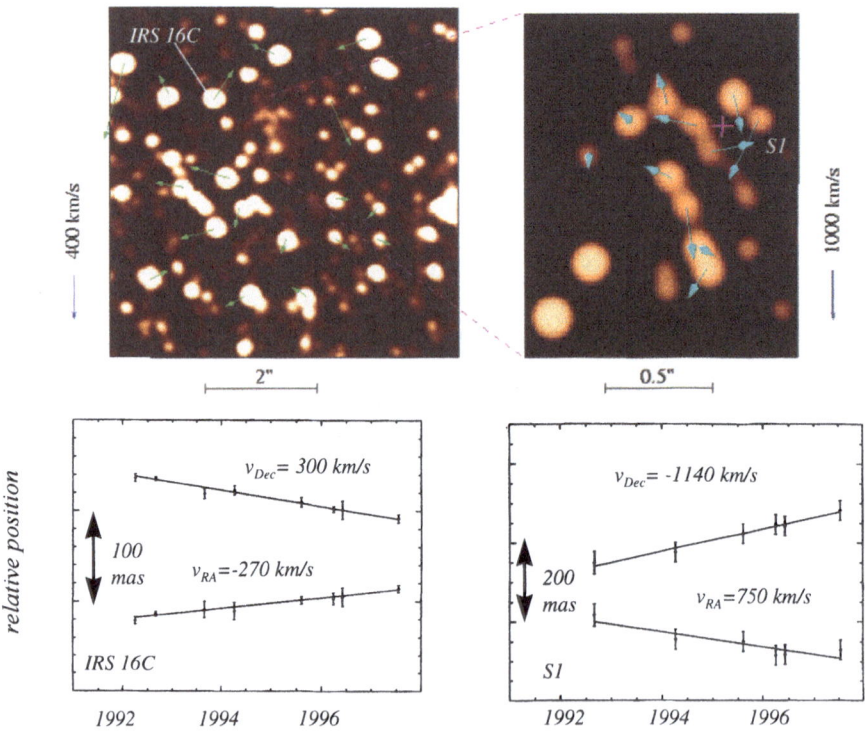

Fig. 2 Stellar proper motions. The upper left inset shows the derived proper motions in the central 6″ as vectors (green), with lengths proportional to the absolute value of the motions. The upper right inset shows the proper motion vectors (blue) in the immediate vicinity of the compact radio source SgrA* (cross). The bottom two insets show the relative positions, given in milli-arcseconds (mas) in right ascension and declination, of two stars (marked on the upper images) as a function of time between 1992 and 1997. The error bars in each epoch are the 1σ errors of the mean of several independent measurements in each epoch. The derived proper motions (linear fits to the data) are for an assumed Sun–Galactic Center distance of 8 kpc.

changes over up to eight epochs exclude such 'Christmas tree effects'.

Because of the rapid increase of stellar density toward the center, contamination by stars in the foreground or background are not very significant when one considers stars outside the central parsec. Projection effects within the central stellar cluster, however, have to be explicitly considered in the analysis (Genzel et al. 1997).

4   Derived mass distribution

A first rough analysis shows that the projected velocity dispersions of a number of stars in a given annulus of projected radius p increase with $p^{-1/2}$ between $p \approx 1$ pc and $p \approx 0.01$ pc, as expected in the potential of a central point mass (a 'Kepler law'). The location of the largest stellar velocities (the dynamic center), the stellar density maximum and the position of SgrA* (now determined relative to the stars to 30 milli-arcseconds, Menten et al. 1997) all agree to within ± 0.004 pc ($0.1''$, Ghez et al. 1997). Between $5'' \geq p \geq 1''$ — where both radial and proper motions, sometimes from the same stars, are available — the mean velocities in all three directions agree within the error bars. This means that anisotropy of the stellar orbits — caused, for instance, by predominantly very elliptical orbits — does not play a significant role in the Galactic Center.

The final distribution of the enclosed mass as a function of true radius (from SgrA*) is shown in Fig. 3 and is the result of applying the so-called Jeans equation as well as projected mass estimators to all available stellar radial and proper motion data (Eckart and Genzel 1997, Genzel et al. 1997). The data are fitted extremely well by the combination of a central point mass $\left(2.61 \left[\pm 0.15_{stat}, \pm 0.35_{stat+sys}\right] \times 10^6 M_\odot\right)$ and a nearly isothermal stellar cluster of core radius ~ 0.38 pc and core density 4×10^6 $M_\odot$ pc^{-3}. The latter is a good fit to the stellar light distribution with a mass to $2\,\mu$m-band luminosity ratio of 2 (indicated as a fat dashed line in Figure 3). The central mass is 'dark', as it has to have a mass to luminosity ratio of 100 or greater. If the central point mass is replaced by a dark cluster, its central density has to be in excess of 2×10^{12} $M_\odot$ pc^{-3} in order to still be consistent with the data, about 500,000 times greater than that of the visible cluster.

5   Nature of the central mass

Basic considerations on the stability of dark clusters composed of white dwarfs, neutron stars, stellar black holes or sub-stellar entities show that a dark cluster of mass $2.6 \times 10^6 M_\odot$ and density $2 \times^{12}$ $M_\odot$ pc^{-3} or greater cannot be stable for more than about 10 million years (Maoz 1995, 1997, Genzel et al. 1997). The majority of the Galactic Center stars, however, are older than 10^8 or 10^9 years. It is also not possible that the dark mass concentration is the core-collapsed state of a dynamically evolving cluster. In that case, the distribution – while very dense in a tiny core – would have a soft, quasi-isothermal envelope,

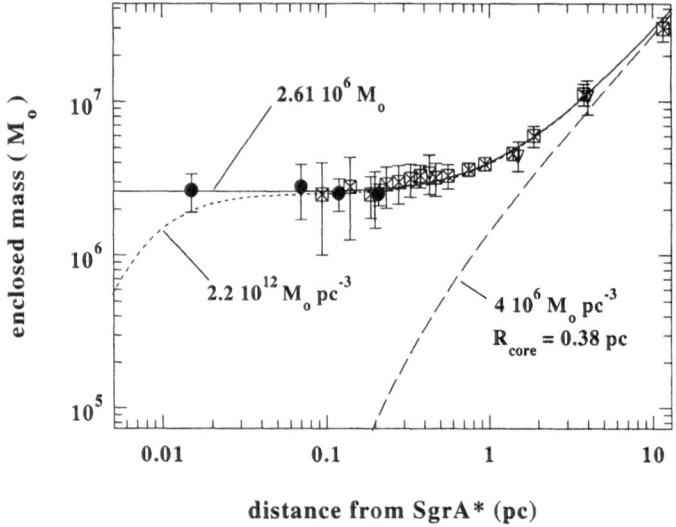

Fig. 3 Mass distribution in the central 10 pc of the Galaxy as obtained from stellar dynamics. Shown as filled circles and crossed rectangles with 1σ error bars are the Jeans equation and projected mass estimator, mass estimates obtained from proper motions and radial motions, respectively, assuming a Sun–Galactic Centre distance of 8 kpc (Genzel et al. 1997). Mass estimates from the proper motions data of Ghez et al.(1997) are in excellent agreement with these values. For comparison, mass estimates from gas velocities (e.g. Serabyn and Lacy 1985, Lacy et al. 1991, Guesten et al. 1987) are given as open triangles. The thick dashed curve represents the mass model for the (visible) stellar cluster $(M/L(2\,\mu m)=2$, $R_{core}=0.38$ pc, $r(R=0)=4\times10^6\ M_\odot$ pc^{-3}, Genzel et al. 1996). The thin continuous curve is the sum of this stellar cluster plus a point mass of $2.61\times10^6\ M_\odot$. The thin dotted curve is the sum of the visible stellar cluster plus an $\alpha=5$ Plummer model of a dark cluster of central density $2.2\times10^{12}\ M_\odot$ pc^{-3} and $R_o=0.0065$ pc (Genzel et al. 1997). This measured enclosed mass distribution is the currently best conclusive evidence for a massive black hole at the center of the Milky Way.

unlike what is observed in the Galactic Center. Finally, if the dark mass were conjectured to consist of a degenerate gas of fermions, the m^{-2} dependence of the Chandrasekhar mass on the mass m of the constituent particles requires that the mass of the fermions cannot be much larger than that of the electron. The only realistic configuration without net electric charge would then be a positron-electron plasma which would, however, rapidly decay through annihilation line radiation. Two further arguments strengthen the conclusion that the dark mass in the Galactic Center must be a black hole. The first comes from the fact that SgrA* itself is known from VLBI measurements to have a proper motion of less than about 20 km/s (Backer 1996). In the very dense Galactic Center core, the fast moving stars near SgrA* and SgrA* approximately should have the same kinetic energy. The large (factor 100) difference in observed motions means that SgrA* must be at least 10^4 times more massive than those stars, or 10^5 $M_{\odot}$, unless its motion is exactly along the line of sight. If one further assumes that the mass of SgrA* is at least as concentrated as its radio emission (radius 1.5×10^{13} cm, Backer 1996), the inferred density of SgrA* is at least $10^{20.5}$ $M_{\odot}$. This lower limit is only five orders of magnitude smaller than the equivalent density of a 2.6×10^6 $M_{\odot}$ black hole within its Schwarzschild radius of $\sim10^{12}$ cm. The second argument is an inversion of the well known dilemma that if SgrA* is a million solar mass black hole, it is presently radiating at a rest mass energy to radiation conversion efficiency of 10^{-5} to 10^{-6}, considering the accretion of stellar wind gas from its environment (Melia 1992). The only possible way out – apart from very large amplitude variability in the accretion – is the argument that in purely radial (Bondi–Hoyle) or in low-density non-radial flows most of the rest mass energy of the accretion flow can be advected into the hole, rather than radiated away (Rees et al. 1982, Melia 1992, Narayan et al. 1995, 1998). This explanation then requires the existence of an event horizon and does not work with any configuration other than a black hole (Narayan et al. 1997). Taking all these arguments together it is hard to escape the conclusion that the core of the Milky Way in fact harbors a massive, but presently inactive central black hole.

Bibliography

[1] Allen, D.A., Hyland, A.R. and Hillier, D.J., MNRAS 244, 706 (1990)

[2] Backer, D.C., in: Unsolved Problems of the Milky Way, eds. L.Blitz and P.Teuben (Kluwer: Dordrecht), 193 (1996)

[3] Balick, B. and Brown, R.L., Ap.J. 194, 265 (1974)

[4] Eckart, A., Genzel, R., Hofmann, R., Sams, B. and Tacconi-Garman, L.E., Ap.J. 445, L26 (1995)

[5] Eckart, A. and Genzel, R., Nature 383, 415 (1996)

[6] Eckart, A. and Genzel, R., MNRAS 284, 576 (1997)

[7] Forrest, W.J., Shure, M.A., Pipher, J.L. and Woodward, C.A., in 'The Galactic Center', ed. D.Backer, AIP Conf. Proc 155, 153 (1987)

[8] Genzel, R. and Townes, C.H. , Ann. Rev. Astr. Ap. 25, 377(1987)

[9] Genzel, R., Thatte, N., Krabbe, A., Eckart, A., Kroker, H. and Tacconi-Garman, L.E., Ap.J. 472, 153 (1996)

[10] Genzel, R., Eckart, A., Ott, T. and Eisenhauer, F., MNRAS 291, 219 (1997)

[11] Ghez, A., Klein, B., Morris, M. and Becklin, E.E., Proc. of IAU Symposium 184, 1998 (23rd General Assembly 1997 in Kyoto on "The Central Regions of the Galaxy and Galaxies")

[12] Guesten, R., Genzel, R., Wright, M.C.H., Jaffe, D.T., Stutzki, J. and Harris, A.I., Ap.J. 472, 153 (1987)

[13] Haller, J.W., Rieke, M.J., Rieke, G.H., Tamblyn, P., Close, L. and Melia, F., Ap.J. 456, 194 (1996)

[14] Kormendy, J. and Richstone, D., Ann. Rev. Astr. Ap. 33, 581 (1995)

[15] Krabbe, A., Genzel, R., Drapatz, S. and Rotaciuc, V., Ap.J. 382, L19 (1991)

[16] Krabbe, A. Genzel, R., Eckart, A., Najarro, F., Lutz, D. et al., Ap.J. Lett. 447, L95 (1995)

[17] Lacy, J.H., Baas, F., Townes, C.H. and Geballe, T.R. Ap.J. 227, L17 (1979)

[18] Lacy, J.H., Townes, C.H. and Hollenbach, D.J., Ap.J. 262, 120 (1982)

[19] Lacy, J.H., Achtermann, J.M. and Serabyn, E., Ap.J. 380, L71 (1991)

[20] Lynden-Bell, D. and Rees, M., MNRAS 152, 461 (1971)

[21] Maoz, E. , Ap.J. 447, L91 (1995)

[22] Melia, F., Ap.J. 387, L25 (1992)

[23] Menten, K.M., Reid, M., Eckart, A. and Genzel, R., Ap.J. 475, L111 (1997)

[24] Myoshi, M., Moran, J.M., Hernstein, J., Greenhill, L., Nakai, N., Diamond, P. and Inoue, M., Nature 373, 127 (1995)

[25] Narayan, R., Yi, I., and Mahadevan, R., Nature 374, 623 (1995)

[26] Narayan, R., Mahadevan, R., Grindlay, J., Popham, R.G. and Gammie, C., Ap.J. 492, 554 (1998)

[27] Rees, M., Phinney, E.S., Begelman, M.C. and Blandford, R.D., Nature 295, 17 (1982)

[28] Rieke, G.H. and Rieke, M.J., Ap.J. 330, L33 (1988)

[29] Sellgren, K., McGinn, M.T., Becklin, E. and Hall, D.N.B., Ap.J. 359, 112 (1990)

[30] Serabyn, E. and Lacy, J. , Ap.J. 293, 445 (1985)

[31] Wollman, E.R., Geballe, T.R., Lacy, J.H., Townes, C.H. and Rank, D.M., Ap.J. 218, L103 (1977)

Broad Iron Lines in Active Galactic Nuclei: A Possible Test of the Kerr Metric?

Jörn Wilms[1], Roland Speith[2], and Christopher S. Reynolds[3]

[1] Institut für Astronomie und Astrophysik, Abt. Astronomie, Waldhäuser Str. 64, D-72076 Tübingen, Germany
[2] Institut für Astronomie und Astrophysik, Abt. Theoretische Astrophysik, Auf der Morgenstelle 10, D-72076 Tübingen, Germany
[3] JILA, University of Colorado, C. B. 440, Boulder, CO 80309-0440, U.S.A.

Abstract. The broad lines in Active Galactic Nuclei (AGN) discovered recently have been interpreted as evidence for emission close to a central black hole. We briefly describe the physical processes leading to the line emission, describe the computational methods used to compute the emerging line profiles, and summarize the qualitative behavior of these lines. We present the observational evidence for the relativistic lines, concentrating on the properties of the line in MCG−6-30-15, where the line profile shows strong indications that a Kerr black hole is present in the object. Finally, we show how future X-ray missions will help in deepening our understanding of the emission of broad iron lines from AGN.

1 Introduction

As reviewed in several chapters of this volume there is ample evidence for the presence of optically thick accretion disks in Active Galactic Nuclei (AGN) and in Galactic black hole candidates. While the high luminosity of AGN is a good indicator for the presence of a deep potential well, the evidence for the geometry of the accretion process has to rely mainly on indirect evidence from the UV, X-ray, and γ-ray spectrum. The ultraviolet excess seen in most AGN, the "big blue bump", and (probably) also the soft X-ray excess below 1 keV are usually assumed to originate in the accretion disk. At energies above 1 keV the spectrum of AGN can be roughly described by a power-law with a photon index of 1.7 and, at least in Seyfert galaxies, an exponential cutoff above 100 or 200 keV. The current physical interpretation of this X-ray and γ-ray power-law component is that of Comptonization of the soft photons in a hot electron plasma, usually called an accretion disk corona, situated geometrically close to the accretion disk [6, 13, 12]. See [26] for a review of the radiation processes around AGN. Although the geometry of the X-ray producing region is still unclear, with possibilities ranging from "standard" thin accretion disks to more complicated accretion geometries as the advection dominated flows and the solutions proposed by Chakrabarti in this volume, there is general agreement that the high temperatures necessary for the production of the hard radiation are only possible in the close vicinity of a black hole, closer than about 100 Schwarzschild radii, where relativistic effects are important. The X-ray and γ-ray radiation from Seyfert galaxies should therefore exhibit signatures that allow us to directly probe this region and perhaps even to find physical processes enabling us to directly measure parameters of the black hole as its mass or its angular momentum.

The availability of high sensitivity X-ray and γ-ray satellites in the past ten years has allowed the observational study of the broad band spectrum of AGN to search for such processes. Recently, extremely broad Iron fluorescence lines have been observed in several Seyfert galaxies. The most convincing interpretation for these lines is that they are produced in a geometrically thin accretion disk close to a central black hole. If this interpretation is correct, the line profiles are the best evidence for the existence of black holes known so far. In this review, we give a brief introduction to the field. In §2 we describe the physical processes leading to line emission close to the black hole, i.e. Compton reflection and fluorescent line emission (§2.1), followed by a description of the computational methods used to calculate the emerging line profiles (§2.2), and a summary of the qualitative behavior of the emitted lines (§2.3). In §3, we describe the observational evidence for the relativistic lines and give a summary of possible future observations. A recommended review of the subject stressing the observational material has recently been published by Fabian [7].

2   Line Emission Close to the Black Hole

2.1   Compton Reflection and Reprocessing

One direct consequence of the Accretion Disk Corona model is that it requires the presence of a hot electron plasma with a temperature of a few $100\,\mathrm{keV}$ in the close vicinity of the cold accretion disk, which has a temperature of less than $10^6\,\mathrm{K}$ ($\sim 0.1\,\mathrm{keV}$). Due to the proximity of the cold material, hard X-rays emitted from the corona interact with the cold material, leading to observable spectral features. In a gas with $kT \lesssim 0.2\,\mathrm{keV}$ only Hydrogen and Helium are fully ionized. Most metals, i.e. elements with a nuclear charge number $Z > 2$, are only moderately ionized [24]. Since the cross-section for photo-absorption is $\sigma_{\mathrm{bf}} \propto E^{-3}$, most of the irradiating soft X-rays (i.e. photons with $E \lesssim 10\,\mathrm{keV}$) get photo-absorbed within the accretion disk. On the other hand, the cross-section for Compton scattering is almost equal to the Thomson cross section σ_{T} (a constant), so that photons with high energies predominantly Compton scatter off the electrons in the disk. The threshold energy above which Compton scattering dominates is about $15\,\mathrm{keV}$. Since the electrons in the accretion disk have low thermal velocities, Compton scattered X-ray photons with $E \gg 15\,\mathrm{keV}$ lose energy. The result of these two processes, photo-absorption and Compton scattering, is a "hump" of radiation in the spectrum emerging from the disk, peaking at about $30\,\mathrm{keV}$ (Fig. 1; see also, [18, 10, 19], and references therein). Such humps have indeed been found in many Seyfert galaxies, proving the presence of cold matter in these objects [21].

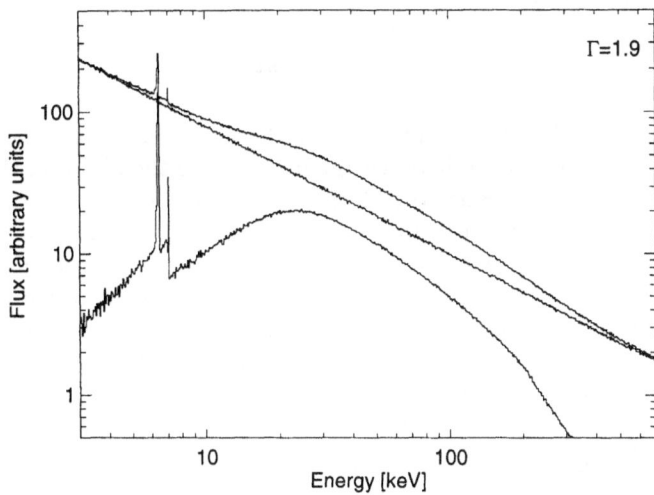

Fig. 1 Reflection spectrum for a cold disk irradiated with a power-law with photon-index $\Gamma = 1.9$. From top to bottom, the plot shows the total emerging spectrum, the incident power-law, and the reflection spectrum. Note the strong Iron K_α line at 6.4 keV and the Iron K_β line and Iron K edge at 7.1 keV. The spectrum was generated with our Monte Carlo code (Wilms, 1996, unpublished), using the cosmic abundances given by Grevesse [11], photo-absorption cross-sections from Verner et al. [29], and fluorescence yields from Kaastra & Mewe [15].

In addition to the reflection hump, the reprocessing of the irradiated X-rays within the accretion disk also leads to the production of emission lines in the X-ray spectrum below 10 keV. The absorption of an X-ray photon by the K-shell of an atom can lead to the emission of a K_α fluorescence photon. For astronomical objects, features of Iron are especially abundant, since Iron has a high cosmic abundance and high fluorescence yield (Fig. 1). Consistent with this picture, Iron features have been found in most Seyfert galaxies [22].

2.2 Radiative Transfer in the Kerr Metric

Since the spectrum is emitted close to the black hole, an observer at infinity will see the spectrum of Fig. 1 "distorted" by relativistic effects, namely Doppler boosting and gravitational red-shift. In this section, we briefly show how to take care of these effects. Due to space limitations, we can only sketch the important physics, for the details we refer to the literature referenced.

The specific flux F_{ν_o} at frequency ν_o as seen by an observer at infinity is defined as the (weighted) sum of the observed specific intensities I_{ν_o} from all parts of the accretion-disk,

$$F_{\nu_o} = \int_\Omega I_{\nu_o} \cos\theta \, d\Omega, \tag{1}$$

where Ω is the solid angle subtended by the accretion disk as seen from the observer and θ is the angle between the direction to the disk and the direction of the observed photon. Since the black hole (=AGN) is assumed to be very far away from the observer (=us), we can safely set $\cos\theta = 1$. Thus, we "only" have to compute the specific intensity I_{ν_o} at infinity from the spectrum emitted on the surface of the accretion disk, I_{ν_e}. In an axisymmetric accretion disk, I_{ν_e} is a function of the radial distance from the point of emission from the black hole r_e and of the inclination angle i_e of the emitted photon, measured with respect to the normal of the accretion disk.

Due to Doppler boosting and gravitational red-shift, the observed frequency ν_o is related to the emitted frequency ν_e by

$$g = \frac{\nu_o}{\nu_e} = \frac{1}{1+z}, \tag{2}$$

where z is the red-shift of the photon. According to Liouville's theorem, the phase-space density of photons, proportional to I/ν^3, is constant along the path of propagation of the photon (the null-geodesic). It is therefore possible to express eq. (1) in terms of the emitted specific flux on the accretion disk:

$$F_{\nu_o} = \int_\Omega \frac{I_{\nu_o}}{\nu_o^3} \nu_o^3 \, d\Omega = \int_\Omega \frac{I_{\nu_e}}{\nu_e^3} \nu_o^3 \, d\Omega = \int_\Omega g^3 I_{\nu_e}(r_e, i_e) \, d\Omega. \tag{3}$$

In other words, the computation of the emerging spectrum breaks down to the computation of the "red-shift" g. In the weak field limit, when $r/M \gg 3$ in geometric units, and in the Schwarzschild metric, g and therefore the line profile emitted by the accretion disk, can be evaluated analytically. Profiles computed this way have been presented, e.g., by Fabian et al. [9] for the Schwarzschild case, and by Chen & Halpern [4] in the weak field limit. In most cases, however, the computation has to be done in the Kerr metric since the accreting black hole will be sped up by the captured material [28].

The "brute force" approach to the computation of g in the Kerr metric is the direct integration of the trajectory of the photon in the Kerr metric [3, 16]. This approach allows the computation of exact line profiles even in the case of very complicated geometries, like thick accretion disks, but is very expensive: The computation of one line profile takes several hours on a typical workstation, and several tens of minutes on a supercomputer (Bromley, Chen & Miller [3] quote a computation time of 15 minutes on a Cray T3D with 128 nodes for the computation of one line profile). It is clear, therefore, that ray-tracing is not suitable for the analysis of X-ray observations, where a direct comparison between the measured data and the theory is to be made using a χ^2 minimization method.

The second way to compute the observed flux was first used by Cunningham [5] who noted that the observed flux from eq. (3) can be expressed by

$$F_{\nu_o} = \int T(i_e, r_e, g) I_{\nu_e}(r_e, i_e) \, dg \, r_e \, dr_e \,, \tag{4}$$

where the integration is carried out over all possible "red-shifts" g and over the whole surface of the accretion disk. This form is well suited for fast numerical evaluation. All relativistic effects are hidden in the *transfer-function* T. We refer to [5], [17], [25], and the references in these works for the technical details. Numerical values for T from the computations of Laor [17] are available in FITS-format as part of the popular X-ray analysis package XSPEC [1]. For detailed studies, a FORTRAN 77 code is available [25]. This code needs about 5 minutes on a DEC Alpha machine (233 MHz) to compute T for one value of i_e. The evaluation of the line profile afterwards takes almost no time.

2.3   The Emerging Line Profile

In Figs. 2 to 4 we illustrate the relativistic effects on the emerging line profile. Due to its astrophysical importance, we chose the Iron fluorescence line at 6.4 keV for our examples. To facilitate the translation to other lines, we indicate the red-shift on the upper abscissa of the figures. All line profiles have been computed with the code of Speith, Riffert & Ruder [25].

In our computations we assumed a geometrically thin, but optically thick Keplerian accretion disk to be the source of the line radiation. We note that our adopted velocity profile is different from the profiles resulting from the thick accretion disks presented by Chakrabarti elsewhere in this volume. The local emissivity of the line on the disk was parameterized as

$$I_{\nu_e}(r_e, i_e) \propto (1 + a\mu)\mu^b \cdot r^{-\alpha} \,, \tag{5}$$

where a, b, and α are free parameters, and where $\mu = \cos i_e$. This parameterization is sufficient for most practical work [2]. For optically thick material, $a, b = 0$, for the optically thin case, $a = 0$, $b = -1$, and for general limb-darkening, $a \neq 0$, $b = 0$. In our computations, the disk was assumed to be optically thick. For realistic accretion disks, the coefficient for the radial emissivity α ranges from 2 to 3 (the comparison with observations, §3, indicates $\alpha \approx 3$).

Common to all line profiles is a characteristic double-horned shape (Fig. 2). This shape is due to the Doppler effect, with the line emitted from material receding from the observer being red-shifted, and the line emitted from material moving towards the observer being blueshifted. Contrary to accretion-disks around normal stars, the emitted line profile is not symmetric: relativistic boosting causes the blue wing of the line to be much stronger than the red wing (it is customary in astronomy to call lower energies "red" and higher energies "blue", even when talking about lines in other energy bands than the optical). In addition to the Doppler boosting, the line is also red-shifted due to the gravitational

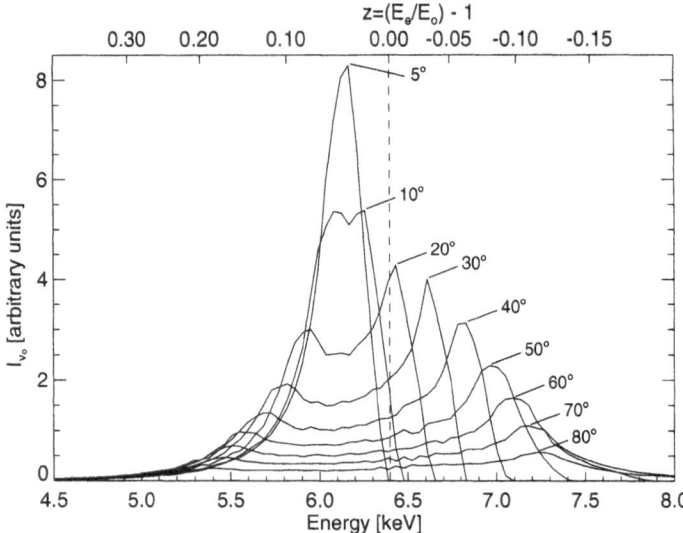

Fig. 2 Line profile as a function of the inclination angle i_o for $a = 0.9981$ and $\alpha = 0.5$.

red-shift. The influence of both effects on the line profile depends on the observers' inclination angle i_o: For a disk seen almost face on (i.e. i_o close to 0°), the gravitational red-shift dominates. With larger and larger i_o, Doppler effects become dominant.

The broadest parts of the profile are due to material emitted very close to the black hole, as is evident from Fig. 3 where line profiles for different emissivity coefficients α are shown. For large α, most of the line emission takes place close to the last stable orbit, so that these profiles are the broadest. Note that for values of $\alpha \gtrsim 2$ the red wing of the profile gets weaker until it is almost undetectable.

For the same emissivity coefficient α, the blue wings of lines emitted from disks around Schwarzschild and Kerr black holes are almost indistinguishable (Fig. 4), the red wings, however, are very different since in the Kerr case the radius of marginal stability, i.e. the inner edge of the accretion disk, is closer to the black hole than in the Schwarzschild case. Therefore, the red wing of the line can extend to much lower energies than in the Schwarzschild case. These effects from regions close to the last stable orbit, are the most promising for measuring general relativistic effects around Kerr black holes [17].

3   Observational Evidence for Broad Iron Lines

3.1   The Case of MCG−6-30-15

The rapid evolution of moderate resolution X-ray detectors in the past decade finally made the discovery of relativistically broadened Iron lines possible. The

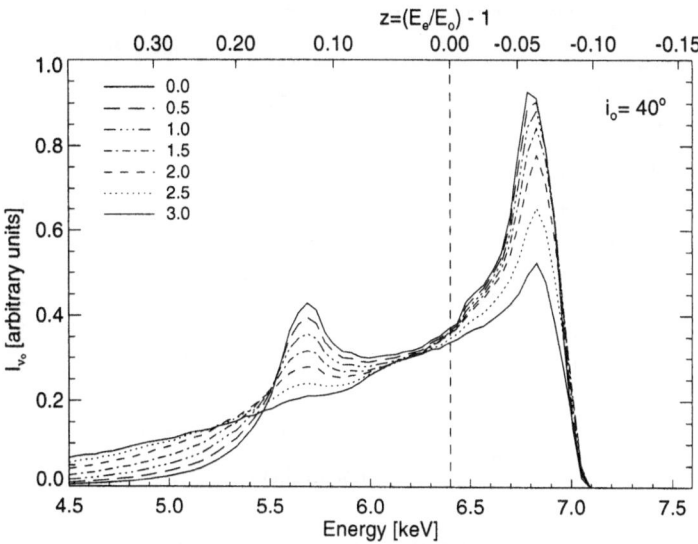

Fig. 3 Line profile as a function of the coefficient of radial emissivity α, where the emitted intensity profile is $I_{\nu_e} \propto r^{-\alpha}$ (eq. (5)), for a black hole with $a = 0.5$. To emphasize the different profiles, the lines have been flux-normalized.

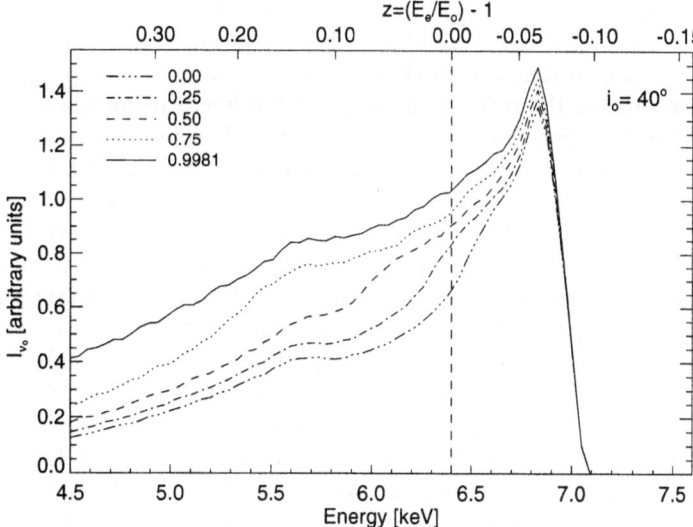

Fig. 4 Line profile for different angular momenta $a = J/M$ of the Kerr black hole for an inclination of $i_o = 40°$ and $\alpha = 3$.

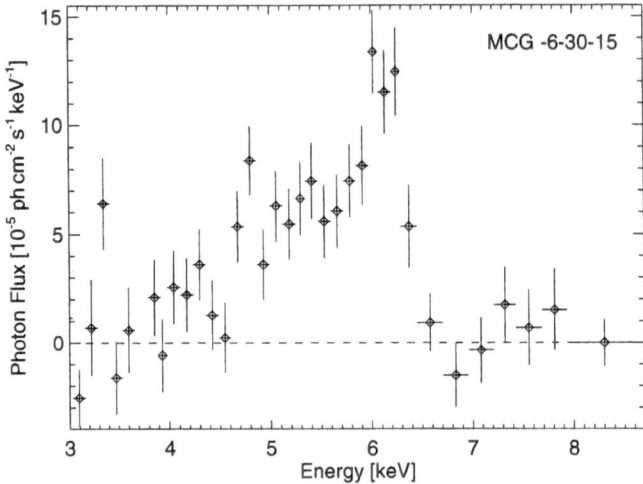

Fig. 5 Average line profile of MCG−6-30-15, as observed with ASCA. The best-fit power-law continuum has been subtracted [27].

best candidate for such a line is the Seyfert galaxy MCG−6-30-15. Here, the Japanese Advanced Satellite for Cosmology and Astrophysics (ASCA) discovered a strongly broadened Iron feature with a full width at zero intensity of 100 000 km/sec [27]. Comparing Fig. 5 with Fig. 2 shows that such a profile has to come from a disk that is seen close to face-on, since the blue-wing of the line is still very close to the rest-frame energy of the line. Fitting the data with the theoretical line models shows that the observed line profile is consistent with that emitted by an accretion disk seen at an inclination of $30 \pm 3°$ [27]. As we showed in the last section, the line profiles from Schwarzschild and Kerr black holes are very similar, with the main difference being in the very red parts of the line. For determining the type of the black hole, therefore, a more careful analysis of the observation of MCG−6-30-15 has to be done. Iwasawa et al. [14] looked at the temporal changes of the Iron line profile during the 4.5 days of the ASCA observation and correlated the profile with the observed variability of MCG−6-30-15. They were able to find three distinct "states" of the line: when MCG−6-30-15 was close to its average flux (Fig. 6b), the profile is similar to the average profile shown in Fig. 5, at times where the flux was very large (Fig. 6a), the line profile is very narrow and centered at 6.4 keV. Finally, when the source intensity was very small (Fig. 6c), the line is very broad and extends down to about 4 keV. This large width is only possible if the line is emitted from material inside six gravitational radii, i.e. inside the marginally stable orbit for a Schwarzschild black hole. Such an emission is only possible if the central object is a Kerr black hole where the disk can extend to smaller radii.

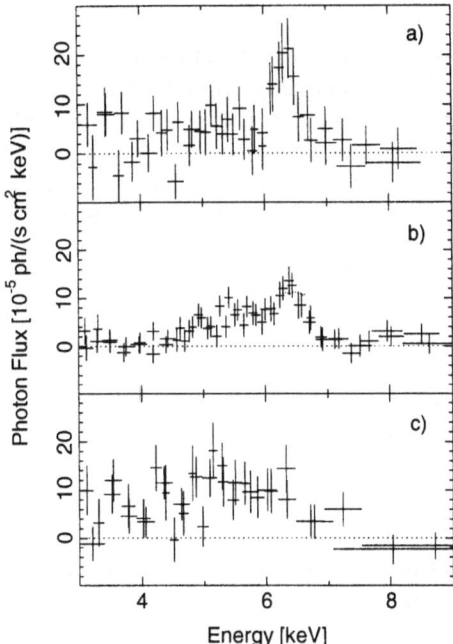

Fig. 6 Variability of the fluorescent iron line in MCG−6-30-15, the different panels show the line profile for phases with high flux (a) down to phases where the continuum flux was very small [14]

If this interpretation of the line variability is true, then the observations present the first *direct* evidence for rotating black holes. Although various objections have been raised against the interpretation of the line as a relativistic line, most objections can be rejected on physical grounds [8][1].

There is strong evidence that the broad Iron line of MCG−6-30-15 is not a special case, but that broad Iron lines are a common phenomenon. In a recent study, Nandra et al. [20] analyzed ASCA observations of 18 Seyfert 1 galaxies. They find evidence for broad lines with a strong asymmetry of the profiles to the red in all 14 objects in which they could detect a line. Nandra et al. [20] were able to explain these lines with relativistic line profiles, but due to the poor signal to noise ratio they could not distinguish between lines from Kerr or Schwarzschild black hole accretion disks.

3.2 The Future: AXAF, XMM, and Such

Although the ASCA results are very exciting and have undoubtfully opened the door to directly observing relativistic effects in AGN, more detailed observations

---

[1] See, however, the objections by Reynolds & Begelman [23] that could weaken the result for the angular momentum of the central black hole.

are needed.

With the currently planned next generation X-ray missions, the American *Advanced X-ray Astrophysics Facility* (AXAF; launch 1998), the Russian Spectrum-X/γ (SXG, launch 1998), the European *X-ray Multiple Mirror mission* (XMM; launch 1999), and the US-Japanese Astro E (launch 2000), we will be able to use X-ray instruments with a much higher energy resolution and larger effective areas than presently available. The huge effective area of XMM will allow us to probe for temporal variability in the line on much smaller time-scales than presently possible: With XMM we might be able to study the time-delay between fluctuations in the continuum and the reaction of the line to these fluctuations, allowing us to directly probe the geometry of the accretion flow. The large energy resolution of AXAF and Astro E will enable us to measure line profiles with a much higher resolution than ever before, which should help us to distinguish without doubt between the current relativistic models for the broad line emission. In the framework of the European EPIC consortium for XMM, two of us (J.W. and C.S.R.) have proposed an uninterrupted 100 ksec observation of MCG−6-30-15. The signal to noise ratio of such an observation will be much higher than that of Fig. 6, making the study of the line variability on short time-scales possible. Our future ability to observe relativistic effects happening on a large scale close to 10^8 M$_\odot$ black holes look very positive indeed.

Acknowledgments J.W. thanks Prof. N. Straumann for his enthusiastic suggestion to write this review and to Prof. F. W. Hehl for his invitation to include the review in these proceedings. We thank J. Dove, I. Kreykenbohm, K. Pottschmidt, T. Rauch, and R. Staubert for helpful discussions.

Bibliography

[1] Arnaud, K. A., 1996, in Astronomical Data Analysis Software and Systems V, ed. J. H. Jacoby, J. Barnes, (San Francisco: Astron. Soc. Pacific), 17

[2] Bao, G., Hadravana, P., Østgaard, E., 1994, ApJ, 435, 55

[3] Bromley, B. C., Chen, K., Miller, W. A., 1997, ApJ, 475, 57

[4] Chen, K., Halpern, J. P., 1989, ApJ, 344, 115

[5] Cunningham, C. T., 1975, ApJ, 202, 788

[6] Dove, J. B., Wilms, J., Begelman, M. C., 1997, ApJ, 487, 747

[7] Fabian, A., 1997, Astron. Geophys., 38, 10

[8] Fabian, A. C., et al., 1995, MNRAS, 277, L11

[9] Fabian, A. C., et al., 1989, MNRAS, 238, 729

[10] George, I. M., Fabian, A. C., 1991, MNRAS, 249

[11] Grevesse, N., Anders, E., 1989, in Cosmic abundances of matter, ed. C. Waddington, (New York: AIP Conf. Proc.), 1

[12] Haardt, F., Maraschi, L., Ghisellini, G., 1994, ApJ, 432, L95

[13] Hua, X.-M., Titarchuk, L., 1995, ApJ, 449, 188

[14] Iwasawa, K., et al., 1996, MNRAS, 282, 1038

[15] Kaastra, J. S., Mewe, R., 1993, A&AS, 97, 443

[16] Karas, V., Vokrouhlický, D., Polnarev, A. G., 1992, MNRAS, 259, 569

[17] Laor, A., 1991, ApJ, 376, 90

[18] Lightman, A. P., White, T. R., 1988, ApJ, 335, 57

[19] Magdziarz, P., Zdziarski, A. A., 1995, MNRAS, 273, 837

[20] Nandra, K., et al., 1997, ApJ, 477, 602

[21] Nandra, K., Pounds, K. A., 1994, MNRAS, 268, 405

[22] Pounds, K. A., et al., 1989, MNRAS, 240, 769

[23] Reynolds, C. S., Begelman, M. C., 1997, ApJ, 488, 109

[24] Shull, J. M., Van Steenburg, M., 1982, ApJS, 48, 95

[25] Speith, R., Riffert, H., Ruder, H., 1995, Comput. Phys. Commun., 88, 109

[26] Svensson, R., 1996, A&AS, 120, C475

[27] Tanaka, Y., et al., 1995, Nature, 375, 659

[28] Thorne, K. S., 1974, ApJ, 191, 507

[29] Verner, D. A., et al. , 1993, Atomic Data Nucl. Data Tables, 55, 233

Accretion and Winds around Galactic and Extragalactic Black Holes

Sandip K. Chakrabarti

S.N. Bose National Centre for Basic Sciences, JD-Block, Sector-III, Salt Lake, Calcutta 700091, India

Abstract. We describe the evolution of models of accretion disks around black holes. We emphasize the importance of centrifugal barrier supported quasi-spherical dense boundary layer (CENBOL) surrounding black hole that is being recognized in recent times. Most of the current observations, such as the hard to soft state transition, quasi-periodic oscillations, power-law spectral slope in soft states, almost constancy of power-law slopes in both hard and soft states, pattern of the rise and fall of intensities in X-ray novae and processes and rates of mass outflow could be easily understood using the advective disk model which includes this barrier.

1 Introduction

Study of modern accretion process on stars and compact objects began with Bondi's solution of spherical flows (Bondi, 1952). This revolutionary work was carried out when 'black hole' phrase was not known and although the Schwarzschild solution of Einstein equations was discovered, the Kerr solution for rotating compact flow was not. The Bondi solution was obtained in Newtonian geometry for a pointlike mass. The general conclusion was that the subsonic flow with specific energy $\mathcal{E} \sim n a_\infty^2 \geq 0$ (where n is the polytropic index of the flow, and a_∞ is the adiabatic sound speed at a large distance) which begins at rest at infinity would pass through a sonic point and becomes supersonic till the star surface. This conclusion would be strictly incorrect if the stars were of finite size, since in the latter case, matter has to stop on a hard surface and therefore must pass through a shock where the supersonic flow becomes subsonic (unless, of course the entire flow is subsonic, see, Chakrabarti & Sahu, 1997; hereafter CS97). In this way, the boundary layer could be studied as a part of the inflow itself. For a black hole accretion, the flow passes through the horizon with the velocity of light, and therefore it must remain supersonic. These conclusions are valid for rotating flows as well (Chakrabarti, 1989; hereafter C89; Chakrabarti, 1996a). For a recent review on spherical flows see, Chakrabarti (1996b, hereafter C96) and CS97 and references therein.

Although the excessive luminosities of quasars and active galaxies in the sixties and seventies were readily interpreted to be due to gravitational energy release of matter accreting on black holes, problems arose in the procedural details. Rapidly inflowing spherical matter is of very low density and advects virtually all the energy through the black hole horizon. Magnetic dissipation could increase the efficiency of emission (Shapiro 1973ab), but the assumptions which went in (for instance, equipartition of gas and magnetic fields) were not at all satisfactory. Shakura & Sunyaev (1973) and Novikov & Thorne (1973) increased the efficiency of emission by assuming the flow to be rotating in Keplerian orbits. This basically rotating matter is of high density and the radiation emitted from this optically thick flow is basically black body. The multicolour black body emission (obtained by summing black body contributions from a large number of annuli) roughly agrees with the observed accretion disk spectra in binary systems as well as in active galaxies (see, e.g., C96b).

However, this cannot be the full story. As early as 1979, observed spectra of the black hole candidates, such as Cyg X-1 (Sunyaev & Trümper, 1979) indicated that the spectrum consists of two distinct components. Whereas the soft X-ray bump in these spectra could be explained by a Keplerian disk, the explanation of the power-law component required additional, and mostly ad hoc, 'Compton clouds' (or, magnetic corona) which are supposed to produce the power-law component by reprocessing the intercepted soft photons emitted by the Keplerian disk. The behaviour of the power-law component was complex: the energy spectral index α ($F_\nu \propto \nu^{-\alpha}$) apparently stays closer to 0.5 when the soft bump is very weak or non-existent (and remains almost constant even when the intensity of the soft bump changes by a factor of several), and closer to 1.5 when the soft bump is very strong. In the first case, most of the power is generated in the hard component and the black hole is said to be in 'hard state', and in the second case, most of the power is generated in the soft component and the black hole is said to be in 'soft state' (see, Tanaka & Lewin, 1995; Ebisawa, Titarchuk & Chakrabarti, 1996; hereafter ETC96 and references therein). Generally, neutron star candidates are not seen in soft states. Even when they in soft states, they do not show constant slope power law component which is insensitive to the luminosity.

The idea that a complete accretion flow must take either the form of a Bondi flow or the form of a Keplerian disk is clearly absurd. On the one hand, the incoming flow must have some angular momentum as it is coming from an orbiting companion (in a binary system) or some orbiting stars (in a galactic nucleus), therefore the flow cannot be Bondi-like. On the other hand, the inner boundary condition on the horizon (that the flow velocity be velocity of light) suggests that the flow must be supersonic, and hence sub-Keplerian at least close to the horizon (see, C96b, Chakrabarti, 1996c; hereafter C96c). Thus a realistic flow *must* be an intermediate solution between the Bondi flow and a Keplerian disk. Far away from the black hole, the flow may be cold, Keplerian (may even be sub-Keplerian if matter is accreted from a large number of stars), but most certainly, close to the black hole flow must be sub-Keplerian. Roughly speaking, the radial component of the equation of motion of infalling matter on a Schwarzschild black hole is given by,

$$v\frac{dv}{dx} + \frac{1}{\rho}\frac{dp}{dx} + \frac{\lambda^2(x)}{x^3} - \frac{1}{2(x-1)^2} = 0.$$

[Here, we chose $2GM/c^2 = 1$ as the unit of length x, c is the unit of velocity and M as the unit of mass. λ is specific angular momentum, $-\frac{1}{2(x-1)^2}$ is pseudo-Newtonian force that mimics the force due to a Schwarzschild Black hole, see, Paczyński & Wiita, (1980). Also, we use x, r and R as radial distance without any distinction.] A flow will deviate from a Keplerian disk only when the inertial force ($v \, dv/dx$) and/or the pressure gradient force ($1/\rho \, dp/dx$) are/is substantial compared to the gravitational and centrifugal forces. (At pressure extrema if the radial velocity vanishes, or, a low pressure flow if there is a velocity extremum, the disk could be momentarily Keplerian. Thus, pressure or velocity does not have to vanish for a Keplerian disk.) Very close to the horizon, the inertial force causes the deviation [$v \sim 1$, $dv/dx < 0$ implies $\lambda < \lambda_{Kep} = (x/2)^{1/2}/(1 - 1/x)$] while deviation farther away is also partly contributed by the pressure force term. The pressure force term is contributed by the radiation pressure if the accretion rate is very high (Paczyński & Wiita, 1980), or, by the ion pressure and possibly magnetic pressure if the accretion rate is very low (Rees et al., 1982). For low accretion rates, the cooling processes are inefficient and the flow advects most of its energy and viscously generated heat towards the black hole just as in a Bondi flow. Such constant energy, inefficiently radiating, steady as well as time dependent accretion solutions on black holes and Neutron stars are abound in the literature (C89; C90; C96c; Molteni, Sponholz & Chakrabarti, 1996; Ryu, Chakrabarti & Molteni, 1997). In flows with shocks, entropy generated At the shock is also advected towards the black hole. When accretion rate is increased, cooling becomes efficient and the disk settles down to a basically Keplerian disk except at the inner part where it still maintains the sub-Keplerian behaviour.

An important ingradiant of the state-of-the-art accretion flow is the centrifugal pressure supported denser region close to a black hole. Roughly speaking, the infall time scale being very short compared to the viscous (transport of angular momentum) time scale, the angular momentum λ remains almost constant close to the black hole particularly for lower viscosity. As a result, the centrifugal force λ^2/x^3 increases much faster compared to the gravity $\sim 1/x^2$ as the flow approaches the black hole. Matter starts piling up behind this centrifugal barrier and becomes denser, with opacity $\tau \sim \dot{m}$, where $\dot{m}$ is the accretion rate in units of the Eddington rate Eddington!rate. Eventually, of course, the gravity wins and matter enters the black hole supersonically since the effective potential is infinitely negative for all possible angular momentum (e.g., Shapiro & Teukolsky, 1983). What this means is that matter with any amount of angular momentum can be made to accrete on a black hole if it is 'pushed' hard enough. This is to be contrasted with the fact that an infinite force is required to push matter to the surface of a Newtonian point mass with even an insignificant angular momentum. This is why a rotating flow has a saddle type sonic point close to a black hole, while the closest sonic point is of unphysical 'center' type for a rotating accretion on a Newtonian compact star.

At the centrifugal pressure supported barrier, matter slows down and its thermal energy increases. In some region of the parameter space this slowing down takes place rather abruptly at a standing shock. Most of the thermal energy of the flow could be extracted through inverse Compton effect if soft photons are injected on this region from the Keplerian disk region. This region is therefore analogous to the boundary layer of a compact star and would be termed as CENBOL (centrifugal pressure supported boundary layer) in the rest of the article. Whereas the boundary layer of a white dwarf is of thickness less than a percentage of its radius, the thickness of the boundary layer (CENBOL) of a black hole is several (typically $10-20$) times larger! If the neutron star is not compact enough, its boundary layer is also of similar size. For compact neutron stars the boundary layer could be very thin (see, C89). Absence of a centrifugal barrier in a Bondi flow causes the flow to be inefficient and one requires sufficient magnetic field to 'jack up' the cooling efficiency. However, CENBOL has just the right properties: the efficiency of its emission is neither almost zero as in a Bondi flow, nor fixed and maximum as in a Keplerian disk. Its size and optical depth are determined by viscosity and accretion rates, thereby giving rise to varieties of spectral properties as are observed.

The present review of the accretion disk model primarily emphasizes the importance of CENBOL as predicted by the advective disks in black hole astrophysics: from the steady and non-steady spectral properties of the black hole candidates, to the formation of the jets and outflows, and to the possibility of nucleosynthesis (which includes enhancement of metalicity in the galaxy and influence on deuterium and lithium abundances). This review also discusses, albeit briefly, how all the other models of the accretion flows could be derived from the exact global solutions of the advective disks.

2 Complete Solution Topologies of the Advective Disks

Advective accretion flows are those which self-consistently include advection velocity as in Bondi flows at the same time include viscosity, heating and cooling processes. For a black hole accretion, these are same as viscous transonic flows (VTF) discussed in detail in Chakrabarti (1990) and C96a. For a neutron star accretion the flow need not be transonic and the advective disks include that possibility as well. Typical hydrodynamic equations which govern vertically averaged advective flows in the pseudo-Newtonian geometry are as follows (C96a):
(a) The radial momentum equation:

$$v\frac{dv}{dx} + \frac{1}{\rho}\frac{dp}{dx} + \frac{\lambda^2_{Kep} - \lambda^2}{x^3} = 0 \tag{1a}$$

(b) The continuity equation:

$$\frac{d}{dx}(\Sigma x v) = 0 \tag{1b}$$

(c) The azimuthal momentum equation:

$$v\frac{d\lambda(x)}{dx} - \frac{1}{\Sigma x}\frac{d}{dx}(x^2 W_{x\phi}) = 0 \tag{1c}$$

(d) The entropy equation:

$$\Sigma v T \frac{ds}{dx} = \frac{h(x)v}{\Gamma_3 - 1}(\frac{dp}{dx} - \Gamma_1 \frac{p}{\rho}) = Q_{nuc}^+ + Q_{vis}^+ - Q^-$$

$$= Q^+ - g(x, \dot{m})q^+ = f(\alpha, x, \dot{m})q^+. \tag{1d}$$

Here, we have included the possibility of nuclear energy release as well. On the right hand side, we wrote Q^+ collectively proportional to the cooling term for simplicity (purely on dimensional grounds). The quantity f is almost zero on the Keplerian disk and about 1 close to the horizon. Here,

$$\Gamma_3 = 1 + \frac{\Gamma_1 - \beta}{4 - 3\beta}; \Gamma_1 = \beta + \frac{(4 - 3\beta)^2(\gamma - 1)}{\beta + 12(\gamma - 1)(1 - \beta)} \tag{2}$$

and $\beta(x)$ is the ratio of gas pressure to total (gas plus magnetic plus radiation) pressure:

$$\beta(x) = \frac{\rho k T / \mu m_p}{\rho k T / \mu m_p + \bar{a} T^4 / 3 + B(x)^2 / 4\pi} \tag{3}$$

Here, $B(x)$ is the strength of magnetic field in the flow, p and ρ are the gas pressure and density respectively, Σ is the density integrated in vertical direction, T is the temperature of the flow (proton and electron), $h(x)$ is the height of the flow chosen to be in vertical equilibrium, $\bar{a}$ is the Stefan's constant, k is the Boltzmann constant, μ is the electron number per particle (and is generally a function of x in case of strong nucleosynthesis effects), m_p is the mass of the proton. Two temperature solutions are important in the case where strong cooling is present (Chakrabarti & Titarchuk, 1995; hereafter CT95). In an optically thick gas, the cooling is governed by black body emission, while in optically thin limit it could be due to bremsstrahlung, Compton effects, synchrotron radiation etc. (see, Rybicki & Lightman, 1979). Except for Compton scattering, other coolings are computed analytically and is very simple to take care of. A novel method to include Compton cooling (first used in CT95) is to fit analytical curves of the numerical results of Sunyaev & Titarchuk (1985) for the cooling function as a function of the optical depth:

$$g(\tau) = (1 - \frac{3}{2}e^{-(\tau_0 + 2)})\cos \frac{\pi}{2}(1 - \frac{\tau}{\tau_0}) + \frac{3}{2}e^{-(\tau_0 + 2)}, \tag{4}$$

where, τ_0 is the total Thomson optical depth of the CENBOL region and by construction $g(\tau_0) = 1$. This is easily translated in radial coordinate for a typical flow model and used in the energy equation 1(d). Similar set of equations can be easily written in Kerr geometry (Chakrabarti, 1996c) and the resulting solution topologies remain identical.

The general procedure of solving this set of simultaneous differential equations is provided in C90 and in C96a in detail. Although the flow deviates from a Keplerian disk to pass through a sonic point, and therefore the sonic point properties are to be obtained a posteriori, it is best to assume the location of the sonic point as well as the angular momentum at that point along with a viscosity parameter. The solutions are integrated outward till they reach a Keplerian disk. This way the shock-free solutions are obtained. Most of the 'shock-free' solutions which pass through the outer saddle type sonic points do pass through shocks and then through the inner sonic points on their way to black holes and neutron stars. (Careless computations usually miss these solutions.) To search for solutions which include shocks, one has to incorporate Rankine-Hugoniot conditions so that global solutions which pass through multiple sonic points may be obtained without difficulty.

An alternate approach may be to start with a Keplerian disk and integrate forward so as to obtain the sonic point and the corresponding angular momentum. This is used in slim disk models (Abramowicz et al. 1988; Narayan & Yi, 1995). But this approach requires a large number of extra unknown parameters to launch the solution from rest and is likely to miss sonic points unless the parameters are fine tuned by several decimal places. By construction, this method always misses the shock solutions (which requires that the flow should pass through two sonic points, and not just one).

2.1   Solution topologies in inviscid flow

In the case of inviscid flow, all possible solutions are shown in Fig. 1a (C89, C90, C96a) where radial Mach number M vs. logarithmic radial distance is plotted. In the central box, the parameter space (spanned by specific energy $\mathcal{E}$ and specific angular momentum λ) is divided into separate regions depending on the nature of the solution topology. Complete solutions in regions O and I have only one sonic point and are analogous to Bondi Flows. Incomplete solutions in O^* and I^* have one 'O' type sonic point as well and are otherwise useless. Constant angular momentum solutions cannot join with a Keplerian disk, but when viscosity is added, solutions in O^* and I^* join with cool Keplerian disks. Complete accretion solution in SA are unstable to form shocks, while those of NSA, NSW and SW are shock-free. Complete wind solutions in SW have shocks, while those of NSA, SA and NSW are shock-free. Low viscosities do not change the conclusions, but high viscosity removes the shocks. The complete solutions from regions $\mathcal{E} \geq 0$ with and without shocks cannot join with a cold Keplerian flow even when viscosity is added, since these flows are not bound. If one writes the net energy (Bernoulli constant) as

$$\mathcal{E} = \frac{1}{2}v^2 + \frac{1}{2}\frac{\lambda^2}{x^2} + \frac{1}{\gamma-1}a_s^2 - \frac{1}{2(x-1)} \tag{5}$$

our arguments will be clearer. (Here we did not include the rest mass energy in $\mathcal{E}$. a_s is the adiabatic sound speed). For a cold Keplerian disk, sound speed $a_s \sim 0$, $v \sim 0$, and $\lambda_K = \frac{1}{2}\frac{x^3}{(x-1)^2}$. At the junction point, where the advective disk meets the Keplerian disk, $\lambda = \lambda_K$ and $\mathcal{E} = \frac{(2-x)}{4(x-1)^2} < 0$ for all $x > 2$. Only when the disk is very hot ($a_s > 0$), or away from the equatorial plane (where

potential energy is smaller) or matter coming out of cold disks and eventually
heated up by, say, magnetic flares, can have specific energy larger than 0 and can
join with the advective disk solutions. Note that these hot, energy conserving
solutions are for strictly inviscid flow. The entire energy of the flow is advected
to the black hole rendering the disk to be non-luminous. It is proposed that
this may be the reason why our galactic center is also faint in X-rays (C96b),
although arguments based on total luminosity is usually not a full proof (SC97).
It is exciting that the same set of equations 1(a-d) shows a rich variety of time
dependent behaviour. This will be discussed in the next Section.

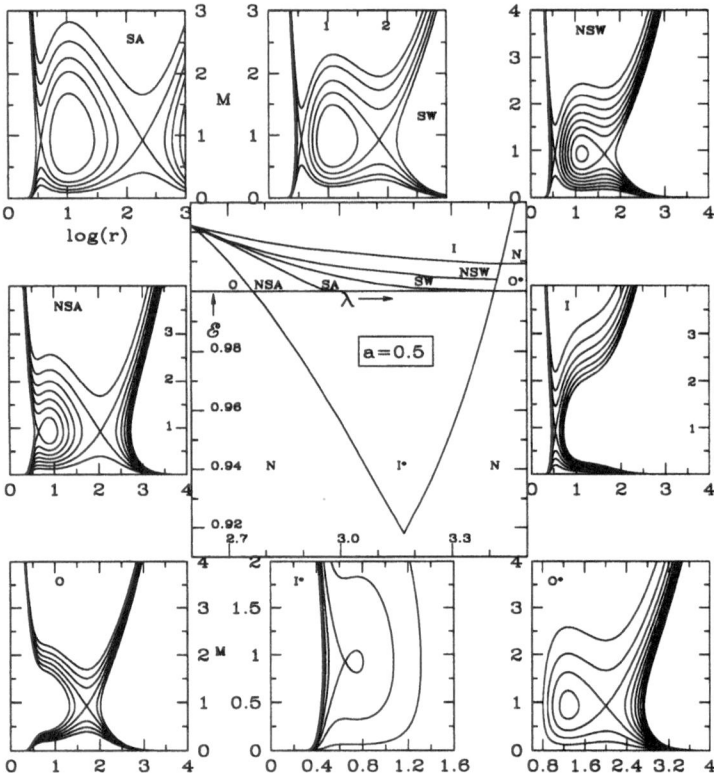

Fig. 1 Classification of the parameter space (central box) in the energy-angular momentum plane in terms of various topology of the black hole accretion. Eight surrounding
boxes show the solutions (Mach number along y axis and logarithmic radial distance
r along x axis) from each of the independent regions of the parameter space. Similar
classification is possible for all adiabatic index $\gamma < 1.5$. For $\gamma > 1.5$, only the inner
sonic point is possible other than an unphysical 'O' type point (C96a). Figure is from
C97c.

The above Figure is drawn for the adiabatic index $\gamma = 4/3$. The classification is universal for all Kerr parameters and is independent of the flow model that is employed as long as $\gamma < 1.5$ (C90) [note that in Chakrabarti (1997a), the sign $<$ was misprinted as $>$]. For $\gamma > 1.5$, two sonic points cannot form and therefore the shocks cannot form, but the centrifugal barrier would still exist.

The solution branch which is supersonic close to the axis is valid for black holes while the solution branches subsonic close to the axis are valid for neutron stars. This is discussed in details in C89 and more recently in CS97. The solution entering through the horizon is unique, since it must pass through the sonic point. This is physically appealing since the horizon properties are independent of any physical parameters such as temperature and pressure etc. The solution touching a neutron star surface is not unique in the same token since any number of subsonic branches from infinity can come close to the axis (either through shocks or without shocks). Of course, ultimately, the one which match with the surface properties of the star will be selected. In a black hole accretion such choices are not present.

2.2 Solution topologies for viscous advective flows

Complete set of topologies of the viscous solutions are presented in C90 for isothermal flows, and they remain identical even when the assumption of isothermality is dropped (C96a). Typical solutions from C90 are shown in Fig. 2(a-d) and the corresponding angular momentum distributions are shown in Fig. 2(e-h). Each solution is identified by only three parameters, namely, the inner sonic point x_{in}, the specific angular momentum at the sonic point λ_{in} and the constant viscosity parameter α. The closed solutions of Fig. 1 open up in presence of viscosity. For low enough viscosity, shock condition may still be satisfied as in Fig. 2a, but as α is increased (2b), λ_{in} is reduced (2c), or x_{in} is reduced, the topologies change completely. The open solution passing through the inner sonic point joins with a Keplerian disk at x_K. For a given cooling process (mainly governed by the accretion rate) x_K strongly depends on viscosity: higher the viscosity, smaller is x_K. This paradoxical property is primarily responsible to the observed nature of the novae outbursts (ETC96). This is exactly what happens if parameters are taken from the region I^* where only inner sonic point is present. The outer sonic point is also present for flows with positive specific energy, and thus, in principle, the solutions passing through the outer sonic point may also join with a Keplerian disk. However, we suspect that in absence of stable shock solutions, flows in 2(b-d) would produce unstable oscillatory behaviour. The region between the Keplerian disk and the black hole is basically freely falling, till close to the horizon ($x \sim \lambda^2$; note that angular momentum is nearly constant close to the black hole) where the centrifugal barrier is formed and matter slows down, heats up and is puffed up. (The geometrical form of the accretion disk takes a very interesting form, somewhat like what Eardley and Lightman (1975) originally proposed.) Highly viscous Keplerian disk stays in the equatorial plane till x_K and then becomes sub-Keplerian (a part of the flow may also become super-Keplerian before becoming sub-Keplerian) as the flow enters through the horizon. If the viscosity monotonically decreases with height, the flow would separate out of a Keplerian disk and form sub-Keplerian halo at a varying distance depending on the viscosity coefficient $\alpha(z)$. Thus, typically a generalized accretion disk would have the shape as shown in Fig. 3. The centrifugal barrier closer to the horizon may or may not be abrupt, depending on the parameters involved. In either case, the flow density, temperature, velocity etc. remain very similar as is shown in Fig. 4, where two solutions, one with and the other without a shock are plotted. Thus the properties of CENBOL is independent of whether a shock actually forms or not. For comparison, a high viscosity flow solution is also presented which deviates from a Keplerian disk closer to the black hole. Figure 2 is drawn for an isothermal advective disk (C90). For a more general energy equation the solution topologies and their interpretations do not change (C96a). In passing, we may mention that not only does the stable shock solutions exist for inviscid and weakly viscous flows, global solutions where the shocks themselves dissipate a large chunk of the flow energy also exist. An infinite number of such one parameter family of dissipative shock solutions are in the literature (Abramowicz & Chakrabarti, 1990). Independent shock solutions in various advective disk models have also been obtained by several authors (e.g., Yang & Kafatos, 1995; Lu et al., 1997; Nobuta & Hanawa, 1995).

3   Complete Set of Numerical Simulation Results

Our understanding of the advective disks is greatly enhanced after a series of numerical simulations were performed to check if the inviscid and viscous solutions were stable or not. Chakrabarti & Molteni (1993), Molteni, Lanzafame & Chakrabarti (1994), Chakrabarti & Molteni (1995), Molteni, Ryu & Chakrabarti (1996), Chakrabarti et al. (1996), Ryu, Chakrabarti & Molteni (1997), Lanzafame, Molteni, & Chakrabarti (1997) found all possible ways advective disks behave when the flow is allowed to be fully time dependent. Shock solutions (from the region SA in Fig. 1) were found to be so stable that the accuracy and performance of codes could be judged by merely comparing the results with theoretical solutions. Figure 5 shows the theoretical and numerical simulation results where the results from three completely different methods (smoothed particle hydrodynamics, total variation diminishing and explicit/implicit code). In two dimensions also shocks formed close to the predicted locations. When the solutions have one sonic point and shocks are not predicted (in regions O and I) shocks do not form (uppermost and lowermost sets of curves). When the solutions have two sonic points but still shocks do not form (in region NSA), the shocks form nevertheless, but they oscillate back and forth thereby changing the size of the CENBOL (Ryu, Chakrabarti & Molteni, 1997). In presence of cooling effects, shocks may oscillate even when stable shocks are present (Molteni, Sponholz & Chakrabarti, 1996), especially when the cooling time scale roughly agrees with the infall time scale. The oscillating shock has the period comparable to the

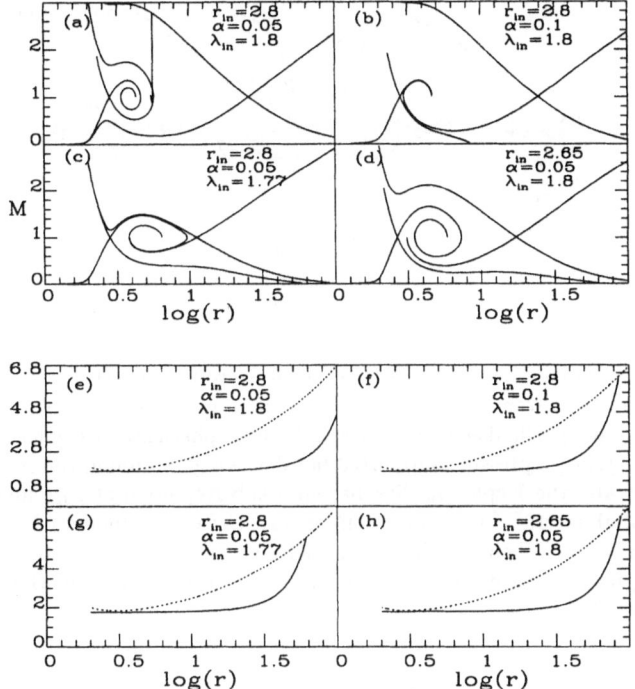

Fig. 2 Mach number variation (a-d) and angular momentum distribution (e-h) of an isothermal viscous transonic flow. Only the topology (a) allows a shock formation. Transition to open (no-shock) topology is initiate by higher viscosity (α) or lower angular momentum (λ_{in}) or inner sonic point location (r_{in}). In (e-h), flow angular momentum (solid) is compared with Keplerian angular momentum (dotted). The location from where the angular momentum is deviated varies with the three parameters Figure is from Chakrabarti, 1996e.

cooling time and is believed to explain the quasi-periodic oscillations observed in the black hole candidates. The viscous flows also show the similar oscillations (Lanzafame, Molteni & Chakrabarti, 1997). We suspect that whenever accretion rates of a black hole change substantially (such as when a black hole changes its state), the oscillations may be set in as a result of competition among various time scales prevalent in the system.

Chakrabarti et al. (1996) provides a collection of numerical simulation results including the formation of an advective disk from a Keplerian one far away from the black hole (and not just near the horizon as in Chakrabarti & Molteni, 1995). Figure 6a shows the ratio of the disk angular momentum to the Keplerian angular

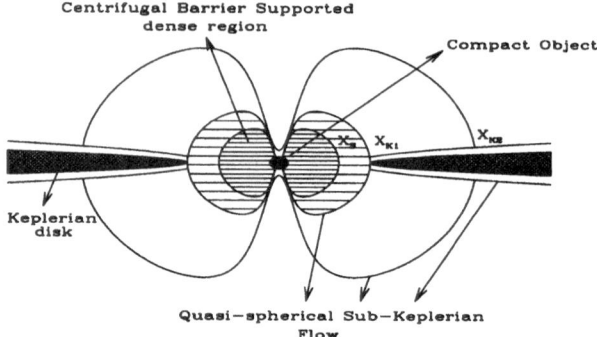

Fig. 3 Schematic diagram of a multi-component accretion flow (C97a). All the components are segregated from the same inflow in different regimes of viscosity. Keplerian disk (cross-hatched) is flanked by (a generally) quasi-spherical sub-Keplerian halo which produces a centrifugal pressure supported hot dense region around the compact object. In the hard state, the Keplerian disk becomes sub-Keplerian at x_{K2} and produces a giant torus of about $10^{2-4}R_g$, which collapses as viscosity is increased and the object goes to softer state. When the shock is absent, $1 < x < x_S \sim x_{K1}$, becomes the centrifugally supported dense region which reprocesses soft photons in the same way as the post-shock flow.

momentum in one of the simulations. The shape is typical of such advective flows (See Fig. 10 of CT95) although the transition from Keplerian to the advective disk is not exactly *smooth*! This is because the derivatives $d\lambda/dr$ in an advective flow is different from that in the Keplerian regime. Figure 6b shows the deviation of angular momentum of a flow which included a standing shock. Apart from a mild kink in the distribution at the shock location, the flow is perfectly smooth, transonic and stable.

4   Complete Set of Spectral Properties of the Advective Disks

Black holes are being fundamentally black, observational evidences must necessarily include 'funny' spectral signatures of radiating matter entering in them. The unique inner boundary condition imposed on the solutions automatically separates the unique solution branch out of many, and therefore it is of no surprise that the spectral properties of the flow entering in a black hole should be different. The problem lies in quantification of this special character. Here we present a few observations and how they may be readily interpreted using the advective disk solutions. The explanations are general (as they are straight from solutions of governing equations), and do not depend on any particular black hole candidate. Here we write accretion rates in units of Eddington rate.

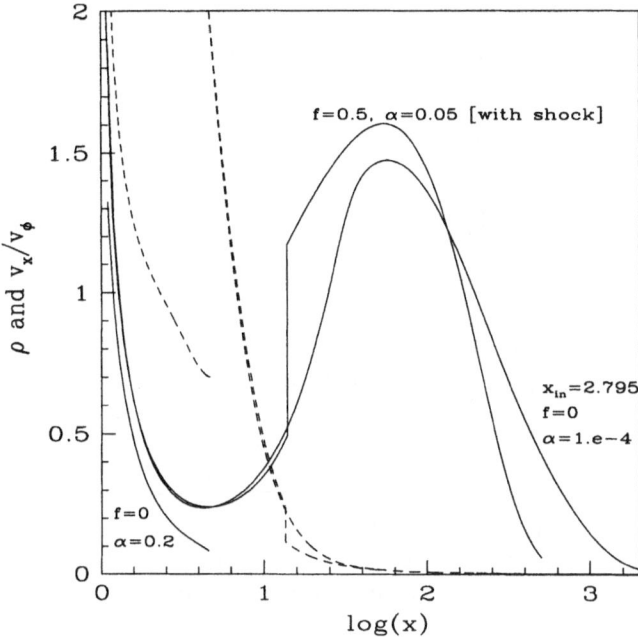

Fig. 4 Ratios v_x/v_ϕ (solid) and densities (dashed) of three illustrative solutions of the advective flows (C96a). Note that the centrifugal barrier close to the hole makes all the three solutions to behave similarly in the region $2 \lesssim x \lesssim 10 - 20$, emission from which strongly determines the spectral properties of the black hole. In a strongly shocked flow the variations in densities and velocities occur in a shorter length scale while in a weakly shocked or shock-free flows the variations occur in an extended region.

4.1   Hard and soft states and triggering of their transitions

Galactic black holes are seen basically in two states. In soft states, more power is in soft X-rays and in hard states more power is in hard X-rays. (The extra-galactic cases such state separation is not obvious, since the observations are poorer, and transition of states may take place in thousands of years. Some of the carefully observed cases the spectral nature was found to be similar to those of the galactic candidates.) The explanation of this apparently puzzling state variation is simple: the Keplerian and sub-Keplerian components redistribute

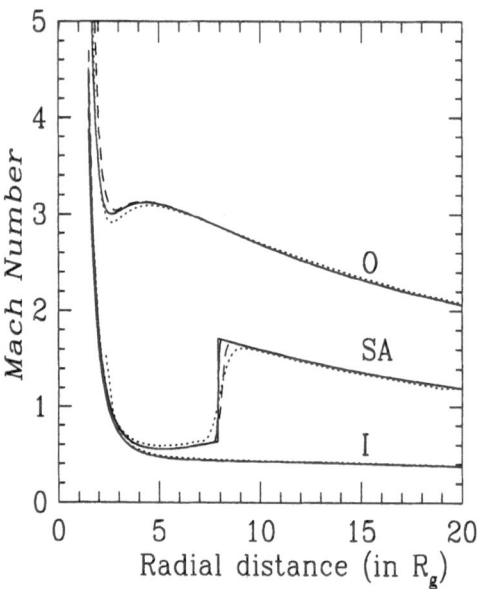

Fig. 5 Comparison of analytical (solid) and numerical results in a one-dimensional accretion flow which may or may not allow a standing shock. The long and short dashed curves are the results of the TVD and SPH simulations respectively while very long dashed curve is using explicit/implicit code. The curves marked 'O' and 'I' are for transonic flows which pass through the outer and the inner sonic points respectively. They are also reproduced perfectly with numerical simulations (from Chakrabarti et al. 1996).

matter among themselves depending on viscosity of the flow. Sudden rise in viscosity would bring more matter to the Keplerian component (with rate $\dot{m}_d$) and sudden fall would bring more matter to sub-Keplerian halo component (with rate $\dot{m}_h$). Disk component $\dot{m}_d$ not only governs the soft X-ray intensity directly coming to the observer, it also provides soft photons to be inverse Comptonized by sub-Keplerian CENBOL electrons. The CENBOL (comprised of matter coming from $\dot{m}_d$ and $\dot{m}_h$) will remain hot and emit power law (energy spectral index, $F_\nu \sim \nu^{-\alpha}$, $\alpha \sim 0.5 - 0.7$) hard X-rays only when its intercepted soft photons from the Keplerian disk (See Fig. 4) are insufficient, i.e., when $\dot{m}_d \ll 1$ to $\dot{m}_d \sim 0.1$ or so, while $\dot{m}_h$ is much higher. For $\dot{m}_d \sim 0.1 - 0.5$ (with $\dot{m}_h \sim 1$),

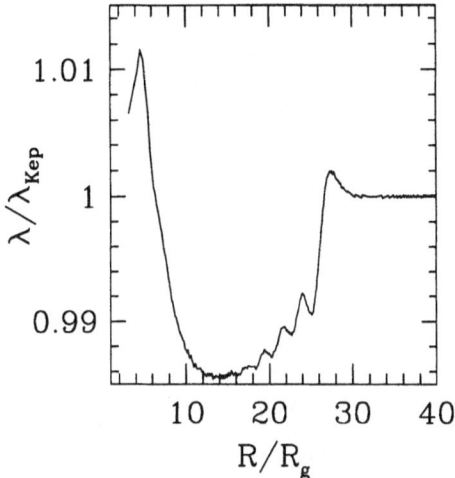

Fig. 6a Ratio of disk angular momentum to the Keplerian angular momentum in a typical time dependent simulation in an advective disk. Note the deviation Keplerian disk at around $R \sim 30$. The flow becomes super-Keplerian close to the hole before becoming sub-Keplerian as it plunges in (Chakrabarti et al. 1996).

CENBOL cools catastrophically and no power law is seen (this is sometimes called a high state). With somewhat larger $\dot{m}_d$, the power law due to the bulk motion of electrons is back at $\alpha \sim 1.5$ (this is sometimes called a very high state). Figure 7 (taken from Chakrabarti, 1997a, hereafter C97a) shows a typical hard to soft state transition as $\dot{m}_d$ is increased. Here, power $EF(E)$ is plotted against the energy E of the emitted photons. The dashed curve drawn for $\dot{m}_d = 1.0$ includes the convergent flow behaviour of the inner part of CENBOL. Details of the solutions are in CT95, C97a, and ETC96.

Such hard/soft transitions are regularly seen in black hole candidates (Dolan et al, 1979; Miyamoto et al, 1991; Ricketts, 1983; Ebisawa et al., 1994; Zhang et al., 1997).

4.2   Constancy of slopes in hard and soft states

Spectra of the advective disk solutions shows a remarkable property: the slope $\alpha \sim const$ in hard states even when $\dot{m}_d$ is increased by a factor of a thousand (CT95). The degree of constancy is increased (C97a) if one assumes that the matter is actually redistributed between the Keplerian and sub-Keplerian halo

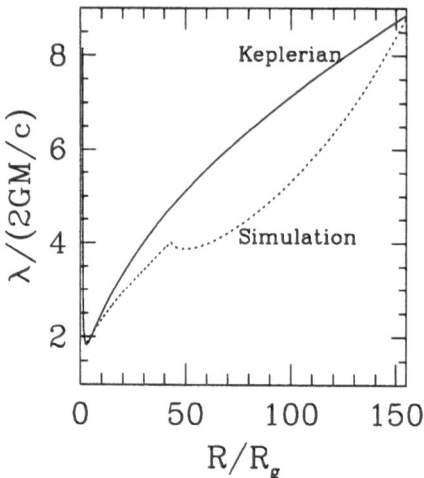

Fig. 6b Numerical simulations of a typical advective disk which forms a standing shock after deviating from a Keplerian disk. While the flow remains generally sub-Keplerian, a kink in the distribution at the shock is produced, which, however, remained stable throughout the simulation (Chakrabarti et al. 1996).

components, rather than assuming that both the components are completely independent. In soft states also, convergent flow calculations show remarkable constancy in the slope at around 1.5 − 1.8 depending on the location of the inner boundary (photon absorption radius). This constancy of slopes is also regularly seen (e.g. Sunyaev et al., 1994, Ebisawa et al., 1994; CT95 ad ETC96 for references; Kuznetsov et al., 1997). Particularly important is the weak power law in the soft state as this is not observed in neutron star candidates. CENBOL around neutron stars may also cool down to produce softs state for the same reason. However, they can go up to high state and not up to 'very high' state where the weak power law due to convergent flow is seen.

In bulk motion Comptonization bulk momentum of the quasi-freely falling electrons (outside the horizon) are transported to the soft photons. In neutron stars, electrons slow down on the hard surface due to radiation forces acting on them, and therefore the effect is negligible. Thus black holes should be definitely identified by spectral signatures alone provided they are seen in soft states (CT95).

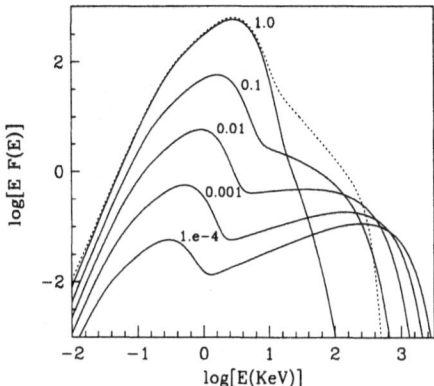

Fig. 7 Spectral evolution of an accretion disk with a strong shock at $X_s = 10$ around a black hole of mass $3.6M_\odot$. The sub-Keplerian halo rate is $\dot{m}_h = 1$ and the Keplerian rates are marked on the curves. The dotted curve is drawn to include the effect of bulk motion Comptonization when $\dot{m}_d = 1$ (Chakrabarti 1997a).

4.3 Variation of inner edge of the Keplerian component

This is trivially achieved in the advective disks (see, CT95; C96a). As viscosity is increased, the location x_K where the disk deviates from Keplerian is *generally* decreased if other two parameters (x_{in} and λ_{in}) are held fixed. Thus, in hard states, not only $\dot{m}_d$ is smaller, the x_K is also larger. As the viscosity increases, x_K becomes smaller in viscous time scale, at the same time more matter is added to the Keplerian component. This behaviour is also seen in black hole candidates (e.g. Gilfanov, Churazov & Sunyaev, 1997).

4.4 Rise and fall of X-ray novae

While in persistent black hole candidates (such as Cyg X-1, LMC X-1, LMC X-3) Keplerian and sub-Keplerian matter may partially redistribute to change states (see §4.1 above), in X-ray novae candidates (e.g., A0620-00, GS2000+25, GS1124-68, V404 Cygni etc.) the net mass accretion rate may indeed decrease, even if some redistribution may actually take place. First qualitative explanation of the change of states in X-ray novae in terms of the advective disk model was put forward by ETC96. The biggest advantage of the advective solution is that it

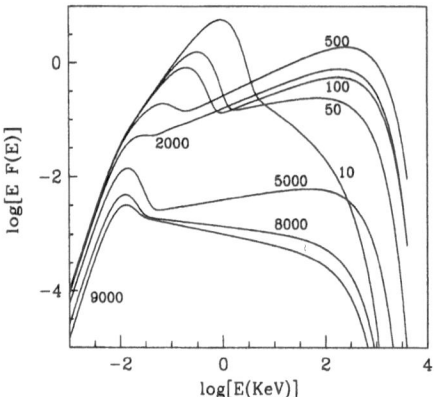

Fig. 8 Typical spectral evolution of an X-ray nova. x_K (marked on the curves) could be very far away as in a low accretion rate, low viscosity disk. We chose: $\dot{m}_h = 1.0$ and $\dot{m}_d = 0.01$. Initially, at the onset of an outburst, the optical intensity goes up as x_K is decreased. Subsequently, the hard X-ray goes up first and then the soft X-ray is intensified. The $x_K = 9000$ and 8000 solutions resemble Novae spectra in quiescence (from Chakrabarti, 1997a).

automatically moves the inner edge of the Keplerian disk as viscosity is varied. Similar to the dwarf novae outbursts, where the Keplerian disk instability is triggered far away (e.g., Cannizzo, 1993) here also the instability may develop and cause the viscosity to increase, and the resulting Keplerian disk with higher accretion rate moves forward. In C97a, several such spectral evolutions have been presented. In Fig. 8, we show one such case, where the increase in viscosity is used to cause the decrease in x_K from 9000 to 10, keeping $\dot{m}_h = 1$ and $\dot{m}_d = 0.01$. Note that as the inner edge goes from 9000 to 5000, the optical (around tens of eV) peaks first, which is followed by hard X-rays (at around hundreds of KeV) till x_K reaches about several hundred Schwarzschild radii. After that the hard X-ray subsides and soft X-ray intensifies. The optical precursor of an X-ray nova GRO J1655-40 have been seen recently (Orosz et al., 1997).

4.5   Quiescent states of X-Ray novae candidates

Advective disks naturally explain quiescent states. As already demonstrated in CT95, C96a, x_K recedes from the black holes as viscosity is decreased. With the decrease of viscosity, less matter goes to the Keplerian component (Chakrabarti & Molteni, 1995), i.e., $\dot{m}_d$ goes down. Since the inner edge of the Keplerian disk does not go all the way to the last stable orbit, optical radiation is weaker in comparison with what it would have been predicted by a Shakura-Sunyaev (1973) model (see, plots for $x_K = 9000$ and 8000 for such spectral behaviour in Fig. 8 above). This behaviour is seen in V404 Cyg (Wagner et al. 1994) and A0620-00 (McClintock et al., 1995). The deviated component from the Keplerian disk almost resembles a constant energy rotating flow described in detail in C89. It is also possible that our own galactic center may have this low viscosity, low accretion rate, global advective disks, as mentioned in C96b.

Recently, a so-called advection dominated model has been used to fit these states (Narayan et al., 1997). In this model highly viscous ($\alpha \sim 0.1 - 0.5$) quasi-spherical flow resulted from Keplerian disk evaporation (which is also in equipartition with magnetic field at all radii!) was used. On the contrary, advective disk solution presented in C96a does not require such evaporation, and the ion torus comes most naturally out of the governing equations only for low viscosity case. The deviation from a Keplerian takes place several thousand Schwarzschild radii (in high viscosity case angular momentum transport rate becomes so high that the flow deviates from a Keplerian disk almost immediately outside the inner sonic point). Advection disks produce the quiescent state like spectra (CT95, C97a) without making any further unwarranted approximations. Detail fits will be presented else where.

4.6 Quasi-periodic oscillations

As mentioned in Section 3, in some large region of the parameter space the solutions of the governing equations 1(a-d) are inherently time-dependent. Just as a pendulum inherently oscillates, the solutions of the advective disks also show oscillations of the CENBOL region. This oscillation is triggered by competitions among various time scales (such as infall time scale, cooling time scales by different processes). Thus, even if black holes do not have hard surfaces, quasi-periodic oscillations could be seen. Although any number of physical processes (such as acoustic oscillations (Taam, Chen & Swank, 1997), diskoseismology (Nowak & Wagoner, 1993), trapped oscillations (Kato, Honma & Matsumoto, 1988) could produce such oscillation frequencies, modulation of $10 - 100$ per cent or above cannot be achieved without bringing in the dynamical participation of the hard X-ray emitting region, namely, the CENBOL. By expanding back and forth (and puffing up and collapsing at the same time), CENBOL intercepts variable amount of soft photons and reprocesses them. Some of the typical observational results are presented in Dotani (1992), Halpern & Marshall (1996); Cui et al. (1997). Recently more complex behaviour has been seen in GRS 1915+105 (Paul et al., 1997; Morgan, Remillard & Greiner, 1997), which may possibly be understood by considering several cooling mechanisms simultaneously. Some chaotic behaviour of x_K under non-linear feed back mechanism cannot be ruled out either. This is being investigated.

5   Spin-offs to Other Branches of Astrophysics

Some of the unexpected spin-offs from the advective disk model is that one can for the first time make quantitative estimate of the mass outflow rate from a black hole accretion process. Similarly, heavy and light element abundances may be changed in the hot CENBOL region. Estimated outflow rate together with modified composition enable one to gauge the importance of black hole nucleosynthesis on the surroundings. Another spin-off is in the gravitational wave astronomy of coalescence of binary compact objects. Briefly, we discuss these issues below.

5.1   Physics of jets: Estimation of the outflow rate from an advective flow

Outflows are common in many astrophysical systems which contain black holes and neutron stars. Difference between stellar outflows and outflows from these systems is that the outflows in these systems have to form from the inflowing material only. One advantage of these compact objects is the formation of CENBOL which behaves like a stellar surface as far as mass loss is concerned. Assuming this (see, Fig. 9) configuration, mass loss is estimated very easily (Chakrabarti, 1997b). The procedure involves first computing the CENBOL temperature T_s from the incoming flow using steady shock condition and then computing the mass loss rate from CENBOL using the same procedure as used on staller surface. The ratio of outflow rate to the inflow rate turns out to be,

$$R_{\dot{m}} = \frac{\dot{M}_{out}}{\dot{M}_{in}} = \frac{\Theta_{out}}{\Theta_{in}} \frac{R}{4} exp(-f) f_0^{3/2} \qquad (6)$$

where, $f = f_0 - \frac{3}{2}$ and $f_0 = (2n+1)R/(2n)$ [$n = (\gamma - 1)^{-1}$ is the polytropic constant.]. Notice that this simple result does not depend on the location of the sonic point or shock (namely the size of the dense cloud) or the outward force causing the mass loss. It is a function of compression ratio R for a given geometry. In a relativistic inflow $n = 3$, $\gamma = 4/3$ and $R = 7$ and the ratio of inflow and outflow becomes,

$$R_{\dot{m}} = 0.052 \frac{\Theta_{out}}{\Theta_{in}} \qquad (16a)$$

and for inflow of an ionized gas $n = 3/2$, $\gamma = 5/3$ and $R = 4$, and the ratio in this case becomes,

$$R_{\dot{m}} = 0.266 \frac{\Theta_{out}}{\Theta_{in}} \qquad (16b)$$

Outflows are usually concentrated near the axis, while the inflow is near the equatorial plane. Assuming a half angle of $10°$ in each case, we obtain,

$$\Theta_{in} = \frac{2\pi^2}{9}; \qquad \Theta_{out} = \frac{\pi^3}{162}$$

and

$$\frac{\Theta_{out}}{\Theta_{in}} = \frac{\pi}{36}.$$

The ratios for $\gamma = 4/3$ and $\gamma = 5/3$ are then

$$R_{\dot{m}} = 0.0045 \quad \text{and} \quad R_{\dot{m}} = 0.023$$

respectively. This is to be compared with the rate $R_{\dot{m}} = 0.004$ found in radiation dominated flow (Eggum, Coroniti & Katz, 1985). Recently, more exact computation of the mass loss rate has been done using exact transonic solutions for the inflow and outflow (Das, Rakshit & Chakrabarti, in preparation). Although the general conclusions remain the same, the details vary. The results would be presented elsewhere.

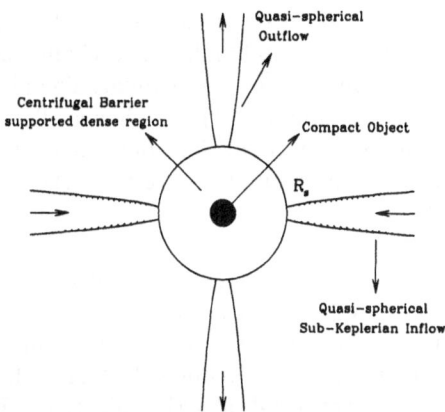

Fig. 9 Schematic diagram of inflow and outflow around a compact object. Rotating matter forms a centrifugal barrier supported dense region around the object which in turn acts like a 'stellar surface' from which the outflowing wind is developed (from Chakrabarti, 1997b).

5.2   Nuclear astrophysics: Nucleosynthesis in advective disks around black holes

Chakrabarti (1986) and Chakrabarti, Jin & Arnett (1987), first pointed out that a considerable nucleosynthesis could take place during the infall and heavier elements may be produced inside the thick disk, a fraction of which could be ejected out through bipolar outflows and jets (see, Horgan, 1988). Although the disk model used was very preliminary, the conclusions were firm and were verified by a large number of independent workers (Arai & Hashimoto, 1992, Hashimoto et al., 1992) using other disk models.

In the decade since these pioneering works were started, the self-consistent advective disk model has been developed. This entices one to look into the nucleosynthesis problem once more, specially when the shocks and CENBOL regions are also included in the computation. Fig. 9 shows the variation of the abundance as the matter enters the advective disk regime. Here only $1M_\odot$ central object and a mass accretion rate of $100\dot{M}_{Edd}$ is used with $\beta = \frac{p_g}{p_{tot}} = 0.03$. The cooling factor $f = 0.5$ and viscosity $\alpha_\Pi = 0.05$ were used which gave $x_K = 480R_g$ are used. The shock is formed at $x_s = 13.9R_g$ (see, Fig. 4 for the full solutions). The dotted curves are drawn when only the supersonic branch through outer sonic point is used, while the solid curves are drawn when the solution takes more stable branch through the shock and finally through the inner sonic point. At the shock, the sudden rise in temperature as well as higher residence time in the post-shock flow causes the abundance to change abruptly although the final product at the horizon remains very similar. The abundances for Neutron (N), Deuterium (D) and Lithium (Li^7) are shown. The initial flow leaving a Keplerian disk was chosen to have a solar abundance. Here 255 isotopes (from neutron, proton, helium to germanium) have been chosen in the network as in Chakrabarti (1986).

In the case of higher viscosity, some heavy elements may be produced and some lighter elements may be destroyed completely. In the low accretion rate case, spallation reaction could become important to produce some excess Li^7 (Jin, 1980), but the main reactants, namely He^4 may themselves be destroyed by photo-dissociation process (Chakrabarti et al., 1987). In fact, both He^4 and D photo-dissociate into protons and neutrons. Neutrons produced in the flow produce a neutron tori (till they decay into protons) in the advective region which, mixing with fresh incoming matter may produce neutron rich isotopes of the galaxies. Heavier elements which are produced in the disk may supply metalicity in the galaxies (Hogan & Applegate, 1987). Jin, Arnett and Chakrabarti (1989) originally concluded that the effect of nucleosynthesis is important only for very low viscosities. This is because they focussed on cooler radiation dominated disks. Using present advective disks, even for $\alpha \sim 0.1 - 0.2$ the nucleosynthesis effect seems to be important for stellar black holes. Detail calculations have been reported elsewhere (Mukhopadhyay & Chakrabarti, submitted). The changes in abundance in the CENBOL region may be important, since the wind is produced from this region (see above) and a part of wind is intercepted by the companion star. For instance, Martin et al. 1994 reportedly detected Li^7 in K-star companions of black holes, but Li^7 itself need not come through winds. The nucleosynthesis inside the disk could affect the energetics very strongly. Some reactions are exothermic and some are endothermic. In future, these considerations are to be made in a self-consistent disk structure.

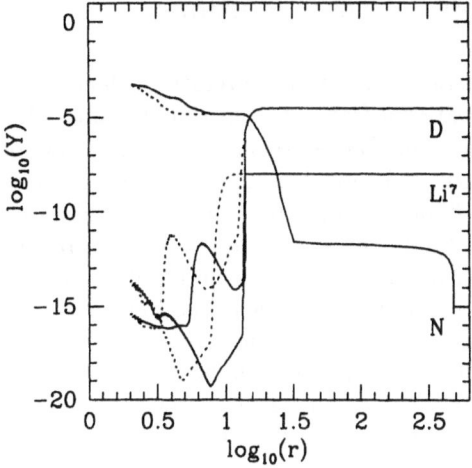

Fig. 10 Hot advective disks allow synthesis of new elements. Here the variation of the abundances of Deutorium (D), Lithium (Li^7) and Neutron (N) in an advective disk are shown as functions of the logarithmic radial distance from the black hole. The dotted curves are drawn for the flow passing through the outer sonic point and the solid curves are when the flow passes through the shock at $13.9r_g$ as is formed in this particular example. Fig. 4 above gives the full shocked solution used here.

5.3   Gravity Wave Astronomy: Effects on gravitational wave emission

Traditionally, coalescence of two compact bodies is studied in the absence of accretion disks. When a supermassive black hole at the galactic centre is surrounded by an advective disk through which a companion star gradually moves in on an instantaneously Keplerian orbit, not only is the angular momentum of the companion lost due to gravitational wave emission, but some angular momentum is changed through the interaction of the disk with the companion. For instance in the super-Keplerian region of the disk, the companion will *gain* angular momentum due to accretion from disk material, while in the sub-Keplerian region it would be the opposite. In either cases, the wave pattern of the emerging gravitational wave would be affected. Through detailed computation it was shown that the disk effect could be 7-10 percent of the main effect (Chakrabarti, 1996d). Similar deviations from standard template is also possible when self-gravitating disk is present, since Keplerian angular momentum distribution of such disks are completely different (Chakrabarti, 1988). The creation of templates with different disk models is in progress (Das & Chakrabarti, in preparation) and would be reported elsewhere.

6   Other Models – Other Solutions

There have been several models in the literature which were brought in to explain different observational features from time to time. Most of these models are far from self-consistent and serve only special purpose for which they were invoked. Since advective disk model discussed here is considered to encompass all possible types of solutions governed by known physical laws, one would expect that this model would have the 'right features in right regime'. As discussed in the Introduction, the Bondi flow and Keplerian disks are both extreme cases: one does not have rotation and the other does not have advection properly included, one is basically energy conserving (roughly valid for low accretion rate) spherical advective flow and the other is highly dissipative (roughly valid for viscous, high accretion rate) disklike flow. Attempts to fill the gap, i.e., to find intermediate structures were made since early eighties. Paczyński and his collaborators advanced two models: (1) thick accretion disks (Paczyński & Wiita, 1980 and references therein): where the flow is still rotation dominated, but it is non-Keplerian. The radiation pressure is high enough to hold matter vertically, preventing it from collapsing. This thick disk model was clearly valid for high accretion rate and the radiation pressure in the funnel was assumed to push matter vertically to form outflows and jets. Ree et al. (1982) pointed out that the low accretion rate flow can support vertical structure due to strong ion pressure while most of the energy is advected. These models were not globally complete as the disk is generally non-accreting. (2) transonic flows (Paczyński & Bishnovatyi-Kogan, 1981; Paczyński & Muchotrzeb 1982): Here the radial motion is also included but global solutions were not obtained. Preliminary non-dissipative solutions indicated (Liang & Thomson, 1980) that unlike Bondi flow, there are two saddle type sonic points in an advective flow. Abramowicz & Zurek (1981) further concluded that the same matter could go through the outer (Bondi) or inner (disklike) sonic points, although it is now known that they have different entropies and are to be used by two different solutions (see, C96a for details.). Matsumoto et al. (1984) were interested to mend the Keplerian disk by trying to let the flow pass through the inner sonic point while Abramowicz et al. (1988), in the so-called slim disk tried to find global solutions of the transonic flow in the high accretion rate (they used $\dot{m} = 800 \dot{M}_{Edd}$) limit without success (although local solutions were proven to be stable). The transonic solution of Fukue (1987) had a standing shock wave. Chakrabarti (1989, 1990, 1996a) found all possible topologies of advective disk solutions in viscous and non-viscous flows, with and with magnetic fields. The recently re-discovered ion tori (known as advection dominated flow) model by Narayan & Yi (1994, 1995) is an ad hoc model, assumed to be something like 'corona without a disk'. This corona comes about by evaporation of the underlying Keplerian disk at low accretion rate. The post-shock and CENBOL regions of advective disk solutions described in (C89, C90, C96a) resemble those of thick disks (but accreting!). High accretion rate solutions of advective disk are the globally correct solutions of 'slim disks' (unlike Abramowicz et al., 1988 model, these flows have angular momentum increasing outwards to join a Keplerian disk) and low accretion rate advective disks are the correct 'advection dominated flows' (unlike Narayan & Yi, 1994, 1995 model, these flows naturally deviate from a Keplerian disk specially for low viscosity). Recently, some attempts are being made by these groups to try to reproduce advective disk solutions in their respective regimes (Chen et al., 1997; Narayan et al., 1997) although the success has been limited because of less general approach of solving the eigenvalue problem that the equations 1(a-d) posit. In their approach the flow is 'let loose' from a Keplerian disk at an arbitrary distance with an arbitrary initial radial and angular velocity components and only one sonic point is assumed (thus by construction this method cannot find shocks). It the advective disk approach (C90, C96a), the flow is allowed to have 'discontinuous' (shock) solutions which joins two branches in the stable manner. Needless to say that all these types of solutions are found to be stable except at least a section of NSA and NSW (see Fig. 1) where the solutions are inherently time-dependent. Observations of the black hole candidates also indicate that advective disk solution behaves in the right manner in right regime. Particularly important is the understanding that the net accretion flow is Keplerian disklike in some region, and advective in some other region. Today, the need for having the admixture of Keplerian and sub-Keplerian components is clearly recognized in most of the observations.

An accretion disk is by definition advective. It was on this philosophy Bondi flow was studied originally. In the intermediate phase, specially, in seventies and early eighties, rotating Keplerian disk took over while advection took the 'back seat'. It is releaving that the more and more observations suggest that the advection effects are important and advective disks are the preferred solutions. This is true both for black holes and neutron stars. The question of whether an advective disk can explain observations is outright irrelevant because this disk represents self-consistent solution of the governing equations which are derived from fundamental laws of nature (such as conservation of energy and momentum; unlike various models where ad hoc components are first thrown in by hand and then justified by tentative arguments). Unexpected solutions such as those including CENBOL emerged which behave like boundary layers of black holes! The same solution produces boundary layer of neutron star as well. These were unbelievable first, but now they are indispensable in most explanations, given that they obviated the need to construct 'Compton Clouds' by ad hoc processes (CT95). The future of the black hole astrophysics is certainly going to be the correct understanding of this advective region of the flow.

Bibliography

[1] Abramowicz, M.A. & Chakrabarti, M.A., 1990, Ap. J., 350, 281

[2] Abramowicz, M.A., Czerny, B., Lasota, J.P., Szuzkiewicz, E. 1988, Ap. J., 332

[3] Abramowicz, M.A., Zurek, W.H. 1981, Ap. J., 246, 314

[4] Arai, K., Hashimoto, M., 1992, Astron. Astrophys. 254, 191.

[5] Bondi, H., 1952, Mon. Not. R. Astron. Soc. 112, 195.

[6] Cannizzo, J. 1993 in *Accretion Disks in Compact Stellar Systems* (Ed.) J. C. Wheeler, World Scientific: Singapore

[7] Chakrabarti, S.K, 1986, in *Accretion Processes in Astrophysics,* (Eds.) J. Audouze & J. Tran Thanh Van, Editions Frontierès, Paris.

[8] Chakrabarti, S.K., 1988, J. Astrophys. Astron., 9, 49.

[9] Chakrabarti, S.K., 1989, Ap. J. 347, 365.

[10] Chakrabarti, S.K., 1990, *Theory of Transonic Astrophysical Flows*, World Scientific Publishers (Singapore).

[11] Chakrabarti, S.K., 1996a, Ap. J. 464, 646.

[12] Chakrabarti, S.K., 1996b, Physics Reports, v.266, No 5 & 6, 229

[13] Chakrabarti, S.K. 1996c, Mon. Not. R. Astron. Soc. 283, 325

[14] Chakrabarti, S.K., 1996d, Phys. Rev. D., 53, 2901.

[15] Chakrabarti, S.K., 1996e, in *Proceedings of the XVIIIth conference of the Indian Association of General Relativity and Gravitation, IMSC report no. 117.*

[16] Chakrabarti, S.K., 1997a, Ap. J., 484, 313.

[17] Chakrabarti, S.K., 1997b, Ap.J. (submitted).

[18] Chakrabarti, S.K., Jin, L., Arnett, W.D., Ap. J., 1987, 313, 674

[19] Chakrabarti, S.K., Molteni, D., 1993, Ap. J. 417, 671.

[20] Chakrabarti, S.K., Molteni, D., 1995, Mon. Not. R. Astron. Soc. 272, 80.

[21] Chakrabarti, S.K., Ryu, D., Molteni, D., Sponholz, H., Lanzafame, G., Eggum, G., 1996, in the *Proceedings of the IAU Asia-Pacific regional meeting* (Eds.) H.M. Lee, S.S. Kim, K.S. Kim, J. Korean Astron. Soc., v. 29, 229.

[22] Chakrabarti, S.K., Sahu, S., 1997, Astron. Astrophys., 323, 382.

[23] Chakrabarti, S.K., Titarchuk, L.G., 1995, Ap. J., 455, 623.

[24] Chen, X.M., Abramowicz, M.A. Lasota, J.-P., 1997, Ap.J., 476, 61.

[25] Crary, D.J. et al. Ap. J., 1996, 462, 71L

[26] Cui, W., Zhang, S.N., Focke, W. Swank, J.H., 1997, Ap. J., 484, 383.

[27] Dolan, J.F., Crannel, C.J., Dennis B.R., Frost, K.J., Orwig, L.E., 1979, Ap. J., 230, 551.

[28] Dotani, Y. 1992 in *Frontiers in X-ray Astronomy*, (Eds.) Y. Tanaka & K. Koyama, p. 152 Universal Academy Press, Tokyo.

[29] Eardley, D.M., Lightman, A.P., 1975, Ap. J. 200, 187.

[30] Ebisawa, K., Ogawa, M., Aoki, T., Dotani, T., Takizawa, M., Tanaka, Y., Yoshida, K., Miyamoto, S. 1994, PASJ, 46, 375.

[31] Ebisawa, K., Titarchuk, L., Chakrabarti, S.K., 1996, Publ. Astron. Soc. Japan, 1996, v. 48, No. 1

[32] Eggum, G. E., Coroniti, F. V., Katz, J. I. 1985, Ap. J., 298, L41

[33] Fukue, J., 1987, Publ. Astron. Soc. Japan, 39, 309.

[34] Gilfanov, M., Churazov, E., Sunyaev, R., 1997, in *Accretion Disks – New Aspects*, (Eds.) E. Meyer-Hofmeister & H. Spruit, Springer (Heidelberg).

[35] Halpern, J., Marshall, H. L. 1996, Ap. J., 464, 760

[36] Hashimoto, M., Eriguchi, Y., Arai, K., Müller, E., 1993, Astron. Astrophys. 268, 131.

[37] Hogan, C.J., Applegate, J.H., 1987, Nature 330, 236.

[38] Horgan,J., 1988, Scientific American, 258, 20.

[39] Jin, L., 1990, Ap. J. 356, 501.

[40] Jin, L., Arnett, W.D., Chakrabarti, S.K., 1989, Ap. J. 336, 572.

[41] Kato, S., Honma, F., Matsumoto, R., 1988, Publ. Astron. Soc. Japan 40, 709.

[42] Kuznetsov, S. et al., 1997, Mon. Not. R. Astron. Soc. (in press).

[43] Lanzafame, G. Molteni, D., Chakrabarti, S. K. 1997, Mon. Not. R. Astron. Soc. (submitted).

[44] Liang, E.P.T., Thompson, K.A., 1980, Ap. J., 240, 271.

[45] Lu, J.F., Yu, K.N., Yuan, F., Young, E.C.M., Astron. Astrophys., 321, 665

[46] Martin, E.L., Rebolo, R., Cesares, J. Charles, P.A., 1994, Ap. J., 435, 791.

[47] Matsumoto, R., Kato, S., Fukue J., Okazaki, A.T., 1984, Publ. Astron. Soc. Japan. 36, 71.

[48] McClintock, J.E., Horne, K. Remillard, R.A., 1995, Ap. J. 442, 358.

[49] Miyamoto, S., Kimura, K., Kitamoto, S., Dotani, T., Ebisawa, K., 1991, Ap. J., 383, 784.

[50] Molteni, D., Lanzafame, G., Chakrabarti, S.K., 1994, Ap. J. 425, 161.

[51] Molteni, D., Ryu, D. Chakrabarti, S. K. 1996, Ap.J. 470, 460.

[52] Molteni,D., Sponholz, H., Chakrabarti, S.K., 1996, Ap. J., 457, 805

[53] Morgan, E.H., Remillard, R. A., Greiner, J. 1997, Ap. J., 482, 993.

[54] Muchotrzeb B., Paczyński, B., 1982, Acta Astron. 32, 1.

[55] Narayan, R., Barret, D. & McClintock, J.E., Ap. J., 1997, 482, 448.

[56] Narayan, R., Kato, S., Honma, F. 1997, Ap. J., 476, 49.

[57] Narayan, R., Yi, I., 1994, Ap. J. 428, L13.

[58] Narayan, R., Yi, I., 1995, Ap. J. 444, 231.

[59] Nobuta K., Hanawa, T., 1994, Publ. Astron. Soc. Japan, 46 (1994) 257.

[60] Novikov, I., Thorne, K.S., 1973, in: *Black Holes*, (eds.) C. DeWitt and B. DeWitt, Gordon and Breach, New York.

[61] Nowak, M., Wagoner, R.V., 1993, Ap. J. 418 (1993) 183.

[62] Orosz, J.A., Remillard, R.A., Bailyn, C.D., McClintock, J.E., 1997, Ap.J., 478, 830.

[63] Paczyński B., Wiita, P.J., 1980, Astron. Astrophys. 88, 23.

[64] Paczyński, B., Bishnovatyi-Kogan, G., 1981, Acta Astron. 31, 283

[65] Paul, B. Agarwal, P.C., Rao, A.R., Vahia, M.N., Yadav, J.S., Seetha, S., Astron. Astrophys. (in press).

[66] Rees, M.J., Begelman, M.C., Blandford, R.D. & Phinney, E.S., 1982, Nature 295, 17.

[67] Ricketts, M., 1983, Astron. Astrophys., 118, L3.

[68] Rybicki, G.B. & Lightman, A.P., 1979, *Radiative Processes in Astrophysics*, John Wiley & Sons, New York.

[69] Ryu, D., Chakrabarti, S.K., Molteni, D., 1997, Ap J., 474, 378.

[70] Shapiro, S.L., 1973a, Ap. J., 180, 531.

[71] Shapiro, S.L., 1973b, Ap. J., 185, 69.

[72] Shapiro, S.L., Teukolsky, S.A. 1983, *Black Holes, White Dwarfs and Neutron Stars — the Physics of Compact Objects* , John Wiley & Sons, New York.

[73] Shakura, N.I., Sunyaev, R.A., 1973, Astron. Astrophys. 24, 337.

[74] Sunyaev, R.A., Trümper, J., 1979, Nature, 279, 506.

[75] Sunyaev, R.A., et al., 1994, Astron. Lett. 20, 777

[76] Sunyaev, R.A., Titarchuk, L.G., 1985, Astron. Astrophys. 143, 374.

[77] Taam, R., Chen, X.M. Swank, J., 1997, Ap. J. 485, L83.

[78] Tanaka, Y., Lewin W.H.G. 1995, *X Ray Binaries*, W.H.G. Lewin, J. Van Paradijs, E.P.J. Van-den Heuvel (eds.) p. 126 Cambridge University Press.

[79] Wagner, R.M., Starrfield, S.G., Hjellming, R.M., Howell, S.B. & Kreidl, T.J., 1994, Ap. J. 401, L97.

[80] Yang, R., Kafatos, M., 1995, Astron. Astrophys. 295, 238.

[81] Zhang, S.N., W. Cui, B.A. Harmon, W.S. Paciesas, R.E. Remillard, J. van Paradijs, Ap. J. 477, L95.

Part III

Classical General Relativity

The Membrane Model of Black Holes and Applications

Norbert Straumann

Institute for Theoretical Physics, University of Zürich, Winterthurer Str. 190, CH-8057 Zürich, Switzerland

Abstract. The "membrane model" of black holes (BHs) is a useful reformulation of the standard relativistic theory of BHs. By splitting the more elegant 4-dimensional physical laws into space and time (3+1 splitting) we can, e.g. in the case of a stationary BH, introduce quantities of an "absolute space" which evolve as functions of an "absolute time". This formulation is much closer to the intuition we have gained from other fields of physics. The 3+1 splitting, brings Maxwell's equations into a form which resembles the familiar form of Maxwell's equations for moving conductors. We can then use the pictures and the experience from ordinary electrodynamics. In a second step, one replaces the boundary conditions at the horizon by physical properties (electric conductivity, etc.) of a fictitious membrane. This procedure is completely adequate as long as one is not interested in fine details very close to the horizon. The details of this boundary layer are, however, completely irrelevant for astrophysical applications.

The following points will be discussed in detail:

- Solutions of Maxwell's equations in a Kerr background
- Space-time splittings
- The horizon as a conducting membrane
- Magnetic energy extraction from a BH
- BHs as current generators or rotators of electric motors
- The Blandford-Znajek process
- Stationary axisymmetric electrodynamics for force-free fields

1 Introduction

In these lectures I give an introduction to what is called *the membrane model of black holes* (BHs). This is not a new theory, but a convenient reformulation of the standard relativistic theory of BHs – as far as physics *outside* the horizon is concerned –, which is much closer to the intuition we have gained from other fields of physics (see [1], for a general reference). While the basic equations look less elegant, it has the advantage that we can understand astrophysical processes near a BH much more easily. For an analogy, imagine you would have to explain how a Tokomak works by using the language and pictures of special relativity (SR), i.e., by using the electromagnetic field tensor and 4-dimensional pictures of plasma flows. I would not know how to do this and how to get, for instance, an understanding of even the simplest plasma instabilities. A closer analogy would be to translate relevant studies of the electrodynamics of pulsars into a 4-dimensional language. The basic equations look beautiful, but it would be hard to understand anything. (You may say that radio pulsars are anyhow not understood.)

In the membrane model (often called *membrane paradigm* [1]) one first splits the elegant 4-dimensional physical laws of general relativity (GR) into space and time (3+1 splitting). For a general situation this can be done in many ways (reflecting the gauge freedom in GR) since there is no canonical fibration of spacetime by level surfaces of constant time. However, for a stationary BH there is a preferred decomposition. Relative to this the dynamical variables (electromagnetic fields, etc) become quantities on an *absolute space* which evolve as functions of an *absolute time*, as we are accustomed to from nonrelativistic physics. We shall see, for example, that the 3+1 splitting brings Maxwell's equations into a form which resembles the familiar form of Maxwell's equations for moving conductors. We can then use the pictures and the experience from ordinary electrodynamics.

In a second step one replaces the boundary conditions at the horizon by physical properties (electric conductivity, etc) of a *fictitious membrane*. This procedure is completely adequate as long as one is not interested in fine details *very close* to the horizon. The details of this boundary layer are, however, completely irrelevant for astrophysical applications. (The situation is similar to many problems in electrodynamics, where one replaces the real surface properties of a conductor and other media by idealized boundary conditions.)

The program of these lectures is as follows. First I will discuss the 3+1 splitting of the spacetime of a stationary rotating BH and of Maxwell's equations outside its horizon. We shall see that this can be achieved very smoothly by using the calculus of differential forms. As an illustration and for later use we shall apply these tools for a discussion of an exact solution of Maxwell's equation on a Kerr background, which describes an asymptotically homogeneous magnetic field. We shall then derive the electromagnetic properties of the fictitious membrane that simulate the boundary conditions at the horizon. Here, I can offer a much simpler derivation than has been given so far in the literature. As an important example of a physical process relatively close to a BH, I will treat in detail the magnetic energy extraction of a hole's rotational energy. Blandford and Znajek have first pointed out the possible relevance of this mechanism for an understanding of active galactic nuclei. It may well play an important role in the formation of energetic jets. The Blandford-Znajek process could also be important for explaining gamma-ray-bursts, because it may energize a Poynting-dominated outflow.

I hope to show you that the physics involved is not very different from that behind the electric generator in Fig. 1.

2   Space-Time Splitting of Electrodynamics

I describe now the 3+1 splitting of the general relativistic Maxwell equations on a stationary spacetime $(M, {}^{(4)}g)$. Most of what follows could easily be generalized to spacetimes which admit a foliation by spacelike hypersurfaces (see, e.g., Ref. [2]), but this is not needed in what follows.

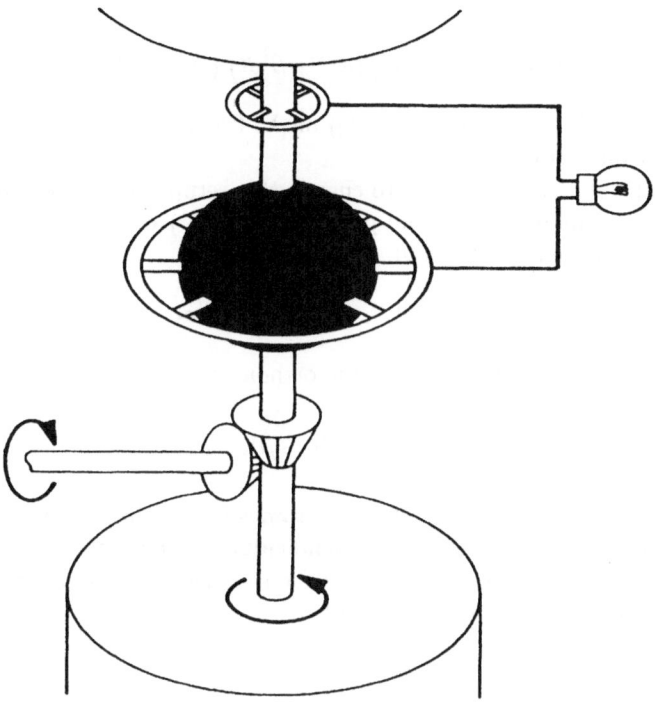

Fig. 1 Electric generator whose physics is similar to the electrodynamics of black holes in external magnetic fields.

Slightly more specifically, we shall assume that globally M is a product $\boldsymbol{R} \times \Sigma$, such that the natural coordinate t of $\boldsymbol{R}$ is adapted to the Killing field k, i.e., $k = \partial_t$. We decompose the Killing field into normal and parallel components relative to the "absolute space" (Σ, g), g being the induced metric on Σ,

$$\partial_t = \alpha\, u + \beta. \tag{1}$$

Here u is the unit normal field and β is tangent to Σ. This is what one calls the decomposition into lapse and shift; α is the *lapse function* and β the *shift vector field*. We shall usually work with adapted coordinates $(x^\mu) = (t, x^i)$, where $\{x^i\}$ is a coordinate system on Σ. Let $\beta = \beta^i \partial_i$ $(\partial_i = \partial/\partial x^i)$, and consider the basis of 1-forms

$$\alpha\, dt, \qquad dx^i + \beta^i dt. \tag{2}$$

One verifies immediately, that this is dual to the basis $\{u, \partial_i\}$ of vector fields. Since u is perpendicular to the tangent vectors ∂_i of Σ, the 4-metric has the form

$$^{(4)}g = -\alpha^2 dt^2 + g_{ij} \left(dx^i + \beta^i dt \right) \left(dx^j + \beta^j dt \right), \tag{3}$$

where $g_{ij} dx^i dx^j$ is the induced metric g on Σ. Clearly, α, β, and g are all time-independent quantities on Σ.

For what follows, I would like to change this setup slightly by using, instead of ∂_i and dx^i, a dual orthonormal pair $\{e_i\}$ and $\{\vartheta^i\}$ on Σ. Instead of (2), we have then the *orthonormal* tetrad

$$\theta^0 = \alpha dt, \qquad \theta^i = \vartheta^i + \beta^i dt, \tag{4}$$

where now $\beta = \beta^i e_i$. This is dual to the orthonormal frame

$$e_0 = u = \frac{1}{\alpha}(\partial_t - \beta), \quad e_i. \tag{5}$$

The tetrad $\{\theta^\mu\}$ describes the reference frames of so-called FIDOs, for *fiducial observers*. Their 4-velocity is thus perpendicular to the absolute space Σ.

Relative to these observers we have for the electromagnetic field tensor F (2-form) the same decomposition as in SR:

$$F = E \wedge \theta^0 + B, \tag{6}$$

where E is the electric 1-form $E = E_i \theta^i$ and B the magnetic 2-form $B = \frac{1}{2} B_{ij}\, \theta^i \wedge \theta^j$. ($E_i$, B_{ij} are the field strengths measured by the FIDOs.)

In a second step we decompose E and B relative to absolute space and absolute time. We have, using (5),

$$E = E_i \theta^i = E_i \left(\vartheta^i + \beta^i dt \right) = \mathcal{E} + i_\beta\, \mathcal{E}\, dt, \tag{7}$$

where

$$\mathcal{E} = E_i \vartheta^i, \qquad i_\bullet: \quad \text{interior product.} \tag{8}$$

Similarly,

$$B = \mathcal{B} + dt \wedge i_\beta\, \mathcal{B}, \quad \mathcal{B} = \frac{1}{2} B_{ij}\, \vartheta^i \wedge \vartheta^j. \tag{9}$$

Together we arrive at the following 3+1 decomposition of F:

$$F = \mathcal{B} + (\alpha\, \mathcal{E} - i_\beta\, \mathcal{B}) \wedge dt. \tag{10}$$

From this the 3+1 splitting of the homogeneous Maxwell equations is readily obtained: $dF = 0$ gives

$$\mathbf{d}\mathcal{B} + dt \wedge \partial_t\, \mathcal{B} + \mathbf{d}(\alpha\, \mathcal{E}) \wedge dt - \mathbf{d}(i_\beta\, \mathcal{B}) \wedge dt = 0.$$

Here $\mathbf{d}$ denotes the differential on Σ. This gives the two equations

$$\mathbf{d}\mathcal{B} = 0, \qquad \mathbf{d}(\alpha\,\mathcal{E}) + \partial_t\,\mathcal{B} = \mathbf{d}(i_\beta\,\mathcal{B})$$

or, with the Cartan identity $L_\beta = \mathbf{d} \circ i_\beta + i_\beta \circ \mathbf{d}$,

$$\mathbf{d}\mathcal{B} = 0, \qquad \mathbf{d}(\alpha\,\mathcal{E}) + (\partial_t - L_\beta)\mathcal{B} = 0. \tag{11}$$

The second equation describes Faraday's induction law in a gravitational field. It will be of crucial importance in later sections. Note, in particular, the coupling of the $\mathcal{B}$-field to the shift through the Lie derivative.

Let us also decompose the representation of F by a potential, $F = dA$. We have, using again (4),

$$A = A_\mu\,\theta^\mu = \alpha A_0\,dt + A_i(\vartheta^i + \beta^i\,dt)$$
$$= (\alpha A_0 + i_\beta\,\mathcal{A})dt + \mathcal{A}, \qquad \mathcal{A} = A_i\,\vartheta^i.$$

Thus

$$A = -\phi\,dt + \mathcal{A}, \tag{12}$$

where

$$\phi = -(\alpha A_0 + i_\beta\,\mathcal{A}). \tag{13}$$

This gives

$$dA = -\mathbf{d}\phi \wedge dt + \mathbf{d}\mathcal{A} + dt \wedge \partial_t\,\mathcal{A},$$

which is of the form (10), with

$$\mathcal{B} = \mathbf{d}\mathcal{A}, \qquad \alpha\,\mathcal{E} = -\mathbf{d}\phi - \partial_t\,\mathcal{A} + i_\beta\,\mathbf{d}\mathcal{A}. \tag{14}$$

Apart from the last term, this is what one is used to.

Now, we turn to the inhomogeneous Maxwell equation

$$d * F = 4\pi\mathcal{S}. \tag{15}$$

We need first the Hodge-dual of (10). We decompose $*F$ similarly to (6):

$$*F = -H \wedge \theta^0 + D, \tag{16}$$

which can be viewed as a definition of H and D. Comparison with (6) shows, that

$$H = B_i\,\theta^i, \qquad B_1 = B_{23}, \text{ etc,}$$
$$D = E_1\,\theta^2 \wedge \theta^3 + E_2\,\theta^3 \wedge \theta^1 + E_3\,\theta^1 \wedge \theta^2. \tag{17}$$

With (4) we find (∗ denotes the Hodge-dual on Σ)

$$
\begin{aligned}
H &= \mathcal{H} + i_\beta\, \mathcal{H} \wedge dt, & \mathcal{H} &= *\mathcal{B}, \\
D &= \mathcal{D} - i_\beta\, \mathcal{D} \wedge dt, & \mathcal{D} &= *\mathcal{E}.
\end{aligned}
\tag{18}
$$

If this is inserted into (16), we obtain

$$
*F = \mathcal{D} - (\alpha \mathcal{H} + i_\beta\, \mathcal{D}) \wedge dt.
\tag{19}
$$

The dual $J = *\mathcal{S}$ of the current 3-form can be decomposed as in SR

$$
J = \rho_{el}\, \theta^0 + j_k\, \theta^k,
\tag{20}
$$

where ρ_{el} is the electric charge density and j^k is the electric current density relative to the FIDOs. We use them to introduce the following quantities on the absolute space

$$
\rho = \rho_{el}\, \vartheta^1 \wedge \vartheta^2 \wedge \vartheta^3, \qquad j = j_k\, \vartheta^k, \qquad \mathcal{J} = *j.
\tag{21}
$$

Using the notation $\eta^\mu := *\theta^\mu$, we can decompose $\mathcal{S}$ as follows

$$
\begin{aligned}
\mathcal{S} &= \rho_{el}\, \eta^0 + j_k\, \eta^k \\
&= \rho_{el}\, \theta^1 \wedge \theta^2 \wedge \theta^3 - (j_1\, \theta^2 \wedge \theta^3 + \ldots) \wedge \theta^0 \\
&= \rho + \rho_{el}(\beta^1\, \vartheta^2 \wedge \vartheta^3 + \ldots) \wedge dt - \alpha\,(j_1\, \vartheta^2 \wedge \vartheta^3 + \ldots) \wedge dt.
\end{aligned}
$$

Thus

$$
\mathcal{S} = \rho + (i_\beta\, \rho - \alpha\, \mathcal{J}) \wedge dt.
\tag{22}
$$

Inserting this and (19) into (15) leads to

$$
\begin{aligned}
d*F &= \mathbf{d}\mathcal{D} + dt \wedge \partial_t\, \mathcal{D} - \mathbf{d}(\alpha\mathcal{H}) \wedge dt - \mathbf{d}(i_\beta\, \mathcal{D}) \wedge dt \\
&= 4\pi\rho + 4\pi(i_\beta\, \rho - \alpha\, \mathcal{J}) \wedge dt,
\end{aligned}
$$

and hence to the following 3+1 split of the inhomogeneous Maxwell equation

$$
\mathbf{d}\mathcal{D} = 4\pi\rho, \qquad \mathbf{d}(\alpha\mathcal{H}) = (\partial_t - L_\beta)\mathcal{D} + 4\pi\alpha\, \mathcal{J}.
\tag{23}
$$

From these laws one obtains immediately the local conservation law of the electric charge (use that $\mathbf{d}$ commutes with L_β):

$$
(\partial_t - L_\beta)\rho + \mathbf{d}(\alpha\, \mathcal{J}) = 0.
\tag{24}
$$

This follows, of course, also from $d\mathcal{S} = 0$ and the decomposition (22).

Integral Formulas

As is well-known from ordinary electrodynamics, it is often useful to write the basic laws (11), (23), and (24) in integral forms. Consider, for instance, the induction law in (11). If we integrate this over a surface area $\mathcal{A}$, which is *at rest* relative to the absolute space, we obtain with Stokes' theorem $(\mathcal{C} := \partial \mathcal{A})$

$$\oint_{\mathcal{C}} \alpha \mathcal{E} = -\frac{d}{dt} \int_{\mathcal{A}} \mathcal{B} + \int_{\mathcal{A}} L_\beta \mathcal{B}.$$

Here, we use $L_\beta \mathcal{B} = \mathbf{d}\, i_\beta \mathcal{B}$ (since $\mathbf{d}\mathcal{B} = 0$) and Stokes' theorem once more, with the result

$$\oint_{\mathcal{C}} \alpha \mathcal{E} = -\frac{d}{dt} \int_{\mathcal{A}} \mathcal{B} + \oint_{\mathcal{C}} i_\beta \mathcal{B}. \tag{25}$$

The left hand side is the electromotive force (EMF) along $\mathcal{C}$. The last term is similar to the additional term one encounters in Faraday's induction law for moving conductors. It is an expression of the coupling of $\mathcal{B}$ to the gravitomagnetic field and plays a crucial role in much that follows. This term contributes also for a stationary situation, for which (25) reduces to

$$\mathrm{EMF}(\mathcal{C}) = \oint_{\mathcal{C}} \alpha \mathcal{E} = \oint_{\mathcal{C}} i_\beta \mathcal{B}. \tag{26}$$

The integral form of the Ampère-Maxwell law is obtained similarly. Integrating the second equation in (23), we obtain with the Cartan identity $L_\beta = \mathbf{d} \circ i_\beta + i_\beta \circ \mathbf{d}$ and Gauss' law (first equation in (23)):

$$\oint_{\mathcal{C}} (\alpha \mathcal{H} + i_\beta \mathcal{D}) = \frac{d}{dt} \int_{\mathcal{A}} \mathcal{D} + 4\pi \int_{\mathcal{A}} (\alpha \mathcal{J} - i_\beta \rho). \tag{27}$$

The integral form of charge conservation is obtained by integrating (24) over a volume $\mathcal{V}$ which is at rest relative to absolute space:

$$\frac{d}{dt} \int_{\mathcal{V}} \rho = -\int_{\partial \mathcal{V}} (\alpha \mathcal{J} - i_\beta \rho) \tag{28}$$

(note that $L_\beta \rho = \mathbf{d}\, i_\beta \rho$).

One could, of course, also derive integral formulas for moving volumes and surface areas (exercise).

Vector Analytic Formulation

The similarity of the basic laws in the 3+1 split with ordinary electrodynamics becomes even closer if we write everything in vector analytic form. I give a dictionary between the two formulations that is valid for any 3-dimensional Riemannian manifold (Σ, g).

The metric g defines natural isomorphisms $\sharp$ and $\flat$ between the sets of 1-forms, $\Lambda^1(\Sigma)$, and vector fields, $\mathcal{X}(\Sigma)$. In addition, the volume form η, belonging to the metric, defines an isomorphism between $\mathcal{X}(\Sigma)$ and the space of 2-forms, $\Lambda^2(\Sigma)$, given by

$$\vec{B} \longmapsto \mathcal{B} = i_{\vec{B}}\,\eta. \tag{29}$$

We have the following commutative diagram, in which $*$ denotes, as always, the Hodge-dual:

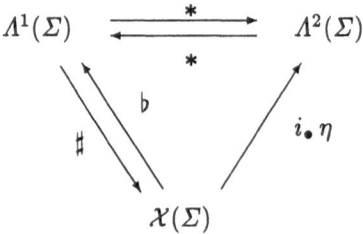

From this one can read off, for instance,

$$i_{\vec{v}}\,\eta = *v \qquad (\vec{v} \in \mathcal{X}(\Sigma),\ v : (\vec{v})^{\flat}). \tag{30}$$

The cross product and the wedge product are related as follows:

$$
\begin{array}{ccc}
\Lambda^1(\Sigma) \times \Lambda^1(\Sigma) & \xrightarrow{\wedge} & \Lambda^2(\Sigma) \\
\updownarrow & & \updownarrow \\
\mathcal{X}(\Sigma) \times \mathcal{X}(\Sigma) & \xrightarrow{\times} & \mathcal{X}(\Sigma)
\end{array}
$$

In particular, we have

$$i_{\vec{v}\times\vec{w}}\,\eta = v \wedge w. \tag{31}$$

With the help of the next commutative diagram one can reduce many of the vector analytic identities to $d \circ d = 0$.

$$
\begin{array}{ccccccccc}
0 & \longrightarrow & \Lambda^0(\Sigma) & \xrightarrow{\mathbf{d}} & \Lambda^1(\Sigma) & \xrightarrow{\mathbf{d}} & \Lambda^2(\Sigma) & \xrightarrow{\mathbf{d}} & \Lambda^3(\Sigma) & \longrightarrow & 0 \\
& & \| & & \flat\uparrow\downarrow\sharp & & \uparrow i_\bullet\eta & & \uparrow\bullet\eta & & \\
0 & \longrightarrow & \Lambda^0(\Sigma) & \xrightarrow{\mathrm{grad}} & \mathcal{X}(\Sigma) & \xrightarrow{\mathrm{curl}} & \mathcal{X}(\Sigma) & \xrightarrow{\mathrm{div}} & \Lambda^0(\Sigma) & \longrightarrow & 0.
\end{array}
$$

We can read off, for example,

$$i_{\mathrm{curl}\,\vec{v}}\,\eta = \mathrm{d}v. \tag{32}$$

For a 1-form w with vector field $\vec{w} = w^{\sharp}$, we have the translation

$$i_{\vec{v}}\,\mathrm{d}w \overset{(32)}{=} i_{\mathrm{curl}\,\vec{v}}\,i_{\mathrm{curl}\,\vec{w}}\,\eta = [(\mathrm{curl}\,\vec{w}) \times \vec{v}]^{\flat}. \tag{33}$$

Here, we made use of the algebraic relation

$$i_{\vec{v}}\, i_{\vec{u}}\, \eta \overset{(30)}{=} i_{\vec{v}} * u = *(u \wedge v) \overset{(31)}{=} (\vec{u} \times \vec{v})^{\flat}. \tag{34}$$

We need also the translation of the Lie derivative of a 1-form w:

$$L_{\vec{v}}\, w = \mathbf{d}\, \underbrace{i_{\vec{v}}\, w}_{(\vec{v},\vec{w})} + i_{\vec{v}}\, \mathbf{d}w \overset{(33)}{=} \mathbf{d}(\vec{v},\vec{w}) + [(\operatorname{curl}\vec{w}) \times \vec{v}]^{\flat}.$$

Thus,

$$L_{\vec{v}}\, w = \{\operatorname{grad}(\vec{v},\vec{w}) + (\operatorname{curl}\vec{w}) \times \vec{v}\}^{\flat}. \tag{35}$$

Similarly, we have for a 2-form $\mathcal{B} = i_{\vec{B}}\, \eta$:

$$L_{\vec{v}}\, \mathcal{B} = L_{\vec{v}}\, i_{\vec{B}}\, \eta = i_{\vec{v}}\, \underbrace{\mathbf{d}\, i_{\vec{B}}\, \eta}_{\operatorname{div}\vec{B}\,\eta} + \underbrace{\mathbf{d}\, i_{\vec{v}}\, i_{\vec{B}}\, \eta}_{(\vec{B}\times\vec{v})^{\flat}}$$

$$\overset{(32)}{=} i_{\{(\operatorname{div}\vec{B})\vec{v} + \operatorname{curl}(\vec{B}\times\vec{v})\}}\, \eta.$$

We have thus the correspondence

$$L_{\vec{v}}\, \mathcal{B} \longleftrightarrow (\operatorname{div}\vec{B})\vec{v} + \operatorname{curl}(\vec{B} \times \vec{v}). \tag{36}$$

Here we have to stress that the right hand side is in general not equal to $L_{\vec{v}}\,\vec{B} = [\vec{v}, \vec{B}]$ (Lie bracket). This comes out as follows:

$$L_{\vec{v}}\, \mathcal{B} = L_{\vec{v}}\, i_{\vec{B}}\, \eta = \underbrace{[L_{\vec{v}}, i_{\vec{B}}]}_{i_{[\vec{v},\vec{B}]}}\, \eta + i_{\vec{B}}\, \underbrace{L_{\vec{v}}\, \eta}_{\operatorname{div}\vec{v}\,\eta}$$

$$= i_{\{(\operatorname{div}\vec{v})\vec{B} + [\vec{v},\vec{B}]\}}\, \eta.$$

The correspondence (36) is thus equivalent to

$$L_{\vec{v}}\, \mathcal{B} \longleftrightarrow L_{\vec{v}}\, \vec{B} + (\operatorname{div}\vec{v})\, \vec{B}. \tag{37}$$

Only for $\operatorname{div}\vec{v} = 0$ do the Lie derivatives $L_{\vec{v}}\,\mathcal{B}$ and $L_{\vec{v}}\,\vec{B}$ correspond to each other!

In Maxwell's equations the Lie derivative $L_{\vec{\beta}}\,\mathcal{B}$ occurs. This will be replaced in the vector analytic translation by $L_{\vec{\beta}}\,\vec{B}$, because $\operatorname{div}\vec{\beta}$ is (for a stationary metric) proportional to the trace of the second fundamental form of the time slices and this vanishes for *maximal slicing*. For the Kerr solution we are, for instance, in this situation (exercise).

Part of what has been said is summarized for convenience in the table below.

Dictionary

calculus of forms	vector analysis	notation
$v \wedge w$	$\vec{v} \times \vec{w}$	$v = (\vec{v})^\flat,\ w = (\vec{w})^\flat$
$i_{\vec{v}}\mathcal{B},\ i_{\vec{v}}\mathcal{D}$	$\vec{B} \times \vec{v},\ \vec{E} \times \vec{v}$	$\mathcal{B} = i_{\vec{B}}\,\eta,\ \mathcal{D} = i_{\vec{E}}\,\eta$
$\mathbf{d}f$	$\operatorname{grad} f$	$f :$ function
$\mathbf{d}v$	$\operatorname{curl} \vec{v}$	
$\mathbf{d}\mathcal{B},\ \mathbf{d}\mathcal{D}$	$\operatorname{div} \vec{B},\ \operatorname{div} \vec{E}$	
$L_{\vec{v}}\,w$	$\operatorname{grad}(\vec{v},\vec{w}) - \vec{v} \times \operatorname{curl} \vec{w}$	$w = (\vec{w})^\flat$
$L_{\vec{v}}\mathcal{B},\ L_{\vec{v}}\mathcal{D}$	$(\operatorname{div} \vec{B})\vec{v} - \operatorname{curl}(\vec{v} \times \vec{B}),\ \vec{B} \longleftrightarrow \vec{E}$	

Summary

For reference, we write down once more the 3+1 split of Maxwell's equations (11) and (23) in Cartan's calculus

$$\mathbf{d}\mathcal{B} = 0, \quad \mathbf{d}(\alpha\,\mathcal{E}) + (\partial_t - L_\beta)\mathcal{B} = 0,$$
$$\mathbf{d}\mathcal{D} = 4\pi\rho, \quad \mathbf{d}(\alpha\mathcal{H}) = (\partial_t - L_\beta)\mathcal{D} + 4\pi\alpha\,\mathcal{J}. \tag{38}$$

The dictionary above allows us to translate these into the vector analytic form:

$$\vec{\nabla} \cdot \vec{B} = 0, \quad \vec{\nabla} \times (\alpha\,\vec{E}) + (\partial_t - L_{\vec{\beta}})\vec{B} = 0,$$
$$\vec{\nabla} \cdot \vec{E} = 4\pi\rho_{el}, \quad \vec{\nabla} \times (\alpha\,\vec{B}) = (\partial_t - L_{\vec{\beta}})\vec{E} + 4\pi\alpha\,\vec{j}. \tag{39}$$

3   Black Hole in a Homogeneous Magnetic Field

As an instructive example and a useful tool we discuss now an exact solution of Maxwell's equations in the Kerr metric, which becomes asymptotically a homogeneous magnetic field. This solution can be found in a strikingly simple manner [3].

For any Killing field K one has the following identity

$$\delta\,\mathbf{d}\,K^\flat = 2\,R(K), \tag{40}$$

where δ denotes the codifferential and $R(K)$ is the 1-form with components $R_{\mu\nu}K^\nu$. In components (40) is equivalent to

$$K_{\mu\ ;\alpha}^{;\alpha} = -R_{\mu\alpha}\,K^\alpha. \tag{41}$$

This form can be obtained by contracting the indices σ and ρ in the following general equation for a vector field

$$\xi_{\sigma;\rho\mu} - \xi_{\sigma;\mu\rho} = \xi_\lambda R^\lambda_{\ \sigma\rho\mu}$$

and by using the consequence $K^{\sigma}_{;\sigma} = 0$ of the Killing equation $K_{\sigma;\rho} + K_{\rho;\sigma} = 0$. For a vacuum spacetime we thus have

$$\delta\, d\, K^{b} = 0 \tag{42}$$

for any Killing field. Hence, the vacuum Maxwell equations are satisfied if F is a constant linear combination of the differential of Killing fields (their duals, to be precise). For the Kerr metric, as for any axially symmetric stationary spacetime, we have two Killing fields k and m, say; in adapted coordinates these are $k = \partial_t$ and $m = \partial_{\varphi}$. The Komar formulae provide convenient expressions for the total mass M and the total angular momentum J of the Kerr BH:

$$M = -\frac{1}{8\pi} \int_{\infty} *dk^{b}, \qquad J = \frac{1}{16\pi} \int_{\infty} *dm^{b} \tag{43}$$

(for $G = 1$).

We try the ansatz

$$F = \frac{1}{2} B_0\,(dm^{b} + 2a\, dk^{b}) \qquad (B_0 = \text{const}), \tag{44}$$

and choose a such that the total electric charge

$$Q = -\frac{1}{4\pi} \int_{\infty} *F \tag{45}$$

vanishes. The Komar formulae (43) tell us that

$$Q = -\frac{1}{8\pi} B_0\,(16\pi J - 2a \cdot 8\pi M), \tag{46}$$

and this vanishes if $a = J/M$ (which is the standard meaning of the symbol a in the Kerr solution).

Clearly, F is stationary and axisymmetric:

$$L_k F = L_m F = 0, \tag{47}$$

because (dropping b from now on)

$$L_k\, dk = d\, L_k\, k = 0 \qquad (L_k\, k = [k, k] = 0), \text{ etc.}$$

For the further discussion we need the Kerr metric. In Boyer-Lindquist coordinates and more or less standard notation it has the form (3), i.e.,

$$^{(4)}g = [-\alpha^2\, dt^2 + g_{\varphi\varphi}(d\varphi + \beta^{\varphi}\, dt)^2] + [g_{rr}\, dr^2 + g_{\vartheta\vartheta}\, d\vartheta^2], \tag{48}$$

with only the component β^{φ} of the shift being $\neq 0$. With the abbreviations

$$\rho^2 := r^2 + a^2 \cos^2\vartheta, \quad \Delta := r^2 - 2Mr + a^2, \quad \Sigma^2 := (r^2 + a^2)^2 - a^2 \Delta \sin^2\vartheta, \tag{49}$$

the metric coefficients are

$$g_{rr} = \frac{\rho^2}{\Delta}, \quad g_{\vartheta\vartheta} = \rho^2, \quad g_{\varphi\varphi} = \sin^2\vartheta \, \frac{\Sigma^2}{\rho^2},$$

$$g_{tt} = -1 + \frac{2Mr}{\rho^2}, \quad g_{t\varphi} = -\frac{2Mra\sin^2\vartheta}{\rho^2}, \tag{50}$$

while the lapse and shift are given by

$$\alpha^2 = \frac{\rho^2}{\Sigma^2} \Delta, \quad \beta^\varphi = -a \frac{2Mr}{\Sigma^2}. \tag{51}$$

This gives asymptotically

$$^{(4)}g = -\left[1 - \frac{2M}{r} + \mathcal{O}\left(\frac{1}{r^2}\right)\right] dt^2 - \left[\frac{4aM}{r}\sin^2\vartheta + \mathcal{O}\left(\frac{1}{r^2}\right)\right] dt\,d\varphi$$

$$+ \left[1 + \mathcal{O}\left(\frac{1}{r}\right)\right] [dr^2 + r^2(d\vartheta^2 + \sin^2\vartheta \, d\varphi^2)]. \tag{52}$$

To establish the connection with our general discussion, we introduce here as orthonormal basis of the absolute space naturally

$$\vartheta^r = \sqrt{g_{rr}}\,dr, \quad \vartheta^\vartheta = \sqrt{g_{\vartheta\vartheta}}\,d\vartheta, \quad \vartheta^\varphi = \sqrt{g_{\varphi\varphi}}\,d\varphi. \tag{53}$$

The shift vector is $\beta = \beta^\varphi\,\partial_\varphi$.

The angular velocity ω of the FIDOs, with 4-velocity $u = \frac{1}{\alpha}(\partial_t - \beta^\varphi\,\partial_\varphi)$, is

$$\omega = \frac{u^\varphi}{u^t} = -\beta^\varphi = -\frac{g_{t\varphi}}{g_{\varphi\varphi}}. \tag{54}$$

The last equality sign implies that the angular momentum of these (Bardeen) observers vanishes: $(u, m) = 0$. The FIDOs are, therefore, sometimes also called ZAMOs (for *zero angular momentum observers*).

Now, we want to discuss in detail the solution (44), which we can express in terms of a potential: $F = dA$, with

$$A = \frac{1}{2} B_0 (m + 2ak). \tag{55}$$

Let us first look at the asymptotics. The 1-forms $k_\mu\,dx^\mu$, $m_\mu\,dx^\mu$ belonging to the Killing fields ($k_\mu = g_{\mu t}$, $m_\mu = g_{\mu\varphi}$) are asymptotically $k \sim -dt$, $m \sim r^2\sin^2\vartheta\,d\varphi$, whence (44) gives

$$F \sim B_0\left[\sin\vartheta\,dr \wedge r\sin\vartheta\,d\varphi + \cos\vartheta\,rd\vartheta \wedge r\sin\vartheta\,d\varphi\right]. \tag{56}$$

This is a magnetic field in the z-direction whose magnitude is B_0.

In (55) we need

$$m + 2ak = (g_{\mu\varphi} + 2a\,g_{\mu t})\,dx^\mu = (g_{t\varphi} + 2a\,g_{tt})\,dt + (g_{\varphi\varphi} + 2a\,g_{\varphi t})\,d\varphi$$

$$\overset{(54)}{=} (-\omega\,g_{\varphi\varphi} + 2a\,g_{tt})dt + (g_{\varphi\varphi} - 2a\,\omega\,g_{\varphi\varphi})d\varphi.$$

Using the notation [1]

$$\tilde{\omega}^2 := g_{\varphi\varphi}, \qquad \tilde{\omega} = \frac{\Sigma}{\rho} \sin\vartheta \tag{57}$$

we thus have

$$m + 2ak = \left[-\omega\tilde{\omega}^2 + 2a\left(\omega^2\tilde{\omega}^2 - \alpha^2\right)\right] dt + \tilde{\omega}^2(1 - 2a\omega)\, d\varphi.$$

Comparing this with (12), we obtain for the potentials

$$\phi = \frac{1}{2} B_0 \left[\omega\tilde{\omega}^2 + 2a\left(\alpha^2 - \omega^2\tilde{\omega}^2\right)\right], \tag{58}$$

$$\mathcal{A} = A_\varphi\, d\varphi, \qquad A_\varphi = \frac{1}{2} B_0\, \tilde{\omega}^2(1 - 2a\omega). \tag{59}$$

The fields $\mathcal{E}$ and $\mathcal{B}$ can now be obtained from (14). We have

$$\mathcal{B} = d\mathcal{A} = A_{\varphi,r}\, dr \wedge d\varphi + A_{\varphi,\vartheta}\, d\vartheta \wedge d\varphi$$

$$= \frac{1}{\Sigma \sin\vartheta} \left[A_{\varphi,\vartheta}\, \underbrace{\vartheta^2 \wedge \vartheta^3}_{*\vartheta^1} - \sqrt{\Delta}\, A_{\varphi,r}\, \underbrace{\vartheta^3 \wedge \vartheta^1}_{*\vartheta^2} \right]. \tag{60}$$

A_φ is explicitly (use (59), (57), (54), and (51))

$$A_\varphi = \frac{1}{2} B_0 \frac{\Sigma^2}{\rho^2} \sin^2\vartheta \left(1 - 4a^2 \frac{Mr}{\Sigma^2}\right).$$

We write this as

$$A_\varphi = \frac{1}{2} B_0\, X, \qquad X = \frac{\sin^2\vartheta}{\rho^2}\left(\Sigma^2 - 4a^2 Mr\right). \tag{61}$$

With this notation, (60) reads

$$\mathcal{B} = \frac{B_0}{2\Sigma \sin\vartheta} \left[X_{,\vartheta} * \vartheta^r - \sqrt{\Delta}\, X_{,r} * \vartheta^\vartheta\right]. \tag{62}$$

The corresponding vector field $\vec{B}$ is thus

$$\vec{B} = \frac{B_0}{2\Sigma \sin\vartheta} \left[X_{,\vartheta}\, \vec{e}_r - \sqrt{\Delta}\, X_{,r}\, \vec{e}_\vartheta\right]. \tag{63}$$

For $\mathcal{E}$ we have, with $\beta = -\omega\, \partial_\varphi$,

$$\alpha\, \mathcal{E} = -d\phi + i_\beta\, d\mathcal{A} = -d\phi + \omega\, dA_\varphi. \tag{64}$$

From this one finds quickly

$$\vec{E} = -\frac{B_0\, a\Sigma}{\rho^2} \left\{ \left[\frac{\partial\alpha^2}{\partial r} + \frac{M \sin^2\vartheta}{\rho^2}\left(\Sigma^2 - 4a^2 Mr\right)\frac{\partial}{\partial r}\left(\frac{r}{\Sigma^2}\right)\right] \vec{e}_r \right.$$

$$\left. + \frac{1}{\sqrt{\Delta}} \left[\frac{\partial\alpha^2}{\partial\vartheta} + r\, \frac{M \sin^2\vartheta}{\rho^2}\left(\Sigma^2 - 4a^2 Mr\right)\frac{\partial}{\partial\vartheta}\left(\frac{1}{\Sigma^2}\right)\right] \vec{e}_\vartheta \right\}. \tag{65}$$

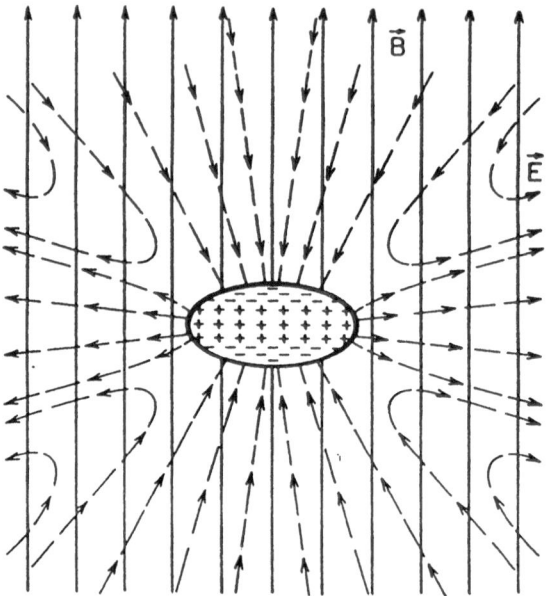

Fig. 2 Field lines of $\vec{E}$ (from Ref. [1]).

The field lines of $\vec{E}$ are shown in Fig. 2.

It is of interest to work out the magnetic flux through the equator of the BH, i.e.,

$$\Phi = \int_{\text{upper h.}} \mathcal{B} = \int_{\text{equator}} \mathcal{A} = 2\pi \, \mathcal{A}_\varphi \big|_{\text{equator}} . \qquad (66)$$

Specializing (61) to $\vartheta = \pi/2$ and $r = r_H$ gives (the horizon is located at $\Delta = 0$)

$$\Phi = 4\pi B_0 \, M(r_H - M) = 4\pi B_0 \, M \sqrt{M^2 - a^2}. \qquad (67)$$

Here, we have used

$$r_H = M + \sqrt{M^2 - a^2}. \qquad (68)$$

Note that this vanishes for an extremal BH ($a = M$). Generically one has, as expected, $\Phi \approx \pi r_H^2 B_0$.

4   The Horizon as a Conducting Membrane

As long as one is not interested in fine details very close to the horizon, one can regard the boundary conditions implied by the horizon as arising from physical properties of a fictitious membrane. This membrane is thought of as endowed with surface charge density σ_H, surface current $\vec{\mathcal{J}}_H$, and surface resistivity R_H. This idea was first pursued by Damour [4] and independently by Znajek [5]. Later it was further developed by Thorne and Mac Donald [6] and other authors (see [1] and references therein).

Below we give a much simplified derivation that the following surface properties for the parallel and perpendicular components hold:

$$
\begin{aligned}
\text{Gauss's law}: \ & E_\perp \longrightarrow 4\pi\,\sigma_H, \\
\text{Ampère's law}: \ & \alpha\,\vec{B}_\| \longrightarrow \vec{B}_H = 4\pi\,\vec{\mathcal{J}}_H \times \vec{n}, \\
\text{charge conservation}: \ & \alpha\,j_\perp \longrightarrow -\partial_t\,\sigma_H - {}^{(2)}\vec{\nabla}\cdot\vec{\mathcal{J}}_H, \\
\text{Ohm's law}: \ & \alpha\,\vec{E}_\| \longrightarrow \vec{E}_H = R_H\,\vec{\mathcal{J}}_H.
\end{aligned}
\tag{69}
$$

The surface resistivity R_H turns out to be equal to the vacuum impedance

$$
R_H = 4\pi = 377 \ \text{Ohm}.
\tag{70}
$$

For the horizon fields $\vec{E}_H$, $\vec{B}_H$ we have, therefore,

$$
\vec{B}_H = \vec{E}_H \times \vec{n},
\tag{71}
$$

as for a plane wave in vacuum.

Toward the horizon the FIDOs move relative to freely falling observers, say, more and more rapidly, approaching the velocity of light. Mathematically, the tetrad $\{\theta^\mu\}$ becomes singular at the horizon, and therefore the components of F relative to $\{\theta^\mu\}$ are ill-behaved. The 2-form F should, of course, remain regular and the laws (69) and (70) are just an expression of this requirement. For illustration, we demonstrate this first for a Schwarzschild BH, and generalize afterwards the argument in a simple manner.

(a) Derivation for a Schwarzschild BH

The procedure is simple: we pass to a regular, not necessarily orthonormal tetrad. The angular part $\theta^2 = r\,d\vartheta$, $\theta^3 = r\sin\vartheta\,d\varphi$ is kept, but instead of $\theta^0 = \alpha\,dt$, $\theta^1 = \alpha^{-1}\,dr$ we use $d\bar{t}$, where $\bar{t}$ is the Eddington-Finkelstein time coordinate

$$
\bar{t} = t + 2M\ln\left(\frac{r}{2M} - 1\right) \longrightarrow d\bar{t} = dt + \alpha^{-2}\frac{2M}{r}\,dr \qquad (r > 2M).
\tag{72}
$$

Since the Eddington-Finkelstein coordinates are regular at the horizon, the same is true for the basis $\{d\bar{t}, dr, \theta^2, \theta^3\}$. We have the relations

$$
\theta^0 = \alpha\,d\bar{t} - \frac{1}{\alpha}\frac{2M}{r}\,dr,
$$

$$
\theta^1 = \frac{1}{\alpha}\,dr,
\tag{73}
$$

which allow us to rewrite the decomposition (6) as follows

$$F = E_1\,\theta^1 \wedge \theta^0 + \ldots + B_3\,\theta^1 \wedge \theta^2 + \ldots$$
$$= E_1\,dr \wedge d\bar{t} + E_2\left(\alpha\,\theta^2 \wedge d\bar{t} - \frac{1}{\alpha}\frac{2M}{r}\theta^2 \wedge dr\right) + E_3\,(\ldots)$$
$$+ B_3\,\frac{1}{\alpha}\,dr \wedge \theta^2 - B_2\,\frac{1}{\alpha}\,dr \wedge \theta^3 + B_1\,\theta^2 \wedge \theta^3$$

or

$$F = E_1\,dr \wedge d\bar{t} + \alpha\,E_2\,\theta^2 \wedge d\bar{t} + \alpha\,E_3\,\theta^3 \wedge d\bar{t} + B_1\,\theta^2 \wedge \theta^3$$
$$+ \frac{1}{\alpha}\left(B_3 + \frac{2M}{r}E_2\right)dr \wedge \theta^2 + \frac{1}{\alpha}\left(-B_2 + \frac{2M}{r}E_3\right)dr \wedge \theta^3. \qquad (74)$$

The regularity of the coefficients in this expansion implies the following behavior when the horizon is approached ($\alpha \downarrow 0$):

(i) radial components: $E_1, B_1 = \mathcal{O}(1)$,

(ii) tangent components: $E_2, E_3, B_2, B_3 = \mathcal{O}\left(\dfrac{1}{\alpha}\right)$,

(iii) $E_2 + B_3, E_3 - B_2 = \mathcal{O}(\alpha)$. \qquad (75)

This shows that

$$\alpha\,\vec{E}_\| = \vec{n} \times \alpha\,\vec{B}_\| + \mathcal{O}(\alpha^2), \qquad (\vec{n} := \vec{e}_r),$$
$$\alpha\,\vec{B}_\| = -\vec{n} \times \alpha\,\vec{E}_\| + \mathcal{O}(\alpha^2), \qquad (76)$$
$$E_n \equiv E_\perp, \quad B_n \equiv B_\perp \text{ remain finite.} \qquad (77)$$

This is basically already what was claimed in (69) and (70). It is useful to introduce the *stretched horizon* $\mathcal{H}^s = \{\alpha = \alpha_H \ll 1\}$ which is arbitrary close to the event horizon. Let

$$\vec{E}_H := (\alpha\,\vec{E}_\|)_{\alpha_H}, \qquad \vec{B}_H := (\alpha\,\vec{B}_\|)_{\alpha_H} \qquad (78)$$

and as above

$$E_n := \vec{E} \cdot \vec{n}, \qquad B_n := \vec{B} \cdot \vec{n}. \qquad (79)$$

These components remain *finite* for $\alpha_H \downarrow 0$ and we have up to $\mathcal{O}(\alpha_H^2)$

$$\vec{E}_H = \vec{n} \times \vec{B}_H, \qquad \vec{B}_H = -\vec{n} \times \vec{E}_H. \qquad (80)$$

The surface charge density σ_H and the surface current density $\vec{J}_H$ on $\mathcal{H}^s$ are defined by

$$\sigma_H := \left(\frac{E_n}{4\pi}\right)_{\mathcal{H}^s}, \qquad \vec{B}_H =: \left(4\pi\,\vec{J}_H \times \vec{n}\right)_{\mathcal{H}^s}. \qquad (81)$$

The second equation and (50) imply Ohm's law in (69)

$$\vec{\mathcal{J}}_H = \frac{1}{R_H}\,\vec{E}_H, \qquad R_H = 4\pi. \tag{82}$$

Finally, we make use of the Ampère-Maxwell law (39) (for $\vec{\beta} = 0$):

$$\partial_t\,\vec{E} = \vec{\nabla} \times (\alpha\,\vec{B}) - 4\pi\,\alpha\,\vec{j}.$$

The normal component on $\mathcal{H}^s$ is

$$\partial_t\,E_n = \left[\vec{\nabla} \times (\alpha\,\vec{B})\right]_n - 4\pi\,\alpha\,\vec{j}_n\,\Big|_{\mathcal{H}^s}\,.$$

Using

$$\left[\vec{\nabla} \times (\alpha\,\vec{B})\right]_n\Big|_{\mathcal{H}^s} = \left[\vec{\nabla} \times (4\pi\,\vec{\mathcal{J}}_H \times \vec{n})\right]_n\Big|_{\mathcal{H}^s} = -\,4\pi\,{}^{(2)}\vec{\nabla} \cdot \vec{\mathcal{J}}_H,$$

where ${}^{(2)}\vec{\nabla}$ denotes the induced covariant derivation on $\mathcal{H}^s$, we obtain

$$\partial_t\,\sigma_H + {}^{(2)}\vec{\nabla} \cdot \vec{\mathcal{J}}_H + (\alpha\,j_n)_{\mathcal{H}^s} = 0, \tag{83}$$

as an expression of charge conservation.

This completes the derivation of (69) and (70) for the Schwarzschild BH. Next, we generalize the discussion to an arbitrary static BH.

(b) Derivation for static BH

We introduce first a parameterization of the exterior metric which was used also in Israel's famous proof of the uniqueness theorem for the Schwarzschild BH.

The starting point is (3) for $\beta = 0$, i.e.,

$${}^{(4)}g = -\alpha^2 dt^2 + g, \qquad \alpha \text{ and } g \text{ independent of } t. \tag{84}$$

Note that $\alpha^2 = -(k|k) \geq 0$, $k = \partial_t$, and that the horizon has to be at $\alpha = 0$. We assume that the lapse function has no critical point, $d\alpha \neq 0$. The absolute space Σ is then foliated by the leaves $\{\alpha = \text{const}\}$. The function

$$\rho := (d\alpha|d\alpha)^{-\frac{1}{2}} \tag{85}$$

is then positive on Σ.

Now we introduce adapted coordinates on Σ. Consider in any point $p \in \Sigma$ the 1-dimensional subspace of $T_p(\Sigma)$ perpendicular to the tangent space of the leave $\{\alpha = \text{const}\}$ through p. This defines a 1-dimensional distribution which is, of course, involutive (integrable). The Frobenius theorem then tells us that we can introduce coordinates $\{x^i\}$ on Σ, such that x^A $(A = 2, 3)$ are constant along the integral curves of the distribution. For x^1 we can choose the lapse function, and thus obtain

$$g = \rho^2 d\alpha^2 + \tilde{g}, \quad \tilde{g} = \tilde{g}_{AB} dx^A dx^B. \tag{86}$$

Here, ρ and $\tilde{g}_{AB}$ depend in general on all three coordinates $x^1 = \alpha$, x^A.

We also need the surface gravity κ on the horizon. A useful formula is (see, e.g., [7])

$$\kappa^2 = -\frac{1}{4}(dk|dk)\big|_H. \tag{87}$$

Now, $k = -\alpha^2 dt$, $dk = -2\alpha\, d\alpha \wedge dt$, whence

$$\kappa = \frac{1}{\rho_H}. \tag{88}$$

From the zeroth law of BH physics we know that $\kappa = \text{const}$. Since we want to assume a regular Killing horizon (generated by k), we conclude

$$0 < \rho_H < \infty, \qquad \rho_H = \text{const}. \tag{89}$$

Combining (84) with (86) we have outside the horizon

$$^{(4)}g = -N\, dt^2 + \frac{\rho^2}{4N}\, dN^2 + \tilde{g}, \qquad N := \alpha^2. \tag{90}$$

The natural FIDO tetrad is

$$\theta^0 = \sqrt{N}\, dt,$$
$$\theta^1 = \frac{\rho}{2\sqrt{N}}\, dN,$$
$$\theta^A \quad (A = 2,3): \quad \text{orthonormal 2-bein for } \tilde{g}. \tag{91}$$

This becomes again singular at the horizon ($N = 0$).

Now we imitate what we did for the Schwarzschild BH. We search for a basis of 1-forms which is well-defined in the neighborhood of the horizon. Guided by (72) we introduce

$$\bar{\theta}^t \equiv dt + \frac{\rho}{2N}\, dN \tag{92}$$

and rewrite (90)

$$^{(4)}g = -N\, (\bar{\theta}^t)^2 + \rho\, \bar{\theta}^t\, dN + \tilde{g}.$$

Thanks to (89) we conclude that

$$\{\bar{\theta}^t,\, dN,\, \theta^A\, (A = 2,3)\} \tag{93}$$

remains a regular basis on the horizon. Outside the horizon we can express (91) in terms of this basis:

$$\theta^0 = \sqrt{N}\,\bar{\theta}^t - \frac{\rho}{2\sqrt{N}}\,dN,$$

$$\theta^1 = \frac{\rho}{2\sqrt{N}}\,dN. \tag{94}$$

We can now proceed as in the derivation of (74), obtaining now

$$F = \frac{\rho\,E_1}{2}\,dN \wedge \bar{\theta}^t + \sqrt{N}E_2\,\theta^2 \wedge \bar{\theta}^t + \sqrt{N}E_3\,\theta^3 \wedge \bar{\theta}^t + B_1\theta^2 \wedge \theta^3$$

$$+ \frac{\rho}{2\sqrt{N}}\,(E_2 + B_3)\,dN \wedge \theta^2 + \frac{\rho}{2\sqrt{N}}\,(B_2 - E_3)\,dN \wedge \theta^3. \tag{95}$$

This implies again the limiting behavior (75) for the normal and parallel components of the electric and magnetic fields.

(c) Derivation for rotating BHs

Finally, I give a similar derivation for rotating BHs. (This was worked out in collaboration with Gerold Betschart in the course of his diploma work.)

I first need the Papapetrou parameterization of a stationary, axisymmetric BH. Since this will be treated in detail in the lectures by Markus Heusler [7], I can be brief.

Since the isometry group is $R \times SO(2)$ we have two commuting Killing fields k and m, say, which are tangent to the orbits belonging to the group action. We assume that k and m satisfy the Frobenius integrability conditions

$$k \wedge m \wedge dk = 0, \qquad k \wedge m \wedge dm = 0. \tag{96}$$

The Frobenius theorem then guarantees that the distribution of subspaces orthogonal to k and m is (locally) integrable. I recall that (96) is implied by the field equations for vacuum spacetimes and also for certain matter models (electromagnetic fields, ideal fluids, but not for Yang-Mills fields).

In this situation, spacetime is (locally) a product manifold, $M = \Sigma \times \Gamma$, where $\Sigma = R \times SO(2)$ and Γ is perpendicular to Σ. Thus the metric splits

$$^{(4)}g = \sigma + g \tag{97}$$

such that σ is an invariant 2-dimensional Lorentz metric on Σ, depending, however, on $y \in \Gamma$, and the fact that (Γ, g) is a 2-dimensional Riemannian space. In adapted coordinates

$$x^\mu : \quad x^0 = t, \quad x^1 = \varphi \text{ for } \Sigma; \quad x^2, x^3 \text{ for } \Gamma, \tag{98}$$

we have

$$k = \partial_t, \qquad m = \partial_\varphi, \tag{99}$$

and

$$\sigma = \sigma_{ab}\, dx^a dx^b, \qquad g = g_{ij}\, dx^i dx^j, \tag{100}$$

where $a, b = 0, 1$ and $i, j = 2, 3$. The metric functions σ_{ab} and g_{ij} depend only on the coordinates x^i of Γ.

The following functions on Γ have an invariant meaning

$$-V := (k|k), \qquad W := (k|m), \qquad X := (m|m), \tag{101}$$

and we have

$$\sigma = -V\, dt^2 + 2W\, dtd\varphi + X\, d\varphi^2. \tag{102}$$

We use also

$$\rho := \sqrt{VX + W^2} = \sqrt{-\sigma}, \qquad A := \frac{W}{X}, \tag{103}$$

in terms of which (102) takes the form

$$\sigma = -\frac{\rho^2}{X}\, dt^2 + X\, (d\varphi + A\, dt)^2. \tag{104}$$

It turns out that the partial trace $R^a_{\ a}$ of the 4-dimensional Ricci tensor is proportional to $^{(g)}\Delta\rho$ and ρ is thus a harmonic function on Γ, whenever $R^a_{\ a}$ vanishes. This is of course the case for vacuum manifolds, but also for the Kerr-Newman solution, and in some other cases [7]. With the help of the Riemann mapping theorem one can then show that ρ is a well-defined coordinate on (Γ, g) (ρ has no critical points). It is then possible to introduce a second coordinate z, such that

$$g = \frac{1}{X} e^{2h} (d\rho^2 + dz^2). \tag{105}$$

In terms of t, φ, and the *Weyl coordinates* ρ, z, $^{(4)}g$ assumes the *Papapetrou parameterization*

$$^{(4)}g = -\frac{\rho^2}{X}\, dt^2 + X\, (d\varphi + A\, dt)^2 + \frac{e^{2h}}{X} (d\rho^2 + dz^2). \tag{106}$$

We emphasize once more, that the functions X, A, and h depend only on the Weyl coordinates ρ and z.

The weak rigidity theorem tells us that on the surface on which

$$\xi = k + \Omega m, \qquad \Omega = -A = -W/X, \tag{107}$$

becomes null, Ω is constant and ξ is a Killing field on this surface, denoted by $H[\xi]$ in what follows. In addition, $H[\xi]$ is a stationary null surface, in particular a Cauchy horizon. (For proofs, see [7].)

Note that ξ, as a 1-form, can be expressed as

$$\xi = -\frac{\rho^2}{X} \, dt. \tag{108}$$

One also knows that X is well-behaved on the horizon

$$X = \mathcal{O}(1), \qquad X^{-1} = \mathcal{O}(1) \quad \text{on } H[\xi] \tag{109}$$

(see [8]). Since $(\xi|\xi) = -\rho^2/X$ the horizon is at $\rho = 0$ (as is well-known from the Kerr solution).

Although ξ is not a Killing field, the weak rigidity theorem implies that the surface gravity is still given by

$$\kappa^2 = -\frac{1}{4}(d\xi|d\xi)\big|_{H[\xi]}. \tag{110}$$

(A priori, the formula holds for $l := k + \Omega_H \, m$, $\Omega_H = \Omega|_{H[\xi]}$, but $d\Omega = 0$ on $H[\xi]$.) From (108) we obtain

$$d\xi = -\frac{2\rho}{X} \, d\rho \wedge dt - \rho^2 \, d\left(\frac{1}{X}\right) \wedge dt.$$

Using the scalar products $(d\rho|d\rho) = Xe^{-2h}$, $(dt|dt) = -X/\rho^2$, $(d\rho|dt) = 0$ gives thus, together with the zeroth law,

$$\kappa = e^{-h}\big|_{H[\xi]} = \text{const} \neq 0. \tag{111}$$

The reader should verify that this gives the correct result for the Kerr solution

$$\kappa = \frac{r_H - M}{2Mr_H}. \tag{112}$$

After these preparations, our argument proceeds similarly as in (b). The natural FIDO tetrad $\{\theta^\mu\}$ for (106) is ($N := \rho^2$)

$$\theta^0 = \sqrt{\frac{N}{X}} \, dt, \quad \theta^1 = \frac{e^h}{2\sqrt{XN}} \, dN, \quad \theta^2 = \frac{e^h}{\sqrt{X}} \, dz, \quad \theta^3 = \sqrt{X}(d\varphi + A \, dt). \tag{113}$$

Clearly, θ^0 and θ^1 are again ill-defined on the horizon $N = 0$. We therefore pass over to a new basis

$$\bar{\theta}^t := dt + \frac{e^h}{2N} \, dN, \qquad \bar{\theta}^\varphi := d\varphi + \Omega \, \frac{e^h}{2N} \, dN, \tag{114}$$

together with dN and dz. In order to check whether this new basis remains valid when the horizon is approached, we express the metric (106) in terms of it. Since

$$\theta^0 = \sqrt{\frac{N}{X}} \, \bar{\theta}^t - \frac{e^h}{2\sqrt{XN}} \, dN, \qquad \theta^3 = \sqrt{X}(\bar{\theta}^\varphi + A \, \bar{\theta}^t), \tag{115}$$

we find readily

$$^{(4)}g = -V(\bar{\theta}^t)^2 + 2W\,\bar{\theta}^t\bar{\theta}^\varphi + X(\bar{\theta}^\varphi)^2 + \frac{e^h}{X}\,\bar{\theta}^t\,dN + \frac{e^{2h}}{X}\,dz^2. \tag{116}$$

The determinant of the metric coefficients in this expression is $\frac{e^{4h}}{4X^2}$ and thus remains regular, thanks to (108) and (110). Since we postulate a regular horizon, it is then clear that $\{\bar{\theta}^t, \bar{\theta}^\varphi, dN, dz\}$ indeed forms a well-defined basis also on the horizon.

It is now straightforward to express the Maxwell 2-form F in terms of this basis. Instead of (96) we obtain now

$$\begin{aligned}
F = &\left\{E_1\frac{e^h}{2X} + \frac{Ae^h}{2\sqrt{N}}(E_3 - B_2)\right\}dN \wedge \bar{\theta}^t + E_3\sqrt{N}\,\bar{\theta}^\varphi \wedge \bar{\theta}^t \\
&\left\{E_2e^h\sqrt{\frac{N}{X}} + B_1e^hA\right\}dz \wedge \bar{\theta}^t + \frac{e^h}{2\sqrt{N}}(B_2 - E_3)\,dN \wedge \bar{\theta}^\varphi \\
&+B_1e^h\,dz \wedge \bar{\theta}^\varphi - \frac{e^{2h}}{2X\sqrt{N}}(E_2 + B_3)\,dz \wedge dN.
\end{aligned} \tag{117}$$

Since $\alpha := \sqrt{\frac{N}{X}}$ is the lapse function, we obtain once more the limiting behavior (75), and thus the basic membrane laws (69) and (70).

5   Magnetic Energy Extraction from a Black Hole

As an interesting, and possibly astrophysically important application of our basic laws (38) and (69) I show now that it is possible, in principle, to extract the rotational energy of a BH with the help of external magnetic fields. In the next section, we will work out some of the details for an ideal gedanken experiment. This will serve as a preparation for an understanding of the Blandford-Znajek process.

Our starting point is Faraday's induction law (25) in integral form, which we write down once more

$$\mathrm{EMF}(\mathcal{C}) = -\frac{d}{dt}\int_A \mathcal{B} + \oint_{\mathcal{C}} i_\beta\,\mathcal{B}. \tag{118}$$

For stationary situations this reduces to

$$\mathrm{EMF}(\mathcal{C}) = \oint_{\mathcal{C}} i_\beta\,\mathcal{B}. \tag{119}$$

In Fig. 3 we consider a stationary rotating BH in an external magnetic field (like in §3). The integral in (119) along the field lines gives no contribution and

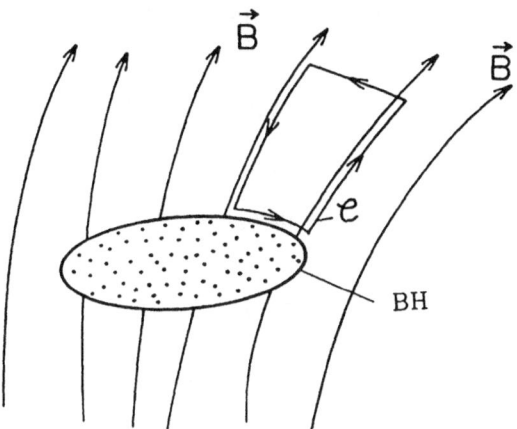

Fig. 3 Arrangement for eq. (119).

far away β drops rapidly ($\sim r^{-2}$). Thus, there remains only the contribution from the horizon (C_H) of the path C in Fig. 3:

$$\text{EMF} = \int_{C_H} i_{\beta_H} \mathcal{B}, \quad \beta_H = -\Omega_H \, \partial_\varphi = -\Omega_H \, \tilde{\omega}_H \, e_\varphi \tag{120}$$

(only the normal component $\vec{B}_\perp$ contributes). I recall that $\Omega_H = a(2Mr_H)^{-1}$, $r_H = M + \sqrt{M^2 - a^2}$.

We shall show in section 6 that it is possible to construct a generator such that (with optimal impedance matching) a *maximal extraction rate* equal to

$$\frac{1}{4} \frac{(\text{EMF})^2}{R(C_H)} \tag{121}$$

becomes possible, where $R(C_H)$ is the horizon ("internal") resistance

$$R(C_H) = \int_{C_H} R_H \, \frac{dl}{2\pi \, \tilde{\omega}}, \quad R_H = 377 \text{ Ohm.} \tag{122}$$

(It should be known from electrodynamics, that we have to divide the surface resistivity R_H by the length of the cross section through which the current is flowing.) We shall also show that an equal amount of energy is dissipated by ohmic heating at the (stretched) horizon.

Let us work this out for the special case of an axisymmetric field: $L_{\partial_\varphi} \, \mathcal{B} = 0 \leftrightarrow \mathbf{d}i_{\partial_\varphi} \, \mathcal{B} = 0$, whence

$$i_{\partial_\varphi} \, \mathcal{B} = -\frac{\mathbf{d}\Psi}{2\pi}. \tag{123}$$

From this we conclude that $\mathcal{B}$ can be expressed in terms of two potentials Ψ and g,

$$\mathcal{B} = \underbrace{\frac{1}{2\pi} \, d\Psi \wedge d\varphi}_{\text{poloidal part}} + \underbrace{g * d\varphi}_{\text{toroidal part}} \,, \tag{124}$$

both of which can be taken to be independent of φ. This is equivalent to the vector formulae

$$\vec{B}^{\text{pol}} = \frac{1}{2\pi \, \tilde{\omega}} \vec{\nabla}\Psi \times \vec{e}_\varphi, \qquad \vec{B}^{\text{tor}} = \frac{g}{\tilde{\omega}} \vec{e}_\varphi \tag{125}$$

(Exercise). Ψ is the magnetic flux function (see Fig. 4), because the poloidal flux inside a tube $\{\Psi = \text{const}\}$ is

$$\int \mathcal{B} = \int d\Psi = \Psi, \qquad \Psi(0) = 0. \tag{126}$$

Ψ is constant along magnetic field lines, as should be clear from Fig. 4. Formally,

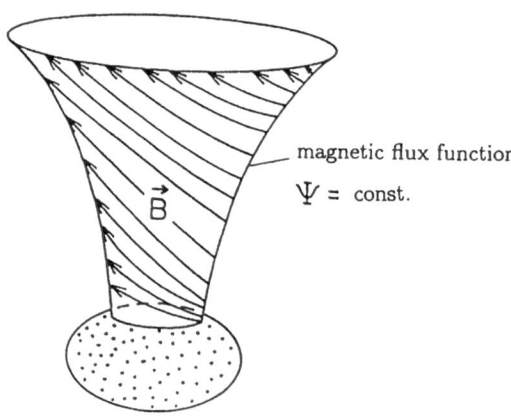

Fig. 4 Axisymmetric magnetic field. The total flux inside the magnetic surface defines the flux function Ψ.

this comes about as follows:

$$
\begin{aligned}
i_{\vec{B}} \, d\Psi &= *(*d\Psi \wedge B) = *(d\Psi \wedge *B) = *(d\Psi \wedge B) \\
&\overset{(124)}{=} g * (d\Psi \wedge *d\varphi) = 0,
\end{aligned}
$$

thus $< d\Psi, \vec{B} > = 0$.

For the closed path $\mathcal{C}$ in Fig. 5 the EMF is by (120)

$$\text{EMF} \equiv \Delta V = \int_{\mathcal{C}_H} i_{\beta_H}\, \mathcal{B} \stackrel{(123)}{=} -\left(-\frac{\Omega_H}{2\pi}\right) \int_{\mathcal{C}_H} d\Psi,$$

i.e.

$$\Delta V = \frac{\Omega_H}{2\pi}\, \Delta\Psi. \tag{127}$$

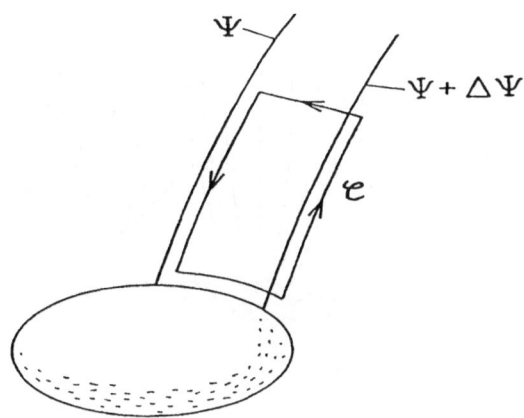

Fig. 5 Integration path for the voltage in eq. (127).

The internal resistance (122) becomes

$$\triangle R_H = R_H\, \frac{\Delta l}{2\pi\, \tilde{\omega}}. \tag{128}$$

Since, on the other hand,

$$\Delta\Psi = 2\pi\, \tilde{\omega}\, B_\perp\, \Delta l, \tag{129}$$

we find, by eliminating Δl,

$$\triangle R_H = R_H\, \frac{\Delta\Psi}{4\pi^2\, \tilde{\omega}^2\, B_\perp}. \tag{130}$$

Inserting (127) and (130) into the expression (121) for the maximal power output gives

$$\frac{1}{4}\, \frac{(\Delta V)^2}{\triangle R_H} = \frac{\Omega_H^2}{16\,\pi}\, \tilde{\omega}^2\, B_\perp\, \Delta\Psi. \tag{131}$$

This, as well as (127), have to be integrated from the pole to some point north of the equator (see Fig. 6). For the exact solution in §3 we know the result for the EMF, if we integrate up to the equator: From (127) and (67) we get

$$\text{EMF} = \frac{1}{2\pi} \, \Omega_H \, 4\pi \, B_0 \, M \, (r_H - M)$$

or $(\Omega_H = a/2Mr_H)$

$$\text{EMF} = a \, B_0 \, \frac{r_H - M}{r_H} \qquad \left(r_H = M + \sqrt{M^2 - a^2} \right). \tag{132}$$

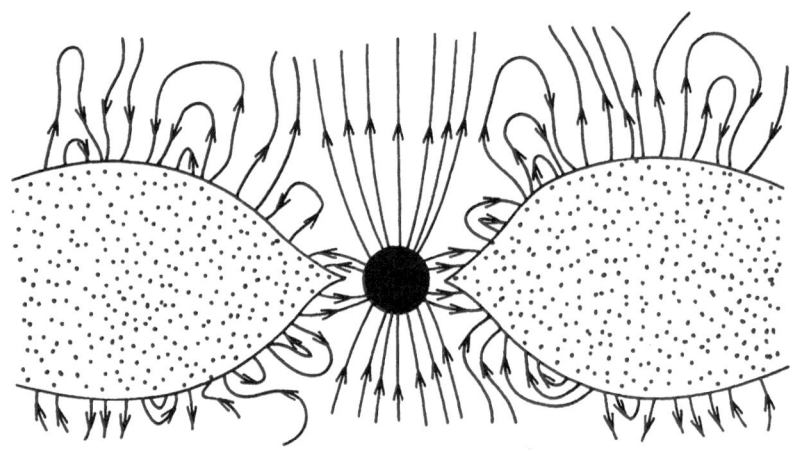

Fig. 6 Plausible structure of the magnetic field close to a supermassive BH which is surrounded by an accretion disk (central engine of a quasar) (adapted from Ref. [1]).

For a general situation, like in Fig. 6, we have roughly

$$\Sigma \, \triangle \Psi = \Psi \sim B_\perp \, \pi \, r_H^2, \qquad \bar{\omega}^2 \sim \, < \bar{\omega}^2 > \, \sim \frac{r_H^2}{2},$$

and we obtain, by (131), for the power output

$$P \sim \frac{1}{128} \left(\frac{a}{M} \right)^2 B_\perp^2 \, r_H^2$$

$$\sim (10^{45} \text{ erg/s}) \left(\frac{a}{M} \right)^2 \left(\frac{M}{10^9 \, M_\odot} \right)^2 \left(\frac{B_\perp}{10^4 \, G} \right)^2. \tag{133}$$

The total EMF is $(V = \Sigma \, \triangle V)$

$$V \sim \frac{1}{2\pi} \, \Omega_H \Psi \sim \frac{1}{2\pi} \, \frac{a}{2Mr_H} \, B_\perp \, \pi \, r_H^2 \simeq \frac{1}{2} \left(\frac{a}{M} \right) M B_\perp \tag{134}$$

(compare this with (132)). Numerically we find

$$V \sim (10^{20} \text{ Volt}) \left(\frac{a}{M}\right) \frac{M}{10^9 \, M_\odot} \frac{B_\perp}{10^4 \, G}. \tag{135}$$

For reasonable astrophysical parameters we obtain magnetospheric voltages $V \sim 10^{20}$ Volts and power output of the magnitude $\sim 10^{45}$ erg/s. This power is what one observes typically in active galactic nuclei, and the voltage is comparable to the highest cosmic ray energies that have been detected.

Note, however, that for a realistic astrophysical situation there is plasma outside the BH and it is, therefore, not clear, how the horizon voltage (135) is used in accelerating particles to very high energies.

Let us estimate at this point the characteristic magnetic field strength than can be expected outside a supermassive BH. A measure for this is the field strength B_E for which the energy density $B_E^2/8\pi$ is equal to the radiation energy density u_E corresponding to the Eddington luminosity

$$L_E = \frac{4\pi M_H m_p c}{\sigma_T} = 1.3 \times 10^{38} \, (\text{erg/s}) \, \frac{M}{M_\odot}. \tag{136}$$

The relation between L_E and u_E is

$$L_E = 4\pi \, r_g^2 \, \frac{c}{4} \, u_E = \pi \, r_g^2 \, c \, u_E \qquad \left(r_g = \frac{GM}{c^2}\right). \tag{137}$$

Thus

$$\frac{1}{8\pi} \, B_E^2 = \frac{4 \, m_p c^2}{\sigma_T \, r_g}, \tag{138}$$

giving $(M_{H,8} \equiv M_H/10^8 \, M_\odot)$

$$B_E = 1.2 \times 10^5 \, M_{H,8}^{-1/2} \text{ Gauss}. \tag{139}$$

For a BH with mass $\sim 10^9 \, M_\odot$ inside an accretion disk acting as a dynamo, a characteristic field of about 1 Tesla (10^4 Gauss) is thus quite reasonable.

6 Rotating BH as a Current Generator

Before we come to realistic possibilities of energy extraction, we analyze in detail an idealized arrangement, sketched in Fig. 7.

We compute first the EMF around a closed path consisting of the following parts: We start from the equator of the (stretched) horizon along a perfectly conducting disk, which is supposed to rotate differentially in such a manner, that all its pieces are at rest relative to the FIDOs (they have thus zero angular momentum). From the boundary of the disk, the path continuous along a wire and through a resistive load (R_L) to the top of a conical conductor. Then we

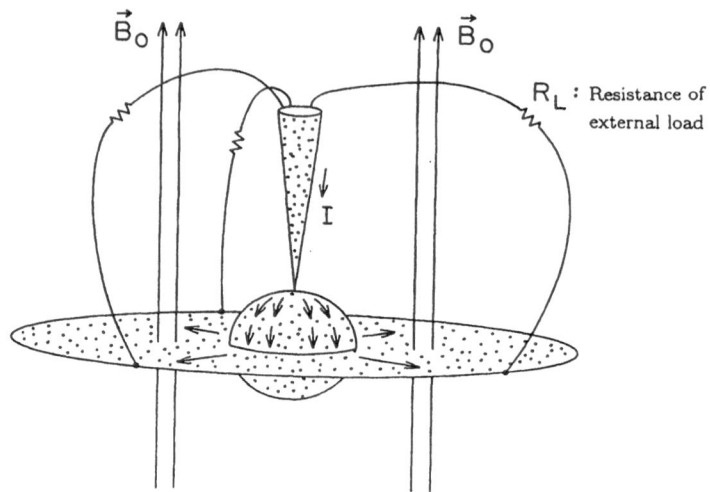

Fig. 7 An idealized current generator (adapted from [1]).

move down to its tip at $\vartheta = \vartheta_0$ close to the north pole of the membrane, and finally down the horizon in the poloidal direction to the starting point. (The conical surface is chosen instead of an infinitely thin wire in order to avoid divergent integrals; it replaces a wire of finite thickness.)

The ideally conducting disk gives no contribution to the EMF. Formally this comes about as follows: The 4-velocity field of the disk u (whose parts move with the FIDOs) is e_0 (see (5)) and thus $-i_u F = E = 0$, for an ideal conductor. Using (7), this implies $\mathcal{E} = 0$.

The contribution V_L to the line integral $\oint \alpha \mathcal{E}$ from the load resistance far from the BH is what we all know from electrodynamics

$$V_L = I\,R_L, \tag{140}$$

where I is the current in the wire.

There remains the contribution V_H to the EMF along the stretched horizon

$$V_H = \int_{\mathcal{C}_H} \alpha\,\mathcal{E} \overset{(\text{Ohm})}{=} \int_{\mathcal{C}_H} R_H * \mathcal{J}_H. \tag{141}$$

The current flows along the northern hemisphere of the horizon as a surface current in the *poloidal* direction, with surface current density

$$\vec{\mathcal{J}}_H = \frac{I}{2\pi\,\tilde{\omega}_H}\,\vec{e}_\vartheta. \tag{142}$$

In order to prove that the surface current $\vec{\mathcal{J}}_H$ has no toroidal component, we apply (119) to a toroidal path $\{\vartheta = \text{const}\}$ on the stretched horizon. Along this

$i_\beta \mathcal{B} = -\mathbf{d}\Psi/2\pi = 0$, and thus

$$\oint \alpha \mathcal{E} = 0 \quad \Longrightarrow \quad i_{\partial_\varphi} \mathcal{E}_H = 0 : \quad \vec{E}_H \cdot \vec{e}_\varphi = 0.$$

Ohm's law in (69) then implies $\vec{\mathcal{J}}_H \cdot \vec{e}_\varphi = 0$.

Using (142) in (141) gives

$$V_H = I \int_{\vartheta_0}^{\frac{\pi}{2}} R_H \frac{\rho_H}{2\pi \, \tilde{\omega}_H} \, d\vartheta$$

or

$$V_H = I \, R_{HT}, \tag{143}$$

where

$$R_{HT} = \int_{\vartheta_0}^{\frac{\pi}{2}} R_H \frac{\rho_H}{2\pi \, \tilde{\omega}_H} \, d\vartheta$$

is the total resistance of the horizon (see (122)). The EMF around our closed path is thus

$$V = V_H + V_L = I(R_{HT} + R_L). \tag{144}$$

This voltage is also equal to the line integral on the right hand side in (119),

$$V = \oint_C i_\beta \mathcal{B}. \tag{145}$$

This receives only contributions from the horizon and the disk (the contribution of the latter was overlooked in Ref. [1]). Using $\beta = -\omega \, \partial_\varphi$, $\mathcal{B}_\perp = B_r \, \vartheta^\vartheta \wedge \vartheta^\varphi$, $\vartheta^\vartheta = \rho \, d\vartheta$, $\vartheta^\varphi = \tilde{\omega} \, d\varphi$, the horizon gives

$$\int_{C_H} i_\beta \mathcal{B} = \Omega_H \int_{\vartheta_0}^{\frac{\pi}{2}} B_\perp \, \tilde{\omega}_H \, \rho_H \, d\vartheta. \tag{146}$$

A similar contribution is obtained along the disk (C_D) and the total voltage is given by

$$V = \Omega_H \int_{\vartheta_0}^{\frac{\pi}{2}} B_\perp \, \tilde{\omega}_H \, \rho_H \, d\vartheta - \int_{r_H}^{r_D} B_\perp \, \omega \, \tilde{\omega} \, \frac{\rho}{\sqrt{\Delta}} \, dr \tag{147}$$

(r_D is the edge of the disk). This "battery" voltage[1], and the total horizon and load resistances R_{HT} and R_L determine the current I according to (144).

---

[1] The part of the integral (145) from the disk is independent of the connecting path, because the induction law gives $\mathbf{d}(\alpha \, \mathcal{E} - i_\beta \, \mathcal{B}) = 0$ and thus $\mathbf{d} \, (i_\beta \, \mathcal{B}) = 0$ inside the disk.

If the magnetic field is axisymmetric, we can use (123) to write the integrand in (145) as follows

$$i_\beta \mathcal{B} = \frac{\omega}{2\pi}\, d\Psi. \tag{148}$$

The voltage is then

$$V = \frac{\Omega_H}{2\pi}\, \Psi(\text{eq}) \left[1 + \int_{horizon}^{edge} \left(\frac{\omega}{\Omega_H} \right) d\left(\frac{\Psi}{\Psi(\text{eq})} \right) \right], \tag{149}$$

where $\Psi(\text{eq})$ denotes the value of the flux function at the equator of the horizon. The two pieces in the square bracket are comparable in magnitude.

We shall see in the next section in detail how the power, dissipated as ohmic losses

$$P_L = I^2\, R_L \tag{150}$$

in the load, is extracted from the hole, but this is clearly at the cost of the mass of the BH:

$$I^2 R_L = -\dot{M}. \tag{151}$$

Let us note that

$$P_L = V^2\, \frac{R_L}{(R_{HT} + R_L)^2}. \tag{152}$$

This becomes maximal for

$$R_{HT} = R_L \qquad \text{(impedance matching)} \tag{153}$$

with

$$P_L^{max} = \frac{V^2}{4\, R_{HT}}. \tag{154}$$

This maximal extraction rate was stated in (121).

Clearly, we can also reverse our gedanken experiment. By applying a voltage, we can use the BH as the rotator of an electric motor (see Fig. 8). You see that some of the physics of BHs is indeed very similar to that of ordinary electric generators and electric motors.

7   Conservation Laws, Increase of Entropy of a BH

The general results which will be obtained in this section will enable us to develop a more profound analysis of the idealized current generator discussed above.

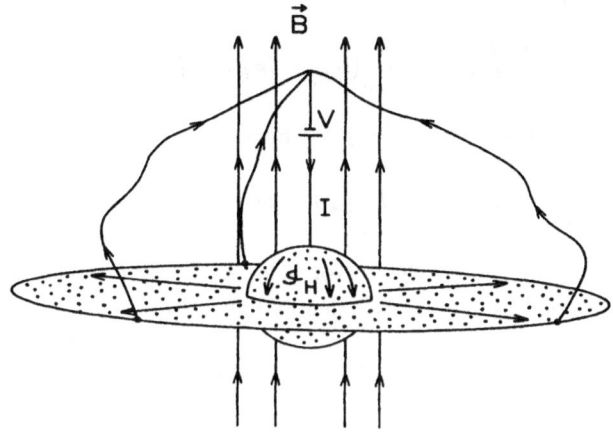

Fig. 8 Black hole playing the role of the rotator of an electric motor.

Contraction of the energy-momentum tensor $T^{\mu\nu}$ with the two Killing fields ∂_t, ∂_φ gives two conserved vector fields, which express the conservation laws of energy and angular momentum in the z-direction. We want to formulate these in the 3+1 splitted form.

To this end, we consider as a preparation a 4-dimensional equation of the type

$$\nabla \cdot J = Q \tag{155}$$

for a vector field J with source term Q. The Hodge dual of the 1-form J^b has the same decomposition as S in (22):

$$*J^b = \rho + (i_\beta \rho - \alpha\,\mathcal{J}) \wedge dt. \tag{156}$$

If $J = J^\mu e_\mu$, then ρ is the 3-form belonging to J^0, and $\mathcal{J}$ is the 2-form corresponding to $\vec{j} = J^k e_k$. Eq. (155) is equivalent to $d\left(*J^b\right) = Q\,\mathrm{vol}_4$ or

$$dt \wedge [(\partial_t - L_\beta)\rho + \mathbf{d}(\alpha\,\mathcal{J})] = Q\,\mathrm{vol}_4. \tag{157}$$

Since $dt = \theta^0/\alpha$, the 3+1 split is

$$(\partial_t - L_\beta)\,\rho + \mathbf{d}(\alpha\,\mathcal{J}) = \alpha\,Q\,\mathrm{vol}_3. \tag{158}$$

If $\operatorname{div}\vec{\beta} = 0$ (as for the Kerr solution), this is equivalent to $(\rho =: \hat{\rho}\,\mathrm{vol}_3)$

$$(\partial_t - L_{\vec{\beta}})\,\hat{\rho} + \vec{\nabla} \cdot \left(\alpha\vec{j}\right) = \alpha\,Q. \tag{159}$$

The integral form

$$\frac{d}{dt} \int_\mathcal{V} \rho = - \int_{\partial\mathcal{V}} (\alpha\, \mathcal{J} - i_\beta\, \rho) + \int_\mathcal{V} \alpha\, Q \,\text{vol}_3 \tag{160}$$

generalizes the conservation law (28). We also write (160) in vector analytic form

$$\frac{d}{dt} \int_\mathcal{V} \hat\rho\, dV = - \int_{\partial\mathcal{V}} (\alpha\vec{j} - \hat\rho\vec\beta) \cdot d\vec{A} + \int_\mathcal{V} \alpha\, Q \, dV. \tag{161}$$

Let us now apply this to the vector fields $k_\mu T^{\mu\nu}$ and $m_\mu T^{\mu\nu}$ ($k = \partial_t$, $m = \partial_\varphi$). The corresponding $\hat\rho$ and $\vec{j}$ are denoted as follows

$$-k \cdot T \longleftrightarrow (\varepsilon_{E_\infty}, \vec{S}_{E_\infty}),$$
$$m \cdot T \longleftrightarrow (\varepsilon_{L_z}, \vec{S}_{L_z}). \tag{162}$$

Eq. (159) gives

$$(\partial_t - L_{\vec\beta})\, \varepsilon_{E_\infty} + \vec\nabla \cdot (\alpha\, \vec{S}_{E_\infty}) = 0 \qquad \text{(Poynting theorem)}, \tag{163}$$
$$(\partial_t - L_{\vec\beta})\, \varepsilon_{L_z} + \vec\nabla \cdot (\alpha\, \vec{S}_{L_z}) = 0 \ \text{(angular momentum conservation)}. \tag{164}$$

The corresponding integral formulas are

$$\frac{d}{dt} \int_\mathcal{V} \varepsilon_{E_\infty} \, dV = - \int_{\partial\mathcal{V}} \left(\alpha\, \vec{S}_{E_\infty} - \varepsilon_{E_\infty} \vec\beta \right) \cdot d\vec{A}, \tag{165}$$
$$\frac{d}{dt} \int_\mathcal{V} \varepsilon_{L_z} \, dV = - \int_{\partial\mathcal{V}} \left(\alpha\, \vec{S}_{L_z} - \varepsilon_{L_z} \vec\beta \right) \cdot d\vec{A}. \tag{166}$$

Now we make use of (5), i.e., $\partial_t = \alpha\, e_0 - \omega\, \partial_\varphi = \alpha\, e_0 - \omega\, \tilde\omega\, \vec{e}_\varphi$, giving us

$$-k \cdot T = -\alpha\, e_0 \cdot T + \omega\, m \cdot T.$$

Here we use the following FIDO decomposition of T:

$$T = \varepsilon\, e_0 \otimes e_0 + e_0 \otimes \vec{S} + \vec{S} \otimes e_0 + \overset{\leftrightarrow}{\mathbf{T}}, \tag{167}$$

and obtain the relations

$$\varepsilon_{E_\infty} = \alpha\, \varepsilon + \omega\, \varepsilon_{L_z}, \qquad \text{(energy density ``at infinity'')}, \tag{168}$$
$$\vec{S}_{E_\infty} = \alpha\, \vec{S} + \omega\, \vec{S}_{L_z}, \quad \text{(energy current density ``at infinity'')}. \tag{169}$$

Application to a Kerr BH

The global conservation laws (165) and (166) are now applied to a Kerr BH. Its mass plays the role of the "energy at infinity" and thus, the energy conservation (165) with (169), gives

$$\frac{dM}{dt} = - \int_{\mathcal{H}^\bullet} \alpha_H\, \vec{S}_{E_\infty} \cdot \vec{n} \, dA = - \int_{\mathcal{H}^\bullet} \left(\alpha_H^2\, \vec{S} + \alpha_H\, \Omega_H\, \vec{S}_{L_z} \right) \cdot \vec{n} \, dA. \tag{170}$$

Similarly, the change of the angular momentum of the BH is by (166)

$$\frac{dJ}{dt} = -\int_{\mathcal{H}^{\bullet}} \alpha_H \, \vec{S}_{L_z} \cdot \vec{n} \, dA; \qquad \vec{S}_{L_z} = \vec{\partial}_{\varphi} \cdot \overset{\leftrightarrow}{\mathbf{T}} \,. \tag{171}$$

Now, we consider the entropy increase of the BH. The first law [7] tells us that

$$T_H \frac{dS_H}{dt} = \frac{dM}{dt} - \Omega_H \frac{dJ}{dt}. \tag{172}$$

On the right hand side we insert (170) and (171), and use also (169), giving us the generally valid formula

$$T_H \frac{dS_H}{dt} = -\int_{\mathcal{H}^{\bullet}} \alpha_H^2 \, \vec{S} \cdot \vec{n} \, dA. \tag{173}$$

Until now we have not specified the matter content. For electrodynamics we have $\vec{S} = \frac{1}{4\pi} \vec{E} \times \vec{B}$, and thus at the horizon $\alpha_H^2 \, \vec{S} = \frac{1}{4\pi} \vec{E}_H \times \vec{B}_H$ (see (78)). In this case we obtain with the laws of Ampère and Ohm

$$T_H \frac{dS_H}{dt} = -\frac{1}{4\pi} \int_{\mathcal{H}^{\bullet}} (\vec{E}_H \times \vec{B}_H) \cdot \vec{n} \, dA = \int_{\mathcal{H}^{\bullet}} \vec{J}_H \cdot \vec{E}_H \, dA = \int_{\mathcal{H}^{\bullet}} R_H \, \vec{J}_H^2 \, dA. \tag{174}$$

All of these familiarly looking expressions for the rate of entropy increase are important and useful.

Let us also evaluate (171) in a similar manner. First, we have

$$\vec{S}_{L_z} = \frac{1}{4\pi} \vec{\partial}_{\varphi} \cdot \left[-\left(\vec{E} \otimes \vec{E} + \vec{B} \otimes \vec{B}\right) + \frac{1}{2}\left(\vec{E}^2 + \vec{B}^2\right) \overset{\leftrightarrow}{\mathbf{g}} \right]. \tag{175}$$

Clearly, only the first term contributes to the integrand in (171). Using this time the laws of Gauss and Ampère, we find

$$\frac{dJ}{dt} = \int_{\mathcal{H}^{\bullet}} \left(\sigma_H \, \vec{E}_H + \vec{J}_H \times \vec{B}_n\right) \cdot \vec{\partial}_{\varphi} \, dA. \tag{176}$$

Note that the first term is absent if $\vec{E}_H$ has no toroidal component.

Finally, we use (172), together with (174) and (176), to obtain the following formula for the change of the mass of the BH

$$\frac{dM}{dt} = \int_{\mathcal{H}^{\bullet}} \left[\vec{J}_H \cdot \vec{E}_H - \vec{\beta}_H \cdot \left(\sigma_H \, \vec{E}_H + \vec{J}_H \times \vec{B}_n\right) \right] dA. \tag{177}$$

Application to the Idealized Current Generator

It is instructive to use these general results for a more detailed analysis of the current generator, discussed in the last section.

We begin by computing the ohmic heating rate of the current flowing through the northern hemisphere (n.H.) of the BH:

$$\int_{n.H.} \vec{E}_H \cdot \vec{J}_H \, dA \;=\; \int_{\vartheta_0}^{\frac{\pi}{2}} E_{H\hat{\vartheta}} \, \mathcal{J}_{H\hat{\vartheta}} \, 2\pi \, \tilde{\omega} \, \rho_H \, d\vartheta$$

$$\overset{(142)}{=} I \int_{\vartheta_0}^{\frac{\pi}{2}} E_{H\hat{\vartheta}} \, \rho_H \, d\vartheta = I \, V_H \overset{(143)}{=} I^2 \, R_{HT}. \qquad (178)$$

According to the general result (174) this rate is equal to $T_H \, dS_H/dt$. Thus,

$$T_H \, \frac{dS_H}{dt} = I^2 \, R_{HT}. \qquad (179)$$

In order to trace the details of the energy flow, we apply the generalized Poynting theorem (165). For a stationary situation this reduces to

$$\int_{\partial \mathcal{V}} \left(\alpha \, \vec{S}_{E_\infty} - \varepsilon_{E_\infty} \, \vec{\beta} \right) \cdot d\vec{A} = 0. \qquad (180)$$

We choose the volume $\mathcal{V}$ such that the boundary $\partial \mathcal{V}$ consists of a horizon part $\mathcal{A}_H$ ($\mathcal{H}^s$ minus the inner edge of the disk), the disk $\mathcal{A}_D$, and a surface $\mathcal{A}_L$ enclosing the load's resistor (see Fig. 9). The second term in (180) does not contribute for $\mathcal{A}_H$ and $\mathcal{A}_D$, and can be ignored for the load, since this is assumed to be located far from the horizon. Using also the relation (169) we have then

$$\int_{\partial \mathcal{V}} \alpha \, \vec{S}_{E_\infty} \cdot d\vec{A} = \int_{\partial \mathcal{V}} (\alpha^2 \vec{S} + \alpha \, \omega \, \vec{S}_{L_z}) \cdot d\vec{A} = 0. \qquad (181)$$

The contribution from the horizon to the first term on the right is

$$-\int_{\mathcal{A}_H} \alpha^2 \vec{S} \cdot \vec{n} \, dA \overset{(173)}{=} T_H \, \frac{dS_H}{dt} \overset{(179)}{=} I^2 R_{HT}, \qquad (182)$$

and the second term gives

$$-\Omega_H \int_{\mathcal{A}_H} \alpha_H \, \vec{S}_{L_z} \cdot \vec{n} \, dA. \qquad (183)$$

At first sight one might think that this is just $\Omega_H \frac{dJ}{dt}$ (see (171)). This is, however, not correct, because there is an additional inflow of angular momentum through the disk. Physically, this is clear: The external magnetic field exerts – through the Lorentz force on the surface current $\vec{J}_D$ of the disk – a torque on the disk, and thus on the BH to which the disk is locked. As in the derivation of (176) we find for (183)

$$-\Omega_H \int_{\mathcal{A}_H} \alpha_H \, \vec{S}_{L_z} \cdot \vec{n} \, dA \;=\; \Omega_H \int_{\mathcal{A}_H} \tilde{\omega}_H (\vec{J}_H \times \vec{B}_n) \cdot \vec{e}_\varphi \, dA$$

$$= -\Omega_H \int_{\vartheta_0}^{\frac{\pi}{2}} \mathcal{J}_{H\hat{\vartheta}} \, B_\perp \, \tilde{\omega}_H \, 2\pi \, \tilde{\omega}_H \, \rho_H \, d\vartheta \overset{(142)}{=} -I \, \Omega_H \int_{\vartheta_0}^{\frac{\pi}{2}} B_\perp \, \tilde{\omega}_H \, \rho_H \, d\vartheta,$$

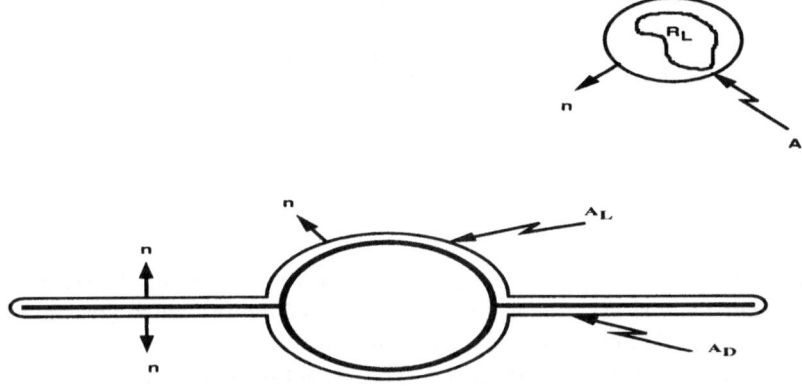

Fig. 9 The boundary $\partial V = \mathcal{A}_H \cup \mathcal{A}_D \cup \mathcal{A}_L$ for (181)

where use has been made of the fact, that $\vec{E}_H$ has no toroidal component. To-gether with (146), we obtain for this horizon contribution

$$-\Omega_H \int_{\mathcal{A}_H} \alpha_H \vec{S}_{L_z} \cdot \vec{n} \, dA = -I \int_{\mathcal{C}_H} i_\beta \, B. \tag{184}$$

Note that the integral on the right is not the total voltage (145). However, we shall see shortly that the disk contributes the remaining part of $-IV$.

Only the last term in (181) gets contributions from the disk (since $\vec{E} = 0$ in the disk). Using the expression (175) for $\vec{S}_{L_z}$, this becomes

$$-\int_{\mathcal{A}_D} \alpha \, \omega \, \vec{S}_{L_z} \cdot \vec{n} \, dA = \int_{\mathcal{C}_D} \alpha \, \omega \, \frac{B_n}{4\pi} \, \vec{\partial}_\varphi \cdot \vec{B} \, 2\pi \tilde{\omega} \, dl. \tag{185}$$

Between the parallel component $\vec{B}_{\parallel}$ and the surface current $\vec{J}_D$ of the disk we have Ampère's relation of ordinary electrodynamics:

$$\alpha \, \vec{B}_{\parallel} = 4\pi \, \vec{J}_D \times \vec{n} = 4\pi \, \frac{I}{2\pi \, \tilde{\omega}} \, \vec{e}_r \times \vec{n}. \tag{186}$$

Making use of this, (185) becomes

$$-\int_{\mathcal{A}_D} \alpha \, \omega \, \vec{S}_{L_z} \cdot \vec{n} \, dA = -I \int_{\mathcal{C}_D} \vec{\beta} \cdot \left(\vec{e}_r \times \vec{B}_n \right) \, dl = I \int_{\mathcal{C}_D} \vec{\beta} \cdot \left(\vec{\beta} \times \vec{B}_n \right) \cdot d\vec{l}$$

or

$$-\int_{\mathcal{A}_D} \alpha \, \omega \, \vec{S}_{L_z} \cdot \vec{n} \, dA = -\int_{\mathcal{C}_D} i_\beta \, B, \tag{187}$$

as already announced.

Finally, the contribution to (181) is what we are used to:

$$\int_{A_L} \alpha \vec{S}_{E_\infty} \cdot d\vec{A} = \int_{A_L} \frac{1}{4\pi} (\vec{E} \times \vec{B}) \, d\vec{A} = I^2 R_L. \tag{188}$$

All together, the Poynting theorem (181) gives

$$\underbrace{I^2 R_{TH} - I \int_{C_H} i_\beta \, \mathcal{B}}_{\text{horizon}} - \underbrace{I \int_{C_D} i_\beta \, \mathcal{B}}_{\text{disk}} = \underbrace{-I^2 R_L}_{\text{load}}. \tag{189}$$

From the derivation it is clear how to interpret this result. First, we note that the left hand side is the energy (measured at infinity) which flows per unit time into the BH, and thus is equal to dM/dt. According to (179) the first term is $T_H \, dS_H/dt$ (ohmic dissipation). The second term in (189) is that part of $\Omega_H \, dJ/dt$ which is due to the torque acting on the surface current density $\vec{J}_H$ (see (171) and (176)). The third term resulted from the disk and gives an additional contribution to the change of the rotational energy, which flows through the disk into the BH. For clarification we note that for a closed surface $\mathcal{A}$ surrounding the disk we have

$$\oint_{\mathcal{A}} \alpha \vec{S}_{E_\infty} \cdot d\vec{A} = 0 = \oint_{\mathcal{A}} \alpha \omega \vec{S}_{L_z} \cdot d\vec{A},$$

since $\vec{S}$ vanishes in the disk. Therefore, the rotational energy which flows through the inner edge of the disk into the BH is (see (171))

$$\Omega_H \int_{\leftarrow|} \alpha \vec{S}_{L_z} \cdot d\vec{A} = - \int_{A_D} \alpha \omega \vec{S}_{L_z} \cdot \vec{n} \, dA \overset{(187)}{=} - \int_{C_D} i_\beta \, \mathcal{B}.$$

The total change of the rotational energy of the BH thus is

$$\Omega_H \frac{dJ}{dt} = -I \oint_C i_\beta \, \mathcal{B} \overset{(145)}{=} -IV$$

$$\overset{(144)}{=} -I^2 (R_{HT} + R_L). \tag{190}$$

The presence of the disk made the discussion a bit complicated, but it is nice to see how the various pieces combine. Schematically, we have

$$\begin{array}{ccccc} \frac{dM}{dt} &=& T_H \frac{dS_H}{dt} &+& \Omega_H \frac{dJ}{dt} \\ \| & & \| & & \| \\ -I^2 R_L & & I^2 R_{HT} & & -I^2 (R_{HT} + R_L). \end{array} \tag{191}$$

The energy (at ∞) dM/dt flows – partly through the disk – down the hole and is, of course, negative because the spin down overcomes the ohmic dissipation. This part of the energy balance results, therefore, in an outflowing energy which is carried without loss by the electromagnetic fields to the load resistors, where it is dissipated at the rate $I^2 R_L$. Netto, we obtain, of course,

$$-\frac{dM}{dt} = I^2 R_L. \tag{192}$$

8 Blandford-Znajek Process

Let us return to Fig. 6, in which a plausible magnetic field structure around a supermassive BH in the center of an active galaxy is sketched. Rotation and turbulence in an accretion disk can generate magnetic fields of the order 1 tesla, as we estimated at the end of section 5.

Close to the BH one expects a force-free electron-positron plasma, which comes about as follows. Imagine first that there are no charged particles in the neighborhood of the hole. In this case unipolar induction generates a quadrupole-like electric field similar to that we found in section 3 (see Fig. 2). Close to the BH the magnitude of this electric field is $E \sim B\,a/M \sim 3 \times 10^6$ (Volt/cm) (a/M). Between the horizon and a few gravitational radii along the magnetic field lines this gives rise to a voltage $V \sim E\,r_H \sim B\,a \sim 10^{20}$ Volt for a BH with mass $M \sim 10^9\,M_\odot$ (see section 5). In this enormous potential stray electrons from the disk or interstellar space will be accelerated along magnetic field lines to ultrarelativistic energies. Inverse Compton scattering with soft photons from the accretion disk leads to γ-quanta which in turn annihilate with soft photons from the disk into electron-positron pairs. These will again be accelerated and by repetition an electron-positron plasma is generated which can become dense enough to annihilate the component of $\vec{E}$ along $\vec{B}$. The electric field is then nearly orthogonal to $\vec{B}$, up to a sufficiently large component which produces occasional electron-positron sparks in order to fill the magnetosphere with plasma. (Similar processes are important in the magnetospheres of pulsars. It is likely that all active pulsars have electron-positron winds.)

Fields with $\vec{E} \cdot \vec{B} = 0$ are called *degenerate*. We assume in what follows that in the neighborhood of the BH, where the $\vec{B}$-field is strong, the electromagnetic field is *force-free*, which means that the ideal MHD condition

$$\rho_e\,\vec{E} + \vec{\jmath} \times \vec{B} = 0 \qquad (193)$$

is satisfied. Clearly, in a force-free plasma $\vec{E}$ is perpendicular to $\vec{B}$. Furthermore, $\vec{E}$ has no toroidal component for an axisymmetric stationary situation. This is an immediate consequence of the induction law: Applying (25) for a stationary and axisymmetric configuration to the path $\mathcal{C}$ in Fig 10, we obtain

$$\oint_{\mathcal{C}} \alpha\,\mathcal{E} = \oint_{\mathcal{C}} i_\beta\,\mathcal{B} \stackrel{(123)}{=} -\frac{1}{2\pi}\oint_{\mathcal{C}} d\Psi = 0 \implies \vec{E}^{\mathrm{tor}} = 0. \qquad (194)$$

Similarly, Ampère's law in integral form (27) reduces to

$$\oint_{\mathcal{C}} \alpha\,\mathcal{H} = 4\pi\oint_{\mathcal{A}} \alpha\,\mathcal{J} = 4\pi I, \qquad (195)$$

where I is the total upward current through a surface $\mathcal{A}$ bounded by $\mathcal{C}$. This gives for the toroidal component of the magnetic field

$$\vec{B}^{\mathrm{tor}} = \frac{2\,I}{\alpha\,\tilde{\omega}}\,\vec{e}_\varphi. \qquad (196)$$

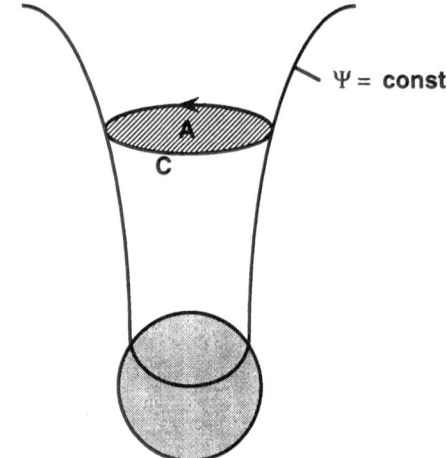

$\Psi = \mathbf{const.}$

Fig. 10 Path for (194) and (195).

Using this in (124) gives

$$B = \frac{1}{2\,\pi}\,\underbrace{\mathbf{d}\Psi \wedge \mathbf{d}\varphi}_{\text{poloidal}} + \underbrace{\frac{2\,I}{\alpha}\,*\,\mathbf{d}\varphi}_{\text{toroidal}}. \tag{197}$$

For later use, we also note the following: The continuity equation (24) reduces to

$$\mathbf{d}\,(\alpha\,\mathcal{J}) = 0. \tag{198}$$

Moreover, $L_{\partial_\varphi}\,(\alpha\,\mathcal{J}) = 0$, we have $\mathbf{d}\,(i_{\partial_\varphi}\,\alpha\,\mathcal{J}) = 0$, thus

$$i_{\partial_\varphi}\,(\alpha\,\mathcal{J}) = \frac{1}{2\,\pi}\,\mathbf{d}I \tag{199}$$

or

$$\alpha\,\mathcal{J}^{\mathrm{pol}} = \frac{1}{2\,\pi}\,\mathbf{d}I \wedge \mathbf{d}\varphi. \tag{200}$$

Clearly, the potential I is the current in (195).

Because $\mathcal{E}$ is poloidal, we can represent the electric field as follows

$$\mathcal{E} = i_{\vec{v}_F}\,B \qquad (\vec{E} = -\vec{v}_F \times \vec{B}), \tag{201}$$

where $\vec{v}_F$ is toroidal. Let us set

$$\vec{v}_F =: \frac{1}{\alpha}\,(\Omega_F - \omega)\,\tilde{\omega}\,\vec{e}_\varphi. \tag{202}$$

For the interpretation of Ω_F note the following: For an observer, rotating with angular velocity Ω, the 4-velocity is $u = u^t \left(\partial_t + \Omega \, \partial_\varphi \right)$. On the other hand, $u = \gamma \left(e_0 + \vec{v} \right)$, where $\vec{v}$ is the 3-velocity relative to a FIDO. Using also $\partial_t = \alpha \, e_0 + \vec{\beta}$ we get $\Omega \, \vec{\partial}_\varphi = \alpha \, \vec{v} - \vec{\beta}$ or

$$\vec{v} = \frac{1}{\alpha} \left(\Omega - w \right) \vec{\partial}_\varphi = \frac{1}{\alpha} \left(\Omega - w \right) \tilde{\omega} \, \vec{e}_\varphi. \tag{203}$$

This has the same form as (202) and, therefore, Ω_F is the angular velocity of the magnetic field lines.

Next, we show that Ω_F is only a function of Ψ. The induction law gives

$$\mathbf{d}(\alpha \, \mathcal{E}) = L_\beta \, \mathcal{B} = -L_{w \, \partial_\varphi} \, \mathcal{B} = -w \, L_{\partial_\varphi} \, \mathcal{B} - dw \wedge i_{\partial_\varphi} \, \mathcal{B}$$

$$= \frac{1}{2 \, \pi} \, dw \wedge d\Psi.$$

On the other hand, (201) and (202) give

$$\mathbf{d}(\alpha \, \mathcal{E}) = \mathbf{d}(\alpha \, i_{\vec{v}_F} \, \mathcal{B}) = \mathbf{d} \left((\Omega_F - w) \, i_{\partial_\varphi} \, \mathcal{B} \right) = -\mathbf{d} \left(\frac{\Omega_F - w}{2 \, \pi} \, d\Psi \right)$$

$$= -\mathbf{d} \left(\frac{\Omega_F - w}{2 \, \pi} \right) \wedge d\Psi.$$

By comparison, we get $d\Omega_F \wedge d\Psi = 0 \implies \Omega_F = \Omega_F(\Psi)$. The calculation above also shows

$$\alpha \, \mathcal{E} = -\frac{\Omega_F - w}{2 \, \pi} \, d\Psi, \tag{204}$$

i.e., $\vec{E}$ is *perpendicular* to the surfaces $\{\Psi = \text{const}\}$.

Specializing to the horizon gives

$$\vec{E}_H = -(\Omega_F - \Omega_H) \, \tilde{\omega} \, \vec{e}_\varphi \times \vec{B}_\perp. \tag{205}$$

The representations (197) of $\mathcal{B}$, and (204) for $\mathcal{E}$ will be important in the final section 9.

Now, we proceed as earlier in section 5 and consider again the closed path $\mathcal{C}$ in Fig. 11.

We already had the relations

$$\Delta V = \frac{1}{2 \, \pi} \, \Omega_H \, \Delta \Psi, \tag{206}$$

and

$$\Delta R_H = R_H \, \frac{\Delta \Psi}{4 \, \pi^2 \, \tilde{\omega}^2 \, B_\perp} \tag{207}$$

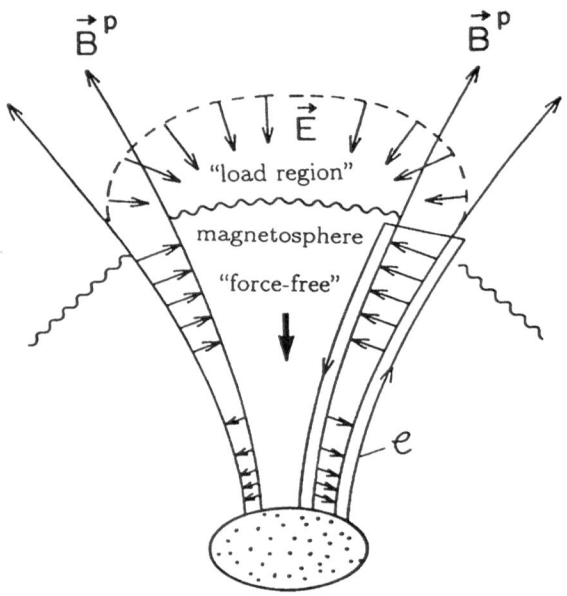

Fig. 11 Magnetosphere of a rotating BH (adapted from [1]).

(see (127) and (130)). As in (143) we obtain for the horizon voltage

$$\triangle V_H = I \triangle R_H. \tag{208}$$

This contribution can alternatively be computed with (204):

$$\triangle V_H = \int_{\mathcal{C}_H} \alpha \, \mathcal{E} = \frac{\Omega_H - \Omega_F}{2 \pi} \triangle \Psi. \tag{209}$$

The total voltage $\triangle V$ of $\mathcal{C}$ is the sum of $\triangle V_H$ and the voltage drop $\triangle V_L$ in the astrophysical load region (see Fig. 11). For the latter we obtain from (204) (since $\omega \simeq 0$)

$$\triangle V_L = \frac{1}{2\pi} \, \Omega_F \, \triangle \Psi \tag{210}$$

and, of course, also

$$\triangle V_L = I \triangle R_L, \tag{211}$$

where $\triangle R_L$ is the resistance of the load region. From this equations, and

$$\triangle V = \triangle V_H + \triangle V_L, \tag{212}$$

we immediately find the relations

$$\frac{\Delta V_L}{\Delta V_H} = \frac{\Omega_F}{\Omega_H - \Omega_F} = \frac{\Delta R_L}{\Delta R_H}, \tag{213}$$

$$I = \frac{\Delta V}{\Delta R_H + \Delta R_L} = \frac{1}{2}(\Omega_H - \Omega_F)\,\tilde{\omega}^2\,B_\perp. \tag{214}$$

The ohmic dissipation at the horizon is as in (179)

$$T_H\,\frac{d(\Delta S_H)}{dt} = I^2\,\Delta R_H = I\,\Delta V_H$$

$$= \frac{(\Omega_H - \Omega_F)^2}{4\pi}\,\tilde{\omega}^2\,B_\perp\,\Delta\Psi. \tag{215}$$

The power ΔP_L deposited in the load is thus

$$\Delta P_L = I^2\,\Delta R_L = I\,\Delta V_L \overset{(210),\,(214)}{=} \frac{\Omega_F\,(\Omega_H - \Omega_F)}{4\pi}\,\tilde{\omega}^2\,B_\perp\,\Delta\Psi. \tag{216}$$

As in (153), ΔP_L becomes maximal for $\Delta R_H = \Delta R_L$, which is the case for (see (213))

$$\Omega_F = \frac{1}{2}\,\Omega_H. \tag{217}$$

Then the estimates (133) and (135) hold.

Whether the crucial condition (217) is approximately satisfied in realistic astrophysical scenarios is a difficult problem for model builders. Therefore, the question remains open whether the process proposed by Blandford and Znajek [9] is important for the energy production in active galactic nuclei. It could, however, play a significant role in the formation of relativistic jets.

In this connection it should be mentioned that the Blandford-Znajek process may play an important role in gamma-ray bursts, as has been suggested, for instance by Mészáros and Rees [10]. One of the proposed models involves the toroidal debris from a disrupted neutron star orbiting around a BH. This debris may contain a strong magnetic field, perhaps amplified by differential rotation, and an axial magnetically-dominated wind may be generated along the rotation axis which would contain little baryon contamination. Energized by the BH via the Blandford-Znajek process, a narrow channel Poynting-dominated outflow with little baryon loading could be formed. The latter property is crucial for explaining the efficient radiation as gamma rays. (For a review of this topic, see for instance [11].)

9 Axisymmetric, Stationary Electrodynamics of Force-Free Fields

In this concluding section we discuss the electrodynamics of force-free fields around a BH a bit more systematically. The main goal is to derive the general relativistic *Grad-Shafranov equation*, whose nonrelativistic limit plays an important role in the electrodynamics of pulsars.

For *stationary* fields, Maxwell's equations (38) become

$$\mathbf{d}\mathcal{B} = 0, \quad \mathbf{d}(\alpha\,\mathcal{E}) = L_\beta\,\mathcal{B}, \tag{218}$$

$$\mathbf{d} * \mathcal{E} = 4\pi\rho, \quad \mathbf{d}(\alpha * \mathcal{B}) = 4\,\pi\,\alpha\,\mathcal{J} - L_\beta * \mathcal{E}. \tag{219}$$

From now on we assume also *axisymmetry*. Then $L_\beta\,\rho = \mathbf{d}\,i_\beta\,\rho = 0$, and hence the continuity equation (24) reduces to

$$\mathbf{d}\,(\alpha\,\mathcal{J}) = 0. \tag{220}$$

I recall the representation (197) of $\mathcal{B}$ in terms of the potential Ψ and I:

$$\mathcal{B} = \frac{1}{2\,\pi}\,\mathbf{d}\Psi \wedge \mathbf{d}\varphi + \frac{2\,I}{\alpha} * \mathbf{d}\varphi. \tag{221}$$

Ψ and I are independent of φ; their physical significance has already been discussed: Using the notation in Fig. 10 we have

$$\Psi = \int_A \mathcal{B}, \qquad I = \int_A \alpha\,\mathcal{J}. \tag{222}$$

At this point we make the simplifying assumption that the ideal MHD condition (193) holds *everywhere*. The fields are thus force-free, and we know from section 8 that the $\mathcal{E}$-field is poloidal and can be represented as (see (204))

$$\alpha\,\mathcal{E} = -\frac{\Omega_F - \omega}{2\,\pi}\,\mathbf{d}\Psi, \tag{223}$$

The representation (200) of the poloidal current

$$\alpha\,\mathcal{J}^{\mathrm{pol}} = \frac{1}{2\,\pi}\,\mathbf{d}I \wedge \mathbf{d}\varphi. \tag{224}$$

was obtained without assuming the ideal MFD condition. But if this is assumed, the toroidal part of $\vec{j} \times \vec{B}$ in (193) has to vanish, which means that $\mathcal{J}^{\mathrm{pol}}$ and $\mathcal{B}^{\mathrm{pol}}$ are proportional to each other. Comparison of (221) and (224) shows that $\mathbf{d}I$ must then be proportional to $\mathbf{d}\Psi$, thus I is a function of Ψ alone:

$$I = I(\Psi). \tag{225}$$

We have shown already in section 8 that Ω_F is also a function of Ψ,

$$\Omega_F = \Omega_F(\Psi). \tag{226}$$

According to (221) and (224) we now have

$$\mathcal{J}^{\mathrm{pol}} = \frac{1}{\alpha}\,\frac{dI}{d\Psi}\,\mathcal{B}^{\mathrm{pol}}. \tag{227}$$

The total current $\vec{\jmath}$ can now be represented as

$$\vec{\jmath} = \rho_e \, \vec{v}_F + \frac{1}{\alpha} \frac{dI}{d\Psi} \, \vec{B}. \tag{228}$$

In view of (227), the poloidal part of this equation is certainly correct, and the toroidal part on the right hand side is chosen such that $\vec{\jmath} \times \vec{B} = \rho_e \, \vec{v}_F \times \vec{B} = -\rho_e \, \vec{E}$ (see (201)), which is just the ideal MFD condition.

Our goal is now to derive – for given functions (225) and (226) – a partial differential equation for the potential Ψ. We shall achieve this by computing the toroidal part of the current in two independent ways. First, we take the toroidal part of (228) and obtain with $B^{\text{tor}} = \frac{2I}{\alpha \tilde{\omega}}$ (see (221))

$$j^{\text{tor}} = \rho_e \, \frac{\Omega_F - \omega}{\alpha} \, \tilde{\omega} + \frac{1}{\alpha} \frac{dI}{d\Psi} \frac{2I}{\alpha \tilde{\omega}}.$$

Here, we also eliminate ρ_e: From (223) and (219) we deduce

$$8\pi^2 \rho = -\mathbf{d} * \left(\frac{\Omega_F - \omega}{\alpha} \, \mathbf{d}\Psi \right), \tag{229}$$

i.e.,

$$8\pi^2 \rho_e = -\vec{\nabla} \cdot \left(\frac{\Omega_F - \omega}{\alpha} \, \vec{\nabla}\Psi \right). \tag{230}$$

We thus arrive at

$$j^{\text{tor}} = -\frac{1}{8\pi^2} \frac{\Omega_F - \omega}{\alpha} \, \tilde{\omega} \, \vec{\nabla} \cdot \left(\frac{\Omega_F - \omega}{\alpha} \, \vec{\nabla}\Psi \right) + \frac{2I}{\alpha^2 \tilde{\omega}} \frac{dI}{d\Psi}. \tag{231}$$

On the other hand, this quantity can also be obtained from Maxwell's equation (219):

$$4\pi\alpha \, \mathcal{J}^{\text{tor}} = [\mathbf{d}(\alpha * B)]^{\text{tor}} + [L_\beta * \mathcal{E}]^{\text{tor}}. \tag{232}$$

For the first term on the right we have

$$2\pi \, [\mathbf{d}(\alpha * B)]^{\text{tor}} = 2\pi \, \mathbf{d}(\alpha * B^{\text{tor}}) \stackrel{(221)}{=} \mathbf{d} \, [\alpha * (\mathbf{d}\Psi \wedge \mathbf{d}\varphi)]. \tag{233}$$

Now, we use the following useful general identity, whose proof is left as an exercise: Let χ be a p-form and $\vec{m}$ a Killing field. Assume also $L_{\vec{m}} \chi = 0$, then

$$\delta \, (m \wedge \chi) = -m \wedge \delta \chi, \tag{234}$$

where δ is the codifferential.

For $\chi = (\alpha/\tilde{\omega}^2) \, \mathbf{d}\Psi$, $\vec{m} = \vec{\partial}_\varphi$, $m = \tilde{\omega}^2 \mathbf{d}\varphi$ this gives

$$\delta \, (\alpha \, \mathbf{d}\Psi \wedge \mathbf{d}\varphi) = \delta \left(\frac{\alpha}{\tilde{\omega}^2} \, \mathbf{d}\Psi \right) \tilde{\omega}^2 \, \mathbf{d}\varphi$$

or

$$\mathbf{d} * [\alpha \, \mathbf{d}\Psi \wedge \mathbf{d}\varphi] = \delta \left(\frac{\alpha}{\tilde{\omega}^2} \, \mathbf{d}\Psi \right) \tilde{\omega}^2 * \mathbf{d}\varphi.$$

Thus, (233) becomes

$$2\pi \, \mathbf{d}(\alpha * \mathcal{B}^{\text{pol}}) = -\vec{\nabla} \cdot \left(\frac{\alpha}{\tilde{\omega}^2} \, \vec{\nabla}\Psi \right) \tilde{\omega}^2 * \mathbf{d}\varphi. \tag{235}$$

Next, we turn to the second term in (232). By (223) we have

$$-L_\beta * \mathcal{E} = L_\beta \left[\frac{\Omega_F - \omega}{2\pi\alpha} * \mathbf{d}\Psi \right].$$

If χ denotes the square bracket, we can write $L_\beta \chi = L_{-\omega \, \partial_\varphi} \chi = -\omega L_{\partial_\varphi} \chi - \mathbf{d}\omega \wedge i_{\partial_\varphi} \chi$, and obtain

$$*L_\beta * \mathcal{E} = -\frac{\Omega_F - \omega}{2\pi\alpha} i_{\vec{\nabla}\omega} \, (\mathbf{d}\Psi \wedge \tilde{\omega}^2 \, \mathbf{d}\varphi). \tag{236}$$

With (235) and (236) we obtain for the Hodge-dual of (232)

$$4\pi\alpha * \mathcal{J}^{\text{tor}} = -\frac{1}{2\pi} \vec{\nabla} \cdot \left(\frac{\alpha}{\tilde{\omega}^2} \, \vec{\nabla}\Psi \right) \tilde{\omega}^2 \, \mathbf{d}\varphi - \frac{\Omega_F - \omega}{2\pi\alpha} \left(\vec{\nabla}\omega \cdot \vec{\nabla}\Psi \right) \tilde{\omega}^2 \, \mathbf{d}\varphi.$$

Since $*\mathcal{J}^{\text{tor}} = j^{\text{tor}} \frac{1}{\tilde{\omega}} \tilde{\omega}^2 \, \mathbf{d}\varphi$, we get

$$\frac{4\pi\alpha}{\tilde{\omega}} j^{\text{tor}} = -\frac{1}{2\pi} \vec{\nabla} \cdot \left(\frac{\alpha}{\tilde{\omega}^2} \, \vec{\nabla}\Psi \right) - \frac{\Omega_F - \omega}{2\pi\alpha} \left(\vec{\nabla}\omega \cdot \vec{\nabla}\Psi \right). \tag{237}$$

Inserting $\vec{\nabla}\omega = \vec{\nabla}(\omega - \Omega_F) + (d\Omega_F/d\Psi) \, \vec{\nabla}\Psi$ leads finally to our second formula for j^{tor}:

$$8\pi^2 j^{\text{tor}} = -\frac{\tilde{\omega}}{\alpha} \vec{\nabla} \cdot \left(\frac{\alpha}{\tilde{\omega}^2} \, \vec{\nabla}\Psi \right) + \frac{\tilde{\omega}}{\alpha^2} \left(\Omega_F - \omega \right) \vec{\nabla}\Psi \cdot \vec{\nabla}(\Omega_F - \omega)$$

$$- \frac{\tilde{\omega}}{\alpha^2} \left(\Omega_F - \omega \right) \frac{d\Omega_F}{d\Psi} (\vec{\nabla}\Psi)^2. \tag{238}$$

Comparison of (232) with (238) gives, after a few steps, the following *generalized Grad-Shafranov equation*:

$$\vec{\nabla} \cdot \left\{ \frac{\alpha}{\tilde{\omega}^2} \left[1 - \frac{(\Omega_F - \omega)^2 \, \tilde{\omega}^2}{\alpha^2} \right] \vec{\nabla}\Psi \right\} + \frac{\Omega_F - \omega}{\alpha} \frac{d\Omega_F}{d\Psi} (\vec{\nabla}\Psi)^2 + \frac{16\pi^2}{\alpha \tilde{\omega}^2} I \frac{dI}{d\Psi} = 0. \tag{239}$$

The integration of the original equations (for axisymmetric stationary situations) is reduced to this single partial differential equation. $I(\Psi)$ and $\Omega_F(\Psi)$ are free functions[2], and if Ψ is a solution of (239) the electromagnetic fields are given

---

[2] These are, of course, restricted by boundary conditions, but we do not discuss this here.

by (221) and (223), while the charge and current distribution can be obtained from (230), (224), and (231).

A limiting case of (239) is known from the electrodynamics of pulsars: Ignoring the curvature of spacetime and using that Ω_F is equal to the angular velocity Ω of the neutron star (to derived shortly), we get the *pulsar equation*:

$$\vec{\nabla} \cdot \left\{ \frac{1}{\tilde{\omega}^2} \left[1 - \Omega^2 \tilde{\omega}^2 \right] \vec{\nabla}\Psi \right\} + \frac{16\,\pi^2}{\tilde{\omega}^2} I \frac{dI}{d\Psi} = 0 \qquad (240)$$

($\tilde{\omega}$ is the radial cylindrical coordinate in flat space).

The equation (239) holds also outside an aligned pulsar, since all the basic equations in section 2 remain valid there. However, the different boundary condition at the surface of the neutron star implies $\Omega_F = \Omega$, as we now show. In the interior of the neutron star the 3-velocity is (see (203))

$$\vec{v} = \frac{1}{\alpha}\,(\Omega - \omega)\,\vec{\partial}_\varphi. \qquad (241)$$

Since the neutron star matter is ideally conducting, the electric field there is

$$\mathcal{E} = i_{\vec{v}}\,\mathcal{B} = -\frac{1}{2\pi\alpha}\,(\Omega - \omega)\,\mathrm{d}\Psi,$$

if the $\mathcal{B}$-field is poloidal (first term in (221)). At the boundary this has to agree with (223), implying $\Omega_F = \Omega$.

Another remark should be made at this point about the interior of the neutron star. We showed in section 8 that $\mathrm{d}\Omega_F \wedge \mathrm{d}\Psi = 0$ implying $\mathrm{d}\Omega \wedge \mathrm{d}\Psi = 0$ inside the star. It will, in general, not be possible to represent Ψ as a function of Ω, since Ψ has to satisfy the Grad-Shafranov equation for $I(\Psi) = 0$ inside the star. Therefore, the rotation must be *rigid*, $\Omega = $ const.

Bibliography

[1] Thorne, K.S., Price, R.H., MacDonald, D.A. (1986): Black Holes: The Membrane Paradigm, Yale Univ. Press.

[2] Durrer, R., Straumann, N. (1988): *Helv. Phys. Acta.* **61**, 1027

[3] Wald, R.M. (1974): *Phys. Rev.* **D** 10, 1680

[4] Damour, T., *Phys. Rev. D* **18**, 3598 (1978); Ph. D. diss., Université de Paris (1979).

[5] Znajek, R.L. (1978): *Mon. Not. Roy. Astron. Soc.* **185**, 833

[6] Thorne, K.S., MacDonald, D.A. (1982): *Mon. Not. Roy. Astron. Soc.* **198**, 339

[7] Heusler, M. (1996): Black Hole Uniqueness Theorems, Cambridge Univ. Press.

[8] Weinstein, G. (1990): *Commun. Pure Appl. Math.* **43**, 903

[9] Blandford, R.D., Znajek, R.L. (1977): *Mon. Not. Roy. Astron. Soc.* **179**, 433

[10] Mészáros, P., Rees, M.J. (1997): *Ap. J.* **482**, L29

[11] Rees, M.J.: Gamma-Ray bursts: Challenges to Relativistic Astrophysics, Los Alamos e-print `astro-ph/9701162`, (Presented on Symposium on Black Holes and Relativistic Stars, Chicago, 14–15 Dec. 1996)

Uniqueness Theorems for Black Hole Space-Times

Markus Heusler

Institute for Theoretical Physics, University of Zürich, Winterthurer Str. 190, CH-8057 Zürich, Switzerland

Abstract. This lecture gives a brief outline of the proof of the black hole uniqueness theorem. Since the latter applies to stationary space-times, we first recall the reduction of the Einstein-Maxwell action in the presence of a Killing field. The resulting coset structure of the field equations gives rise to the Mazur identity, being the key to the uniqueness proof for the Kerr-Newman metric.

1 Introduction

All stationary, asymptotically flat black hole solutions (with non-degenerate horizon) of the Einstein-Maxwell (EM) equations are parameterized by the Kerr-Newman metric. The proof of this celebrated result, conjectured by Israel, Penrose and Wheeler in the late sixties (see [52] for a historical account) has been completed during the last three decades (see, e.g. [17] and [42]). Some open gaps, notably the electrovac staticity theorem (Sudarsky and Wald 1992, 1993) and the topology theorem [30], have been closed recently (see also [18]). The uniqueness theorem implies that all stationary electrovac black hole space-times are characterized by their mass, angular momentum and electric charge. This beautiful feature – together with the striking analogy between the laws of black hole physics and the laws of equilibrium thermodynamics – provided support for the expectation that the stationary black hole solutions of other self-gravitating matter fields are also parameterized by a set of asymptotic flux integrals (no-hair conjecture).

However, during the last eight years, a variety of new black hole configurations which violate the generalized no-hair conjecture were found (see, e.g. Volkov and Gal'tsov 1989, Künzle and Masood-ul-Alam 1990, Bizon 1990, Droz et al 1991, Heusler et al 1991, 1992, 1993, Breitenlohner et al 1992, Lavrelashvili and Maison 1993, Greene et al 1993). While these counterexamples consist in spherically symmetric solutions which are not characterized by their masses and charges, more recent studies have revealed that static black holes need not even be spherically symmetric ([78], [59]) and, moreover, that non-rotating black holes need not be static ([8]). The rich spectrum of stationary black hole configurations demonstrates that the matter content is by far more critical to the uniqueness properties than originally expected.

In fact, the classical uniqueness theorem, to which the present lecture is devoted, applies only to vacuum and electrovac space-times. The proof of the theorem is, at least in the axisymmetric case, heavily based on the fact that the EM equations in the presence of a Killing field form a sigma-model, effectively coupled to three-dimensional gravity ([26], [73]). Since this property is not shared by models with non-Abelian gauge fields ([7]), it is, with hindsight, not too surprising that the Einstein-Yang-Mills (EYM) system admits black holes with "hair". There exist, however, other black hole solutions which are likely to be subject to a generalized no-hair theorem. These solutions appear in theories with self-gravitating massless scalar (moduli) and Abelian vector fields. The expectation that uniqueness results apply to a variety of these models arises from the fact that their dimensional reduction (with respect to the Killing symmetry) yields a sigma-model with symmetric target space (see, e.g. [6], [24], [31], and references therein).

The uniqueness proof for the stationary electrovac black hole solutions comprises various steps, not all of which are established in an equally reliable manner (see, e.g. Chruściel 1994, 1996a). The purpose of this lecture is to provide an introduction to the reasoning, rather than the technical details of the uniqueness proof. We shall do so by first discussing the Kaluza-Klein reduction of the Einstein-Maxwell action in the presence of a Killing symmetry. We then recall Hawking's rigidity theorem, some properties of Killing horizons, and the staticity and circularity theorems. Finally, we demonstrate the uniqueness of the Reissner-Nordström solution among the static, and the Kerr-Newman solution among the circular electrovac black hole configurations. In order to keep track of the reasoning, we start by briefly introducing the various issues involved in the uniqueness program (see also Fig. 1).

2   The Corner-Stones of the Uniqueness Theorem

At the basis of the reasoning lies Hawking's *strong rigidity theorem* (Hawking 1972, Hawking and Ellis 1973). It relates the global concept of event horizons to the independently defined – and logically distinct – local notion of Killing horizons: Requiring that the fundamental matter fields obey well behaved hyperbolic equations, and that the stress-energy tensor satisfies the weak energy condition, the theorem asserts that the event horizon of a *stationary* black hole spacetime is a *Killing horizon*. This also implies that stationary black hole spacetimes are either *non-rotating* or axisymmetric. The uniqueness proofs for the Reissner-Nordström and the Kerr-Newman metric are, however, based on the requirement that the domain of outer communications (DOC) is either *static* or *circular*. Hence, in both cases, one has to prove beforehand that the Frobenius integrability conditions for the Killing fields are satisfied as a consequence of the symmetry properties and the field equations.

The *circularity theorem*, due to Kundt and Trümper (1966) and Carter (1969), implies that the metric of a vacuum or electrovac spacetime can, without loss of generality, be written in the well-known Papapetrou $(2+2)$-split. [It has become clear only recently that stationary and axisymmetric EYM configurations need *not* be circular (Heusler 1995, 1996a), although the circularity property can be established for a variety of other matter models ([39]).]

Stationary, asymptotically flat black hole spacetime

(Killing field k^μ)

STRONG RIGIDITY THM (1st part)

Event horizon = Killing horizon $H[\xi]$

(null generator Killing field ξ^μ)

non-rotating, $k^\mu|_{H[\xi]} = \xi^\mu$

rotating, $k^\mu|_{H[\xi]} \neq \xi^\mu$

NO ERGO-REGION THM

STRONG RIGIDITY THM

DOC strictly stationary

$(k^\mu k_\mu \leq 0)$

DOC axisymmetric

($\exists$ Killing field m^μ)

STATICITY THM

CIRCULARITY THM

DOC static

$\exists$ coordinate t: $g = \sigma^{(1)} + g^{(3)}$

$\sigma^{(1)} = -V\,dt^2$

DOC circular

$\exists$ coordinates t, φ: $g = \sigma^{(2)} + g^{(2)}$

$\sigma^{(2)} = -V\,dt^2 + 2W\,dt\,d\varphi + X\,d\varphi^2$

STATIC UNIQUENESS THM

(originally by means of Israel's thm,
later by using the positive energy thm)

CIRCULAR UNIQUENESS THM

(originally by means of Robinson's thm,
later by using sigma-model identities)

Schwarzschild (Reissner-Nordström)

Kerr (Kerr-Newman) metric

Fig. 1 The various issues involved in the uniqueness proof

The *staticity theorem*, establishing that the stationary Killing field in a non-rotating, electrovac black hole spacetime is hyper-surface orthogonal, is more involved than the circularity problem. First, one has to establish *strict* stationarity, that is, one needs to exclude ergo-regions. This problem, first discussed by Hajicek (1973, 1975) and Hawking and Ellis (1973), was solved only recently by Sudarsky and Wald (1992, 1993), assuming a foliation by maximal slices (Chruściel and Wald 1994a). If ergo-regions are excluded, it still remains to prove that the stationary Killing field satisfies the Frobenius integrability condition. In the vacuum case, this was achieved by Hawking (1972), who was able to generalize a theorem due to Lichnerowicz (1955) to black hole space-times. As already mentioned, Sudarsky and Wald (1992, 1993) eventually succeeded in solving the staticity problem for *electrovac* black holes, by using the generalized version of the first law of black hole physics. (Like the circularity theorem, the staticity theorem is easily extended to scalar fields ([39]), whereas the generalization to non-Abelian gauge fields requires additional assumptions.)

The main task of the uniqueness problem is to show that the static electrovac black hole space-times (with non-degenerate horizon) are described by the Reissner-Nordström metric, whereas the circular ones (i.e., the stationary and axisymmetric ones with integrable Killing fields) are represented by the Kerr-Newman metric.

In the static case it was Israel (1967, 1968) who, in his pioneering work, was able to show that both static vacuum and electrovac black hole space-times are *spherically symmetric*. Israel's ingenious method, based on integral identities and Stokes' theorem, triggered a series of investigations devoted to the uniqueness problem (see, e.g. Müller-zum-Hagen et al 1973, 1974, Robinson 1974, 1977). Later on, Simon (1985), Bunting and Masood-ul-Alam (1987) and Masood-ul-Alam (1992) were able to find new proofs of the Israel theorem, which were based on the positive energy theorem (see Schoen and Yau 1979, 1981, and Witten 1981 for the proof of the latter).

The uniqueness theorem for the Kerr metric relies heavily on the Ernst formulation of the Einstein vacuum equations (Ernst 1986; see also Ehlers 1959, Neugebauer and Kramer 1969, Geroch 1971, 1972). The key to the proof consists in Carter's observation that the field equations reduce to a two-dimensional boundary value problem (Carter 1971, 1973b). An amazing identity due to Robinson (1975) then establishes that all vacuum solutions with the same boundary and regularity conditions are identical. The uniqueness problem for the stationary and axisymmetric case *with* electromagnetic fields remained open until Mazur (1982, 1984a,b) and, independently, Bunting (1983) were able to obtain a generalization of the Robinson identity in a systematic way: The Mazur identity is based on the observation that the Ernst equations describe a nonlinear sigma-model with coset space G/H, where G is a connected Lie group and H is a maximal compact subgroup of G. In the electrovac case one finds $G/H = SU(1,2)/S(U(1) \times U(2))$. Within this approach, the Robinson identity turns out to be the explicit form of the Mazur identity for the vacuum case, $G/H = SU(1,1)/U(1)$.

The uniqueness theorem presented in this lecture applies exclusively to spacetimes having Killing horizons with non-vanishing surface gravity. The multi black hole solutions of Papapetrou (1945) and Majumdar (1947) illustrate that stationary EM black holes with degenerate Killing horizons need not belong to the Kerr-Newman family. The uniqueness of the Papapetrou-Majumdar solution amongst the stationary electrovac black hole configurations with degenerate horizons is not yet completely established (see Chruściel and Nadirashvili 1995, [43] for recent progress).

3 Stationary Electrovac Space-Times

For physical reasons, the black hole equilibrium states are expected to be stationary. In order to describe these configurations as isolated systems, spacetime is also assumed to be asymptotically flat. As for the matter content, it is required that the DOC is either empty or contains no other fields than electromagnetic ones. Hence, the uniqueness theorems are statements about *asymptotically flat black hole solutions of the stationary EM equations in the absence of additional matter fields*. In the presence of a Killing symmetry, the EM equations exhibit some distinguished features, to which this section is devoted.

3.1 Reduction of the Einstein-Hilbert Action

By definition, a stationary spacetime (M, g) admits an asymptotically time-like Killing field, that is, a vector field k with $L_k g = 0$, L_k denoting the Lie derivative with respect to k. At least locally, M has the structure $\Sigma \times G$, where $G \approx \mathbb{R}$ denotes the one-dimensional group generated by the Killing symmetry, and Σ is the three-dimensional quotient space M/G. A stationary spacetime is called *static*, if the integral trajectories of k are orthogonal to Σ.

With respect to the adapted time coordinate t, defined by $k \equiv \partial_t$, the metric of a stationary spacetime is parametrized in terms of a three-dimensional (Riemannian) metric $\bar{g}$, a scalar field σ and a one-form $a \equiv a_i dx^i$ on $(\Sigma, \bar{g})$:

$$g = -\sigma(dt + a) \otimes (dt + a) + \frac{1}{\sigma}\bar{g}. \tag{1}$$

Note that a is the connection of a fiber bundle with base space Σ and fiber G. The derivative of a is closely related to the dual of the twist one-form, ω, assigned to k: In terms of the Killing one-form $\tilde{k} \equiv -\sigma(dt + a)$, one has

$$2\omega \equiv *(\tilde{k} \wedge d\tilde{k}) = -\sigma^2 \bar{*} da \tag{2}$$

where $\bar{*}$ denotes the Hodge dual with respect to the three-dimensional Riemannian metric $\bar{g}$. (In the following we shall use the symbol k for both the Killing field and the one-form $\tilde{k}$.) It is worth noticing that the "field strength", $f \equiv da$, is gauge invariant, since a transforms like an Abelian gauge potential under coordinate transformations.

It is a straightforward task to compute the Ricci scalar for the decomposition (1). Using the symbols $\bar{R}$ and $\bar{\Delta}$ for the Ricci scalar and the Laplacian with respect to $\bar{g}$, one finds (Exercises 1 & 2)

$$R\sqrt{-g} = \sqrt{\bar{g}}\left(\bar{R} + \bar{\Delta}\ln\sigma - \frac{1}{2\sigma^2}\langle d\sigma, d\sigma\rangle + \frac{\sigma^2}{2}\langle da, da\rangle\right), \tag{3}$$

where here and in the following $\langle \, , \, \rangle$ denotes the inner product with respect to $\bar{g}$ (i.e., $\bar{*}\langle\beta, \beta\rangle \equiv \beta \wedge \bar{*}\beta$ for an arbitrary p-form β). The above formula shows that the Einstein-Hilbert action of a stationary spacetime reduces to the action for a scalar field σ and an Abelian vector field a, which are coupled to three-dimensional gravity. The fact that this coupling is *minimal* is a consequence of the particular choice for the conformal factor in front of the three-metric $\bar{g}$ in (1).

The vacuum field equations are, therefore, equivalent to the three-dimensional Einstein-matter equations obtained from variations of the effective action

$$S_{\text{eff}} = \int \bar{*} \left(\bar{R} - \frac{1}{2\sigma^2} \langle d\sigma , d\sigma \rangle + \frac{\sigma^2}{2} \langle da , da \rangle \right) , \tag{4}$$

with respect to $\bar{g}_{ij}$, σ and a.

3.2   The Vacuum Target Space and the Ernst Potential

It is of crucial importance that the one-form a, parametrizing the non-static part of the metric, enters the effective action (4) only via the field strength, $f \equiv da$. For this reason, the off-diagonal Einstein equation, that is, the variational equation for a, assumes the form of a source-free Maxwell equation,

$$d\bar{*} \left(\sigma^2 da \right) = 0 . \tag{5}$$

Hence, there exists (locally) a function Y, such that

$$dY \equiv -\bar{*} \left(\sigma^2 da \right) , \tag{6}$$

where the definition (2) shows that Y is the potential for the twist one-form, $2\omega = dY$. In order to write the effective action (4) in terms of the twist potential Y, rather than the one-form a, one uses (6) and a Lagrangian multiplier to impose the constraint $d \left(\sigma^{-2} \bar{*} dY \right) = 0$ (since $d^2 a = 0$). A short computation shows that this effects in replacing $\sigma^2 \langle da , da \rangle$ by $-\sigma^{-2} \langle dY , dY \rangle$. Thus, the action (4) for the stationary vacuum Einstein equations becomes

$$S_{\text{eff}} = \int \bar{*} \left(\bar{R} - \frac{\langle d\sigma , d\sigma \rangle + \langle dY , dY \rangle}{2\,\sigma^2} \right) , \tag{7}$$

where we recall that $\langle \, , \, \rangle$ is the inner product with respect to the three-metric $\bar{g}$, defined in (1).

The action (7) describes a harmonic mapping into a two-dimensional target space, effectively coupled to three-dimensional gravity. In terms of the complex potential $\mathcal{E}$ (Ernst 1968), one has

$$S_{\text{eff}} = \int \bar{*} \left(\bar{R} - 2 \frac{\langle d\mathcal{E} , d\bar{\mathcal{E}} \rangle}{(\mathcal{E} + \bar{\mathcal{E}})^2} \right) , \quad \mathcal{E} \equiv \sigma + iY . \tag{8}$$

The stationary vacuum Einstein equations are obtained from variations with respect to the three-metric $\bar{g}$ [(ij)-equations] and the Ernst potential $\mathcal{E}$ [(0μ)-equations]. One easily finds

$$\bar{R}_{ij} = \frac{2}{(\mathcal{E} + \bar{\mathcal{E}})^2} \mathcal{E}_{,i} \bar{\mathcal{E}}_{,j} , \tag{9}$$

and

$$\bar{\Delta}\mathcal{E} \ = \ \frac{2}{\mathcal{E} + \bar{\mathcal{E}}} \, \langle \mathrm{d}\mathcal{E} \, , \, \mathrm{d}\mathcal{E} \rangle \, . \tag{10}$$

(In the static case, $Y = 0$, (10) reduces to the Poisson equation for the gravitational potential $\ln \sigma$ with respect to the metric $\bar{g}$.) Introducing the complex potential ε according to

$$\varepsilon \ = \ \frac{1 - \mathcal{E}}{1 + \mathcal{E}} \, , \tag{11}$$

reveals that the target space is the complex unit disc with standard metric or, equivalently, the pseudo-sphere, PS^2, in stereo-graphic coordinates $\mathrm{Re}(\varepsilon)$, $\mathrm{Im}(\varepsilon)$ (Exercise 3):

$$\frac{\langle \mathrm{d}\mathcal{E} \, , \, \mathrm{d}\bar{\mathcal{E}} \rangle}{(\mathcal{E} + \bar{\mathcal{E}})^2} \ = \ \frac{\langle \mathrm{d}\varepsilon \, , \, \mathrm{d}\bar{\varepsilon} \rangle}{(1 - |\varepsilon|^2)^2} \, . \tag{12}$$

As $SO(2,1)$ acts isometrically on PS^2 with isotropy group $SO(2)$, the target space is the coset $SO(2,1)/SO(2) \approx SU(1,1)/U(1)$.

3.3 Stationary Maxwell Fields

A stationary gauge potential A, say, can be decomposed into parallel and orthogonal components to the three-manifold Σ. Introducing the function ϕ on Σ and the one-form $\bar{A}$ on Σ, we may write

$$A \ = \ \phi \, (\mathrm{d}t + a) + \bar{A} \, , \tag{13}$$

where a is the non-static part of the metric defined in (1). (In general, a gauge potential A is called stationary if it is invariant – up to gauge transformations – under the group generated by the Killing field k (Forgács and Manton 1980). That is, $L_k A = D\mathcal{V}$ for some function $\mathcal{V}$, where $D\mathcal{V} = \mathrm{d}\mathcal{V}$ in the Abelian case under consideration; see, e.g. Heusler and Straumann 1993a,b.) By virtue of the above decomposition, the Abelian field strength becomes

$$F \ = \ \mathrm{d}\phi \wedge (\mathrm{d}t + a) + (\bar{F} + \phi f) \, , \tag{14}$$

where $\bar{F} \equiv \mathrm{d}\bar{A}$, and we recall that $f \equiv \mathrm{d}a$. (The Maxwell equations, $\mathrm{d} * F = 0$, with respect to the metric (1) are derived in Exercise 4.)

It is not hard to write the matter Lagrangian, $F \wedge *F$, in terms of the fields ϕ, $\bar{A}$ and the metric (1). Using the result (4) for the vacuum action, one easily finds that the Einstein-Hilbert-Maxwell action,

$$S_{\mathrm{EM}} \ = \ \int (*R - 2F \wedge *F) \, , \tag{15}$$

gives rise to the effective action (Exercise 5)

$$S_{\text{eff}} = \int \bar{*} \left(\bar{R} - \frac{1}{2\sigma^2} \langle d\sigma \rangle^2 + \frac{\sigma^2}{2} \langle da \rangle^2 + \frac{2}{\sigma} \langle d\phi \rangle^2 - 2\sigma \langle d\bar{A} + \phi \, da \rangle^2 \right), \quad (16)$$

where $\langle \, \rangle^2$ is a shorthand for the inner product $\langle \, , \, \rangle$ with respect to the metric $\bar{g}$.

In addition to the scalars σ and ϕ, the above action contains the metric one-form a and the magnetic one-form $\bar{A}$. Like in the vacuum case, the former enters the effective action only via the field strength $f \equiv da$, and gives, therefore, rise to the conservation law

$$d\bar{*} \left[\sigma^2 f - 4\sigma \phi (\bar{F} + \phi f) \right] = 0. \quad (17)$$

In a similar way, the magnetic gauge potential, $\bar{A}$, enters the effective action only via the field strength $\bar{F}$. Since, in the Abelian case, $\bar{F} = d\bar{A}$, the Maxwell equation for $\bar{A}$ assumes the form of a conservation law as well,

$$d\bar{*} \left[\sigma (\bar{F} + \phi f) \right] = 0. \quad (18)$$

The closed one-forms $\bar{*}[\dots]$ defined by the above equations give rise to two scalar potentials, Y and ψ, say. By virtue of (18), the magnetic potential ψ is defined by

$$d\psi \equiv \sigma \bar{*} (\bar{F} + \phi f). \quad (19)$$

Using this in the twist equation (17), the latter can be written in the symmetric form $d(\sigma^2 \bar{*} f - 2\phi d\psi + 2\psi d\phi) = 0$, which suggests the definition

$$dY \equiv -\sigma^2 \bar{*} f + 2\phi d\psi - 2\psi d\phi = 2 \left[\omega + \phi d\psi - \psi d\phi \right], \quad (20)$$

where we have also used the definition (2) of ω.

It is worth noticing that the decomposition (14) of F and the definition (19) imply that the electromagnetic potentials are obtained from F and $k = \partial_t$ by $d\phi = -F(k, \cdot)$ and $d\psi = (*F)(k, \cdot)$; see Exercise 4. We also emphasize that the generalized twist potential, Y, still exists for non-Abelian self-gravitating gauge fields, since (17) remains valid in this case. However, as F is no longer an exact differential-form, the conservation law (18) ceases to exist for Yang-Mills fields, and so does the magnetic potential ψ. This is, in fact, the only difference in the Kaluza-Klein reduction of the EM and the EYM system (see, e.g. Brodbeck and Heusler 1997).

In order to pass from the one-forms a and $\bar{A}$ to the scalar potentials Y and ψ, one applies again the Lagrange multiplier method. Using the definitions (19) and (20), as well as the constraints $d\bar{F} = 0$ and $df = 0$ in the effective action (16), yields (Exercise 6)

$$S_{\text{eff}} = \int \bar{*} \left(\bar{R} + 2 \frac{\langle d\phi \rangle^2 + \langle d\psi \rangle^2}{\sigma} - \frac{\langle d\sigma \rangle^2 + \langle dY - 2\phi d\psi + 2\psi d\phi \rangle^2}{2\sigma^2} \right), \quad (21)$$

which reduces to the vacuum action (7) for $\phi = 0$, $\psi = 0$. Hence, the stationary EM system is described by a non-linear sigma-model with four-dimensional target space, minimally coupled to three-dimensional gravity. The EM equations are obtained from variations of the effective action (21) with respect to the electromagnetic potentials ϕ and ψ, the gravitational potentials σ and Y, and the three-metric $\bar{g}$.

3.4   The Coset Formulation for Electrovac Space-Times

So far we have considered the dimensional reduction of the EM system in the presence of a time-like Killing field. It is clear that the method can be applied in a similar way to electrovac space-times admitting a *space-like* Killing symmetry. In this case, one has to define the scalar potentials and the metric $\bar{g}$ with respect to the space-like Killing field. (In particular, it will be crucial that the second term in the effective action (21) then has the same sign as the third term.) For an arbitrary Killing field ξ, say, one finds

$$S_{\text{eff}} = \int \bar{*} \left(\bar{R} - 2 \frac{\langle d\phi \rangle^2 + \langle d\psi \rangle^2}{N} - \frac{\langle dN \rangle^2 + \langle dY - 2\phi d\psi + 2\psi d\phi \rangle^2}{2\, N^2} \right), \quad (22)$$

where the electromagnetic potentials are obtained from the four-dimensional field strength F and the Killing field ξ by

$$d\phi = -i_\xi F, \quad d\psi = i_\xi * F. \quad (23)$$

(Any two-form β can be assigned the one-form $i_\xi \beta$ with components $\xi^\mu \beta_{\mu\nu}$.) The gravitational scalars, N and Y, are related to the norm and the twist of ξ, respectively:

$$N = (\xi, \xi), \quad dY = 2\,(\omega + \phi d\psi - \psi d\phi), \quad (24)$$

where $2\omega \equiv *(\xi \wedge d\xi)$. The inner product $\langle\ ,\ \rangle$ is taken with respect to the three-metric $\bar{g}$, which becomes pseudo-Riemannian if ξ is space-like. In the stationary and axisymmetric case, to be considered below, the Kaluza-Klein reduction will be performed with respect to the *space-like* Killing field. The presence of the stationary symmetry will then imply that the inner products in (22) have a fixed sign, despite the fact that $\bar{g}$ is not a Riemannian metric in this case.

We have already mentioned that the action (22) describes a harmonic mapping into a four-dimensional target space, effectively coupled to three-dimensional gravity. The target space can be parametrized in terms of the complex potentials $\mathcal{E}$ and Λ, defined by (Ernst 1968)

$$\mathcal{E} = -N - (\phi^2 + \psi^2) + iY, \quad \Lambda = -\phi + i\psi. \quad (25)$$

In terms of the Ernst potentials, the effective action (22) assumes the form

$$S_{\text{eff}} = \int \bar{*} \left(\bar{R} - 2 \frac{|\, d\Lambda\, |^2}{N} - \frac{1}{2} \frac{|\, d\mathcal{E} + 2\bar{\Lambda} d\Lambda\, |^2}{N^2} \right), \quad (26)$$

where $|d\Lambda|^2 \equiv \langle d\Lambda, \overline{d\Lambda} \rangle$. The field equations are obtained from variations with respect to the three-metric $\bar{g}$ and the Ernst potentials (Exercise 7). In particular, the latter turn out to be subject to the equations

$$\bar{\Delta}\mathcal{E} = -\frac{\langle d\mathcal{E}, d\mathcal{E} + 2\bar{\Lambda}d\Lambda \rangle}{N(\mathcal{E}, \Lambda)}, \quad \bar{\Delta}\Lambda = -\frac{\langle d\Lambda, d\mathcal{E} + 2\bar{\Lambda}d\Lambda \rangle}{N(\mathcal{E}, \Lambda)}, \qquad (27)$$

which generalize the vacuum result (10). [According to (25) one has $-N(\mathcal{E}, \Lambda) = \Lambda\bar{\Lambda} + \frac{1}{2}(\mathcal{E} + \bar{\mathcal{E}})$.]

The isometries of the target manifold are found by solving the respective Killing equations (Neugebauer and Kramer 1969). This reveals the coset structure of the target space and provides a parametrization of the latter in terms of the Ernst potentials. For vacuum gravity, we have already argued that the coset space, G/H, is $SU(1,1)/U(1)$, whereas one finds $G/H = SU(2,1)/S(U(1,1) \times U(1))$ for the EM equations with a time-like Killing field. If the dimensional reduction is performed with respect to a space-like Killing field, then $G/H = SU(2,1)/S(U(2) \times U(1))$.

The explicit representation of the coset manifold in terms of the above Ernst potentials, $\mathcal{E}$ and Λ, is given by the hermitian matrix Φ, with components

$$\Phi_{AB} = \eta_{AB} + 2\,\mathrm{sig}(N)\,\bar{v}_A v_B, \quad \text{where } \eta = \mathrm{diag}(-1, +1, +1), \qquad (28)$$

and where the vector v is defined by (Kinnersley 1973, 1977, Kinnersley and Chitre 1977, 1978)

$$(v_0, v_1, v_2) = \frac{1}{2\sqrt{|N|}}(\mathcal{E} - 1, \mathcal{E} + 1, 2\Lambda). \qquad (29)$$

It is straightforward to verify that, in terms of Φ, the effective action (22) assumes the $SU(2,1)$ invariant form (Exercise 8)

$$\mathcal{S}_{\text{eff}} = \int \bar{*}\left(\bar{R} - \frac{1}{4}\mathrm{Tr}\langle J, J \rangle\right), \quad \text{with } J \equiv \Phi^{-1}d\Phi, \qquad (30)$$

where $\mathrm{Tr}\langle J, J \rangle \equiv \langle J_B^A, J_A^B \rangle \equiv \bar{g}^{ij}(J_i)_B^A(J_j)_A^B$. The equations of motion following from the above action are the three-dimensional Einstein equations (obtained from variations with respect to $\bar{g}$) and the sigma-model equations (obtained from variations with respect to Φ) (Exercise 9):

$$\bar{R}_{ij} = \frac{1}{4}\mathrm{Tr}\{J_i J_j\}, \quad d\bar{*}J = 0. \qquad (31)$$

3.5   The Structure of the Mazur Identity

In the presence of a second Killing field, the above equations experience further, considerable simplifications, which will be discussed later. In the remainder of this section we will, however, not yet assume the existence of an additional Killing symmetry. The structure of the Mazur identity (Mazur 1982, 1984), being the key to the uniqueness theorem for the Kerr-Newman metric, is a consequence of the coset structure of the field equations, which only requires the existence of one Killing field. (The second Killing field is, of course, of crucial importance to the boundary value formulation of the field equations and the integration of the Mazur identity.)

In order to obtain the general form of the Mazur identity, we consider two arbitrary hermitian matrices, Φ_1 and Φ_2, say. Our aim is to compute the Laplacian (with respect to an arbitrary metric, $\bar{g}$, say) of the relative difference between Φ_2 and Φ_1,

$$\Psi \equiv \Phi_2 \Phi_1^{-1} - \mathbb{1}. \tag{32}$$

We define the current matrices J_1 and J_2 and their difference, $J_\triangle$, according to

$$J_1 = \Phi_1^{-1} \bar{\nabla} \Phi_1, \quad J_\triangle = J_2 - J_1, \tag{33}$$

where $\bar{\nabla}$ denotes the covariant derivative with respect to the metric under consideration. Using $\bar{\nabla}\Psi = \Phi_2 J_\triangle \Phi_1^{-1}$, one immediately finds for the Laplacian of Ψ,

$$\bar{\Delta}\Psi = \langle \bar{\nabla}\Phi_2 , J_\triangle \rangle \Phi_1^{-1} + \Phi_2 \langle J_\triangle , \bar{\nabla}\Phi_1^{-1} \rangle + \Phi_2 \left(\bar{\nabla} J_\triangle \right) \Phi_1^{-1} .$$

Now using the fact that $\Phi = \Phi^\dagger$, we have $\bar{\nabla}\Phi_2 = J_2^\dagger \Phi_2$ and $\bar{\nabla}\Phi_1^{-1} = -\Phi_1^{-1} J_1^\dagger$. Taking the matrix trace of the above identity finally yields

$$\mathrm{Tr}\left\{ \bar{\Delta}\Psi \right\} = \mathrm{Tr} \langle \Phi_1^{-1} J_\triangle^\dagger , \Phi_2 J_\triangle \rangle + \mathrm{Tr} \left\{ \Phi_2 \left(\bar{\nabla} J_\triangle \right) \Phi_1^{-1} \right\} . \tag{34}$$

Before we use this identity to prove the uniqueness theorem for the Kerr-Newman metric, we have to recall some basic properties of stationary black hole space-times.

4   Stationary Black Holes

Space-times with event horizons exhibit a variety of interesting local and global properties. Particularly intriguing features are the mass variation formula by Bardeen, Carter and Hawking (1973), and the area increase theorem, which suggest a relationship between the physics of (stationary) black holes and the laws of (equilibrium) thermodynamics. (See, e.g. Wald 1984 for the area theorem, Sudarsky and Wald 1992, 1993, Iyer and Wald 1994, Heusler and Straumann 1993 for some recent results on mass variation formulas, and the Lecture by G. Neugebauer for an up-to-date review on black hole thermodynamics.)

Of particular relevance to the uniqueness theorems for stationary black holes is the strong rigidity theorem (Hawking 1972, Hawking and Ellis 1973), which yields a subdivision of the stationary electrovac black hole configurations into non-rotating and axisymmetric ones (see Fig. 1). In addition, the theorem guarantees the Killing property of the horizon, which, in turn, implies that the surface gravity of a stationary black hole is constant (zeroth law).

The Frobenius integrability conditions provide the link between the outcome of the rigidity theorem and the requirements which are needed to prove the uniqueness of the Reissner-Nordström and the Kerr-Newman metric: In the non-rotating case one must establish staticity, whereas circularity is required in the stationary and axisymmetric situation (see, e.g. Carter 1987). [As already mentioned in the introduction, the integrability properties of the Killing fields are sensitive to the matter model. In particular, the Abelian nature of the gauge fields is of decisive importance in this context (see, e.g. Heusler 1996a).]

4.1 Killing Horizons

We start by recalling the definition of a Killing horizon: Consider a Killing field ξ, say, and the set of points with $N \equiv (\xi, \xi) = 0$. A connected component of this set which is a null hyper-surface, $(dN, dN) = 0$, is called a *Killing horizon*, $H[\xi]$.

An immediate consequence of the above definition is the fact that ξ and dN are proportional on $H[\xi]$. (Note that $(\xi, dN) = 0$, since $L_\xi N = 0$, and that two orthogonal null vectors are proportional.) This suggests the following definition of the *surface gravity*, κ,

$$dN = -2\kappa\xi \quad \text{on } H[\xi]. \tag{35}$$

It is an interesting fact that the surface gravity plays a similar role in the theory of stationary black holes as the temperature does in ordinary thermodynamics. Since the latter is constant for a body in thermal equilibrium, the result that

$$\kappa = \text{constant on } H[\xi] \tag{36}$$

is usually called the zeroth law of black hole physics (Bardeen et al 1973).

The zeroth law can be established by different means, depending on the specific assumptions (see Kay and Wald 1991, Rácz and Wald 1992, 1996, and Heusler 1996b for a compilation of the methods). A rather simple, purely geometrical proof can be obtained for static or circular space-times (Exercise 10). If, however, the Killing fields are not required to be hyper-surface orthogonal, then Einstein's equations and the dominant energy condition are needed to establish that κ is uniform over $H[\xi]$. The idea of this (original) proof is the following: First, one establishes the relations (see, e.g. Wald 1984)

$$R(\xi, \xi) = 0 \quad \text{on } H[\xi], \tag{37}$$

$$\xi \wedge d\kappa = -\xi \wedge R(\xi) \quad \text{on } H[\xi], \tag{38}$$

where $R(\xi, \xi) \equiv R_{\mu\nu}\xi^\mu\xi^\nu$, and $R(\xi)$ is the one-form with components $[R(\xi)]_\mu \equiv R_{\mu\nu}\xi^\nu$. In combination with Einstein's equations, and the fact that ξ is null on the horizon, (37) implies that the component $T(\xi,\xi)$ of the stress-energy tensor vanishes on the horizon. Thus, the one-form $T(\xi)$ is perpendicular to ξ and, therefore, space-like or null on $H[\xi]$. On the other hand, the dominant energy condition requires that $T(\xi)$ is time-like or null. Hence, $T(\xi)$ is null on the horizon and, therefore, also proportional to ξ. Using Einstein's equations again yields

$$\xi \wedge R(\xi) = 0 \quad \text{on } H[\xi].\tag{39}$$

Now one uses (38) to conclude from (39) that $\xi \wedge d\kappa$ vanishes on the horizon. Hence, for any vector field tangent to the horizon, τ^μ, say, one has $\tau^\mu \kappa_{,\mu} = 0$, which proves the zeroth law (36).

4.2   The Strong Rigidity Theorem

The power of the strong rigidity theorem lies in the fact that it relates the global concept of event horizons to the local notion of Killing horizons. Under certain conditions, the theorem asserts that the event horizon of a stationary black hole spacetime *is* a Killing horizon. Moreover, if the stationary Killing field, k, does not coincide with the horizon Killing field, ξ, then the rigidity theorem guarantees that spacetime admits at least one axial Killing field. Stationary electrovac black hole space-times are, therefore, *either non-rotating* (that is, $\xi = k$) *or axisymmetric*.

Among other assumptions, the proof of the rigidity theorem requires that spacetime is analytic, the fundamental matter fields obey well behaved hyperbolic equations and the stress-energy tensor fulfills the weak energy condition. Unfortunately, the analyticity assumption has, for instance, no justification if the domain of outer communications admits regions where the stationary Killing field becomes null or space-like. (See Chruściel 1996, Chruściel and Galloway 1996 for recent progress concerning the rigidity theorem.)

5   Staticity and Circularity

The integrability theorems for the Killing fields provide the link between the strong rigidity theorem and the assumptions on which the classical uniqueness theorems are based. In the non-rotating case one has to show that a stationary domain of outer communications (with Killing field k, say) is static,

$$\omega_k \equiv \frac{1}{2} * (k \wedge dk) = 0,\tag{40}$$

whereas in the stationary and axisymmetric situation (with Killing fields k and m, say) one must establish the circularity conditions:

$$(m,\omega_k) \equiv -\frac{1}{2} * (m \wedge k \wedge dk) = 0,\tag{41}$$

and similarly for $k \leftrightarrow m$. For vacuum space-times, the staticity and circularity theorems were proven by Carter (1973b). While Carter also succeeded in establishing the electrovac circularity theorem in the early seventies (see, e.g. Carter 1987), it took, however, some effort until the corresponding staticity issue was settled by Sudarsky and Wald (1992, 1993).

The task is to establish (40) and (41) by using the invariance properties of the matter fields with respect to the Killing symmetries. The link between the relevant components of the stress-energy tensor and the geometrical conditions (40) and (41) is provided by the Ricci identity for Killing fields, $\Delta \xi = -2R(\xi)$, and the Einstein equations (Exercise 11). By virtue of these, one obtains the useful relation $d\omega = 8\pi * [\xi \wedge T(\xi)]$ between the derivative of the twist (of an arbitrary Killing field ξ) and the stress-energy one-form $T(\xi)$ (Exercise 12). Hence, the differentiated Frobenius integrability conditions (40) and (41) assume the simple form

$$d\omega_k = 8\pi * [k \wedge T(k)] = 0,\qquad(42)$$

$$d(m, \omega_k) = -8\pi * [m \wedge k \wedge T(k)] = 0,\qquad(43)$$

and similarly for $k \leftrightarrow m$. [In order to obtain the second equation one uses $d(m, \omega_k) = di_m\omega_k = (L_m - i_md)\omega_k$, and the consequence $L_m\omega_k = 0$ of the commutation property of the Killing fields (Carter 1970). A powerful proof of $[k, m] = 0$ which does not require the existence of an axis was given by Szabados (1987).]

The above formulas show that the Frobenius conditions (40), (41) imply the properties (42), (43). The task is, however, to establish the *converse* direction. In the stationary and axisymmetric case, this is not too hard, since the vanishing of the one-form $d(m, \omega_k)$ implies that the *function* (m, ω_k) is constant. Since, in addition, (m, ω_k) vanishes on the rotation axis, the circularity condition (41) follows if the stress-energy tensor has the property (43).

In the non-rotating case the situation is more difficult, since $d\omega_k = 0$ does not automatically imply that the twist one-form itself vanishes. However, using the identity (Exercise 13)

$$d\left(\omega_k \wedge \frac{k}{\sigma}\right) = d\omega_k \wedge \frac{k}{\sigma} - 2\frac{(\omega_k, \omega_k)}{\sigma^2} * k,\qquad(44)$$

Stokes' theorem shows that $d\omega_k = 0$ does in fact imply $\omega_k = 0$, provided that the domain of outer communications is *strictly* stationary, that is, if $\sigma \equiv -(k, k) \geq 0$. In order to see this, one uses asymptotic flatness and the general properties of Killing horizons to conclude that $\omega_k \wedge \frac{k}{\sigma}$ vanishes at infinity and at (each component of) the horizon. This shows that a strictly stationary domain of outer communications with non-rotating Killing horizon is static, if $d\omega_k = 0$, that is, if $k \wedge T(k) = 0$. In particular, no restrictions concerning the connectedness of the horizon enter the above argument. (The original proof of the vacuum staticity theorem was based on the fact that $d\omega_k = 0$ implies the local existence of a potential, and was therefore subject to stronger topological restrictions.) Hence, under the (weak) assumptions used above, it is sufficient to establish the properties (42) and (43) in order to conclude that the Frobenius integrability conditions (40) and (41) are fulfilled.

In order to prove the *circularity* theorem for the EM system, it therefore remains to establish (43). The expression for the Maxwell stress-energy tensor yields

$$* [k \wedge T(k)] = \frac{1}{4\pi} (i_k F) \wedge (i_k * F) . \tag{45}$$

The symmetry condition $L_k A = dV$ (for some V) implies $-i_k F = -i_k dA = di_k A - L_k A = d(i_k A - V) \equiv d\phi$. Since $0 = *L_k F = L_k * F$, the Maxwell equation $d * F = 0$ implies $di_k * F = 0$, and hence $i_k * F = d\psi$, for some locally defined scalar magnetic potential ψ. Thus,

$$d\omega_k = 8\pi * [k \wedge T(k)] = -2 d\phi \wedge d\psi , \tag{46}$$

which enables one to introduce a generalized twist potential Y, defined by

$$dY = 2 (\omega_k + \phi d\psi - \psi d\phi) . \tag{47}$$

This is, of course, the same expression as we have already obtained earlier from the Kaluza-Klein decomposition of self-gravitating Maxwell fields [see (14), (20), (23) and (24)]. Since the potentials ϕ and ψ are invariant under the action of the axial Killing field m, $L_m \phi = L_m \psi = 0$, (46) yields the desired result:

$$d(m, \omega_k) = -i_m d\omega_k = 2i_m (d\phi \wedge d\psi) = 2 [(L_m \phi) d\psi - (L_m \psi) d\phi] = 0 .$$

It is worthwhile noticing that the circularity theorem does not hold for non-Abelian gauge fields. Although the twist potential, Y, can still be introduced, it is not possible to conclude from the Yang-Mills equations and the symmetry conditions that $d(m, \omega_k)$ vanishes: For gauge fields with arbitrary gauge groups one finds

$$d\omega_k = 2 d [\text{Tr} \{\phi i_k * F\}] , \tag{48}$$

and, with $d(m, \omega_k) = -i_m d\omega_k$,

$$d(m, \omega_k) = 2 d [\text{Tr} \{\phi (*F)(k, m)\}] , \tag{49}$$

which does not vanish automatically, unless the gauge field is Abelian. (Only in the Abelian case one has $(*F)(k, \cdot) = d\psi$, implying that $(*F)(k, m) = i_m d\psi = L_m \psi = 0$.)

As for the *staticity* theorem, we recall that the proof follows from Stokes' theorem and the identity (44), provided that $d\omega_k = 0$, i.e., $k \wedge T(k) = 0$ can be established from the matter equations. Whereas this is trivial in the vacuum case, the Maxwell equations and the symmetry conditions do not automatically imply that $k \wedge T(k)$ vanishes. Hence, what is needed is a more powerful identity, including the electromagnetic potentials. While identities of this kind have been constructed, they turned out to be useless to the staticity problem, since they do not yield semi-definite integrands. In fact, a proof of the electrovac staticity theorem along the lines presented in this section is still outstanding (see, e.g. Carter 1987 or Heusler 1996b). As already mentioned, the problem was solved by different means a couple of years ago (Sudarsky and Wald 1992, 1993).

6   Uniqueness of the Reissner-Nordström Metric

Since the staticity theorem holds for both vacuum and electrovac black hole space-times with non-rotating horizons, it remains to show that the Reissner-Nordström metric is the unique asymptotically flat black hole solution to the *static* EM equations. This was first achieved by Israel in 1967 (vacuum) and 1968 (electrovac), and later, using the positive energy theorem, by Simon (1985), Bunting and Masood-ul-Alam (1987), and Masood-ul-Alam (1992). Here we give a brief outline of Israel's original reasoning, where, for the sake of simplicity, we restrict ourselves to the vacuum case (Israel 1967).

6.1   The Israel Theorem

The celebrated Israel theorem establishes that all static black hole solutions of Einstein's vacuum equations (with non-degenerate horizon) are spherically symmetric. Israel (1967) was able to obtain this result by considering a particular foliation of the static three-dimensional hyper-surface Σ: Requiring that the metric function S is an admissible coordinate, the spacetime metric is written in the form

$$g = -S^2 \, dt \otimes dt + \rho^2 \, dS \otimes dS + \tilde{g}, \tag{50}$$

where both ρ and the metric $\tilde{g}$ depend on S and the coordinates of the two-dimensional surfaces with constant S. With respect to the tetrad fields $\theta^0 = S dt$ and $\theta^1 = \rho dS$ one finds (Exercise 14)

$$G_{00} + G_{11} = \frac{1}{\rho}\left[\frac{K}{S} - \frac{\partial K}{\partial S} - \frac{\rho}{2}K^2\right] - \frac{2\tilde{\Delta}\sqrt{\rho}}{\sqrt{\rho}} - \left[\frac{\left(\tilde{\nabla}\rho, \tilde{\nabla}\rho\right)}{2\rho^2} + \overset{\circ}{K}_{ab}\overset{\circ}{K}{}^{ab}\right], \tag{51}$$

$$G_{00} + 3G_{11} = \frac{1}{\rho}\left[3\frac{K}{S} - \frac{\partial K}{\partial S}\right] - \tilde{R} - \tilde{\Delta}\ln\rho - \left[\frac{\left(\tilde{\nabla}\rho, \tilde{\nabla}\rho\right)}{\rho^2} + 2\overset{\circ}{K}_{ab}\overset{\circ}{K}{}^{ab}\right], \tag{52}$$

where $K_{ab} = (2\rho)^{-1} \partial \tilde{g}_{ab} / \partial S$ is the extrinsic curvature of the embedded surface $S = $ constant in Σ, and $\overset{\circ}{K}_{ab} \equiv K_{ab} - \frac{1}{2}\tilde{g}_{ab}K$ is the trace-free part of K_{ab}. Using the "Poisson" equation

$$\frac{\partial \rho}{\partial S} = \rho^2 \left(K - \rho S R_{00} \right),\tag{53}$$

and the vacuum Einstein equations, $G_{00} = G_{11} = R_{00} = 0$, one obtains the following inequalities from (51) and (52):

$$\frac{\partial}{\partial S} \left(\frac{\sqrt{\tilde{g}}}{\sqrt{\rho}} \frac{K}{S} \right) \leq -2 \frac{\sqrt{\tilde{g}}}{S} \tilde{\Delta}\sqrt{\rho},\tag{54}$$

$$\frac{\partial}{\partial S} \left(\frac{\sqrt{\tilde{g}}}{\rho} [KS + \frac{4}{\rho}] \right) \leq - S \sqrt{\tilde{g}} (\tilde{\Delta} \ln \rho + \tilde{R}),\tag{55}$$

where equality holds if and only if $\overset{\circ}{K}_{ab} = \tilde{\nabla}\rho = 0$. The strategy is now to integrate the above estimates over a space-like hyper-surface extending from the horizon ($S = 0$) to a two-sphere at space-like infinity, S_∞^2 ($S = 1$). Using the Gauss-Bonnet theorem, one has $\int_\Sigma S\tilde{R}dS \wedge \tilde{\eta} = \int_0^1 SdS \int_S \tilde{R}\tilde{\eta} = \frac{1}{2}8\pi$. Since $\int_S (\tilde{\Delta}\sqrt{\rho})\tilde{\eta} = 0$, one finds

$$\left[\int_S \frac{K}{\sqrt{\rho}S} \tilde{\eta} \right]_0^1 \leq 0, \quad \left[\int_S \frac{KS + 4\rho^{-1}}{\rho} \tilde{\eta} \right]_0^1 \leq -4\pi.\tag{56}$$

In order to evaluate these inequalities at infinity ($S = 1$), one uses asymptotic flatness to conclude that

$$\rho^{-1} \to \frac{M}{r^2}, \quad K \to \frac{2}{r}, \quad \text{as } S \to 1,\tag{57}$$

(Exercise 14). At the horizon ($S = 0$), one takes advantage of the G_{11} vacuum equation and the fact that the curvature invariant $R_{\alpha\beta\gamma\delta}R^{\alpha\beta\gamma\delta}$ is required to remain finite. This yields (Exercise 14)

$$K_{ab} \to 0, \quad \frac{K}{S} \to \frac{1}{2}\rho\tilde{R}, \quad \text{as } S \to 0.\tag{58}$$

By virtue of the above limits one now obtains $8\pi\sqrt{M} - 4\pi\sqrt{\rho_H}$ for the evaluation of the first, and $-4\mathcal{A}\rho_H^{-2}$ for the second integrand in (56), where $\mathcal{A}$ denotes the area of the horizon. Hence, the inequalities (56) yield the estimates

$$M \leq \frac{\rho_H}{4}, \quad \frac{\mathcal{A}}{4\pi\rho_H} \geq \frac{\rho_H}{4},\tag{59}$$

where we recall that equality holds if and only if both $\overset{\circ}{K}_{ab}$ and $\tilde{\nabla}\rho$ vanish. Since the Komar expression for the total mass of a static *vacuum* black hole spacetime yields (Exercise 16)

$$M = \frac{\mathcal{A}}{4\pi\rho_H}, \qquad (60)$$

we conclude that equality must hold in the above estimates. [Also note that ρ_H^{-1} is the surface gravity of the horizon, $\rho_H^{-1} = \kappa$ (Exercise 15).] Equality in the estimates (56) implies equality in (54), (55), and hence

$$K_{ab} - \frac{1}{2}\tilde{g}_{ab}K = 0, \qquad \tilde{\nabla}\rho = 0. \qquad (61)$$

Using this in (51) and (53) shows that both ρ and K depend only on S. Equation (52) then implies that $\tilde{R}$ is constant on the surfaces of constant S. Explicitly one finds $\rho = 4c(1 - S^2)^{-2}$, $K = c^{-1}S(1 - S^2)$ and $\tilde{R} = \frac{1}{2}c^{-2}(1 - S^2)^2$, where c is a constant of integration. Hence, defining $r(S)$ by the relation $\tilde{R} = 2/r^2$, yields (with $c = M$)

$$S^2 = 1 - \frac{2M}{r}, \quad \rho^2 dS^2 = (1 - \frac{2M}{r})^{-1} Dr^2, \quad \tilde{g} = r^2 d\Omega^2, \qquad (62)$$

which is the familiar form of the Schwarzschild metric.

6.2 Uniqueness and the Positive Energy Theorem

The Uniqueness of the Reissner-Nordström metric was also established by Israel (1968), who was able to generalize the above ideas to the EM system (see also Müller zum Hagen et al 1973, 1974). A more recent proof, using the positive energy theorem – and avoiding the Israel slicing and the connectedness requirement for the horizon – was given by Simon (1985), Bunting and Masood-ul-Alam (1987), and Masood-ul-Alam (1992). The reader who is interested in this elegant and powerful approach to the static uniqueness theorem is referred to the original literature and to Heusler (1996b) for a brief outline of the ideas.

7 Uniqueness of the Kerr-Newman Metric

The circularity theorem implies that stationary and axisymmetric electrovac black hole space-times admit a foliation by two-surfaces orthogonal to the Killing fields k and m. This can be used to reduce the field equations (31) to a set of differential equations on a fixed, two-dimensional background. The Mazur identity then shows that the Kerr-Newman metric is the unique solution to the corresponding boundary value problem.

7.1   Reduction to a Boundary Value Problem

We have demonstrated that the EM equations in the presence of a Killing field describe a harmonic mapping into a coset space, effectively coupled to three-dimensional gravity. For the Kaluza-Klein reduction with respect to the axial Killing field $m = \partial_\varphi$ with norm X, say, the metric assumes the form (1),

$$g = X(\mathrm{d}\varphi + a) \otimes (\mathrm{d}\varphi + a) + \frac{1}{X}\bar{g}, \tag{63}$$

where now $\bar{g}$ is a pseudo-Riemannian three-metric. According to (31) the EM equations are

$$\bar{R}_{ij} = \frac{1}{4}\mathrm{Tr}\{J_i J_j\}, \quad \mathrm{d}\bar{\ast}J = 0, \quad \text{with } J \equiv \Phi^{-1}\mathrm{d}\Phi, \tag{64}$$

where the hermitian 3×3 matrix Φ comprises the electromagnetic potentials ϕ and ψ, and the gravitational potentials X and Y. We emphasize again that now all scalar potentials are defined with respect to the space-like Killing field, m. As already mentioned, this implies that the coset space becomes $SU(2,1)/S(U(2) \times U(1))$, which will be of crucial importance to the signs in the Mazur identity.

In the stationary and axisymmetric case under consideration, there exists, in addition to m, an asymptotically time-like Killing field k. Since k and m fulfill the Frobenius integrability conditions, the spacetime metric can be written in the Papapetrou (2+2)-split. For the Kaluza-Klein metric (63) this simply implies that $\bar{g}$ is a *static* pseudo-Riemannian three-metric, $\bar{g} = -\rho^2\mathrm{d}t^2 + \tilde{g}$, and that a is orthogonal to the two-dimensional Riemannian manifold $\tilde{\Sigma}$ with metric $\tilde{g} = \tilde{g}_{ab}\mathrm{d}x^a\mathrm{d}x^b$, i.e., $a = a_t\mathrm{d}t$. Moreover, as k and m are Killing fields, all quantities depend only on the two coordinates on $(\tilde{\Sigma}, \tilde{g})$. With respect to the resulting metric (Papapetrou 1953),

$$g = X\left(\mathrm{d}\varphi + a_t\mathrm{d}t\right)^2 + \frac{1}{X}\left(-\rho^2\mathrm{d}t^2 + \tilde{g}\right), \tag{65}$$

the Einstein-Maxwell equations (64) become a set of partial differential equations on the two-dimensional Riemannian manifold $(\tilde{\Sigma}, \tilde{g})$:

$$\tilde{\Delta}\rho = 0, \tag{66}$$

$$\tilde{R}_{ab} - \frac{1}{\rho}\tilde{\nabla}_b\tilde{\nabla}_a\rho = \frac{1}{4}\mathrm{Tr}\left\{J_a J_b\right\}, \tag{67}$$

$$\tilde{\nabla}^a\left(\rho J_a\right) = 0. \tag{68}$$

Here we have used the standard reduction of the Ricci tensor $\bar{R}$ with respect to the static three-metric $\bar{g} = -\rho^2\mathrm{d}t^2 + \tilde{g}$ (Exercise 17). We have also used the fact that J has no components orthogonal to $\tilde{\Sigma}$, implying $J_t = 0$ and $\bar{\ast}J = -\rho\mathrm{d}t\wedge\tilde{\ast}J$.

[This follows from $J = \Phi^{-1}\mathrm{d}\Phi$ and the circumstance that Φ is a matrix valued function on $(\bar{\Sigma}, \tilde{g})$.] It is worthwhile noticing that the sigma-model equations (68) are a consequence of the the Bianchi identity for the Einstein tensor of the metric $\tilde{g}$ and the remaining equations, (66) and (67); see Exercise 18.

The last simplification of the field equations is obtained from the fact that ρ can be chosen as one of the coordinates on $(\bar{\Sigma}, \tilde{g})$. This follows from the fact that ρ is harmonic (with respect to the Riemannian two-metric $\tilde{g}$) and non-negative, and that the domain of outer communications of a stationary black hole is simply connected (Chruściel and Wald 1994b). The coordinates ρ and z are called Weyl coordinates, where z is the conjugate harmonic function. Using Morse theory (see, e.g. Milnor 1963), Carter (1973b) was able to exclude critical points of ρ in the domain of outer communications, $\rho \geq 0$. A more recent, and very elegant proof for the existence of Weyl coordinates was given by Weinstein (1990), taking advantage of the Riemann mapping theorem (or, more precisely, Caratheodory's extension of it (see, e.g. Behnke and Sommer 1976).

Since $\tilde{g}$ can be chosen to be conformally flat with respect to Weyl coordinates, we end up with the metric

$$g = -\frac{\rho^2}{X}\mathrm{d}t^2 + X\left(\mathrm{d}\varphi + a_t \mathrm{d}t\right)^2 + \frac{1}{X}e^{2h}\left(\mathrm{d}\rho^2 + \mathrm{d}z^2\right), \tag{69}$$

the sigma-model equations

$$\left(\rho\, J_\rho\right)_{,\rho} + \left(\rho\, J_z\right)_{,z} = 0, \tag{70}$$

and the remaining field equations (Exercise 19)

$$h_{,\rho} = \frac{\rho}{2}\operatorname{Tr}\left\{J_\rho J_\rho - J_z J_z\right\}, \quad h_{,z} = \rho\operatorname{Tr}\left\{J_\rho J_z\right\}, \tag{71}$$

for the function $h(\rho, z)$. It is not hard to verify that (70) is the integrability condition for (71). [This is, of course, a consequence of the Bianchi identity, as discussed above.] Since (68) is conformally invariant, the metric function $h(\rho, z)$ does not appear in (70). Therefore, the stationary and axisymmetric EM equations reduce to a boundary value problem for the matrix Φ on a fixed, two-dimensional background. Once the solution of (70) is found, the remaining metric function $h(\rho, z)$ is obtained from (71) by quadrature.

7.2   The Ernst Equations and the Kerr-Newman Solution

In order to derive the Kerr-Newman metric, and to prove its uniqueness, it is sufficient to consider the non-linear partial differential equation (70) for the matrix Φ. (Recall that $J = \Phi^{-1}\mathrm{d}\Phi$.) Writing out the components of this equation yields the Ernst equations and a set of additional, redundant differential relations between the potentials. [The latter are, nevertheless, very useful, since their integration over a space-like hyper-surface yields interesting relations between the total mass, the angular momentum and the electric charge (Heusler 1997b).]

As it is rather tedious, although straightforward, to write out the components of (70) in terms of the Ernst potentials $\mathcal{E}$ and Λ, one better uses (27) to obtain the explicit form of the Ernst equations. Using $\bar{g} = -\rho^2 dt^2 + \tilde{g}$, one has $\bar{\Delta}\mathcal{E} = \tilde{\Delta}\mathcal{E} + \rho^{-1}\langle d\rho, d\mathcal{E}\rangle$, where $\langle\,,\,\rangle$ now refers to the Riemannian two-metric $\tilde{g}$. Equations (27) now yield the following conformally invariant equations on $(\tilde{\Sigma}, \tilde{g})$ (Exercise 20):

$$\tilde{\Delta}\varepsilon + \frac{\langle d\rho, d\varepsilon\rangle}{\rho} + \frac{2\langle d\varepsilon, \bar{\varepsilon}d\varepsilon + \bar{\lambda}d\lambda\rangle}{1 - |\varepsilon|^2 - |\lambda|^2} = 0, \tag{72}$$

$$\tilde{\Delta}\lambda + \frac{\langle d\rho, d\lambda\rangle}{\rho} + \frac{2\langle d\lambda, \bar{\varepsilon}d\varepsilon + \bar{\lambda}d\lambda\rangle}{1 - |\varepsilon|^2 - |\lambda|^2} = 0, \tag{73}$$

where the Laplacian $\tilde{\Delta}$ refers to $\tilde{g}$. The complex potentials ε and λ are defined by

$$\varepsilon \equiv \frac{1-\mathcal{E}}{1+\mathcal{E}}, \quad \lambda \equiv \frac{\Lambda}{1+\mathcal{E}}. \tag{74}$$

(Note that this is not the same definition of ε as in the vacuum case (11), since there we have considered the dimensional reduction with respect to the time-like Killing field.)

In order to control the boundary conditions, it is convenient to introduce prolate spheroidal coordinates, x and y, defined in terms of ρ and z by

$$\rho^2 = \mu^2 (x^2 - 1)(1 - y^2), \quad z = \mu xy, \tag{75}$$

where μ is a constant. The domain of outer communications, that is, the upper half-plane $\rho \geq 0$, corresponds to the semi-strip $\{(x,y)|x \geq 1, |y| \leq 1\}$. The boundary $\rho = 0$ consists of the horizon ($x = 0$) and the northern ($y = 1$) and southern ($y = -1$) segments of the rotation axis. In terms of x and y, the Riemannian metric $\tilde{g}$ becomes

$$\tilde{g} = e^{2h} (d\rho^2 + dz^2) = e^{2h}\mu^2 (x^2 - y^2) \left(\frac{dx^2}{x^2 - 1} + \frac{dy^2}{1 - y^2}\right). \tag{76}$$

The Ernst equations (see Exercise 20) become

$$[(x^2 - 1)\varepsilon_{,x}]_{,x} + [(1 - y^2)\varepsilon_{,y}]_{,y}$$

$$= -2\frac{(x^2 - 1)\varepsilon_{,x}(\bar{\varepsilon}\varepsilon_{,x} + \bar{\lambda}\lambda_{,x}) + (1 - y^2)\varepsilon_{,y}(\bar{\varepsilon}\varepsilon_{,y} + \bar{\lambda}\lambda_{,y})}{1 - |\varepsilon|^2 - |\lambda|^2}, \tag{77}$$

and similarly for $\varepsilon \leftrightarrow \lambda$. They admit the simple solution

$$\varepsilon = px + i\,qy, \quad \lambda = \lambda_0, \quad \text{where } p^2 + q^2 + \lambda_0^2 = 1, \tag{78}$$

with real constants p, q and λ_0.

The norm X, the twist potential Y and the electromagnetic potentials ϕ and ψ (all defined with respect to the axial Killing field) are obtained from the above solution by using the definitions (74) and (25). The off-diagonal element of the metric, $a = a_t dt$ [see (63)] is then computed by integrating the twist equation (20). [Note that the Hodge dual in (20) now refers to the decomposition (63) with respect to the axial Killing field.] Finally, the metric function h is obtained from the Ernst potentials by integrating the equations (71).

The solution derived in the above way is the "conjugate" of the Kerr-Newman metric. In order to obtain the Kerr-Newman solution itself, one has to consider the quantities $\hat{X}$, $\hat{a}_t$ and $\hat{h}$, defined by

$$\hat{X} = X\left(a_t^2 - \frac{\rho^2}{X^2}\right), \quad \hat{a}_t = -a_t\left(a_t^2 - \frac{\rho^2}{X^2}\right)^{-1}, \quad e^{2\hat{h}} = e^{2h}\frac{\hat{X}}{X}. \quad (79)$$

Since g is invariant under the simultaneous transformation $(X, a_t, h) \rightarrow (\hat{X}, \hat{a}_t, \hat{h})$ and $t \rightarrow \hat{t} = \varphi$, $\varphi \rightarrow \hat{\varphi} = -t$, the quantities $\hat{X}$, $\hat{a}_t$ and $\hat{h}$ are indeed solutions to the field equations. This additional step in the derivation of the Kerr-Newman metric is necessary because the Ernst potentials were defined with respect to the axial Killing field, ∂_φ. If, on the other hand, one uses the stationary Killing field, ∂_t, then the Ernst equations become singular at the boundary of the ergo-sphere, that is, the region outside the horizon where ∂_t is space-like. (A discussion of this point and a derivation of the Kerr-Newman solution can, for instance, be found in Heusler 1996b, Chaps. 4 & 5.)

In terms of Boyer-Lindquist coordinates,

$$r = m\left(1 + px\right), \quad \cos\vartheta = y, \quad (80)$$

one eventually finds the familiar form of the Kerr-Newman metric:

$$g = \frac{1}{\Xi}\left[-(\Delta - \alpha^2 \sin^2\vartheta)\, dt^2 + 2\alpha \sin^2\vartheta\, (\Delta - (r^2 + \alpha^2))\, dt\, d\varphi\right.$$
$$\left. + \sin^2\vartheta\, ((r^2 + \alpha^2)^2 - \Delta\alpha^2 \sin^2\vartheta)\, d\varphi^2\right] + \Xi\left[\frac{1}{\Delta}dr^2 + d\vartheta^2\right], \quad (81)$$

where α is defined by $a_t \equiv \alpha \sin^2\vartheta$. The expressions for the quantities Δ and Ξ show that the Kerr-Newman metric is parametrized in terms of the total mass M, the electric charge Q, and the angular momentum $J = \alpha M$:

$$\Delta = r^2 - 2Mr + \alpha^2 + Q^2, \quad \Xi = r^2 + \alpha^2 \cos^2\vartheta. \quad (82)$$

Finally, the electromagnetic vector potential, A, becomes

$$A = \frac{Q}{\Xi}r\{dt - \alpha \sin^2\vartheta\, d\varphi\}. \quad (83)$$

7.3   The Uniqueness Proof

The uniqueness of the Kerr-Newman metric is obtained by integrating the Mazur identity (34) and using Stokes' theorem. Since the components of Φ are functions on the two-dimensional Riemannian manifold $(\tilde{\Sigma}, \tilde{g})$, the Mazur identity (34) with respect to the pseudo-Riemannian metric $\bar{g} = -\rho^2 dt^2 + \tilde{g}$ becomes

$$\text{Tr}\left\{ \tilde{\nabla}\left(\rho \tilde{\nabla}\Psi \right) \right\} = \rho \, \text{Tr}\langle \Phi_1^{-1} J_{\triangle}^{\dagger} , \, \Phi_2 J_{\triangle}\rangle + \text{Tr}\left\{ \Phi_2 \, \tilde{\nabla}\left(\rho J_{\triangle}\right) \Phi_1^{-1}\right\}, \qquad (84)$$

where we recall that Ψ is the relative difference of two configurations, $\Psi \equiv \Phi_2 \Phi_1^{-1} - \mathbb{1}$, and $J_{\triangle}$ denotes the difference of their currents, $J_{\triangle} \equiv J_2 - J_1$. (Here we have again used $\langle\ ,\ \rangle$ for the inner product with respect to the two-dimensional metric $\tilde{g}$.)

Let us now assume that Φ_1 and Φ_2 are two *solutions* of the field equations (68), $\tilde{\nabla}(\rho J_1) = \tilde{\nabla}(\rho J_2) = 0$. Then the last term on the RHS of the Mazur identity vanishes, and Stokes' theorem yields

$$\int_{\partial S} \rho \, \ast\!\text{d}\, (\text{Tr}\,\Psi) = \int_{S} \rho \, \text{Tr}\,\langle \Phi_1^{-1} J_{\triangle}^{\dagger} , \, \Phi_2 J_{\triangle}\rangle\, \bar{\eta} \geq 0, \qquad (85)$$

where S is the semi-strip $\{(x,y)|x \geq 1, |y| \leq 1\}$. The important observation consists in the fact that the integrand on the RHS is non-negative: First, the inner product is definite, since $\tilde{g}$ is a Riemannian metric. Second, the factor ρ is non-negative in S, since S is the image of the upper half-plane, $\rho \geq 0$. Third, the current $J_{\triangle}$ is space-like, since the matrices Φ depend only on the coordinates of $(\tilde{\Sigma}, \tilde{g})$. Last, the hermitian matrices Φ are positive, since the embedding of the symmetric space $SU(p,q)/S(U(p) \times U(q))$ in $SU(p,q)$ can be represented in the form $gg^{\dagger}$. (See, e.g. Eichenherr and Forger 1980, Boothby 1975, and Kobayashi and Nomizu 1969 for this point.) Hence,

$$\text{Tr}\,\langle \Phi_1^{-1} J_{\triangle}^{\dagger} , \, \Phi_2 J_{\triangle}\rangle = \text{Tr}\,\langle \mathcal{M} , \, \mathcal{M}^{\dagger}\rangle \geq 0, \qquad (86)$$

where $\mathcal{M} \equiv g_1^{-1} J_{\triangle}^{\dagger} g_2$. We therefore conclude that two solutions, Φ_1 and Φ_2, of the field equations (68) are identical in the semi-strip S, if $\rho\, \text{d}(\text{Tr}\,\Psi) = 0$ on the boundary ∂S, and if Φ_1 and Φ_2 coincide at least in one point.

In order to prove the uniqueness of the Kerr-Newman solution, it remains to show that

$$\rho \, d(\text{Tr}\,\Psi) = 0 \quad \text{on } \partial S, \qquad (87)$$

provided that Φ_1 and Φ_2 are two solutions of the field equations with the *same mass, angular momentum and electric charge*. In order to establish this, one needs the general asymptotic behavior and the regularity conditions on the horizon and the rotation axis of the norm X, the twist potential Y and the electromagnetic potential $\Lambda = -\phi + i\psi$. Using the formula (Exercise 21)

$$\text{Tr}\,\Psi = \frac{(\triangle X)^2 + |\triangle \Lambda|^2 \left[|\triangle \Lambda|^2 + 2(X_1 + X_2)\right] + \left[\triangle Y + i\,(\Lambda_1 \bar{\Lambda}_2 - \Lambda_2 \bar{\Lambda}_1)\right]^2}{X_1\, X_2}$$

$$(88)$$

we shall now argue that the boundary condition (87) is indeed satisfied. (Here $\triangle X \equiv X_2 - X_1$, etc.)

Let us start by considering the asymptotic behavior. In a stationary and axisymmetric spacetime, the norm X, the twist Y and the electromagnetic potential Λ (all defined with respect to the axial Killing field) behave like

$$X = (1 - y^2) \left[\mu^2 x^2 + \mathcal{O}(x)\right], \quad Y = 2 J y (3 - y^2) + \mathcal{O}(x^{-1}),$$

$$\Lambda = -(1 - y^2)\frac{Q J}{M} \left[(\mu x)^{-1} + \mathcal{O}(x^{-2})\right] + i Q \left[y + \mathcal{O}(x^{-2})\right], \qquad (89)$$

where M, Q and J are the total mass, the electric charge and the total angular momentum, respectively, and $\mu \equiv M^2 - (J/M)^2 - Q^2$. For $x \to \infty$, the leading contribution in the numerator of (88) is $(X_2 - X_1)^2 = \mathcal{O}(x^2)$, whereas the leading term in the denominator is $(1 - y^2)^2 \mu^4 x^4$. (Note that $\mu_1 = \mu_2$, since Φ_1 and Φ_2 are required to be solutions with the same set of asymptotic charges.) Now using $\rho = \mathcal{O}(x)$ we find $\rho \, d(\text{Tr}\, \Psi) \to 0$ as $x \to \infty$.

On the horizon, that is for $x = 1$, the potentials must behave regularly,

$$X = \mathcal{O}(1), \quad X^{-1} = \mathcal{O}(1), \quad Y_{,x} = \mathcal{O}(1), \quad Y_{,y} = \mathcal{O}(1),$$

$$\Lambda_{,x} = \mathcal{O}(1), \quad \Lambda_{,y} = \mathcal{O}(1). \qquad (90)$$

This implies that $d(\text{Tr}\, \Psi)$ remains finite for $x = 1$ which, together with $\rho = 0$, enables one to conclude that $\rho \, d(\text{Tr}\, \Psi)$ vanishes on the horizon.

Finally, in the vicinity of the rotation axis, that is for $y \to \pm 1$, regularity requires

$$X = \mathcal{O}(1 - y^2), \quad (1 - y^2) X^{-1} X_{,y} = \mp 2 + \mathcal{O}(1 - y^2),$$

$$Y = \pm 4 J + \mathcal{O}(1 - y^2), \quad Y_{,x} = \mathcal{O}((1 - y^2)^2),$$

$$\Lambda = \pm i Q + \mathcal{O}(1 - y^2), \quad \Lambda_{,x} = \mathcal{O}(1 - y^2). \qquad (91)$$

Using this, one can show that X, Y and Λ are completely determined by the asymptotic conditions (89) on the *entire* axis: One finds $X = 0$, $Y = \pm 4J$ and $\Lambda = \pm iQ$. The differences $X_2 - X_1$, $Y_2 - Y_1$ and $\Lambda_2 - \Lambda_1$ are, therefore, of $\mathcal{O}(1 - y^2)$ for $y \to \pm 1$, and so is X. Hence, $d(\text{Tr}\, \Psi)$ remains finite on the axis, implying that $\rho \, d(\text{Tr}\, \Psi)$ vanishes for $y = \pm 1$.

This concludes the proof of the stationary and axisymmetric uniqueness theorem, due to Robinson (1975) in the vacuum case, and to Mazur (1982) and Bunting (1983) for electrovac space-times. The theorem establishes that the Kerr-Newman metric (81) with vector potential (83) and parameters M, $\alpha = J/M$ and Q is the only electrovac black hole solution with $M^2 > \alpha^2 + Q^2$, vanishing magnetic charge, non-degenerate event horizon and stationary and axisymmetric, asymptotically flat domain of outer communications.

Bibliography

[1] Bardeen, J.M., Carter, B., Hawking, S.W. (1973): The Four Laws of Black Hole Mechanics. Commun. Math. Phys. **31**, 161–170

[2] Behnke, H., Sommer, F. (1976): *Theorie der Analytischen Funktionen einer Komplexen Veränderlichen*, (Springer, Berlin)

[3] Bizon, P. (1990): Colored Black Holes. Phys. Rev. Lett. **64**, 2844–2847

[4] Boothby, W.M. (1975): *An Introduction to Differentiable Manifolds and Riemannian Geometry*, (Academic Press, New York)

[5] Breitenlohner, P., Forgács, P., Maison, D. (1992): Gravitating Monopole Solutions. Nucl. Phys. B **383**, 357–376

[6] Breitenlohner, P., Maison, D., Gibbons, G. (1988): 4-Dimensional Black Holes from Kaluza-Klein Theories. Commun. Math. Phys. **120**, 295–334

[7] Brodbeck, O., Heusler, M. (1997): Stationary Perturbations and Infinitesimal Rotations of Static Einstein-Yang-Mills Configurations with Bosonic Matter. Phys. Rev. D **56**, 6278–6283

[8] Brodbeck, O., Heusler, M., Straumann, N., Volkov, M. (1997): Rotating Solitons and Non-rotating, Non-static Black Holes. Phys. Rev. Lett. **79**, 4310–4313

[9] Bunting, G.L. (1983): *Proof of the Uniqueness Conjecture for Black Holes*, (PhD Thesis, Univ. of New England, Armidale, N.S.W.)

[10] Bunting, G.L., Masood-ul-Alam, A.K.M. (1987): Nonexistence of Multiple Black Holes in Asymptotically Euclidean Static Vacuum Space-Times. Gen. Rel. Grav. **19**, 147–154

[11] Carter, B. (1969): Killing Horizons and Orthogonally Transitive Groups in Space-Time. J. Math. Phys. **10**, 70–81

[12] Carter, B. (1970): The Commutation Property of a Stationary, Axisymmetric System. Commun. Math. Phys. **17**, 233–238

[13] Carter, B. (1971): Axisymmetric Black Hole has only Two Degrees of Freedom. Phys. Rev. Lett. **26**, 331–332.

[14] Carter, B. (1973a): Rigidity of a Black Hole. Nature (Phys. Sci.) **238**, 71–72

[15] Carter, B. (1973b): in *Black Holes*, eds. C. DeWitt & B.S. DeWitt (Gordon & Breach, New York)

[16] Carter, B. (1987): in *Gravitation in Astrophysics*, eds. B. Carter & J.B. Hartle (Plenum, New York)

[17] Chruściel, P.T. (1994): in *Differential Geometry and Mathematical Physics*, eds. J. Beem & K.L. Duggal (Am. Math. Soc., Providence)

[18] Chruściel, P.T. (1996a): Uniqueness of Stationary, Electro-Vacuum Black Holes Revisited. Helv. Phys. Acta **69**, 529–552

[19] Chruściel, P.T. (1996b): On Rigidity of Analytic Black Holes. gr-qc/**9610011**

[20] Chruściel, P.T., Galloway, G.J. (1996): Nowhere Differentiable Horizons. gr-qc/**9611032**

[21] Chruściel, P.T., Nadirashvili, N.S. (1995): All Electrovac Majumdar–Papapetrou Space-Times with Non-Singular Black Holes. Class. Quantum Grav. **12**, L17–L23

[22] Chruściel, P.T., Wald, R.M. (1994a): Maximal Hypersurfaces in Stationary Asymptotically Flat Space-Times. Commun. Math. Phys. **163**, 561–604

[23] Chruściel, P.T., Wald, R.M. (1994b): On the Topology of Stationary Black Holes. Class. Quantum Grav. **11**, L147–L152

[24] Clément, G., Gal'tsov, D.V. (1996): Stationary BPS Solutions to Dilaton-Axion Gravity. Phys. Rev. D **54**, 6136–6152

[25] Droz, S., Heusler, M., Straumann, N. (1991): New Black Hole Solutions with Hair. Phys. Lett. B **268**, 371–376

[26] Ehlers, J. (1959): *Les Theories Relativistes de la Gravitation*, (CNRS, Paris)

[27] Eichenherr, H., Forger, M. (1980): More about Non-Linear Sigma-Models on Symmetric Spaces. Nucl. Phys. B **164**, 528–535

[28] Ernst, F.J. (1968): New Formulation of the Axially Symmetric Gravitational Field Problem (I & II). Phys. Rev. **167**, 1175–1178; Phys. Rev. **168**, 1415–1417

[29] Forgács, P., Manton, N.S. (1980), Space–Time Symmetries in Gauge Theories. Commun. Math. Phys. **72**, 15–35

[30] Galloway, G.J. (1996): A "Finite Infinity" Version of Topological Censorship. Class. Quantum Grav. **13**, 1471–1478

[31] Gal'tsov, D.V., Letelier, P.S. (1997): Interpolating Black Holes in Dilaton-Axion Gravity. Class. Quantum Grav. **14**, L9–L14

[32] Geroch, R.P. (1971): A Method for Generating Solutions of Einstein's Equations. J. Math. Phys. **12**, 918–924

[33] Geroch, R.P. (1972): A Method for Generating New Solutions of Einstein's Equations. J. Math. Phys. **13**, 394–404

[34] Greene, B.R., Mathur, S.D., O'Neill, C.M. (1993): Eluding the No-Hair Conjecture: Black Holes in Spontaneously Broken Gauge Theories. Phys. Rev. D **47**, 2242–22259

[35] Hawking, S.W. (1972): Black Holes in General Relativity. Commun. Math. Phys. **25**, 152–166

[36] Hawking, S.W., Ellis, G.F.R. (1973): *The Large Scale Structure of Space Time*, (Cambridge Univ. Press, Cambridge)

[37] Hajicek, P. (1973): General Theory of Vacuum Ergo-spheres. Phys. Rev. D **7**, 2311–2316

[38] Hajicek, P. (1975): Stationary Electrovac Space-Times with Bifurcate Horizon. J. Math. Phys. **16**, 518–527

[39] Heusler, M. (1993): Staticity and Uniqueness of Multiple Black Hole Solutions of Sigma Models. Class. Quantum Grav. **10**, 791–799

[40] Heusler, M. (1995): The Uniqueness Theorem for Rotating Black Hole Solutions of Self–gravitating Harmonic Mappings. Class. Quantum Grav. **12**, 2021–2035

[41] Heusler, M. (1996a): No-Hair Theorems and Black Holes with Hair. Helv. Phys. Acta **69**, 501–528

[42] Heusler, M. (1996b): *Black Hole Uniqueness Theorems*, (CLN in Physics, Cambridge Univ. Press, Cambridge)

[43] Heusler, M. (1997a): On the Uniqueness of the Papapetrou-Majumdar Metric. Class. Quantum Grav. **14**, L129–L134

[44] Heusler, M. (1997b): Bogomolnyi Type Equations for a Class of Nonrotating Black Holes. Phys. Rev. D **56**, 961–973

[45] Heusler, M., Straumann, N. (1993a): The First Law of Black Hole Physics for a Class of Non–Linear Matter Models. Class. Quantum Grav. **10**, 1299–1321

[46] Heusler, M., Straumann, N. (1993b): Mass Variation Formulae for Einstein–Yang–Mills–Higgs and Einstein–Dilaton Black Holes. Phys. Lett. B **315**, 55–66

[47] Heusler, M., Droz, S., Straumann, N. (1991): Stability Analysis of Self-Gravitating Skyrmions. Phys. Lett. B **271**, 61–67

[48] Heusler, M., Droz, S., Straumann, N. (1992): Linear Stability of Einstein-Skyrme Black Holes. Phys. Lett. B **285**, 21–26

[49] Heusler, M., Straumann, N., Zhou, Z-H. (1993): Self-Gravitating Solutions of the Skyrme Model and their Stability. Helv. Phys. Acta **66**, 614–632

[50] Israel, W. (1967): Event Horizons in Static Vacuum Space-Times. Phys. Rev. **164**, 1776–1779

[51] Israel, W. (1968): Event Horizons in Static Electrovac Space-Times. Commun. Math. Phys. **8**, 245–260

[52] Israel, W. (1987): in *300 Years of Gravitation*, eds. S.W. Hawking & W. Israel (Cambridge Univ. Press, Cambridge)

[53] Iyer, V., Wald, R.M. (1994): Some Properties of Noether Charge and a Proposal for Dynamical Black Hole Entropy. Phys. Rev. D **50**, 846–864

[54] Kay, B.S., Wald, R.M. (1991): Theorems on the Uniqueness and Thermal Properties of Stationary, Nonsingular, Quasifree States on Space-Times with a Bifurcate Horizon. Phys. Rep. **207**, 49–136

[55] Kinnersley, W. (1973): Generation of Stationary Einstein-Maxwell Fields. J. Math. Phys. **14**, 651–653

[56] Kinnersley, W. (1977): Symmetries of the Stationary Einstein-Maxwell Field Equations (I). J. Math. Phys. **18**, 1529–1537

[57] Kinnersley, W., Chitre, D.M. (1977): Symmetries of the Stationary Einstein-Maxwell Field Equations (II). J. Math. Phys. **18**, 1538–1542

[58] Kinnersley, W., Chitre, D.M. (1978): Symmetries of the Stationary Einstein-Maxwell Field Equations (III & IV). J. Math. Phys. **19**, 1926–1931; J. Math. Phys. **19**, 2037–2042.

[59] Kleihaus, B., Kunz, J. (1997): Static Black Hole Solutions with Axial Symmetry. Phys. Rev. Lett. **79**, 1595–1598

[60] Kobayashi, S., Nomizu, K. (1969): *Foundations of Differential Geometry*, (Interscience Publishers, New York)

[61] Künzle, H.P., Masood-ul-Alam, A.K.M. (1990): Spherically Symmetric Static SU(2) Einstein-Yang-Mills Fields. J. Math. Phys. **31**, 928–935

[62] Kundt, W., Trümper, M. (1966): Ann. Physik **192**, 414–418

[63] Lavrelashvili, G., Maison, D. (1993): Regular and Black Hole Solutions of Einstein-Yang-Mills-Dilaton Theory. Nucl. Phys. B **410**, 407–422

[64] Lichnerowicz, A. (1955): *Théories Relativistes de la Gravitation et de l'Electromagnétisme*, (Masson, Paris)

[65] Majumdar, S.D. (1947): A Class of Exact Solutions of Einstein's Field Equations. Phys. Rev. **72**, 390–398

[66] Masood-ul-Alam, A.K.M. (1992): Uniqueness Proof of Static Black Holes Revisited. Class. Quantum Grav. **9**, L53–L55

[67] Mazur, P.O. (1982): Proof of Uniqueness of the Kerr-Newman Black Hole Solution.
J. Phys. A: Math. Gen. **15**, 3173–3180

[68] Mazur, P.O. (1984a): Black Hole Uniqueness from a Hidden Symmetry of Einstein's Gravity. Gen. Rel. Grav. **16**, 211–215

[69] Mazur, P.O. (1984b): A Global Identity for Nonlinear Sigma-Models. Phys. Lett. A **100**, 341–344

[70] Milnor, J. (1963): *Morse Theory*, (Princeton Univ. Press, Princeton)

[71] Müller zum Hagen, H, Robinson, D.C., Seifert, H.J. (1973): Black Holes in Static Vacuum Space-Times. Gen. Rel. Grav. **4**, 53–78

[72] Müller zum Hagen, H, Robinson, D.C., Seifert, H.J. (1974): Black Holes in Static Electrovac Space–Times. Gen. Rel. Grav. **5**, 61–72

[73] Neugebauer, G., Kramer, D. (1969): Eine Methode zur Konstruktion stationärer Einstein-Maxwell-Felder. Ann. Physik (Leipzig) **24**, 62–71

[74] Papapetrou, A. (1945): A Static Solution of the Gravitational Field for an Arbitrary Charge-Distribution. Proc. Roy. Irish Acad. **51**, 191–204

[75] Papapetrou, A. (1953): Eine Rotationssymmetrische Lösung in der Allgemeinen Relativitätstheorie. Ann. Physik **12**, 309–315

[76] Rácz, I., Wald, R. M. (1992): Extension of Space-Times with Killing Horizons. Class. Quantum Grav. **9**, 2643–2656

[77] Rácz, I., Wald, R. M. (1996), Global Extensions of Space-Times Describing Asymptotic Finite States of Black Holes. Class. Quantum Grav. **13**, 539–552

[78] Ridgway, S.A., Weinberg, E.J. (1995): Static Black Hole Solutions without Rotational Symmetry. Phys. Rev. D **52**, 3440–3456

[79] Robinson, D.C. (1974): Classification of Black Holes with Electromagnetic Fields. Phys. Rev. **10**, 458–460

[80] Robinson, D.C. (1975): Uniqueness of the Kerr Black Hole. Phys. Rev. Lett. **34**, 905–906

[81] Robinson, D.C. (1977): A Simple Proof of the Generalization of Israel's Theorem. Gen. Rel. Grav. **8**, 695–698

[82] Schoen, R., Yau, S.-T. (1979): On the Proof of the Positive Mass Conjecture in General Relativity. Commun. Math. Phys. **65**, 45–76

[83] Schoen, R., Yau, S.-T. (1981): Proof of the Positive Mass Theorem. Commun. Math. Phys. **79**, 231–260

[84] Simon, W. (1985): A Simple Proof of the Generalized Israel Theorem. Gen. Rel. Grav. **17**, 761–768.

[85] Sudarsky, D., Wald, R.M. (1992): Extrema of Mass, Stationarity and Staticity, and Solutions to the Einstein-Yang-Mills Equations. Phys. Rev. D **46**, 1453–1474

[86] Sudarsky, D., Wald, R.M. (1993): Mass Formulas for Stationary Einstein-Yang-Mills Black Holes and a Simple Proof of Two Staticity Theorems. Phys. Rev. D **47**, R5209–R5231

[87] Szabados, L.B. (1987): Commutation Properties of Cyclic and Null Killing Symmetries. J. Math. Phys. **28**, 2688–2691

[88] Volkov, M.S., Gal'tsov, D.V. (1989): Non-Abelian Einstein-Yang-Mills Black Holes. JETP Lett. **50**, 346–350

[89] Wald, R.M. (1984): *General Relativity*, (Univ. of Chicago Press, Chicago)

[90] Weinstein, G. (1990): On Rotating Black Holes in Equilibrium in General Relativity. Commun. Pure Appl. Math. **43**, 903–948

[91] Witten, E. (1981): A New Proof of the Positive Energy Theorem. Commun. Math. Phys. **80**, 381–402

Black Hole Hair: A Review

Darío Núñez, Hernando Quevedo, and Daniel Sudarsky

Instituto de Ciencias Nucleares, Universidad Nacional Autónoma de México,
Apartado Postal 70–543, México D.F. 04510, Mexico

Abstract. The issue of black hole hair (*i.e.*, the feasibility of stationary black hole solutions not completely specified by the conserved charges defined at asymptotic infinity)
has received renewed attention in the last few years due, in part, to the unexpected
discovery of such solutions in various theories. In the present work, we give a brief
review of these developments emphasising in the point of view of the authors.

1   Introduction

As it is well known, for vacuum spacetimes, or for those whose stress energy
tensor is described by the electro-magnetic one, the variety of stationary black
hole solutions of Einstein's equations is very small [1]. Such solutions are completely characterized by three parameters (M_{ADM} = ADM mass, J = angular
momentum, Q = electric (or magnetic) charge) all of which are associated with
general conservation laws defined at asymptotic infinity, i°. That is, it is enough
to "measure" those parameters at large distances from the black hole in order to
determine, with infinite accuracy, the stationary black hole we are dealing with
(although, in practice, it is not clear how we would know for certain that we are
facing such a type of black hole).

Moreover, by the time that a proof for such a complete characterization of
the black hole was practically established, it was possible to show [2], [3] that,
with the introduction of other types of fields, new kinds of stationary black holes
do not appear. Probably the most important of such results, by its simplicity
and generality, is the demonstration, due to Bekenstein [2], showing that the
introduction of a scalar field with "convex" potential does not produce new
types of black holes. Clearly, the scalar field ϕ has to satisfy the Klein–Gordon
type equation, the relativistic equation for a scalar field, in the corresponding
spacetime:

$$\nabla_a \nabla^a \phi - \frac{\partial V}{\partial \phi} = 0, \tag{1}$$

where $V = V(\phi)$ is the scalar potential of the corresponding Lagrangian and
$a = 0, 1, 2, 3$ are spacetime indices.

The spacetime is taken such that it corresponds to a stationary black hole
with a bifurcate Killing horizon (see, Wald [4]). Let ξ^a be the stationary Killing
field and t the Killing parameter that we take as one of the coordinates. We
foliate the spacetime by means of spatial hypersurfaces Σ_t (with t = const.)
such that they intersect themselves at the bifurcation surface S.

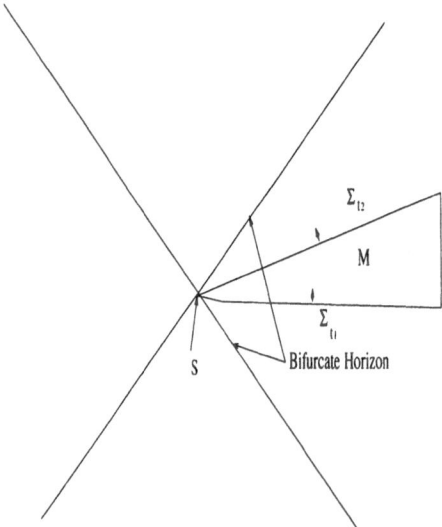

Fig. 1 Spacetime diagram for an hypothetical static black hole configuration in the Einstein theory with a scalar field. The shaded zone, M, is the integration region. We can see in the figure the bifurcate Killing horizon.

Take $\tilde{M}$, the region bounded by the 3-dimensional hypersurfaces $\Sigma_{t_1}, \Sigma_{t_2}$, by S, the bifurcation surface, and by asymptotic infinity. Multiplying Eq. (1) by ϕ and performing an integration over $\tilde{M}$, we obtain

$$0 = \int_{\tilde{M}} (\phi \nabla_a \nabla^a \phi - \phi \frac{\partial V}{\partial \phi}) d\tilde{M} = \int_{\partial \tilde{M}} \phi \nabla^a \phi dS_a - \int_{\tilde{M}} (\nabla_a \phi \nabla^a \phi + \phi \frac{\partial V}{\partial \sigma}) d\tilde{M}. \tag{2}$$

The integration over $\partial \tilde{M}$ has three contributions: From Σ_{t_2}, from Σ_{t_1} and from asymptotic infinity (there is no contribution from S because it has zero measure). Σ_{t_2} and Σ_{t_1} are equal (due to the invariance with respect to $t \to t + \Delta t$) and with opposite signs (the normals point outside and inside of $\tilde{M}$, respectively). The contribution from asymptotic infinity is zero because ϕ and $\nabla^a \phi$ must decrease fast enough in order to be compatible with asymptotic flatness. Finally, we write the metric as $g^{ab} = -c\xi^a \xi^b + \xi^a v^b + v^a \xi^b + h^{ab}$, where h^{ab} is a (spatial) metric on Σ.

The stationarity of the solution implies that $\xi^a \nabla_a \phi = 0$. Thus we have

$$0 = \int_{\tilde{M}} (g^{ab} \nabla_a \phi \nabla_b \phi + \phi \frac{\partial V}{\partial \phi}) d\tilde{M} = \int (h^{ab} \nabla_a \phi \nabla_b \phi + \phi \frac{\partial V}{\partial \phi}) d\tilde{M}. \tag{3}$$

The first term in the last integrand is non negative, and so is the second one, because the convexity condition corresponds to $\phi \frac{\partial V}{\partial \phi} \geq 0$ which holds, in particular, for the usual mass term $V = \frac{m^2}{2} \phi^2$. Obviously, the only possibility is then $\phi = \text{const} = 0$ (we are assuming, without loss of generality, that the minimum of the potential is at $\phi = 0$).

This result, together with the uniqueness theorems in Einstein vacuum and Einstein–Maxwell theory (see [16] for a review on that subject) and with the heuristic arguments stating that any field that was in the exterior of a black hole, would be either "radiated to infinity" or "sucked into the black hole", generated an almost universal agreement about the validity of Wheeler's conjecture [5]: "Black holes have no hair", understanding by hair any field that would be outside the black hole (besides the electromagnetic field).

The first suggestions on the limitation of such arguments came from ideas associated with quantum effects [6], analogous to the Aharonov–Bohm effect[7], where there happens to exist quantum observables, such as phase changes, in spite of the fact that the classical fields vanish. In the present work we are not interested in this kind of quantum hair, but in the classical hair, which corresponds to cases where there do exist classical observables which are not associated with conservation laws in spacetimes with stationary black holes.

2   The discovery of hair

The first examples of black holes with hair, that is, stationary black holes which require not only the values of conserved quantities defined at asymptotic infinity for their complete specification, were found in 1990 [8] within the Yang–Mills theory coupled to Einstein's relativity, the Einstein–Yang–Mills (EYM) theory.

It is not surprising that the examples where they have been found correspond to configurations with high symmetry: Those hairy black holes are not only stationary, but static as well, and besides, they have spherical symmetry.

The specific configurations are described in the following way: The most general metric describing the exterior field of a black hole with such features can be written as:

$$ds^2 = -\mu e^{2\delta} dt^2 + \mu^{-1} dr^2 + r^2 (d\theta + \sin^2 \theta d\varphi^2), \qquad (4)$$

where $\mu = \mu(r), \delta = \delta(r)$, and the black hole horizon corresponds to the place where the Killing vector, $(\frac{\partial}{\partial t})^a$ becomes null, that is $\mu(r_H) = 0$. It is usual to write $\mu(r) = 1 - 2m(r)/r$ in such a way that the horizon is at r_H if $m(r_H) = r_H/2$. Such a metric will then correspond to a static, spherically symmetric black hole with regular horizon at $r = r_H$ iff $\forall r > r_H \ m(r) > r/2, m(r_H) = r_H/2$, and $\delta(r)$ is finite $\forall r \geq r_H$.

The configuration is asymptotically flat if $\mu \to 1$ and $\delta \to \delta_0$ (finite), when $r \to \infty$. (By means of a rescaling of the t coordinate, it can be reached that $\delta \to 0$, when $r \to \infty$).

The value of the ADM mass, M_{ADM}, of such configuration is then

$$M_{\text{ADM}} = \lim_{r \to \infty} m(r), \tag{5}$$

The hairy black holes found in [8] correspond to the gauge theory based on the $SU(2)$, and the configuration of the gauge fields is described by the ansatz:

$$A = \tau_1 \omega d\theta + (\tau_3 \cot\theta + \tau_2 \omega) \sin\theta d\theta, \tag{6}$$

where τ_i are the Pauli matrices. The result obtained is that for each value of r_H, and for each integer n, there is a black hole characterized by the metric given by Eq. (4), with horizon at $r = r_H$, and the gauge field of the form described by Eq. (6), with the function ω passing through 0 exactly n times while varying r in the interval $[r_H, \infty)$. The corresponding M_{ADM} value turns out to be a growing function of r_H and of n as well.

Clearly, it is a solution for a black hole with hair, because for a given value of M_{ADM}, the corresponding metric is not completely specified, in particular, the value of r_H can be determined only if, besides the value of M_{ADM}, the integer n is specified.

These spacetimes have a metric which is practically indistinguishable from the Schwarzschild one for $r >> r_H$, and it is only when getting closer to the horizon that important differences are found. All these solutions turn out to be unstable, and for this reason some researchers have ignored them, considering that they have no importance. We think that this point of view is not justified. Now it is known that all black holes, with the exception of those with surface temperature $\kappa = 0$ evaporate, due to the Hawking radiation.

It seems possible to understand the existence of such solutions (under certain additional technical assumptions, see [9]), as well as their instability, based on the following topological argument. Let us take Γ, the phase space of the EYM theory, corresponding to asymptotically flat initial data on a hypersurface Σ, with $S^2 \times \mathcal{R}^+$ topology, and internal boundary S, with S^2 topology.

We restrict ourselves to those $\Gamma_A \subset \Gamma$ in which the area of S has a prefixed value A, and we will study the behavior of the function M_{ADM} in Γ. The minimum of M_{ADM} at Γ_A corresponds to the Schwarzschild solution, and we label this configuration as P_1.

Clearly, by means of a large gauge transformation, we can obtain another physically identical configuration, that is, another minimum P_2 of M_{ADM} at Γ_A. Now let us consider the integral curves of $-\nabla^a M_{\text{ADM}}$ at Γ_A, that is, the curves obtained by taking at each point the direction in which M_{ADM} decreases the fastest, and then divide Γ_A in sectors according to the final destinations where these integral curves end. We define $\Gamma_A(1) \subset \Gamma_A$ as the set covered by integral curves ending at P_1, and $\Gamma_A(2) \subset \Gamma_A$ as the set covered by integral curves ending at P_2, etc.

The points P_2, P_3, etc. are physically equivalent to P_1 and, in particular, correspond to the same metric. It is interesting that they allow us to establish

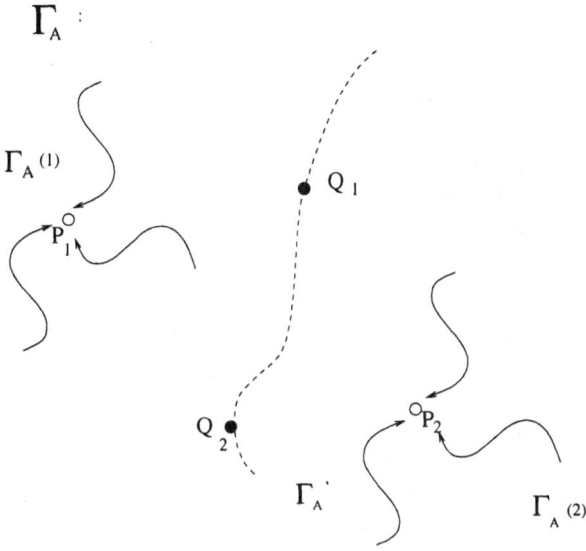

Γ_A :

Γ_A (1)

Q_1

P_1

Q_2

P_2

Γ_A

Γ_A (2)

Fig. 2 Representation of the phase space of the EYM theory, Γ_A, and the integral curves of $-\nabla^a M$.

(based on the assumption that Γ_A is connected), the existence of a boundary between the sectors $\Gamma_A(1), \Gamma_A(2)$, etc. We call such boundary Γ'_A.

Now, it is clear that $-\nabla^a M_{\mathrm{ADM}}$ is tangent to Γ'_A, and so their integral curves must take us to a minimum of M_{ADM} at Γ'. This point, Q_1, is clearly an extreme of M_{ADM} at Γ_A, and the value of M_{ADM} at Q_1 has to be greater than its value at P_1. Thus we end with a static black hole with mass greater than the mass of the Schwarzschild black hole for the corresponding value of the area at the horizon.

This argument has the feature that it can be repeated infinite times: Besides the point Q_1, there exists another minimum of M_{ADM} at Γ'_A, Q_2, obtained from Q_1 by a large gauge transformation and, given the fact that $-\nabla^a M_{\mathrm{ADM}}$ is tangent to Γ'_A, we can perform exactly the same type of division of Γ'_A in sectors $\Gamma'_A(1), \Gamma'_A(2)$, etc., and establish (under the assumption that Γ'_A is connected) the existence of a new boundary between them, Γ''_A, etc. In this way, the argument presented above explains the existence of the infinite series of static black holes that we mentioned before.

On the other side, the relationship that apparently exists between the extrema of the canonical energy $\mathcal{E}$, with fixed value of the area and angular momentum, allows a further application of the above arguments in order to construct infinite series similar to the last one, which would represent rotating black holes with color [8]. In fact, in this theory it is possible to represent the canonical energy (for stationary configurations) as $\mathcal{E} = M_{\mathrm{ADM}} + QV$, where Q is analogous to the electric charge and V to the potential. Then, the first law of thermodynamics may be written as

$$\delta M_{\mathrm{ADM}} = \frac{1}{8\pi}\kappa\delta A + \Omega\delta J - V\delta Q, \qquad (7)$$

and the arguments given above predicts the existence of an infinite series of colored black holes which is associated with each of the special solutions contained in the Kerr–Newman solution. So far, these solutions have not been found.

Other examples of black holes with hair have been found in theories like:
1) Non Abelian Einstein–Proca theory [10] in which a mass term is added to the EYM theory that destroys the gauge symmetry.
2) Einstein–Yang–Mills–Dilaton theory [11] in which a scalar field is added to the EYM theory that does not break the gauge symmetry.
3) Einstein-Yang-Mills-Higgs theory [10] in which a double scalar field is added that preserves the gauge symmetry, but leads to a spontaneous symmetry breaking of the state with minimal energy. This case is of special interest because at present this theory leads to the most accurate physical description of all the non gravitational interactions existing in nature.
4) Einstein-Skyrme theory [12]. This theory tries to develop a phenomenological model for hadronic physics.

All the solutions that have been found so far are static, spherically symmetric (except for the example discussed in [13]) and classically stable only in cases 1 and 4, [14].

The discovery that black holes may have hair arises several questions, for instance:

1. In which theories there exist black holes with hair? Or, in other words, is it possible to determine a priori, without solving the field equations, if there are solutions representing black holes with hair?

2. Which are the physical properties of these black holes? That is, the values and relationships between the quantities which were necessary to completely describe the old solutions: M_{ADM}, Q, J, K, Ω, A, and the values of the new quantities necessary to specify a particular solution in theories which allow the existence of hair.

3. What happened to the physical argument put forward to support the "no hair conjecture" that suggested that all matter fields present in a black hole spacetime would eventually be either radiated to infinity, or "sucked" into the black hole, except when those fields were associated with conserved charges defined at asymptotic infinity?

4. What is the physical relevance of the hair of black holes?

In the following sections we briefly investigate these questions.

3   In which theories there exist black holes with hair?

A partial answer to this question follows from the generalization of the argument used in the description of black holes "with color" [9]. In those theories with discrete symmetries that guarantee the "repetition" of the values of the canonical energy in the corresponding phase space (with a fixed value of the area of the internal boundary and other quantities like the angular momentum), we can expect different extrema of the canonical energy associated with points called "mountain path" which, as we have seen, are associated with static or stationary black holes.

Obviously, this argument is purely heuristic and cannot lead to concrete conclusions. In particular, if the phase space is disconnected the previous argument is no more valid (i.e., the boundary Γ' turns out to be an empty set). An example of this situation is that of a scalar field in a potential with double minimum (for instance, $V(\phi) = \lambda(\phi^2 - a^2)^2$) for which, on the other hand, Bekenstein's theorem does not apply because the potential is not convex. In fact, there are topological arguments that in this case suggest the existence of black holes with hair [15]. However, it is possible to prove explicitly that in this type of theories there are no static spherically symmetric black holes with hair [16], [17], [18]. In the first of these proofs one assumes that such a non trivial configuration exists and proves that it is possible to construct a variation that changes the value of M_{ADM} without changing the horizon area. This contradicts the results given in [9], where it is shown that a static black hole solution corresponds to inital data that extremizes the M_{ADM} at fixed area of the internal boundary. Nevertheless, a direct proof of this result consists in analyzing the specific expression of the conservation law for the energy–momentum tensor $\nabla_\mu T^{\mu\nu} = 0$ of configurations corresponding to this type of black holes. To this end, we consider the line element (4) for which Einstein's equations may be written as

$$\mu' = 8\pi T_t^t + \frac{(1-\mu)}{r}, \qquad (8)$$

$$\delta' = \frac{4\pi}{\mu}(T_t^t - T_r^r), \qquad (9)$$

where $'$ denotes the derivative with respect to r. Furthermore, the r–component of the equation $\nabla_\mu T^{\mu\nu} = 0$ becomes

$$e^\delta (e^{-\delta} T_r^r)' = \frac{1}{2\mu r}[(T_t^t - T_r^r) + \mu(2T - 3(T_t^t + T_r^r))]. \qquad (10)$$

In the case of a theory with scalar fields, the components of the energy-momentum tensor for static spherically symmetric configurations are:

$$T^r_r = \frac{1}{2}\mu \sum_i (\phi'_i)^2 - V(\phi_i), \tag{11}$$

$$T^t_t = T^\theta_\theta = T^\varphi_\varphi = -\frac{1}{2}\mu \sum_i (\phi'_i)^2 - V(\phi_i). \tag{12}$$

On the horizon $\mu = 0$ and therefore $T^r_r(r_H) < 0$. On the other side, we have that $T^t_t - T^r_r \leq 0$ for any matter distribution which satisfies the weak energy condition. Finally, $2T - 3(T^t_t + T^r_r) = -2[\mu \sum_i (\phi'_i)^2 + V] \leq 0$.

Consequently, the right hand side of Eq. (10) is negative semidefinite and hence $e^{-\delta}T^r_r$ is a decreasing function of r (or a constant). Since $e^{-\delta}T^r_r(r_H) \leq 0$ it follows that there cannot be non trivial solutions with $\lim_{r\to\infty} T^r_r = 0$ as it is required for asymptotically flat solutions. This means that the only possibility is $\phi_i = $ const which corresponds to a zero of $V(\phi_i)$. Consequently, in a theory with matter described by scalar fields only, static spherically symmetric black holes have no hair. Obviously, this result applies to any theory in which matter satisfies the weak energy condition and the condition $2T - 3(T^t_t + T^r_r) \equiv 6T^\theta_\theta - T \leq 0$ is fulfilled.

More recently, the study of this question has produced interesting developments in two directions. On the one hand, several no hair theorems have been proved for theories involving scalar fields with non minimal couplings [18, 19, 20], and for nonlinear sigma models [21]. On the other hand, new hairy solutions have been found some of which challenge our preconceptions. One such example is provided by static black hole solutions in theories involving Yang-Mills fields that are not spherically symmetric, but only axisymmetric [13].

Another interesting case is the configuration discovered by Bekenstein [22] in the theory of conformally coupled potentialless scalar fields which would be the only case of hair associated exclusively with scalar fields. The interpretation of this particular configuration as a static black hole solution of the Einstein Conformally-Invariant Field equations has been challenged in [23] where it is argued that in this case the energy-momentum tensor is ill defined at the horizon.

4   Which are the physical properties of black holes with hair?

This question can be investigated by means of the new state functions which describe black holes with hair, i.e. the functions analogous to "Smarr's generalized formula" [24]: $M_{\text{ADM}} + VQ - 2\Omega J = \frac{\kappa A}{4\pi}$ for the case of the Einstein–Maxwell theory. However, there are still no satisfactory analogous functions for more general theories and, therefore, this question is still open. However, there has been some advance in this respect obtained by proving the following inequalities:

1) $\kappa \le 4\pi M/A$ for any stationary black hole if the matter satisfies the strong energy condition [25].

2) $\kappa \le 4\pi M/A$ for any stationary black hole if the matter satisfies the dominant energy condition [26].

3) $\kappa \le \sqrt{\pi/A}$ for any static spherically symmetric black hole if the matter satisfies the dominant energy condition [27].

4) $M - \frac{\kappa A}{4\pi} \ge \Omega J - UQ \ge 0$ for stationary black holes in Einstein-Yang-Mills theory [25].

5   On the "no hair conjecture"

Today we know that the non linear character of the matter fields plays an important role in the theories in which black holes with hair have been found. This nonlinearity represents the interaction of the matter fields with themselves; and this interaction may be understood as a kind of "glue" which connects the fields in the region close to the black hole (where the tendency is that the matter fields are "sucked" into the black hole) with the fields situated in the region far away (where the tendency is that the matter fields are radiated to infinity). Due to this interaction, a state of equilibrium is reached between these two different tendencies.

This intuitive picture suggests that the hair of a black hole must be present in both regions simultaneously. In particular, it would be not possible to restrict the region where the matter fields are present to an arbitrarily small region around the horizon. In fact, this can be proved for the case of spherically symmetric black holes [28] in theories with matter fields which satisfy the weak energy condition and $T < 0$. To this end, we write Eq. (10) in the following form:

$$ e^{\delta}(e^{-\delta} r^4 T_r^r)' = \frac{1}{2\mu}[(3\mu - 1)(T_r^r - T_t^t) + 2\mu T]. \tag{13} $$

As in the case of a scalar field, in these theories we have that $T_r^r(r_H) \le 0$. On the other side, the weak energy condition guarantees the fulfilment of the inequality $T_r^r - T_t^t \ge 0$. Thus the right hand side of the equation (13) is nowhere positive except in those regions where $3\mu - 1 > 0$ (remember that $\mu(r_H) = 0$). The supposition that there are no other charges defined at asymptotic infinity associated with the matter fields, indicates that these fields decrease faster than $1/r^2$, and since the components of $T_{\mu\nu}$ are at least quadratic in the fields, the energy–momentum tensor decreases faster that r^{-4}. This implies that the function $e^{-\delta} r^4 T_r^r \le 0$ at r_H and must reach the value of zero when $r \to \infty$. Moreover, since this is a decreasing function at least up to the radius r_1 for which $3\mu - 1 = 0$, for it to grow towards zero it is necessary that the region with $T_{\mu\nu} \ne 0$ must extend at least up the radius r_1. On the other side, this radius is characterized by

$$r_1 = 3m(r_1) > 3m(r_H) = \frac{3}{2}r_H. \tag{14}$$

This means that the region with hair (the "hairosphere") must extend at least up to the radius $\frac{3}{2}r_H$. The conjecture that this results applies for all stationary black holes is what we call the "no short hair conjecture" and is a subject of further investigation.

Among the consequences of the existence of the "hairosphere" it is important to mention the fact that it rules out the possibility of a static realistic shell with finite thickness (made out of matter satisfying the energy conditions mentioned above) laying completely within the hairosphere [29]. This result indicates certain analogy between the hairosphere and the ergosphere.

Another important aspect of the existence of an upper bound for the hairosphere follows when we consider theories with massive matter fields (like in theories 1, 3 and 4). In these cases, one could expect that the region where the matter fields are non zero is characterized by the distance $\mathcal{M}^{-1}$, with $\mathcal{M}$ being the scale of the characteristic mass of the corresponding theory. This fact combined with the result presented above leads to an upper bound of the size of the horizon:

$$\frac{3}{2}r_H < r_{hair} < \mathcal{M}^{-1}. \tag{15}$$

Recent numerical investigations support this conjecture [14].

6   What is the physical significance of the hair of black holes?

We do not know the correct answer to this question, but it would be very surprising if, despite the fact that nature shows us the way to theories which present this type of interesting solutions, nevertheless, these solutions do not play any role in the description of some physical phenomena. In some sense, this reminds us of the situation caused by the existence of several families of fundamental constituents of elementary particle physics [30], and that induced Rabi [31] to ask himself *Who ordered this?* at the time when the *muon* was discovered, the unstable and heavier partner of the electron.

Bibliography

[1] Heusler, M. *Uniqueness Theorems for Black Hole Space–Times*, in these Proceedings.

[2] J. D. Bekenstein, *Phys. Rev. Lett.* **28**, 452 (1972); *Phys.Rev* **D5**, 1239 (1972); 2403 (1972).

[3] C. Teitelboim, *Phys. Rev.* **D5**, 2941 (1972).

[4] R. M. Wald, *General Relativity*, (The University of Chicago Press, Chicago, 1984).

[5] This conjecture was originally proposed by J.A. Wheeler, see B. Carter, in *Black Holes*, Proceedings of 1972 session of Ecole d'été de physique théorique, edited by C. De Witt, and B. S. De Witt, Gordon and Breach, New York, 1973.

[6] J. M. Bardeen, C. Carter, and S.W. Hawking, *Comm. Math. Phys.* **31**, 161 (1973).

[7] Y. Aharonov, and D. Bohm, *Phys. Rev.* **115**, 485 (1959).

[8] P. Bizon, *Phys. Rev Lett.*, **64**, 28844 (1990); M. S. Volkov and D. V. Gal'tsov, *Sov. J. Nucl. Phys.* **51**, 1171 (1990); H. P. Kunzle and A. K. M. Masood-ul-Alam, *J. Math. Phys.* **31**, 928 (1990).

[9] D. Sudarsky and R. M. Wald, *Phys Rev* **D46**, 1453 (1992).

[10] B. R. Greene, S. D. Mathur, C. M. O'Neill, *Phys. Rev.* **D47**, 2242 (1993).

[11] G. Lavrelashvili and D. Maison, *Nucl. Phys.* **B410**, 407 (1993).

[12] P. Bizon and T. Chamj,*Phys. Lett.* **B 297**, 55 (1992); M. Heusler, S. Droz, and N. Straumann, *Phys. Lett.* **B268**, 371 (1991); **B271**, 61 (1991); **B258**, 21 (1992).

[13] S. A. Ridgway and E. J. Weinberg, *Gen. Rel. Grav.* **27**, 1017 (1995); O. Brodbeck, M. Heusler, N. Straumann, M. Volkov, *Phys. Rev. Lett.* **79**, 4310 (1997).

[14] T. Torii, K. Maeda, and T. Tachizawa, *Phys. Rev.* **D51**, 1510 (1995).

[15] D. Sudarsky, in *Recent Developments in Gravitation and Mathematical Physics: Proceedings of the First Mexican School on Gravitation and Mathematical Physics* (Guanajuato, Mexico, 1994) edited by A. Macías, T. Matos, O. Obregón, and H. Quevedo (World Scientific, Singapore, 1996) p. 146.

[16] M. Heusler, *J. Math. Phys.* **33**, 3497 (1992).

[17] J. D. Bekenstein, *Phys. Rev.* **D51**, R6608 (1995).

[18] A. Mayo, and J. Bekenstein, *Phys. Rev.* **D54**, 5059 (1996).

[19] T. Zannias, *J. Math. Phys.* **36**, 6970 (1995); B. C. Xanthopoulos and T. Zannias, *J. Math. Phys.* **32**, 1875 (1991).

[20] A. Saa, *Phys. Rev.* **D53**, 7373 (1996); **37**, 2349 (1996).

[21] M. Heusler, *Class. Quant. Grav.* **12**, 2036 (1995).

[22] J. Bekenstein, *Ann. Phys. (NY)* **82**, 535 (1974); *Ann. Phys. (NY)* **91**, 72 (1975); N. Bocharova, K. Bronikov and V. Melnikov, *Vestn. Mosk. Univ. Fiz. Astron.*, **6**, 706 (1970).

[23] D. Sudarsky and T. Zannias, *Phys. Rev.* **D58** (1998) in press.

[24] B. Carter, *in General Relativity, an Einstein Centenary Survey* ed. by S. W. Hawking and W. Israel (Cambridge University Press: Cambridge), 1979.

[25] D. Sudarsky and R.M. Wald, *Phys. Rev.* **D47**, R5209 (1993).

[26] M. Heusler, *Class. Quantum Grav.* **12**, 779 (1995).

[27] M. Visser, *Phys. Rev.* **D46**, 2445 (1992).

[28] D. Núñez, H. Quevedo and D. Sudarsky, Phys. Rev. Lett **76**, 571 (1996).

[29] We thank T. Jacobson for pointing out this item.

[30] See, for instance: L. D. Okun, *Leptons and Quarks* (North Holland, Amsterdam, 1982).

[31] For a historical review see for instance, V. C. W. Davies, *The Forces of Nature* (Cambridge University Press: Cambridge) 1986. Heinz R. Pagels, *The Cosmic Code* (Bantam Books, New York) 1982.

Local Version of the Area Theorem (on a Question by G. 't Hooft)

Domenico Giulini

Institut für Theor. Physik, University of Zürich, Winterthurer Str. 190, CH-8057 Zürich, Switzerland

Introduction. During a lecture Prof. 't Hooft posed the question whether one could, in principle, have a multi-black-hole process where the area of at least one black-hole decreased, while the total sum of areas increased. However, if stated carefully, the Area Theorem in fact asserts that *locally* the area cannot decrease. Since black holes cannot bifurcate in the future, the only possibility for a final black hole to be smaller than any initial one is that it was created in the meantime. Hence, given that the Area Theorem holds, we answer the question by the following

Assertion 1. *Consider two Cauchy surfaces, Σ and Σ', with Σ' to the future of Σ. Suppose there is a black hole at time Σ' whose area is smaller than any black hole area at time Σ. Then all null generators of the future event horizon intersecting the surface of this black hole must have past end points between Σ and Σ'.*

Below we will explain and justify this statement in a more precise fashion. In doing this we shall take the opportunity to recall the arguments that lead to the formulation of the Area Theorem. In particular, we wish to point out that standard text-book proofs, as e.g. given in [2], [5], or [4], do not appropriately deal with the fact that the future event horizon, and hence the surfaces of black holes, cannot be smooth (C^1) in general. Hence the application of *differential* geometric methods needs extra justification, usually invoking additional smoothness assumptions. In our discussion we *assume* piecewise C^2-smoothness of the black hole surfaces on the initial Cauchy slice. Since this need not be satisfied in general, better arguments have to be devised to cover the most general case. These might involve suitably smooth approximations of the horizon, or the restriction to just topological and measure-theoretic arguments. But we are presently unaware of such a proof in the literature. However, given the widely believed connection of the Area Theorem with thermodynamic properties of black holes on one side (see Neugebauer's lecture), and the widely expressed hope that this connection may be of heuristic value in understanding certain aspects of quantum gravity on the other (see the lectures by 't Hooft and Kiefer), it seems well motivated to call for a proof of the Area Theorem without additional differentiability assumptions.

Notation, Facts and Assumptions. We assume the space-time (M, g) to be strongly asymptotically predictable (in the sense of Wald 1984) and globally hyperbolic. (It would be sufficient to restrict to a globally hyperbolic portion, as in Thm. 12.2.6 of Wald 1984.) $\mathcal{I}^+$ (scri-plus) denotes future null infinity, $J^-(\mathcal{I}^+)$ its causal past and $B := M - J^-(\mathcal{I}^+)$ the black-hole region. Its boundary, $\partial B =: H$, is the future-event-horizon. H is a closed, imbedded, achronal three-dimensional C^{1-} submanifold of M (proposition 6.3.1 in Hawking and Ellis 1973), where C^{1-} denotes Lipschitz continuity. H is generated by null geodesics without future end points. Past end points occur only where null geodesics, necessarily coming from $J^-(\mathcal{I}^+)$, join onto H. Such points are called "caustics" of H. At a caustic H is not C^1 and has therefore no (continuous) normal. Once a null geodesic has joined onto H it will never encounter a caustic again, never leave H and not intersect any other generator. See Box 34.1 in [4] for a lucid discussion and partial proofs of these statements. Hence there are two different processes through which the area of a black hole may increase: First, new generators can join the horizon and, second, the already existing generators can diverge.

Let Σ be a suitably smooth (below we choose C^2) Cauchy surface, then $\mathcal{B} := B \cap \Sigma$ is called a black-hole region at time Σ and $\mathcal{H} := H \cap \Sigma = \partial \mathcal{B}$ the (future-event-) horizon at time Σ. A connected component $\mathcal{B}_i$ of $\mathcal{B}$ is called a black-hole at time Σ. Its surface is $\mathcal{H}_i = \partial \mathcal{B}_i$, which is a two-dimensional, imbedded C^{1-} submanifold of Σ. In general $\mathcal{H}$ may contain all kinds of caustic sets, like dense ones and/or those of non-zero measure, which are not easily dealt with in full generality. Below we shall avoid this problem by adding the hypothesis of piecewise C^2-smoothness.

By exp we denote the exponential map $TM \to M$. Recall that $\exp_p(v) := \gamma(1)$, where γ is the unique geodesic with initial conditions $\gamma(0) = p \in M$ and $\dot{\gamma}(0) = v \in T_p(M)$. For each p it is well defined for v in some open neighbourhood of $0 \in T_p(M)$. One has $\gamma(t) = \exp_p(tv)$. We shall assume the Lorentzian metric g of M to be C^2, hence the connection (i.e. the Christoffel Symbols) is C^1 and therefore the map $\exp : TM \to M$ is also C^1. The last assertion is e.g. proven by [3].

Local formulation of Area Theorem. We consider two C^2 Cauchy surfaces with Σ' to the future of Σ. The corresponding black-hole regions and surfaces are denoted as above, with a prime distinguishing those on Σ'. We make the *assumption* that $\mathcal{H}$ is piecewise C^2, i.e. each connected component $\mathcal{H}_i$ of $\mathcal{H}$ is the union of open subsets $\mathcal{H}_i^k$ which are C^2 submanifolds of M and whose measure exhaust that of $\mathcal{H}_i$: $\mu(\mathcal{H}_i - \bigcup_k \mathcal{H}_i^k) = 0$, where μ is the measure on $\mathcal{H}$ induced from the metric g.

For each point $p \in \mathcal{H}_i^k$ there is a unique future- and outward-pointing null direction perpendicular to $\mathcal{H}_i^k$, which we generate by some future directed $l(p) \in T_p(M)$. We can choose a C^1-field $p \mapsto l(p)$ of such vectors over $\mathcal{H}_i^k$. The geodesics $\gamma_p : t \mapsto \gamma_p(t) := \exp_p(tl(p))$ are generators of H without future end point. This implies that each γ_p cuts Σ' in a unique point $p' \in \mathcal{H}'$ at a unique parameter value $t = \tau(p)$. By appropriately choosing the affine parametrisations of γ_p as p varies over $\mathcal{H}_i^k$ we can arrange the map τ to be also C^1. Hence $p \rightarrow m(p) := \tau(p)l(p)$ is a null vector field of class C^1 over $\mathcal{H}_i^k$. We can now define the map

$$\varPhi_i^k : \mathcal{H}_i^k \rightarrow \mathcal{H}', \quad p \mapsto \varPhi_i^k(p) := \exp_p(m(p)), \tag{1}$$

which satisfies the following

Lemma. $\varPhi_i^k$ is (i) C^1, (ii) injective, (iii) non-contracting.

(i) follows from the fact that the functions m and $\exp$ are C^1. Injectivity must hold, since otherwise some of the generators of H through $\mathcal{H}_i^k$ would cross in the future. By non-contracting we mean the following: Let μ and μ' be the measures on $\mathcal{H}$ and $\mathcal{H}'$ induced by the space-time metric g. Then $\mu[U] \leq \mu'[\varPhi(U)]$ for each measurable $U \subset \mathcal{H}_i^k$. Assuming the weak energy condition, this is a consequence of the *nowhere* negative divergence for the future geodesic congruence $p \mapsto \gamma_p$ (lemma 9.2.2 in Hawking and Ellis 1973), as we will now show.
Proof of (iii): Set $H_i^k := \bigcup_{p,t} \exp_p(tl(p))$, $\forall p \in \mathcal{H}_i^k$ and $\forall t \in R_+$, which is a C^1-submanifold of M. Let l be the unique (up to a constant scale) future directed null geodesic (i.e. $\nabla_l l = 0$) vector field on H_i^k parallel to the generators. Then $0 \leq \nabla_\mu l^\mu = \pi_\nu^\mu \nabla_\mu l^\nu$, where π denotes the map given by the g-orthogonal projection $T(M)|_{H_i^k} \rightarrow T(H_i^k)$, followed by the quotient map $T(H_i^k) \rightarrow T(H_i^k)/\mathrm{span}\{l\}$. Note that tangent spaces of C^1-cross-sections of H_i^k at point p are naturally identified with $T_p(H_i^k)/\mathrm{span}\{l(p)\}$. Since $\pi_\nu^\mu l^\nu = 0$, we also have $\pi_\nu^\mu \nabla_\mu k^\nu \geq 0$ for $k = \lambda l$ and *any* C^1-function $\lambda : H_i^k \rightarrow \mathbf{R}_+$. Hence this inequality is valid for any future pointing C^1-vector-field k on H_i^k parallel to the generators. Given that, let then $t \mapsto \phi_t$ be the flow of k and $A(t) := \mu_t[\phi_t(U)] := \int_{\phi_t(U)} d\mu_t$, then $\dot{A}(t) = \int_{\phi_t(U)} \pi_\nu^\mu(t)\nabla_\mu k^\nu(t) d\mu_t \geq 0$, where $\pi(t)$ projects onto $T(\phi_t(\mathcal{H}_i^k))/\mathrm{span}\{l\}$, $k(t) = \frac{d}{dt'}|_{t'=t}\phi_{t'}$ and μ_t = measure on $\phi_t(\mathcal{H}_i^k)$. Now choose k such that $\phi_{t=1} = \varPhi_i^k$. Then $\mu'[\varPhi_i^k(U)] - \mu[U] = \int_0^1 dt \dot{A}(t) \geq 0$ ∎

Consequences. From proposition 9.2.5 of [2] it is known that black holes cannot bifurcate in the future. Hence all surface elements $\mathcal{H}_i^k$ of the i-th black-hole at time Σ are mapped via $\varPhi_i^k$ into the surface of a single black-hole at time Σ', whose area therefore cannot be less then the area of $\mathcal{H}_i$. Note that this does not exclude that the number N' of black-holes at time Σ' might be bigger than their number N at time Σ. But it implies that this can only be achieved by an intermediate formation of K *new* black-holes $\mathcal{B}'_1 \ldots \mathcal{B}'_K$, where $K \geq N' - N$. That these black-holes are 'new', i.e. not present at time Σ, means that all generators of H which intersect $\mathcal{H}'_1 \cup \cdots \cup \mathcal{H}'_K$ must have past endpoints somewhere between Σ and Σ'. This proves Assertion 1.

There is another interesting consequence of our analysis: Consider a configuration of two black holes which merge between Σ and Σ'. We assume $\mathcal{H}_1$, $\mathcal{H}_2$ and $\mathcal{H}'$ to be homeomorphic to two-spheres. Suppose $\mathcal{H}_1$ did not contain any caustics, i.e., that $\mathcal{H}_1$ was a C^1-submanifold of Σ. Then we can construct a map

$\Phi_1 : \mathcal{H}_1 \to \mathcal{H}'$ analogous to the construction of Φ_i^k above, but now defined on *all* of $\mathcal{H}_1$. The C^1-condition on $\mathcal{H}_1$ now implies that Φ_1 is C^0. Φ_1 is also injective for the same reason as given for Φ_i^k. Since some generators that cut $\mathcal{H}'$ come from $\mathcal{H}_2$, the map Φ_1 cannot be surjective. If p' is a point of $\mathcal{H}'$ not in the image of Φ_1, we have a continuous injective map $S^2 \cong \mathcal{H}_1 \to \mathcal{H} - \{p'\} \cong R^2$. But this is impossible since such a map does not exist. One way to see this is through a theorem in topology, due to Borsuk and Ulam (proven e.g. in chapter 9 of Amstrong 1983), which says that any continuous map $S^2 \to R^2$ identifies some pair of antipodal points. In particular, it cannot be injective. Hence we obtain a contradiction to the assumption that $\mathcal{H}_1$ was C^1. The same applies of course to $\mathcal{H}_2$, or any other black hole that is going to merge at some later time. Note that it does not matter how far back in time Σ actually is. Thus, under the assumption of spherical topologies for the surfaces of the black holes (which should not be essential), we have shown the following

Assertion 2. *At no time before merging can the surface of a black hole that is going to merge be without caustics.*

Bibliography

[1] Amstrong M.A. (1987): *Basic Topology* (Springer Verlag, New York)

[2] Hawking S.W., Ellis G.F.R. (1973): *The large scale structure of space-time* (Cambridge University Press)

[3] Lang S. (1985): *Differential Manifolds* (Springer-Verlag, Berlin)

[4] Misner C.W., Thorne K.S., Wheeler J.A. (1973): *Gravitation* (W.H. Freeman and Company, San Francisco)

[5] Wald R.M. (1984): *General Relativity* (The University of Chicago Press, Chicago and London)

Black Holes as Exact Solutions of the Einstein–Maxwell Equations of Petrov Type D

Alberto García[1] and Alfredo Macías[2]

[1] Centro de Investigacion y de Estudios Avanzardos del I.P.N., Departamento de Física, Apartado Postal 14-740, México D.F. 07000, Mexico.
[2] Departamento de Física, Universidad Autonoma Metropolitana, Av. Michocan La Purisima S/N, Apartado Postal 55-534, México D.F. 09340, Mexico

Abstract. In this review, we derive from a single canonical metric all electrovac solutions with cosmological constant for an algebraically general electromagnetic field, aligned along the Debever–Penrose directions of the Weyl tensor of Petrov type D. Among them, because of their interpretation as black holes, the Reissner–Nordström static solution and the Kerr–Newman stationary axisymmetric solution are singled out.

1 Introduction

During the last three decades extraordinary progress has been made solving the Einstein–Maxwell equations in the presence of two commuting Killing vectors. One of the most important family of solutions is that related to charged black holes, i.e. to the static Reissner–Nordström black hole and the rotating Kerr–Newman black hole. In this report we shall derive the full class of type D metrics with general Maxwell fields, starting from a single line element containing the above mentioned solutions as particular cases.

In order to give a self–contained description of the gravitational fields we are dealing with, we shall briefly revisit the null–tetrad formalism we will use (for a more detailed review see Kramer et al. [8]) . The null–tetrad frame is given as $\{e_a \equiv \partial_a = e_a{}^\mu \partial_\mu\} = \{m, \bar{m}, l, k\}$, while the associated coframe reads

$$e^1 = \bar{m}_\mu \, dx^\mu \,, \quad e^2 = m_\mu \, dx^\mu \,, \quad e^3 = - \, k_\mu \, dx^\mu \,, \quad e^4 = - \, l_\mu \, dx^\mu \,, \qquad (1.1)$$

where $e_1{}^\mu = m^\mu$, $e_2{}^\mu = \bar{m}^\mu$, $e_3{}^\mu = l^\mu$, and $e_4{}^\mu = k^\mu$, such that $k^\mu l_\mu = -1$, $m^\mu \bar{m}_\mu = 1$, and all the remaining scalar products between them vanish. Thus, the metric reads

$$g = g_{ab} \, e^a \otimes e^b = 2e^1 \otimes e^2 - 2e^3 \otimes e^4 \,. \qquad (1.2)$$

The connection one–forms $\Gamma^a{}_b = \Gamma^a{}_{bc} \, e^c$ associated with the tetrad (1.1) are determined by means of the first Cartan structure equations

$$de^a + \Gamma^a{}_b \wedge e^b = 0 \,, \qquad dg_{ab} = \Gamma_{ab} + \Gamma_{ba} = 0 \,. \qquad (1.3)$$

Consequently, in four–dimensions there are only six independent connection one–forms which can be chosen as Γ_{41}, Γ_{32}, $\Gamma_{12} + \Gamma_{34}$, and their complex conjugated partners $\Gamma_{42} = \bar{\Gamma}_{41}$, $\Gamma_{31} = \bar{\Gamma}_{32}$, $-\Gamma_{12} + \Gamma_{34} = \bar{\Gamma}_{12} + \bar{\Gamma}_{34}$. Here the bar denotes complex conjugation. The expressions of Γ_{ab} in terms of the Newman–Penrose coefficients (see below) are given by

$$\Gamma_{41} = \sigma e^1 + \rho e^2 + \tau e^3 + \kappa e^4 , \tag{1.4}$$
$$\Gamma_{32} = -\mu e^1 - \lambda e^2 - \nu e^3 - \pi e^4 , \tag{1.5}$$
$$\Gamma_{12} + \Gamma_{34} = -2\beta e^1 - 2\alpha e^2 - 2\gamma e^3 - 2\epsilon e^4 , \tag{1.6}$$
$$-\Gamma_{12} + \Gamma_{34} = -2\bar{\alpha} e^1 - 2\bar{\rho} e^2 - 2\bar{\gamma} e^3 - 2\bar{\epsilon} e^4 . \tag{1.7}$$

The coefficients of $\Gamma_{41} = \Gamma_{41a}\, e^a$ and Γ_{32} have certain geometrical meanings, which are assigned to the real (null) congruences k and l associated with them,

$$\Gamma_{411} = \sigma = -k_{a;b}\, m^a m^b , \quad \Gamma_{412} = \rho = -k_{a;b}\, m^a \bar{m}^b = -(\Theta + i\omega) , \tag{1.8}$$

where σ represents the shear of k. For $\sigma = 0$ the congruence is shear–free. Θ describes the expansion of k, and ω the twist of k. For $\Theta = 0$ the congruence is divergence–free and for $\omega = 0$ the congruence is twist–free.

$$\Gamma_{413} = \tau = -k_{a;b} m^a l^b , \quad \Gamma_{414} = \kappa = -k_{a;b} m^a k^b . \tag{1.9}$$

The parameter κ represents the geodetic property of k. For $\kappa = 0$, the congruence is geodetic. For the l congruence one has

$$\Gamma_{321} = \mu = -l_{a;b}\, \bar{m}^a m^b = -(\Theta + i\omega) \quad \text{(expansion \& twist)}, \tag{1.10}$$
$$\Gamma_{322} = \lambda = -l_{a;b}\, \bar{m}^a \bar{m}^b \quad \text{(shear)}, \tag{1.11}$$
$$\Gamma_{323} = \nu = -l_{a;b}\, m^a l^b \quad \text{(geodetic property)}, \tag{1.12}$$
$$\Gamma_{411} = \pi = -l_{a;b}\, \bar{m}^a k^b . \tag{1.13}$$

The geometrical meaning of α, β, γ and ϵ is more involved.

The second structure equations

$$d\Gamma^a{}_b + \Gamma^a{}_s \wedge \Gamma^s{}_b = \frac{1}{2} R^a{}_{bcd} e^c \wedge e^d , \tag{1.14}$$

determine the 20 independent components of the Riemann tensor $R^a{}_{bcd}$, which can be uniquely expressed in terms of irreducible pieces with respect to the general Lorentz group, i.e. the Weyl tensor C_{abcd} (10 components), the piece built from the traceless Ricci tensor S_{ab} (9 components), and the scalar curvature R.

The Weyl tensor is related with the Riemann tensor as follows

$$C_{abcd} = R_{abcd} + \frac{1}{6} R \left(g_{ac} g_{bd} - g_{ad} g_{bc} \right)$$
$$- \frac{1}{2} \left(g_{ac} R_{bd} - g_{bc} R_{ad} + g_{bd} R_{ac} - g_{ad} R_{bc} \right) , \tag{1.15}$$

while the traceless Ricci tensor reads

$$S_{ab} = R_{ab} - \frac{1}{4}g_{ab}R = \kappa\left(T_{ab} - \frac{1}{4}g_{ab}T\right).$$

(1.16)

The invariant classification of the Weyl tensor — the *Petrov classification* — is more easily accomplished by using the self–dual Weyl tensor

$$^*C_{abcd} = C_{abcd} + \frac{i}{2}\varepsilon_{abrs}C^{rs}{}_{cd},$$

(1.17)

which allows the representation

$$\begin{aligned}
^*C_{abcd} = {} & 2\Psi_0 U_{ab}U_{cd} + 2\Psi_1\left(U_{ab}W_{cd} + W_{ab}U_{cd}\right)\\
& + 2\Psi_2\left(U_{ab}V_{cd} + V_{ab}U_{cd} + W_{ab}W_{cd}\right)\\
& + 2\Psi_3\left(V_{ab}W_{cd} + W_{ab}V_{cd}\right) + 2\Psi_4 V_{ab}V_{cd},
\end{aligned}$$

(1.18)

with

$$W_{ab} = m_a\bar{m}_b - m_b\bar{m}_a - k_a l_b + k_b l_a,$$

(1.19)

$$V_{ab} = k_a m_b - k_b m_a, \quad U_{ab} = -l_a\bar{m}_b + l_b\bar{m}_a.$$

$U, V, W, \bar{U}, \bar{V}$ and $\bar{W}$ represent the bivector basis.

By solving the eigenvalue equations $\frac{1}{4}{}^*C_{abcd}X^{cd} = \lambda X_{ab}$, where X^{cd} is an eigenbivector, one establishes the following Petrov types: *I, II, D, III, N*, and *O*, to which one associates the existence of the corresponding null eigenvectors $(1,1,1,1), (2,1,1), (2,2), (3,1), (4)$, and (0) respectively. Each entry corresponds to the number of coincidences (common alignments) of the four null directions. For the classification of the Ricci tensor see Kramer et. al. [8]

On the search for exact solutions in Einstein's theory, it is possible to combine different procedures in looking for solutions of a certain Petrov type with pre-determined geometrical properties of the null eigencongruences, e.g., or within given Killing symmetries.

In the null–tetrad formalism, the Einstein equations are contained in the second Cartan structure equations, which read

$$\begin{aligned}
\mathcal{A}\ :\ & d\Gamma_{41} + \Gamma_{41} \wedge (\Gamma_{21} + \Gamma_{43}) = \frac{1}{2}R_{41cd}\,e^c \wedge e^d\\
= {} & \left(\Psi_1 - \frac{1}{2}S_{14}\right)e^1 \wedge e^2 + \frac{1}{2}S_{11}\,e^1 \wedge e^3 - \Psi_0\,e^1 \wedge e^4 + \left(\Psi_2 + \frac{1}{12}R\right)e^2 \wedge e^3\\
& - \frac{1}{2}S_{44}\,e^2 \wedge e^4 - \left(\Psi_1 + \frac{1}{2}S_{14}\right)e^3 \wedge e^4,
\end{aligned}$$

(1.20)

$$\begin{aligned}
\mathcal{B}\ :\ & d\Gamma_{32} - \Gamma_{32} \wedge (\Gamma_{21} + \Gamma_{43}) = \frac{1}{2}R_{32cd}\,e^c \wedge e^d\\
= {} & -\left(\Psi_3 - \frac{1}{2}S_{32}\right)e^1 \wedge e^2 - \frac{1}{2}S_{33}\,e^1 \wedge e^3 - \Psi_4\,e^2 \wedge e^3 + \left(\Psi_2 + \frac{1}{12}R\right)e^1 \wedge e^4\\
& + \frac{1}{2}S_{22}\,e^2 \wedge e^4 + \left(\Psi_3 + \frac{1}{2}S_{32}\right)e^3 \wedge e^4,
\end{aligned}$$

(1.21)

$$\mathcal{C} \; : \; d\left[\Gamma_{21} + \Gamma_{43}\right] + 2\Gamma_{32} \wedge \Gamma_{41} = \frac{1}{2}\left(R_{21cd} + R_{43cd}\right) e^c \wedge e^d$$

$$= -\left(2\Psi_2 - S_{12} - \frac{1}{12}R\right) e^1 \wedge e^2 + S_{13}\, e^1 \wedge e^3 - 2\Psi_1\, e^1 \wedge e^4 + 2\Psi_3\, e^2 \wedge e^3$$

$$-S_{24}\, e^2 \wedge e^4 - \left(2\Psi_2 + S_{12} - \frac{1}{12}R\right) e^3 \wedge e^4. \tag{1.22}$$

The plan of the paper is as follows. In Sec. 2 we revisit the basic metric and its corresponding field equations. In Sec. 3 we present five different families of static electrovacuum solutions. In Sec. 4, the full class of stationary axisymmetric solutions is obtained. In Sec. 5 the results are discussed.

2   The metric and the basic equations

The starting point in the search for stationary axisymmetric exact solutions of the Einstein–Maxwell equations with cosmological constant is the metric

$$g = \frac{1}{H^2}\left[\frac{\Delta}{P}dx^2 + \frac{P}{\Delta}\left(d\tau + Nd\sigma\right)^2 + \frac{\Delta}{Q}dy^2 - \frac{Q}{\Delta}\left(d\tau + Md\sigma\right)^2\right], \tag{2.1}$$

where $H = H(x,y)$, $P = P(x)$, $Q = Q(y)$, $N = N(y)$, $M = M(x)$, and $\Delta = M - N$.

The coordinates $\{x^\mu\} = \{x, y, \tau, \sigma\}$ are assumed, at this level, to range from $-\infty$ to $+\infty$. Since we are assuming stationarity for ∂_τ and axisymmetry for ∂_σ, one has to impose the signature requirements $\Delta/P > 0$ and $\Delta/Q > 0$. Regularity and elementary flatness condition on the rotation axes constrain the range of σ.

The coframe we choose reads

$$\frac{1}{\sqrt{2}}\frac{1}{H}\left[\sqrt{\frac{\Delta}{P}}dx \pm i\sqrt{\frac{P}{\Delta}}\left(d\tau + Nd\sigma\right)\right] = \begin{cases} e^1 \\ e^2 \end{cases}$$

$$\frac{1}{\sqrt{2}}\frac{1}{H}\left[\pm\sqrt{\frac{\Delta}{Q}}dy + \sqrt{\frac{Q}{\Delta}}\left(d\tau + Md\sigma\right)\right] = \begin{cases} e^3 \\ e^4 \end{cases}. \tag{2.2}$$

We assume the electromagnetic field strength F_{ab} to be aligned along its eigendirections e^3 and e^4 and, accordingly, it has nonvanishing components F_{12} and F_{34}. Thus, the electromagnetic two–form

$$\omega = \frac{1}{2}\left(F_{\mu\nu} + {}^*F_{\mu\nu}\right)dx^\mu \wedge dx^\nu = \frac{1}{2}\left(F_{ab} + {}^*F_{ab}\right)e^a \wedge e^b \tag{2.3}$$

$$= d\left(A_\mu dx^\mu + {}^*A_\mu dx^\mu\right) \tag{2.4}$$

acquires the simple form

$$\omega = (F_{12} + F_{34}) \left(e^1 \wedge e^2 - e^3 \wedge e^4 \right)$$
$$= \frac{1}{H^2} (\mathcal{E} + i^*\mathcal{B}) \left[(d\tau + M d\sigma) \wedge dy + i (d\tau + N d\sigma) \wedge dx \right], \quad (2.5)$$

where $\mathcal{E} = F_{34}$, $i^*\mathcal{B} = F_{12}$, and $*$ is the Hodge dual operation. In terms of the field quantities, the electromagnetic invariant is given by

$$F = \frac{1}{4} F_{\mu\nu} F^{\mu\nu} + \frac{1}{4} {}^*F_{\mu\nu} F^{\mu\nu} = -\frac{1}{2} (\mathcal{E} + i^*\mathcal{B})^2 . \quad (2.6)$$

The integrability of ω requires that $d\omega = 0$, which is equivalent to

$$\partial_x \left(\frac{\mathcal{E} + i^*\mathcal{B}}{H^2} \right) = i\partial_y \left(\frac{\mathcal{E} + i^*\mathcal{B}}{H^2} \right), \quad (2.7)$$

$$M\partial_x \left(\frac{\mathcal{E} + i^*\mathcal{B}}{H^2} \right) + \left(\frac{\mathcal{E} + i^*\mathcal{B}}{H^2} \right) M_x - iN\partial_y \left(\frac{\mathcal{E} + i^*\mathcal{B}}{H^2} \right) - i \left(\frac{\mathcal{E} + i^*\mathcal{B}}{H^2} \right) N_y = 0, \quad (2.8)$$

where $M_x = \partial_x M$ and $N_y = \partial_y N$. Inserting (2.7) and the derivatives of $\Delta = M - N$ into (2.8), we obtain the Maxwell equations to be solved

$$\partial_x \ln \left[\Delta \frac{\mathcal{E} + i^*\mathcal{B}}{H^2} \right] - i\frac{N_y}{\Delta} = 0,$$

$$\partial_y \ln \left[\Delta \frac{\mathcal{E} + i^*\mathcal{B}}{H^2} \right] - i\frac{M_x}{\Delta} = 0. \quad (2.9)$$

The integrability of (2.9) is guaranteed if

$$\partial_x \frac{M_x}{\Delta} - \partial_y \frac{N_y}{\Delta} = 0, \quad (2.10)$$

or, after performing the derivatives, by

$$\Delta(M_{xx} - N_{yy}) - (M_x)^2 - (N_y)^2 = 0. \quad (2.11)$$

Equivalently, the equation (2.9) for the electromagnetic field quantities can be written as a one–form equation

$$d\ln \left[\frac{\Delta}{H^2} (\mathcal{E} + i^*\mathcal{B}) \right] - i \left(\frac{M_x}{\Delta} dy + \frac{N_y}{\Delta} dx \right) = 0. \quad (2.12)$$

The only nonvanishing complex Weyl coefficient is Ψ_2 with

$$4\frac{\Delta}{H^2}\Psi_2 = \frac{1}{3} \left\{ P_{xx} - 3\frac{M_x}{\Delta} P_x - 2P \left[\frac{M_{xx}}{\Delta} - 2(\frac{M_x}{\Delta})^2 + (\frac{N_y}{\Delta})^2 \right] \right.$$
$$\left. + Q_{yy} + 3\frac{N_y}{\Delta} Q_y + 2Q \left[\frac{N_{yy}}{\Delta} + 2(\frac{N_y}{\Delta})^2 - (\frac{M_x}{\Delta})^2 \right] \right\}$$
$$+ i \left\{ \frac{N_y}{\Delta} \left(P_x - 2\frac{M_x}{\Delta} P \right) + \frac{M_x}{\Delta} \left(Q_y + 2\frac{N_y}{\Delta} Q \right) \right\}, \quad (2.13)$$

since all the other Weyl invariants can be expressed as powers of this quantity. Thus, Ψ_2 is by itself an invariant quantity. Moreover the null directions e^3 and e^4 are oriented along the double null eigenvectors of the Weyl tensor and, in addition, along the eigendirection of the electromagnetic field. Consequently, we are dealing with an aligned electromagnetic field in the sense of the coincidence of the proper null directions. Furthermore, the nonvanishing components of the traceless Ricci tensor are

$$S_{12} = S_{34} = -F_{34}{}^2 + F_{12}{}^2 = -\left(\mathcal{E}^2 + {}^*\mathcal{B}^2\right). \tag{2.14}$$

Thus, the Einstein field equations to be solved are:

$$D(MN) := \Delta(M_{xx} - N_{yy}) - (M_x)^2 - (N_y)^2 = 0. \tag{2.15}$$

$$S_{11} : 4\Delta H_{xx} - (M_{xx} + N_{yy})H = 0, \tag{2.16}$$

$$S_{33} : 4\Delta H_{yy} + (M_{xx} + N_{yy})H = 0, \tag{2.17}$$

$$\mathrm{Re}S_{13} : 2\Delta H_{xy} - M_x H_y + N_y H_x = 0, \tag{2.18}$$

$$\mathrm{Im}S_{13} : 2M_x H_x + 2N_y H_y - (M_{xx} + N_{yy})H = 0, \tag{2.19}$$

$$S = S_{12} : P_{xx} - 2P_x \left(\frac{H_x}{H} + \frac{M_x}{\Delta}\right) - 2P \left[\frac{M_{xx}}{\Delta} - (\frac{M_x}{\Delta})^2 - (\frac{N_y}{\Delta})^2 - 2\frac{M_x}{\Delta}\frac{H_x}{H}\right]$$
$$- Q_{yy} + 2Q_y \left(\frac{H_y}{H} - \frac{N_y}{\Delta}\right) - 2Q \left[\frac{N_{yy}}{\Delta} + (\frac{M_x}{\Delta})^2 + (\frac{N_y}{\Delta})^2 - 2\frac{N_y}{\Delta}\frac{H_y}{H}\right]$$
$$+ 4\frac{\Delta}{H^2} \left(\mathcal{E}^2 + {}^*\mathcal{B}^2\right) = 0, \tag{2.20}$$

$$R : H^2 P_{xx} - 6HH_x P_x - 4P \left[HH_{xx} - 3(H_x)^2\right] + H^2 Q_{yy} - 6HH_y Q_y$$
$$- 4Q \left[HH_{yy} - 3(H_y)^2\right] + 4\lambda\Delta = 0. \tag{2.21}$$

The complex rotation of the null eigendirections e^3 and e^4 is given by

$$Z(e^3) = Z(e^4) = H\sqrt{\frac{Q}{2\Delta}} \left[\frac{H_y}{H} + \frac{1}{2}\frac{N_y}{\Delta} + \frac{i}{2}\frac{M_x}{\Delta}\right] = H\sqrt{\frac{Q}{2\Delta}} [\theta + i\widetilde{\omega}]. \tag{2.22}$$

Therefore, one can subclassify the solutions of the above system of equations by means of the expansion θ and the twist $\widetilde{\omega}$ of the principal null congruences e^3 and e^4.

In the next sections we shall derive all stationary axisymmetric type D solutions with aligned electromagnetic field, in the sense mentioned above, with cosmological constant. These solutions are characterized by the invariants Ψ_2, $\mathcal{E}$, ${}^*\mathcal{B}$, and $R = -4\lambda$. In general Ψ_a and $\mathcal{E}({}^*\mathcal{B})$ depend on the choice of the tetrad basis. In the present case, the Weyl and the electromagnetic invariants are expressed in terms of Ψ_2 and $\mathcal{E}({}^*\mathcal{B})$ as $\overset{2}{C} = \Psi_2^2$ and $\mathcal{F} = -(\mathcal{E} + i{}^*\mathcal{B})$, respectively.

3   Static gravitational fields

Let us begin with the simplest family of static solutions. Assuming $N = -1/2$, $M = 1/2$ ($\Delta = 1$), the principal null directions e^3 and e^4 are geodetic, shear–free and twist–free. In the branch with $H = H(x)$, the principal directions are additionally without expansion, while for $H = H(y)$, $H_y \neq 0$ the principal null congruences are expanding.

The equations for H (2.16)–(2.19) are

$$H_{xx} = 0, \quad H_{yy} = 0, \quad H_{xy} = 0, \quad \rightarrow \quad H = \alpha + \beta x + \gamma y, \tag{3.1}$$

where α, β and γ are integration constants. The electromagnetic field equation (2.12) integrates to

$$(\mathcal{E} + i\,{}^*\mathcal{B}) = H^2(e + i\,{}^*g), \tag{3.2}$$

with constants e and ${}^*g$.

Introducing new coordinates $d\phi = d\tau - \frac{1}{2}d\sigma$ and $dt = d\tau + \frac{1}{2}d\sigma$, the metric (2.1) reduces to

$$g = \frac{1}{H^2}\left[\frac{dx^2}{P} + Pd\phi^2 + \frac{dy^2}{Q} - Qdt^2\right]. \tag{3.3}$$

The remaining S– and R–equations (2.20) and (2.21), respectively, should be integrated separately.

3.1   Bertotti–Robinson metric

The simplest solution corresponds to a spacetime filled with a homogeneous electromagnetic field. The metric (3.3) splits into two 2–surfaces of constant curvature. By putting $H = 1$ (one can always achieve this simplification by performing scale transformations), the S– and R–equations, (2.20) and (2.21), take the form

$$P_{xx} + Q_{yy} + 4\lambda = 0, \tag{3.4}$$

$$P_{xx} - Q_{yy} + 4(e^2 + {}^*g^2) = 0. \tag{3.5}$$

Thus, substracting (3.4) and (3.5), one obtains

$$P = p_0 + p_1 x - (\lambda + e^2 + {}^*g^2)x^2, \tag{3.6}$$

$$Q = q_0 + q_1 y - (\lambda - e^2 - {}^*g^2)y^2, \tag{3.7}$$

where p_i and q_i are constants. As mentioned above, the metric can be written in the form of two 2–surfaces of constant curvature:

$$g = g_2 + \tilde{g}_2, \tag{3.8}$$

$$g_2 = \frac{dx^2}{P} + P\,d\phi^2, \quad P = p_0 + p_1 x + p_2 x^2, \tag{3.9}$$

$$\tilde{g}_2 = \frac{dy^2}{Q} - Q\,dt^2, \quad Q = q_0 + q_1 y + q_2 y^2. \tag{3.10}$$

Since these two metrics shall be encountered in subsequent sections, we will rewrite them in terms of more suitables coordinates. For g_2, the only nonvanishing component of the Riemann tensor is given by

$$R_{1212} = \frac{P_{xx}}{4} = \frac{1}{2} P_2 = const. \tag{3.11}$$

By means of shift and scale transformations, we arrive at the well–known canonical form of this metric:

$$p_2 = -\alpha^2 : \quad g_2 = \frac{dx^2}{1-\alpha^2 x^2} + (1-\alpha^2 x^2)d\phi^2$$

$$= \frac{1}{\alpha^2}\left(d\theta^2 + \sin^2\theta d\phi^2\right), \quad (\alpha x = \cos\theta) \tag{3.12}$$

$$p_2 = 0 : \quad g_2 = \frac{4}{p_1^2}\left[(d\sqrt{p_1 x})^2 + (\sqrt{p_1 x})^2 d\phi^2\right]$$

$$= \frac{4}{p_1^2}\left(d\theta^2 + \theta^2 d\phi^2\right), \quad (p_1 x = \theta^2) \tag{3.13}$$

$$p_2 = \alpha^2 : \begin{cases} p_0 - \frac{1}{4}\frac{p_1^2}{\alpha^2} > 0 : \quad g_2 = \frac{dx^2}{1+\alpha^2 x^2} + (1+\alpha^2 x^2)d\phi^2 \\ \qquad\qquad\qquad = \frac{1}{\alpha^2}\left(d\theta^2 + \cosh^2\theta d\phi^2\right), \quad (\alpha x = \sinh\theta) \\ p_0 - \frac{1}{4}\frac{p_1^2}{\alpha^2} < 0 : \quad g_2 = \frac{dx^2}{\alpha^2 x^2 - 1} + (\alpha^2 x^2 - 1)d\phi^2 \\ \qquad\qquad\qquad = \frac{1}{\alpha^2}\left(d\theta^2 + \sinh^2\theta d\phi^2\right), \quad (\alpha x = \cosh\theta). \end{cases} \tag{3.14}$$

These canonical expressions for the g_2–metric can be gathered in the form

$$g_2 = 4\frac{d\zeta d\bar\zeta}{(1+\epsilon\alpha^2\zeta\bar\zeta)^2} \quad \epsilon = -1, 0, 1, \tag{3.15}$$

with

$$\epsilon = 1: \quad \alpha\zeta = \tan(\theta/2)\exp i\phi, \tag{3.16}$$

$$\epsilon = 0\,(=\alpha^2): \quad p\zeta = \theta\exp i\phi, \tag{3.17}$$

$$\epsilon = -1: \quad \alpha\zeta = \tanh(\theta/2)\exp i\phi. \tag{3.18}$$

Analogously, for the $\tilde{g}_2$–metric we have

$$q_2 = -\beta^2 : \quad \tilde{g}_2 = \frac{dy^2}{1-\beta^2 y^2} - (1-\beta^2 y^2)dt^2 = \frac{1}{\beta^2}\left(dz^2 - \sin^2 z dt^2\right), \tag{3.19}$$

$$q_2 = 0 : \quad \tilde{g}_2 = \frac{4}{q_1^2}\left[(d\sqrt{q_1 y})^2 + (\sqrt{q_1 y})^2 dt^2\right] = \frac{4}{q_1^2}\left(dz^2 - z^2 dt^2\right), \tag{3.20}$$

$$q_2 = \beta^2 : \tilde{g}_2 = \frac{dy^2}{\beta^2 y^2 \mp 1} - (\beta^2 y^2 \mp 1)dt^2 = \frac{1}{\beta^2},\left(dz^2 - \left(\frac{\sinh^2 z}{\cosh^2 z}\right)dt^2\right), \tag{3.21}$$

which can be gathered in the form of

$$\tilde{g}_2 = 4\frac{dudv}{(1+\nu\beta^2 uv)^2}, \quad \nu = -1, 0, 1, \tag{3.22}$$

with

$$\nu = 1, \quad \beta u = \tan(z/2)\exp t, \quad \beta v = \tan(z/2)\exp(-t), \tag{3.23}$$
$$\nu = 0 \,(= \beta^2), \quad qu = z\exp t, \quad qv = z\exp(-t), \tag{3.24}$$
$$\nu = -1, \quad \beta u = \tanh(z/2)\exp t, \quad \beta v = \tanh(z/2)\exp(-t). \tag{3.25}$$

Returning back to the Bertotti–Robinson solution [1], one can write the corresponding metric in the form

$$g = 4\,\frac{d\zeta\,d\bar{\zeta}}{(1+\epsilon\alpha^2\zeta\bar{\zeta})^2} + 4\,\frac{du\,dv}{(1+\nu\beta^2 uv)^2}, \tag{3.26}$$

where $\epsilon, \nu = -1, 0, 1$. Nevertheless, since

$$\alpha^2(\epsilon = \pm 1) = \pm\left(\lambda + e^2 + {}^*g^2\right), \quad \epsilon = 0\,(= \alpha^2): \quad -\lambda = e^2 + {}^*g^2, \tag{3.27}$$
$$\beta^2(\nu = \pm 1) = \pm\left(\lambda - e^2 - {}^*g^2\right), \quad \nu = 0\,(= \beta^2): \quad \lambda = e^2 + {}^*g^2, \tag{3.28}$$

one has to take care of the values over which λ ranges.

We have the following possible Bertotti–Robinson metrics

1. $g = g_2(\epsilon = 1) + \tilde{g}_2(\nu = -1)$ with $|\lambda| < e^2 + {}^*g^2$.

   For $\lambda = 0$, $\alpha^2 = \beta^2 = e^2 + {}^*g^2$, this particular solution of the B–R class is commonly written as

$$g = \frac{1}{\alpha^2}\left[d\theta^2 + \sin^2\theta d\phi^2 + dz^2 - \sinh^2 z dt^2\right], \tag{3.29}$$

   and is the single homogeneous conformally flat solution of the Einstein–Maxwell equations with homogeneous general electromagnetic field.

2. $g = g_2(\epsilon = 1) + \tilde{g}_2(\nu = 1)$ with $\lambda > e^2 + {}^*g^2$.

3. $g = g_2(\epsilon = -1) + \tilde{g}_2(\nu = -1)$ with $\lambda < -(e^2 + {}^*g^2)$.

4. $g = g_2(\epsilon = 0) + \tilde{g}_2(\nu = -1)$ with $\lambda = -(e^2 + {}^*g^2)$.

5. $g = g_2(\epsilon = 1) + \tilde{g}_2(\nu = 0)$ with $\lambda = e^2 + {}^*g^2$.

The class of Bertotti–Robinson solutions is characterized by

$$\Psi_2 = -\frac{1}{3}\lambda, \qquad R = -4\lambda, \qquad \mathcal{E} + i\,{}^*\mathcal{B} = e + i\,{}^*g. \tag{3.30}$$

If λ vanishes, the Bertotti–Robinson metric becomes conformally flat. For vanishing electromagnetic field, i.e. $e^2 + {}^*g^2 = 0$, one arrives at the de Sitter space.

3.2   Reissner–Nordström class of solutions

By putting $\beta = 0$ in (3.1), we have $H = \alpha + \gamma y \rightarrow H = y$. Then one is dealing with a class of metrics with diverging (geodetic, shear–free, twist–free) principal null directions of the Weyl tensor. The R–equation (2.21) reads

$$y^2 P_{xx} + y^2 Q_{yy} - 6y Q_y + 12Q + 4\lambda = 0 \tag{3.31}$$

which is a separable equation:

$$P_{xx} = 2p_2 \quad \rightarrow \quad P = p_0 + p_1 x + p_2 x^2, \tag{3.32}$$

$$y^2 Q_{yy} - 6y Q_y + 12Q = -4\lambda - 2y^2 p_2. \tag{3.33}$$

The equation for Q is an inhomogeneous Euler equation. The homogeneous part can be solved by $Q = y^k$, and $(k-3)(k-4) = 0$. The general solution reads

$$Q = -\frac{\lambda}{3} - p_2 y^2 + q_3 y^3 + q_4 y^4. \tag{3.34}$$

Substitution of P and Q into the S–equation (2.20) leads finally to:

$$P = p_0 + p_1 x + p_2 x^2, \tag{3.35}$$

$$Q = -\frac{\lambda}{3} - p_2 y^2 + q_3 y^3 + (e^2 + {}^*g^2) y^4. \tag{3.36}$$

The electromagnetic field (3.2) is given by

$$\mathcal{E} + i\,{}^*\mathcal{B} = y^2 (e + i\,{}^*g), \tag{3.37}$$

$$\omega = -(e^2 + {}^*g)\, d(y\,dt + ix\,d\phi), \tag{3.38}$$

and it is characterized by the Weyl curvature coeficient

$$2\Psi_2 = q_3 y^3 + 2(e^2 + {}^*g^2) y^4. \tag{3.39}$$

Therefore, the metric of the Reissner–Nordström class of solutions [12] can be written as

$$g = \frac{4}{y^2} \frac{d\zeta\, d\tilde{\zeta}}{(1 + \epsilon \alpha^2 \zeta \tilde{\zeta})^2} + \frac{1}{y^2} \left(\frac{dy^2}{Q} - Q\, dt^2 \right). \tag{3.40}$$

Performing a scale transformation together with a change of the variable $y = 1/r$, it is possible to bring the metric (3.40) into the form

$$g = r^2 \frac{d\zeta\, d\tilde{\zeta}}{(1 + \epsilon \zeta \tilde{\zeta})^2} + \frac{dr^2}{Q} - Q\, dt^2, \tag{3.41}$$

$$\epsilon = 1, 0, -1, \tag{3.42}$$

$$Q = \epsilon + \frac{q_3}{r} + \frac{(e^2 + {}^*g^2)}{r^2} - \frac{\lambda}{3} r^2. \tag{3.43}$$

The electromagnetic field is now given by

$$\mathcal{E} + i\,{}^*\mathcal{B} = \frac{1}{r^2}(e + i\,{}^*g)\,, \tag{3.44}$$

$$\omega = -(e + i\,{}^*g)\left\{\frac{1}{r}dt + i\begin{pmatrix} \cos\theta \\ \theta^2 \\ \cosh\theta \end{pmatrix}d\phi\right\} = d\,(A_\mu dx^\mu + {}^*A_\mu dx^\mu)\,. \tag{3.45}$$

Consequently, the electromagnetic potential one–form A reads

$$A = -\frac{e}{r}dt + {}^*g\begin{pmatrix} \cos\theta \\ \theta^2 \\ \cosh\theta \end{pmatrix}d\phi\,. \tag{3.46}$$

The function in the parentheses corresponds to the different possible values of $\epsilon = 1, 0, -1$ respectively. So, the metric (3.41) reduces to

$$g = r^2\left\{d\theta^2 + \begin{pmatrix} \sin^2\theta \\ \theta^2 \\ \sinh^2\theta \end{pmatrix}d\phi^2\right\} + \frac{dr^2}{Q} - Q dt^2\,. \tag{3.47}$$

Assigning values to the remaining constants, it is staightforward to find different static solutions.

3.3   The proper Reissner–Nordström solution: the RN black hole

Choosing $\epsilon = 1$ in (3.47), we arrive at the spherical symmetric asymptotically flat Reissner–Nordström solution [15]

$$g = r^2\left(d\theta^2 + \sin^2\theta d\phi^2\right) + \frac{dr^2}{Q} - Q\,dt^2\,, \tag{3.48}$$

with

$$Q = 1 - \frac{2m}{r} + \frac{e^2}{r^2} \tag{3.49}$$

$$A_0 = -\frac{e}{r} = \varphi \tag{3.50}$$

$$\Psi_2 = -\frac{2m}{r^3} + 2\frac{e^2}{r^4}\,. \tag{3.51}$$

This solution is singled out amongst the Reissner–Nordström class since it represents the gravitational field of a static charged mass. It is the only static solution which can be thought of as a static black hole, endowed with mass m and charge e. Switching off the charge, one arrives at the Schwarzschild solution.

3.4 Anti–Reissner–Nordström class of solutions

In a similar way, for $H = x$, it is straightforward to obtain from (2.20) and (2.21) the metric

$$g = \frac{1}{x^2}\left[\frac{dx^2}{P} + Pd\sigma^2\right] + \frac{1}{x^2}\left[\frac{dy^2}{Q} - Qd\tau^2\right],\tag{3.52}$$

where

$$P = -\frac{\lambda}{3} + \epsilon x^2 + 2nx^3 - \left(e^2 + {}^*g^2\right)x^4,\tag{3.53}$$

$$Q = 1 - \epsilon y^2,\tag{3.54}$$

$$\omega = d\left[(e + i\,{}^*g)(yd\tau + ixd\sigma)\right],\tag{3.55}$$

which does not allow for an interpretation as a black hole. The Weyl component Ψ_2 is given by

$$\Psi_2 = nx^3 - e^2x^4.\tag{3.56}$$

3.5 Levi–Civita metric

The principal null congruences of the Weyl tensor of this metric structure are (geodetic, shear–free) diverging and with twist [9]. By scale and shift transformations, the function H (3.1) can be rewritten as

$$H = x + y.\tag{3.57}$$

Then the R–equation (2.21) acquires the form

$$R := (x + y)^2\,(P_{xx} + Q_{yy}) - 6(x + y)\,(P_x + Q_y) + 12\,(P + Q) + 4\lambda = 0.\tag{3.58}$$

Applying the operator $\partial_x\partial_x\partial_y\partial_y$ to R, one arrives at the separable equation

$$P_{xxxx} + Q_{yyyy} = 0\tag{3.59}$$

whose solution reads

$$P = p_0 + p_1x + p_2x^2 + p_3x^3 + \nu x^4,\tag{3.60}$$

$$Q = q_0 + q_1y + q_2y^2 + q_3y^3 - \nu y^4,\tag{3.61}$$

where p_i, q_i and ν are integration constants.

Inserting P from (3.60) and Q from (3.61) and their derivatives into the R–equation (3.58) as well as into the S–equation (2.20), one finds:

$$P = \gamma - \frac{\lambda}{6} + p_1x + p_2x^2 + p_3x^3 - (e^2 + {}^*g^2)x^4,\tag{3.62}$$

$$Q = -\left(\gamma + \frac{\lambda}{6}\right) + p_1y - p_2y^2 + p_3y^3 + (e^2 + {}^*g^2)y^4.\tag{3.63}$$

This solution is characterized by

$$2\Psi_2 = (x+y)^3 \left[p_3 - 2(e^2 + {}^*g^2)(x-y) \right] , \quad R = -4\lambda , \tag{3.64}$$

with the electromagnetic field (3.2) given as follows:

$$\mathcal{E} + i\,{}^*\mathcal{B} = (x+y)^2 (e + i\,{}^*g) , \tag{3.65}$$

$$\omega = -(e + i\,{}^*g)\{ydt + ixd\phi\} , \tag{3.66}$$

$$A = -eydt + {}^*gxd\phi . \tag{3.67}$$

According to the work by Kinnersley–Walker [7], this solution can be thought of as the field produced by the motion of two accelerated charges.

4 Stationary axisymmetric class of solutions

In this section we present the most important classes of type D solutions of the Einstein–Maxwell equations with two commuting Killing vectors, namely the NUT–Carter $B(+)$ [3] and the Plebanski classes [4].

4.1 NUT–Carter $B(+)$ metrics

We now consider the class of metrics such that

$$N = n = const. , \qquad M = n + m \exp(2\alpha x) , \tag{4.1}$$

as solution to the $D(MN)$–equation (2.15). The equations for H (2.16)–(2.19) become

$$H_{xx} - \alpha^2 H = 0 = H_{yy} - \alpha^2 H , \qquad H_x - \alpha H = 0 , \tag{4.2}$$

hence

$$H = \exp(\alpha x)\,(a \cos \alpha y + b \sin \alpha y) \cong h \exp(\alpha x) \cos \alpha (y - y_0) . \tag{4.3}$$

Substitution in the R–equation (2.21) yields

$$-4\lambda \frac{m}{h^2} \left[1 + \tan^2 \alpha (y - y_0) \right] = P_{xx} - 6\alpha P_x + 8\alpha^2 P + Q_{yy} + 6\alpha \tan \alpha (y - y_0) Q_y$$
$$+ 4Q\alpha^2 \left[1 + 3\tan^2 \alpha (y - y_0) \right] . \tag{4.4}$$

Equation (4.4) separates into P and Q. For P we have

$$P_{xx} - 6\alpha P_x + 8\alpha^2 P = 8\alpha^2 p_2 , \tag{4.5}$$

with solution

$$P = p_1 \exp(4\alpha x) + p_0 \exp(2\alpha x) + p_2 . \tag{4.6}$$

The equation for Q acquires a very simple form after introducing a new coordinate $\tilde{y} = l \tan \alpha (y - y_0)$ and a new function $Q = \tilde{Q}(l^2 + \tilde{y}^2)^{-2}$, i.e.

$$\tilde{Q}_{\tilde{y}\tilde{y}} = -4\lambda \frac{m}{h^2\alpha^2} (l^2 + y^2) - 8l^2 p_2 . \tag{4.7}$$

Equation (4.7) integrates to

$$\tilde{Q} = q_0 + q_1 \tilde{y} - 4l^2 p_2 \tilde{y}^2 - 2\lambda \frac{m}{h^2\alpha^2} l^2 \tilde{y}^2 - \frac{\lambda}{3} \frac{m}{h^2\alpha^2} \tilde{y}^4 . \tag{4.8}$$

The derived metric can be brought into a more transparent form by performing the following identifications:

$$\exp(-2\alpha x) = A_0 \tilde{x} , \tag{4.9}$$

$$\frac{4\alpha^2 l^2 h^2}{m} \frac{P \exp(-4\alpha x)}{A_0^2} = P' , \tag{4.10}$$

$$\frac{h^2\alpha^2}{m} \tilde{Q} = Q' , \tag{4.11}$$

$$\frac{A_0}{2\alpha l^2 h^2} (d\tau + n d\sigma) = d\phi' , \tag{4.12}$$

$$\frac{m}{l^2 h^2 \alpha} d\sigma = dt' . \tag{4.13}$$

Dropping primes and tildes, we can write the NUT–Carter $B(+)$ metric as

$$g = (l^2 + y^2) \left[\frac{dx^2}{P} + P \, d\phi^2 \right] + \frac{l^2 + y^2}{Q} dy^2 - \frac{Q}{l^2 + y^2} [dt + 2lx \, d\phi]^2 , \tag{4.14}$$

where

$$P = p_0 + p_1 x - \epsilon x^2 , \tag{4.15}$$

$$Q = e^2 + g^2 - \epsilon l^2 + \lambda l^4 - 2m y + (\epsilon - 2\lambda l^2) y^2 - \frac{\lambda}{3} y^4 . \tag{4.16}$$

The NUT–Carter $B(+)$ solution is endowed with the mass parameter m, the NUT parameter l, the electric charge e, the magnetic charge g, and the cosmological constant λ. The parameter $\epsilon = 1, 0, -1$ determines the geometry of the g_2–metric sector, i.e. $g_2 = dx^2/P + P d\phi^2$. Moreover, the only non–vanishing Weyl coefficient reads:

$$\Psi_2 = -\frac{1}{(y - il)^2} \left[\frac{m - i\left(\epsilon l - \frac{4}{3}\lambda l^3\right)}{y - il} - \frac{e^2 + {}^*g^2}{y^2 + l^2} \right] . \tag{4.17}$$

4.2   Anti–NUT–Carter B(-) class

For completeness, we present the counterpart of the NUT–Carter $B(+)$ solutions, which can be easily derived by following a similar integration procedure as the one presented above. Letting $M \to m_0$, $N \to \exp(2y)$, $H \to \exp(y)\cos x$, and introducing the transformation $x \to \arctan(x/l)$, $y \to -(1/2)\ln y$, $P \to (l^2 + x^2)^{-2}P$, and $Q \to (2ly)^{-2}Q$, it is possible to write the corresponding metric in the form

$$g = \frac{P}{\delta}\left(d\sigma + 2ly\,d\tau\right)^2 + \frac{\delta}{P}\,dx^2 + \delta\left(\frac{dy^2}{Q} - Q\,d\tau^2\right), \qquad (4.18)$$

with

$$\delta = l^2 + x^2, \qquad (4.19)$$

$$P = -e^2 - g^2 - \epsilon l^2 + \lambda l^4 + 2n\,x + (\epsilon - 2\lambda l^2)\,x^2 - \frac{\lambda}{3}\,x^4 \qquad (4.20)$$

$$Q = 1 - \epsilon\,y^2, \qquad (4.21)$$

$$\omega = d\left[(e + i\,^*g)\left(\frac{d\tau}{l + ix} + \frac{l - ix}{l + ix}\,y\,d\tau\right)\right], \qquad (4.22)$$

since this solution is not asymptotically flat, a black hole interpretation is not appropriate. For this metric, the Weyl coefficient reads

$$\Psi_2 = \frac{1}{(x + il)^2}\left[\frac{n - i\left(\epsilon l - \frac{4}{3}\lambda l^3\right)}{x + il} - \frac{e^2 + \,^*g^2}{x^2 + l^2}\right]. \qquad (4.23)$$

4.3   Plebanski class of metrics

The most general family of solutions of the class under study arises from the assumptions $M_x \neq 0$, $N_y \neq 0$. In this case the principal null congruencies of the Weyl tensor are (geodetic, shear–free) diverging and with twist. The starting point for the integration of the field equations is the $D(MN)$ equation (2.15):

$$D(MN) := (M - N)\,(M_{xx} - N_{yy}) - M_x^2 - N_y^2 = 0. \qquad (4.24)$$

Applying $\partial_x\partial_y$ to $D(MN)$, one obtains

$$N_y M_{xxx} + M_x N_{yyy} = 0, \qquad (4.25)$$

which separates into

$$M_{xxx} + \alpha M_x = 0, \quad \to M_{xx} + \alpha M = 2\beta, \qquad (4.26)$$
$$N_{yyy} - \alpha N_y = 0, \quad \to N_{yy} + \alpha N = 2\gamma. \qquad (4.27)$$

For a further integration, one has to consider the various branches, i.e. $\alpha = 0$, $\alpha = k^2$, and $\alpha = -k^2$, where k is an arbitrary constant.

For $\alpha = 0$, the Plebanski metric is obtained in the standard x, y coordinates. For $\alpha = k^2$, the metric is given in terms of trigonometric and hyperbolic functions which allow, for the electromagnetic cases studied, coordinate transformations yielding the standard Plebanski form. If a rigidly rotating perfect fluid is present, then the resulting metric is a charged version of the Wahlquist perfect fluid solution [17]. For $\alpha = -k^2$, again we are facing a trigonometric–hyperbolic representation of the Plebanski metric. For perfect fluid one arrives at a class of metrics presented in [5]. Therefore, the only case one has to study in the electrovacuum case with cosmological constant λ is the case with $\alpha = 0$.

The integrals $M = m_0 + m_1 x + \beta x^2$ and $N = n_0 + n_1 y + \gamma y^2$, fulfilling equation (4.24), are

$$M = \alpha_0 + \frac{1}{8\beta}\left(n_1^2 + m_1^2\right) + m_1 x + \beta x^2 \,, \tag{4.28}$$

$$N = \alpha_0 - \frac{1}{8\beta}\left(n_1^2 + m_1^2\right) + n_1 y - \beta y^2 \,. \tag{4.29}$$

Performing shift and scale transformations, one can bring M and N into the form

$$M = \rho_0 + \beta x^2 \qquad \rightarrow M = x^2 \,, \tag{4.30}$$
$$N = \rho_0 - \beta y^2 \qquad \rightarrow N = -y^2 \,. \tag{4.31}$$

Then the equations (2.16)–(2.19) reduce to

$$H_{xx} = H_{yy} = 0\,, \quad xH_x - yH_y = 0\,, \quad (x^2 + y^2)H_{xy} - xH_y - yH_x = 0\,. \tag{4.32}$$

They have the following general solution:

$$H = \nu + \mu xy \qquad \rightarrow \qquad H = 1 + \mu xy \,. \tag{4.33}$$

The electromagnetic field equation (2.12) reads

$$d\ln\frac{\Delta}{H^2}\,(\mathcal{E} + i^*\mathcal{B}) - 2i\left(\frac{xdy}{x^2 + y^2} - \frac{ydx}{x^2 + y^2}\right) = 0\,. \tag{4.34}$$

Thus

$$d\ln\frac{\Delta}{H^2}\,(\mathcal{E} + i^*\mathcal{B}) - d\ln\frac{x + iy}{x - iy} = 0\,, \tag{4.35}$$

or

$$\mathcal{E} + i^*\mathcal{B} = \frac{H^2}{\Delta^2}(x + iy)^2(e + i^*g)\,, \quad \Delta = x^2 + y^2\,, \tag{4.36}$$

from which it is straightforward to obtain

$$\mathcal{E} = \frac{H^2}{\Delta}\left[e(x^2 - y^2) - 2xy^*g\right]\,, \quad {}^*\mathcal{B} = \frac{H^2}{\Delta}\left[{}^*g(x^2 - y^2) + 2xye\right]\,. \tag{4.37}$$

The quantity $\mathcal{E}^2 + {}^*\mathcal{B}^2$, entering in the S–equation (2.20) acquires, the simple form

$$\mathcal{E}^2 + {}^*\mathcal{B}^2 = \frac{H^4}{\Delta^2}\left(e^2 + {}^*g^2\right)\,. \tag{4.38}$$

The R–equation (2.21) is now given explicitely by

$$0 = (1 + \mu xy)^2 [P_{xx} + Q_{yy}] - 6\mu (1 + \mu xy) [yP_x + xQ_y] \qquad (4.39)$$
$$+ 12\mu^2 [y^2 P + x^2 Q] + 4\lambda(x^2 + y^2).$$

Applying $\partial_x \partial_x \partial_y \partial_y$ to (4.40), one gets

$$\partial_x \partial_x [x^2 P_{xx} - 6xP_x + 12P] + \partial_y \partial_y [y^2 Q_{yy} - 6yQ_y + 12Q] = 0. \qquad (4.40)$$

Therefore (4.40) splits into

$$x^2 P_{xx} - 6xP_x + 12P = 2\nu x^2 + 6\alpha_1 x + 12\alpha_0, \qquad (4.41)$$
$$y^2 Q_{yy} - 6yQ_y + 12Q = -2\nu y^2 + 6\beta_1 y + 12\beta_0, \qquad (4.42)$$

which are inhomogeneous Euler equations and integrate as follows:

$$P = \alpha_0 + \alpha_1 x + \nu x^2 + p_3 x^3 + p_4 x^4, \qquad (4.43)$$
$$Q = \beta_0 + \beta_1 y - \nu y^2 + q_3 y^3 + q_4 y^4. \qquad (4.44)$$

Substituting the functions P in (4.43) and Q in (4.44) as well as their derivatives into the S– and R–equations (2.20) and (2.21) respectively, one arrives at the structural functions

$$P = \gamma - \frac{e^2 + {}^*g^2}{2} + \alpha x + \nu x^2 + \mu \beta x^3 - \left[\frac{\lambda}{3} + \mu^2 \left(\gamma + \frac{e^2 + {}^*g^2}{2} \right) \right] x^4, \quad (4.45)$$

$$Q = \gamma + \frac{e^2 + {}^*g^2}{2} + \beta y - \nu y^2 + \mu \alpha y^3 - \left[\frac{\lambda}{3} + \mu^2 \left(\gamma - \frac{e^2 + {}^*g^2}{2} \right) \right] y^4, \quad (4.46)$$

which, together with

$$H = 1 + \mu xy, \qquad M = x^2, \qquad N = -y^2, \qquad \Delta = x^2 + y^2, \qquad (4.47)$$

are the building blocks of the Plebanski metric

$$g = \frac{1}{H^2} \left[\frac{\Delta}{P} dx^2 + \frac{P}{\Delta} (d\tau - y^2 d\sigma)^2 + \frac{\Delta}{Q} dy^2 - \frac{Q}{\Delta} (d\tau + x^2 d\sigma)^2 \right], \qquad (4.48)$$

with

$$\omega = d \left\{ \frac{1}{\Delta} [-(ey + {}^*gx)d\tau - xy(ex - {}^*gy)d\sigma] \right.$$
$$\left. + \frac{i}{\Delta} [(ex - {}^*gy)d\tau - xy(ey + {}^*gx)d\sigma] \right\}.$$

$$(4.49)$$

It is characterized by the Weyl component

$$2\Psi_2 = \left(\frac{1 + \mu xy}{y + ix} \right)^3 \left[(\beta - i\alpha) + 2 \left(e^2 + {}^*g^2 \right) \frac{1 - \mu xy}{y - ix} \right]. \qquad (4.50)$$

We suggested to name (4.48) the Plebanski metric since it contains as particular cases the Plebanski–Demianski solution [13] and the Plebanski–Carter [A] metric [3, 14].

4.4   Plebanski–Demianski metric

For $\mu = -1$, $\gamma \to \gamma - \lambda/6 - {}^*g^2/2 + e^2/2$, $\alpha \to 2l$, $\nu = -\epsilon$, $\beta \to -2m$, $x \to p$, $y \to q$, $\sigma \to -\sigma$, the structural functions acquire now the form

$$P = \left(\gamma - \frac{\lambda}{6} - {}^*g^2\right) + 2lp - \epsilon p^2 + 2mp^3 - \left(\gamma + \frac{\lambda}{6} + e^2\right) p^4 , \qquad (4.51)$$

$$Q = \left(\gamma - \frac{\lambda}{6} + e^2\right) - 2mq + \epsilon q^2 - 2lq^3 - \left(\gamma + \frac{\lambda}{6} - {}^*g^2\right) q^4 , \qquad (4.52)$$

$$H = 1 - pq, \quad \Delta = p^2 + q^2 , \qquad (4.53)$$

and the metric (4.48) reduces to the well known Plebanski–Demianski form:

$$g = \frac{1}{H^2} \left[\frac{\Delta}{P} dp^2 + \frac{P}{\Delta} \left(d\tau + q^2 d\sigma\right)^2 + \frac{\Delta}{Q} dq^2 + \frac{Q}{\Delta} \left(d\tau - p^2 d\sigma\right)^2 \right] . \qquad (4.54)$$

The only non–vanishing Weyl coefficient reads

$$\Psi_2 = \left(\frac{1 - xy}{y + ix}\right)^3 \left[-(m + il) + (e^2 + {}^*g^2) \frac{1 + xy}{y - ix} \right] . \qquad (4.55)$$

4.5   The Plebanski–Carter[A] metric

For $\mu = 0$, $\gamma \to \gamma - {}^*g^2/2 + e^2/2$, $\alpha = 2l$, $\nu = -\epsilon$, $\beta = -2m$, $x \to p$, $y \to q$, $\sigma \to -\sigma$, the structural functions acquire the form

$$P = \gamma - {}^*g^2 + 2lp - \epsilon p^2 - \frac{\lambda}{3} p^4 , \qquad (4.56)$$

$$Q = \gamma + e^2 - 2mq + \epsilon q^2 - \frac{\lambda}{3} q^4 , \qquad (4.57)$$

$$\Delta = p^2 + q^2 , \qquad (4.58)$$

and the metric (4.48) reduces to the Plebanski–Carter[A] form

$$g = \frac{\Delta}{P} dp^2 + \frac{P}{\Delta} \left(d\tau + q^2 d\sigma\right)^2 + \frac{\Delta}{Q} dq^2 + \frac{Q}{\Delta} \left(d\tau - p^2 d\sigma\right)^2 . \qquad (4.59)$$

The corresponding Weyl component reads

$$\Psi_2 = -\frac{(m + il)(q - ip) - e^2 - {}^*g^2}{(q + ip)^3 (q - ip)} . \qquad (4.60)$$

4.6   The Kerr–Newman metric

By putting in the Plebanski–Carter [A] metric $\epsilon = 1$, $\lambda = 0$, $l = 0$, ${}^*g = 0$, and $\gamma = a^2$, as well as perfoming the following coordinate transformations $p =$

$-a\cos\theta$, $q = r$, $\tau = t + a\phi$, and $\sigma = (1/a)\phi$, and defining, as it is conventionally accepted,

$$P = a^2 \sin^2\theta, \quad \Delta := Q = a^2 + e^2 - 2mr + r^2, \quad \Sigma := p^2 + q^2 = r^2 + a^2 \cos^2\theta, \tag{4.61}$$

one arrives at the Kerr–Newman metric in Boyer–Lindquist [2] coordinates

$$g = -\frac{1}{\Sigma}\left(\Delta - a^2\sin^2\theta\right)dt^2 + \frac{2}{\Sigma}a\sin^2\theta\left(r^2 + a^2 - \Delta\right)dtd\phi$$
$$+\frac{1}{\Sigma}\sin^2\theta\left[\left(r^2 + a^2\right)^2 - a^2\Delta\sin^2\theta\right]d\phi^2 + \Sigma d\theta^2 + \frac{\Sigma}{\Delta}dr^2. \tag{4.62}$$

The Weyl component Ψ_2 is now given by

$$\Psi_2 = -\frac{m\left(r + ia\cos\theta\right) - e^2}{\left(r - ia\cos\theta\right)^3\left(r + ia\cos\theta\right)}. \tag{4.63}$$

This solution corresponds to the standard black hole solution with charge and angular momentum [6, 10, 11].

5   Discussion

At the beginning of the *golden age*, black holes were thought to be just what their name suggests, i.e. holes in space, down which things can fall, out of which nothing can emerge. Later, the picture changed. Black holes were regarded not as mere quiescent holes in spacetime but rather as dynamical objects. A black hole should be able to rotate, and as it rotates it should create a tornado–like swirling motion in the curved spacetime around itself. The greatest surprise to emerge from the *golden age* was the result of general relativity that all properties of a black hole are precisely predictable from just three parameters, i.e. its mass, its angular momentum, and its electric charge. From those three parameters one should be able to compute, for instance, the shape of the hole horizon, the strength of its gravitational pull, and the details of the swirl of the spacetime around it, among other properties. Penrose introduced global concepts and showed how they could be used to establish results about spacetime singularities that did not depend on any exact symmetry or details of the matter content of the universe. Bekenstein introduced the black holes thermodynamics, which culminate in the area theorem. Nowadays, it is the issue of the quantum mechanics of the black holes which is under intensive investigation [16].

We have derived the full class of solutions of the Einstein–Maxwell equations with cosmological constant for the Petrov type D. These solutions are endowed with several free parameters and they contain a broad spectrum of features, for instance spherical symmetry, flat symmetry, or even hyperbolic symmetry, but also asymptotical or non–asymptotical flatness. Moreover, the electromagnetic field can be homogeneous, inhomogeneous, axially distributed, or it decreases asymptotically.

However, if we take into account the uniqueness theorems of black holes, one has to conclude that from the broad spectrum of type D solutions only the proper Reissner–Nordström family (endowed with mass and charge) and the Kerr–Newman family (characterized by mass, charge, and angular momentum) are solutions which can be interpreted as black holes, i.e. as an exterior field of a collapsing star surrounded by an event horizon.

Acknowledgments

We would like to thank Friedrich W. Hehl for useful discussions and literature hints and for his hospitality at the University of Cologne during the elaboration of this review. This work was partially supported by CONACyT, grants No. 3544–E9311, No. 3692P–E9607, and by the joint German–Mexican project KFA–Conacyt E130–2924 and DLR–Conacyt 6.B0a.6A.

Solutions	m	n	e	$^*g$	a	b	λ	$\widetilde{\omega}$	$\widetilde{\theta}$
Plebański–Demiański	×	×	×	×	×	×	×	×	×
Carter A	×	×	×		×	×	×	×	×
Plebański	×	×	×	×	×		×	×	×
Kerr–Newman	×		×		×			×	×
NUT-Carter B(+)	×	×	×	×			×		×
Carter B(+)	×	×	×		×		×		×
anti NUT-Carter B(−)	×	×	×	×	×		×	×	
Carter B(−)	×	×	×		×		×	×	
C–Levi-Civita	×	×	×	×		×	×		
Reissner–Nordström	×		×	×			×		
anti Reissner–Nordström	×		×	×			×		
Schwarzschild	×						×		
Bertotti–Robinson			×	×					
de Sitter							×		

Table 1 Stationary axisymmeric type D electrovacuum solutions: Classification according to the 7 parameters m (mass), n (NUT-parameter), e (electric charge), $^*g$ (magnetic charge), a (angular momentum), b (acceleration), λ (cosmological constant), and the two quantities $\widetilde{\omega}$ (twist), $\widetilde{\theta}$ (expansion).

Bibliography

[1] B. Bertotti, *Phys. Rev* **116** (1959) 1331. I. Robinson, *Bull. Acad. Polon. Sci., Ser. Math. Astr. Phys.* **7** (1959) 351.

[2] R.H. Boyer and R.W. Lindquist, *J. Math. Phys.* **8** (1967) 265.

[3] B. Carter, *Commun. Math. Phys.* **10** (1968) 280.

[4] A. García, *J. Math. Phys.* **25** (1984) 1951.

[5] A. García, *J. Math. Phys.* **31** (1991) 1951.

[6] R. P. Kerr, *Phys. Rev. Lett.* **11** (1963) 237.

[7] W. Kinnersley and M. Walker, *Phys. Rev.* **D2** (1970) 1359.

[8] D. Kramer, H. Stephani, M. MacCallum, and E. Herlt: *Exact Solutions of the Einstein Field Equations.* (Deutscher Verlag der Wissenschaften, Berlin 1980).

[9] T. Levi–Civita, *Rend. Acc. Lincei* **26** (1917) 307; **27** (1918) 3, 183, 220, 240, 283, 343; **28** (1919) 3, 101.

[10] E. Newman, L. Tamburino, and T. Unti, *J. Math. Phys.* **4** (1963) 915.

[11] E.T. Newman, E. Couch, K. Chinnapared, A. Exton, A. Prakash, and R. Torrence, *J. Math. Phys.* **6** (1965) 918.

[12] G. Nordström, *Proc. Kon. Ned. Akad. Wet.* **20** (1918) 1238.

[13] J. Plebański and M. Demiański, *Ann. Phys.* (N.Y.) **98** (1976) 98.

[14] J. Plebański, *Ann. Phys.* (N.Y.) **90** (1975) 196.

[15] H. Reissner, *Annalen der Physik* **50** (1916) 106.

[16] K.S. Thorne, in: *Black Holes & Time Warps* (W.W. Norton, New York 1994).

[17] H.D. Wahlquist, *Phys. Rev.* **172** (1968) 1291.

On the Construction of Time-Symmetric Black Hole Initial Data

Domenico Giulini

Institute for Theoretical Physics, University of Zürich, Winterthurer Str. 190, CH-8057 Zürich, Switzerland

Abstract. We review the 3+1 - split which serves to put Einstein's equations into the form of a dynamical system with constraints. We then discuss the constraint equations under the simplifying assumption of time-symmetry. Multi-Black-Hole data are presented and more explicitly described in the case of two holes. The effect of different topologies is emphasized.

Notation. Space-time is a manifold M with Lorentzian metric g of signature $(-, +, +, +)$. Greek indices are $\in \{0, 1, 2, 3\}$ and latin indices are $\in \{1, 2, 3\}$. Indices from the beginning of the alphabet, like $\alpha, \beta, \ldots$ and $a, b, \ldots$, refer to orthonormal frames and indices from the middle, like $\lambda, \mu, \ldots$ and $l, m, \ldots$ to coordinate frames. The symbol $\circ$ denotes the composition of maps. The relation $:=$ $(=:)$ defines the left (right) hand side. The torsion and curvature tensors for the connection ∇ are defined by $T(X, Y) := \nabla_X Y - \nabla_Y X - [X, Y]$ and $R(X, Y)Z := \nabla_X \nabla_Y Z - \nabla_Y \nabla_X Z - \nabla_{[X,Y]} Z$ respectively. The covariant components of the Riemann and Ricci tensors are defined by $R_{\alpha\beta\gamma\delta} := g(e_\alpha, R(e_\gamma, e_\delta)e_\beta)$ and $R_{\alpha\gamma} := g^{\beta\delta} R_{\alpha\beta\gamma\delta}$ respectively.

1 The 3+1 – Split

In this article we discuss the vacuum Einstein equations

$$G^{\mu\nu} := R^{\mu\nu} - \tfrac{1}{2} g^{\mu\nu} R = 0, \tag{1}$$

which form a system of ten quasi-linear second order differential equations for the ten functions $g_{\mu\nu}$. However, the four equations $G_{\mu 0} = 0$ do not involve the second time derivatives and hence constrain the set of initial data. To see this, recall that the twice contracted second Bianchi identity gives $\nabla_\mu G^{\mu\nu} = 0$, or expanded

$$\partial_0 G^{0\nu} = -\partial_k G^{k\nu} - \Gamma^\mu_{\mu\lambda} G^{\lambda\nu} - \Gamma^\nu_{\mu\lambda} G^{\mu\lambda}. \tag{2}$$

Since the right hand side contains at most second time derivatives the assertion follows. The ten Einstein equations therefore split into four *constraints* and six evolution equations $G^{ik} = 0$. That four equations constrain the initial data rather than guiding the evolution results in four dynamically undetermined functions among the ten $g_{\mu\nu}$. The task is to parameterize the $g_{\mu\nu}$ in such a way that four dynamically undetermined functions can be cleanly separated from the other six. How this can be done via the 3+1 split is explained below. The four dynamically undetermined quantities will be the famous *lapse* (one function) and *shift* (three functions). It follows directly from (2) that the constraints will be preserved under this evolution.

The splitting of the Einstein equations will be formulated in a geometric fashion. We initially think of (M, g) as given and satisfying the Einstein equations. Then we write down the evolution law for the intrinsic and extrinsic geometry of a spacelike 3-manifold Σ as it moves through M. Together with the constraints they are equivalent to all Einstein equations. Finally this procedure is turned upside down by taking the evolution equations for Σ's geometry as starting point. Only after their integration can we construct the ambient space-time.

1.1   3+1 Split Geometry

The topology of space-time (or the portion thereof) which we want to decompose into space and time must be a product $M \cong \Sigma \times \mathbf{R}$. We foliate M by a one-parameter family of embeddings $e_t : \Sigma \to M$, $t \in \mathbf{R}$. For fixed t the image of e_t in M is called Σ_t, or the t'th leaf of the foliation. All leaves are assumed spacelike. Hence there is a normalized timelike vector field, n, normal to all leaves. We choose one of the two possible orientations and thereby introduce the notions of future and past: A timelike vector X is future pointing iff $g(X, n) < 0$ (recall signature convention). The tangent-bundle $T(M)$ can now be split into the orthogonal sum of the subbundle of spacelike vectors, $S(M)$, and the normal bundle, $N(M)$. The associated projection maps are given by

$$\mathsf{S} : T(M) \to S(M), \quad X \mapsto X + n\, g(n, X), \tag{3}$$

$$\mathsf{N} : T(M) \to N(M), \quad X \mapsto -n\, g(n, X), \tag{4}$$

which can be naturally continued to the cotangent bundle by setting $\mathsf{S}^*(\omega) := \omega \circ \mathsf{S}$ and then factorwise on tensor products and linearly on the whole tensor-bundle. Thus we obtain a split of the whole tensor bundle, where from now on the projection maps are simply called S and N for all tensors. Tensors in the image of S are called *spatial*. It is easy to verify that

$$h := \mathsf{S}g = g + n^\flat \otimes n^\flat, \tag{5}$$

where $n^\flat := g(n, \cdot)$. Note that the restriction h_t of h to $T(\Sigma_t)$ is just the induced Riemannian metric on Σ_t. Identifying for the moment Σ and Σ_t via e_t this leads to $h_t = e_t^* g$ (Exercise). For what follows it is however crucial to regard spatial tensors as tensors over M and not over Σ. Otherwise covariant (or Lie-) derivatives in directions off Σ would not make sense.

If X, Y are any spatial vector fields we can write

$$\nabla_X Y = \mathsf{S}\nabla_X Y + \mathsf{N}\nabla_X Y = D_X Y + n\, K(X, Y), \tag{6}$$

where we defined the spatial covariant derivative, D, and the extrinsic curvature, K, by

$$D_X := S \circ \nabla_X, \tag{7}$$

$$K(X,Y) := -g(\nabla_X Y, n) = -g(\nabla_Y X, n) = g(\nabla_X n, Y). \tag{8}$$

The second equality in (8) – and hence the symmetry of K – follows from the vanishing torsion of ∇ and the fact that $[X,Y]$ is spatial. It is easy to prove that K is indeed a tensor and that D defines a connection on the tangent bundle of each leaf Σ_t. Extension via the Leibnitz rule leads to a unique connection on the bundle of spatial tensors, which can be directly defined by (7) with the extended meaning of S described above. In fact it is just the Levi-Civita connection compatible with the metric h. To see this, we compute $D_X h = S \nabla_X (g + n^b \otimes n^b) = 0$, since $\nabla_X g = 0 = S n^b$, so that D is compatible with h. Vanishing torsion is also immediate: $D_X Y - D_Y X - [X,Y] = S(\nabla_X Y - \nabla_Y X - [X,Y]) = 0$, by $[X,Y] = S[X,Y]$ and the vanishing torsion of ∇.

Let $\{e_0, e_1, e_2, e_3\}$ be an orthonormal frame adapted to the foliation, i.e. $e_0 = n$, and $\{e^0, e^1, e^2, e^3\}$ its dual. Then from (5) with $n^b = e^0$ we have

$$g = -e^0 \otimes e^0 + h = -e^0 \otimes e^0 + \sum_{a=1}^{3} e^a \otimes e^a. \tag{9}$$

The family of embeddings $t \mapsto e_t$ defines a vector field, $\partial/\partial t =: \partial_t$, which is easily characterized by its action on any smooth function f:

$$\partial_t f := \left. \frac{d}{dt} \right|_{t=0} f \circ e_t. \tag{10}$$

This vector field can be decomposed into normal and tangential components

$$\partial_t = \alpha n + \beta = \alpha e_0 + \beta^a e_a, \tag{11}$$

with uniquely defined function α and spatial vector field β. They are called the *lapse* (function) and *shift* (vector field) respectively.

Let now $\{x^\mu\}$ be an adapted local coordinate system on M so that $x^0 = t$ and hence spatial fields $\partial_k := \partial/\partial x^k$. The flow lines of ∂_t are then the lines of constant spatial coordinates x^k. Hence (α, β) are interpreted as normal and tangential components of the 4-velocity – measured in units of t – with which the points of constant spatial coordinates move. To express the metric g in terms of these coordinates we use an obvious matrix notation and write

$$\begin{pmatrix} \partial_t \\ \partial_k \end{pmatrix} = \begin{pmatrix} \alpha & \beta^a \\ 0 & A_k^a \end{pmatrix} \begin{pmatrix} e_0 \\ e_a \end{pmatrix}, \tag{12}$$

$$\begin{pmatrix} e^0 & e^a \end{pmatrix} = \begin{pmatrix} dt & dx^k \end{pmatrix} \begin{pmatrix} \alpha & \beta^a \\ 0 & A_k^a \end{pmatrix}. \tag{13}$$

Introducing (13) into (9) yields the 3+1 split form of the metric g:

$$g = -\alpha^2 \, dt \otimes dt + h_{ik} \, (dx^i + \beta^i dt) \otimes (dx^k + \beta^k dt), \tag{14}$$

where $h_{ik} = h(\partial_i, \partial_k) = \sum_a A_i^a A_k^a$ and $\beta^i A_i^a = \beta^a$. For the measure 4-form one easily obtains $e^0 \wedge e^1 \wedge e^2 \wedge e^3 = \alpha \sqrt{\det\{h_{ik}\}} d^4 x$.

In the ambient space-time the notion of time-derivative of spatial tensors is introduced via the Lie-derivative along the time flow generated by ∂_t. But in order to render this an operation within the space of spatial tensor fields we must include a spatial projection. Using (11) we define the "doting" by

$$\dot{h} := SL_{\partial_t} h = \alpha L_n h + SL_\beta h, \qquad (15)$$

where we also used $L_{\alpha n} h = \alpha L_n h$ and that $L_n h$ is already spatial. This is true for any covariant spatial tensor and any smooth function α. To prove this, we first remark that by Leibnitz' rule it suffices to prove it for a general spatial 1-form ω. The first assertion now follows from $L_{\alpha n} \omega = (i_{\alpha n} \circ d + d \circ i_{\alpha n}) \omega = \alpha i_n d\omega = \alpha L_n \omega$. The second statement follows from the general formula $i_n \circ L_v = i_{[n,v]} + L_v \circ i_n$, showing that for $v = n$ the left hand side annihilates any spatial tensor field. This identity also shows why we need the projector in the second expression on the right hand side of (15), since for $v = \beta$ it shows that we would need $[n, \beta] \propto n$ for $L_\beta h$ to be spatial. But this is generally false, as one easily shows that $[n, \beta^k \partial_k] \propto n \Leftrightarrow \partial_t(\beta^k) = 0$.

We proceed by showing that $L_n h$ is just twice the extrinsic curvature:

$$K = \tfrac{1}{2} L_n h. \qquad (16)$$

To prove this relation, we take any spatial vector fields X, Y and compute:
$L_n h(X, Y) = \nabla_n(h(X, Y)) - h([n, X], Y) - h(X, [n, Y]) = h(\nabla_X n, Y) + h(X, \nabla_Y n)$
$= 2K(X, Y)$, where we used the metricity of ∇, $(\nabla_n h)(X, Y) = (\nabla_n g)(X, Y)$
$= 0$, and its vanishing torsion. Hence we arrive at

$$K = \frac{1}{2\alpha} \left(\dot{h} - SL_\beta h \right). \qquad (17)$$

The projected Lie-derivative can be expressed in terms of the spatial covariant derivative in the usual way. In components with respect to a spatial coordinate frame this reads $(SL_\beta h)_{ik} = D_i \beta_k + D_k \beta_i$.

1.2   Constraints and Equations of Motion

Using the splitting formula (6) for the connection ∇ in terms of D and K we can derive the so-called Gauss-Codazzi and Codazzi-Mainardi equations by a straightforward manipulation. In components with respect to $\{e_a\}$ and with $R^{(3)}$ denoting the curvature of D, they read respectively:

$$R_{abcd} = R^{(3)}{}_{abcd} + K_{ac} K_{bd} - K_{ad} K_{bc}, \qquad (18)$$

$$R_{0abc} = D_c K_{ab} - D_b K_{ac}. \qquad (19)$$

From here it is easy to write down the constraints by noting that in orthonormal frames one has $\sum_{a,b} R_{abab} = R + 2R_{00} = 2G_{00}$, i.e. the 00 component of the

Einstein tensor just depends on the spatial components of ∇'s curvature. In fact, it is the sum of the spatial sectional curvatures of ∇. Further, $G_{0b} = R_{0b} = \sum_a R_{0aba}$. Hence we have the constraints, now written in components with respect a coordinate frame,

$$G_{\mu\nu}n^\mu n^\nu = \tfrac{1}{2}(R^{(3)} - K_{ik}K^{ik} + (K_j^j)^2) = 0, \tag{20}$$

$$G_{\mu i}n^\mu = D^k(K_{ik} - h_{ik}K_j^j) = 0. \tag{21}$$

To obtain the dynamical equations one starts again from the defining equation of the curvature and manipulates the expression for R_{0a0b}. Observing that $\nabla_n(g(e_a, \nabla_{e_b} n)) = L_n K_{ab}$ one arrives at

$$R_{0a0b} = -L_n K_{ab} + K_{ac}K_b^c + a_a a_b + D_a a_b, \tag{22}$$

where $a := \nabla_n n$. Note also that $a^b = L_n n^b$ (Exercise: Prove it). Despite appearance, the last term in (22) is also symmetric.[1] Now, $R_{ab} = -R_{0a0b} + \sum_c R_{cacb}$, so that with (18) we have

$$R_{ab} = R^{(3)}{}_{ab} + L_n K_{ab} + K_{ab}K_c^c - 2K_{ac}K_b^c - a_a a_b - D_a a_b. \tag{23}$$

This is almost the evolution equation we wish to obtain. As in (15) we have $\dot{K} = \alpha L_n K + S L_\beta K$, and since we want to write down the final equation in a coordinate basis, we can simplify the terms involving a by noting that $a_i = (L_n n^b)(\partial_i) = n^b([\partial_i, \tfrac{1}{\alpha}(\partial_t - \beta)]) = \partial_i \alpha / \alpha$. Hence[2]

$$\dot{K}_{ik} = \alpha(2K_{ij}K_k^j - K_{ik}K_j^j + R_{ik} - R^{(3)}{}_{ik}) + L_\beta K_{ik} + D_i D_k \alpha, \tag{24}$$

where in the vacuum case we consider here one sets $R_{ik} = 0$. Note that in a coordinate frame dotting just means taking the partial derivative of the components, i.e., $L_{\partial_t} h_{ij} = \partial h_{ij}/\partial t$.

The dynamical formulation is now complete. The constraints are given by eqs. (20)(21) and the six evolution equations of second order are written as twelve equations of first order, given by (15) and (24). The six dynamical components of g are the h_{ij}, whereas there are no evolution equations for the four functions α, β. The initial value problem thus takes the following form: 1.) choose a 3-manifold Σ with local coordinates $\{x^i\}$, 2.) find a Riemannian metric h_{ij} and a symmetric

---

[1] This is due to n being hypersurface-orthogonal. To see this, we first note the identity $(L_n - \tfrac{1}{2}i_n \circ d)dn^b \wedge n^b = da^b \wedge n^b$ (Exercise: Prove it). Now, hypersurface-orthogonality of $n \Leftrightarrow dn^b \wedge n^b = 0 \Rightarrow da^b \wedge n^b = 0 \Leftrightarrow S da^b = 0 \Leftrightarrow D_{[a}a_{b]} = 0$.

[2] Be aware that some authors define the extrinsic curvature with opposite sign, for example E. Seidel in his lecture. Hence the discrepancy between our eqns. (17)(24) with his (5)(6) respectively. In our convention, which agrees with Hawking & Ellis, a positive K_i^i implies volume *expansion* under deformations in normal direction. Also note that "dotting" does not commute with index raising. Hence notations like $\dot{K}^{ij}$ are ambiguous. For example, denoting the index raising operation by a superscript $\sharp$, (16) immediately gives $(L_n(K^\sharp) - (L_n K)^\sharp)^{ik} = -4K_j^i K^{jk}$.

covariant tensor field K_{ij} on Σ which satisfy (20)(21), 3.) choose any convenient functions $\alpha(t, x^k), \beta^i(t, x^k)$, 4.) evolve h_{ij} and K_{ij} via (15)(24) by using the choices made in the previous step, 5.) take the solution curve $h_{ij}(t, x^k)$ and the functions from 3.) to construct the space-time metric according to (14), where $x^0 = t$. The g so constructed solves Einstein's equations. An important theorem guarantees that for suitably specified data a maximal evolution (M, g) exists which is unique up to diffeomorphisms (Choquet-Bruhat and Geroch 1969). See also Choquet-Bruhat and York (1980) for a review and further references.

Regarding step 2.), we remark that all topologies Σ allow *some* initial data, i.e., there are no topological obstructions to (20)(21) (Witt 1986). This might change if geometrically *special* data are sought (see below). To illustrate step 3.), we mention the so-called maximal slicing condition on α. To derive it, we compute $L_n(h^{ij}K_{ij}) = -2K^{ij}K_{ij} + h^{ij}L_nK_{ij} = -R_{00} - K^{ij}K_{ij} + \Delta\alpha/\alpha$, where $\Delta = D^iD_i$ and where we used $R_{00} = \sum_a R_{0a0a}$ and (22) to replace L_nK. Hence

$$L_{\partial_t}(K_i^i) = (\Delta - K^{ij}K_{ij} - R_{00})\alpha + L_\beta(K_i^i), \tag{25}$$

where we left in R_{00} for generality. Note that the strong energy condition implies $R_{00} \geq 0$ through the Einstein equations. The hypersurface $\Sigma \subset M$ is called *maximal* if the trace of its extrinsic curvature – the so-called *mean curvature* – is zero, i.e., $K_i^i = 0$. This is equivalent to Σ being a stationary point of all the 3-dimensional volume functionals for domains in Σ and fixed boundaries. [3] (Exercise: Prove this using (16).) Now, given a maximal slice $\Sigma \subset M$, (25) gives the following simple condition on α if the evolution is to preserve maximality: $\mathcal{O}\alpha = 0$ with elliptic operator $\mathcal{O} = \Delta - K^{ij}K_{ij} - R_{00}$. (Exercise: Assuming the strong energy condition, prove that any smooth function α in the kernel of $\mathcal{O}$ cannot have a positive local maximum or negative local minimum on Σ.) In the vacuum case one can use (20) to write $\mathcal{O} = \Delta - R^{(3)}$, i.e., purely in terms of the intrinsic geometry of Σ, where clearly $R^{(3)} \geq 0$.

The maximal slicing condition plays an important rôle in numerical evolution schemes, since – by definition – the evolving maximal slices $\Sigma_t \subset M$ approach slowest the regions of strongest spatial compression. In this sense they have the tendency to avoid singularities. For further information see section 2.3 of E. Seidel's lecture. Finally we note that since not all topologies Σ allow for metrics with $R^{(3)} \geq 0$, there exist topological obstructions to *maximal* initial data sets (Witt 1986).

2 Time-Symmetric Initial Data

Suppose a hypersurface $\Sigma \subset M$ has vanishing extrinsic curvature, $K = 0$. From (6) we then have $\nabla_X Y = D_X Y$ for all vector fields X, Y tangent to Σ. In

---

[3] The standard terminology is that such stationary points are called "maximal" if the ambient geometry is Lorentzian and "minimal" if it is Riemannian, irrespectively of whether they really are true maxima or minima respectively. True extrema are called stable maximal (minimal) surfaces.

particular, if $\gamma : I \to \Sigma$ is a curve with tangent vector field γ' over γ, then $\nabla_{\gamma'} \gamma' = D_{\gamma'} \gamma'$ and γ is a geodesic in Σ iff it is a geodesic in M. Submanifolds for which this is true are called *totally geodesic*. This is a stronger condition than maximality. In general, constant mean curvature data play an important rôle in the solution theory for the constraints (see York 1973, Ó Murchadha and York 1974). Here we shall vastly shortcut the general procedure by imposing the condition that Σ be totally geodesic. One can then show that the maximal development, M, from these data allows an isometry fixing Σ pointwise and exchanging the two components of $M - \Sigma$. Hence such data are called *time symmetric*. For such cases the constraints reduce to the simple condition that (Σ, h) has vanishing Ricci-scalar:

$$R^{(3)}(h) = 0, \tag{26}$$

where for later convenience we explicitly indicated the metric as argument of $R^{(3)}$. A general idea for solving (26) is to prescribe h up to an overall conformal factor Φ, and let (26) determine the latter. So setting $h = \Phi^4 h'$, with fourth power just for convenience, we have by the conformal transformation law for the Ricci-scalar

$$R^{(3)}(\Phi^4 h') = -8\Phi^{-5}(\Delta_{h'} - \tfrac{1}{8} R^{(3)}(h'))\Phi =: -8\Phi^{-5} C_{h'} \Phi = 0, \tag{27}$$

where $\Delta_{h'}$ is the Laplacian for the metric h'. We are interested in C^2 solutions satisfying $\Phi > 0$ and where (Σ, h) has no boundaries at finite distances, i.e. Σ should be topologically complete in the metric topology defined by the distance function induced by h. The last condition is equivalent to (Σ, h) being geodesically complete (theorem of Hopf-Rinow-DeRahm, see e.g. Spivak 1979). In addition, we shall only be interested in manifolds whose ends are asymptotically flat. Allowing the manifold Σ to have more ends or to be otherwise topologically more complicated allows for a greater variety of solutions. Note that to each of n asymptotically flat ends there corresponds an ADM-mass of which $n - 1$ are independent (see below).

Brill Waves. One may ask whether simple asymptotically flat solutions to $C_{h'} \Phi = 0$ exist on $\Sigma = \mathbf{R}^3$. There are no (regular!) black-hole solutions with this simple topology, but there are solutions representing localized gravitational waves of non-zero total ADM energy (Araki 1959). In the axisymmetric case they were investigated in detail by Brill (1959). Solutions of this kind are collectively called "Brill waves". One takes (from now on in the usual shorthand suppressing the $\otimes$)

$$h' = \exp(\lambda\, q(z, \rho))(dz^2 + d\rho^2) + \rho^2\, d\varphi^2, \tag{28}$$

where the profile-function q must for $r \to \infty$ fall off like r^{-2} and like r^{-3} in its first derivatives in order for h to turn out asymptotically flat. q characterizes the geometry in the meridial cross section ($z\rho$-plane) of the toroidal gravitational wave. Regularity on the axis also requires q and $\partial_\rho q$ to vanish for $\rho = 0$. The parameter $\lambda \in \mathbf{R}_+$ is sometimes introduced to independently parameterize the overall amplitude. Equation (26) for $\Phi(z, \rho)$ takes the particularly simple form

$$\left(\Delta_{\mathrm{f}} + \tfrac{1}{4}\lambda\Delta^{(2)}q \right) \Phi = 0, \tag{29}$$

where Δ_{f} is the *flat* Laplacian and $\Delta^{(2)} = \partial^2/\partial z^2 + \partial^2/\partial \rho^2$. Given q, everywhere positive solutions for Φ exist provided λ is below some critical value depending on the choice q (Araki 1959). To see uniqueness, assume the existence of two solutions Φ_1 and Φ_2 and set $h_i = \Phi_i^4 h'$, $i = 1, 2$. Then $\Phi_3 := \Phi_1/\Phi_2$ is also C^2, positive and tends to 1 at infinity. But (27) immediately implies $C_{h_2}\Phi_3 = -\tfrac{1}{8}\Phi_3^5 R^{(3)}(h_1) = 0$, and since also $R^{(3)}(h_2) = 0$ this is equivalent to $\Delta_{h_2}\Phi_3 = 0$. Hence $\Phi_3 = 1$ due to the fact that the only bounded harmonic functions are the constant ones.

3   Black-Hole Data

A substantial variety of time-symmetric black-hole data can already be obtained by solving (27) when h' is *flat*, i.e., where the 3-metric, h, on the spatial slice at the moment of time-symmetry is conformally flat. One can obtain manifolds with any number of asymptotically flat ends, and then reduce this number by a process which is best described by calling it "plumbing" (see below). We shall devote the rest of this paper to the description of such solutions and techniques. Note that for flat h' we are left with the simple harmonic equation involving only the flat Laplacian:

$$\Delta_{\mathrm{f}}\Phi = 0. \tag{30}$$

In general it is difficult to infer from given initial data whether they correspond to a spacetime with black holes, i.e. with event horizons. However, in the examples to follow it is easy to see that that there will be apparent horizons, since for time symmetric data apparent horizons correspond precisely to minimal surfaces $S \subset \Sigma$ [4]. Proposition 9.2.8 of Hawking and Ellis (1973) now implies the existence of an event horizon whose intersection with Σ is on, or outside, the outermost apparent horizon for any regular predictable spacetime that develops from data satisfying the strong energy condition. Concerning the topology of apparent horizons we remark the following: Using the formula for the second variation of the area functional and the theorem of Gauß -Bonnet,

---

[4] The condition on S being an apparent horizon is that the congruences of outgoing null rays from S must have zero divergence. Analytically this translates into $\mathrm{tr}_2(\kappa) = \pm(\mathrm{tr}(K) - K(\nu, \nu))$ where κ, ν are respectively the extrinsic curvature and normal of S in Σ. The upper sign is valid for *past* apparent horizons, and the lower one for *future* apparent horizons. tr_2 is the 2-dimensional trace using the induced metric of S and tr the 3-dimensional trace using h. For time-symmetric initial data $(K = 0)$ this condition states that κ is traceless and hence S minimal in Σ.

one shows that for ambient metrics with non-negative Ricci scalar any connected component of an orientable stable minimal surface of finite volume must be a topological 2-sphere (Gibbons 1972). Allowing also for non-orientable apparent horizons, one deduces from this that for metrics h of Σ with $R^{(3)}(h) \geq 0$ a connected component of an apparent horizon is either S^2 or RP^2, the latter being the (non-orientable) 2-dimensional real projective space. If Σ is orientable $RP^2 \subset \Sigma$ is one-sided, as in the example below.

3.1 Schwarzschild Data

We start by noting that the most general non-trivial solution of (30) on $\Sigma = \mathbf{R} - \{0\}$ is given by $\Phi(\vec{x}) = 1 + \frac{m}{2r}$ with $r = \|\vec{x}\|$ and $m \in \mathbf{R}_+$. We cannot have any higher multipole moments because then Φ necessarily has zeros on Σ. Just removing $\Phi^{-1}(0)$ from Σ does not work since these points are at finite distance so that the resulting space would not be (geodesically) complete. This is also the reason why m must be positive. Hence we obtain for the metric h in polar coordinates

$$h = \left(1 + \frac{m}{2r}\right)^4 (dr^2 + r^2\, d\Omega^2), \tag{31}$$

with $d\Omega^2 = d\theta^2 + \sin^2\theta\, d\varphi^2$. Now, it is easy to verify that the following two diffeomorphisms, I and $\tilde{I}$, of Σ are involutive (i.e., square to the identity) isometries:

$$I(r, \theta, \varphi) := \left(\frac{m^2}{4r}, \theta, \varphi\right), \tag{32}$$

$$\tilde{I}(r, \theta, \varphi) := \left(\frac{m^2}{4r}, \pi - \theta, \varphi + \pi\right). \tag{33}$$

The map I is called an inversion on the sphere $r = m/2$, whereas $\tilde{I}$ is that inversion plus an additional antipodal map on the spheres of constant r. We shall sometimes refer to them as inversions of the first and second kind respectively. $\tilde{I}$ has no fixed points while I fixes each point of the sphere $S = \{\vec{x} \mid r = \frac{m}{2}\}$ (which, as set, is also left invariant by $\tilde{I}$). As fixed point set of an isometry S must be totally geodesic[5], hence minimal and therefore an apparent horizon. Its surface area is $A = 16\pi m^2$, and it separates the two isometric regions $r > m/2$ and $r < m/2$. The metric (31) corresponds to the spatial part of the Schwarzschild metric of mass m in isotropic coordinates, which cover both asymptotically flat regions (I and III) on the Kruskal manifold. Using this isotropic form, one can read off $\alpha = (1 - m/2r)/(1 + m/2r)$, $\beta = 0$ and verify that with this choice the static form of (24) with $K = 0$ is satisfied (Exercise).

---

[5] Proof: Consider the unique geodesic γ starting on and tangentially to S. It cannot leave S since if it would, its image under I would be a *different* geodesic with the same initial conditions, which contradicts the uniqueness theorem for ODE's.

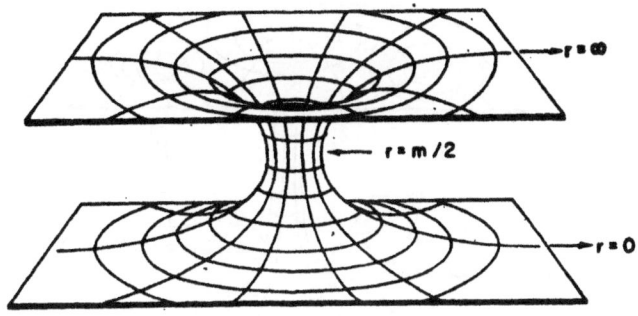

Fig. 1 The Schwarzschild Throat

The manifold Σ has two isometric ends and we can get rid of one by suitably identifications. For this we take the quotient $\tilde{\Sigma}$ of Σ with respect to the free action of $\tilde{I}$. The freeness guarantees that the quotient will be a manifold, and, by being an isometry, the metric descends to a smooth metric on the quotient. $\tilde{\Sigma}$ can be pictured by cutting Σ along S, throwing away one piece, and identifying opposite points on the inner boundary S on the retained piece. Hence topologically $\tilde{\Sigma}$ is the real projective space, $\mathbf{R}P^3$, minus a point. The projection of S into $\tilde{\Sigma}$ is a totally geodesic, one-sided (i.e. non-orientable) surface diffeomorphic to $\mathbf{R}P^2$. $\tilde{\Sigma}$ is orientable, smooth, complete and with one end which is isometric to, and hence has the same ADM mass as, either end in Σ. This demonstrates how the introduction of more ends or other topological features makes it possible to define non-trivial black-hole data. One may also combine Brill waves with a black hole to model a single distorted black hole. This is further discussed in section 2.2 of E. Seidel's lecture.

Multi-Schwarzschild Data. Taking $\Sigma = \mathbf{R}^3 - \{\vec{c}_1, \cdots, \vec{c}_n\}$ the generalization of (31) is easily obtained with n poles of strengths $a_i \in \mathbf{R}_+$ at "positions" $\vec{c}_i$:

$$\Phi(\vec{x}) = 1 + \sum_{i=1}^{n} \frac{a_i}{r_i}, \qquad (34)$$

where $r_i := \|\vec{x} - \vec{c}_i\|$. For each i we can introduce inverted polar coordinates $r_i' = a_i^2/r_i$ to probe the region $r_i \to 0$ by letting $r_i' \to \infty$. Doing this shows that the metric is asymptotically of the form (31) with certain mass parameters $m = m_i$ given below. The same is true for the region $r \to \infty$ with mass M. Hence one obtains $n + 1$ asymptotically flat ends. The internal masses and the overall mass are given by ($r_{ji} := \|\vec{c}_j - \vec{c}_i\|$)

$$m_i = 2a_i(1 + \chi_i), \quad \text{where} \quad \chi_i := \sum_{j \neq i} \frac{a_j}{r_{ji}}, \quad \text{and} \quad M = 2\sum_i a_i. \qquad (35)$$

Fig. 2 Multi-Schwarzschild

In terms of the parameters a_i, r_{ij} the binding energy takes the simple form

$$\Delta M := M - \sum_{i=1}^{n} m_i = -2 \sum_{i=1}^{n} a_i \chi_i = -2 \sum_{i=1}^{n} \sum_{j \neq i} \frac{a_i a_j}{r_{ij}} < 0. \qquad (36)$$

Note that there are as many independent masses as there are generators of the second homology group of Σ. These generators may be represented by stable minimal surfaces associated to each internal end. Their surfaces areas clearly satisfy $A_i > 16\pi(2a_i)^2$, since the right hand side represents the minimal area in the strictly smaller metric (31) for just one hole with parameter $m = 2a_i$. But there is also an upper bound for the area, given by the recently proven Riemannian Penrose inequality[6] (Huisken and Ilmanen 1997), which in our context reads $A \leq 16\pi(2a_i)^2(1 + \chi_i)^2$. Assuming the existence of an event horizon (see above), the Area Theorem (see my other contribution to this volume) implies that the area of the hole in the i'th end cannot evolve below A_i, which, using the first inequality above, implies in particular that the energy which is bound in the final hole is greater than $2a_i = m_i/(1 + \chi_i)$. The difference of the (conserved) ADM mass m_i to the mass of the final hole is therefore bounded above by $m_i \chi_i/(1 + \chi_i)$. In other words, the fraction of energy being radiated is less than $\chi_i/(1 + \chi_i)$. This still allows for total conversion into radiation if one chooses $\chi_i \to \infty$.

It would be of course more interesting to express (36) in terms of physical variables, like the individual masses m_i, and more geometrically defined distance functions than r_{ij}, like e.g. the proper geodesic distance of the minimal surfaces in the i'th and j'th throat. Note that for small mass-to-separation ratios we

---

[6] The proof of [18] applies to all asymptotically flat Riemannian 3-Manifolds whose Ricci scalar satisfies $R \geq 0$. They prove that the area A of the outermost stable minimal surface bounding an end and the ADM mass m of that end satisfy $A \leq 16\pi m^2$. It implies the positive mass theorem for data with $R \geq 0$.

may in a first approximation replace a_i by $\frac{1}{2}m_i$ and r_{ij} by the geodesic distance of the i'th and j'th apparent horizons and get the familiar Newtonian formula. But there will be corrections the precise form of which depend on ones definition of "distance between two holes". Whereas here mass is unambiguously defined for each hole (by ADM), there is no natural definition of distance. Perhaps the easiest intrinsically defined distance is the one given above. For the multi-Schwarzschild manifold it has the disadvantage that the minimal surfaces are not easy to locate analytically and one has to resort to numerical methods (see Brill and Lindquist 1963 for early attempts).

The location of minimal surfaces is interesting for a variety of reasons. It somewhat simplifies in the case of just two holes, which is automatically axisymmetric. Then the variational principle for the minimal surfaces reduces to a geodesic principle for curves in the $z\rho$-half-plane (cylindrical coordinates). The appropriately parameterized solution curves just describe a motion of a point particle in the potential $-\frac{1}{2}\rho^2\varPhi^8$ (Čadež 1974). However, general analytic solution still do not exist. Numerical studies by [3] for equal masses ($a_1 = a_2 =: a$) show the very interesting behaviour above the critical value $a/r_{12} \simeq 1/1.53$, where *two* more minimal surfaces appear, each of which enclosing the previous two. Initially they coincide, but for increasing a/r_{12} they separate with the inner one rapidly increasing in area whereas the outermost staying almost constant. See also [13] for a related discussion.

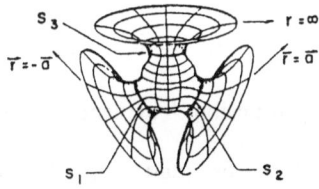

Fig. 3 Two nearby black holes

For the data discussed below the difficulty of determining location and size of minimal surfaces is absent, but somewhat as trade-off the concept of individual mass now becomes slightly more problematic.

Different Topologies for Multi-Hole Data. There are other generalizations of the single hole case. The ones we discuss now will preserve the existence of involutive isometries like (32-33), but now for *each* apparent horizon. The manifolds they exist on have two or even just one end. The construction is somewhat involved (Lindquist 1963) and uses the method of images to construct solutions to (30). This method was introduced by [21] for the time symmetric case and later generalized to more general situations (e.g. Bowen and York 1980, Bowen 1984). (There is also a recent alternative proposal by [6].) For the general understanding it will be sufficient to explain the construction for just two holes. Note that the ADM definition of mass cannot be applied to the individual hole if it does not have an asymptotically flat end associated to it. But there exist alternative proposals for mass due to Lindquist (1963) and Penrose (1982) which can be employed here. (See also the general review by Penrose (1984).) But it should be pointed out that these definitions do not always apply in more general situations. For example, for the applicability of Penrose's mass definition within time-symmetric hypersurfaces the metric of this hypersurface must be conformally flat (Tod 1983, Beig 1991).

3.2   Two Hole Data

Just as in electrostatics, we shall use the method of images to construct special solutions to (30). This is done by placing image masses in an auxiliary, fictitious space so as to enforce special properties of Φ. The properties which will be enforced here are such that the inversions (32)(33) on 2-spheres become isometries.

We start by drawing two 2-spheres $S_i := S(a_i, \vec{c}_i)$, $i = 1, 2$, with radii a_i and centered at $\vec{c}_i$. The spheres are non-intersecting and outside each other, so that $r_{12} > a_1 + a_2$. On $\mathbf{R}^3 - \{\vec{c}_i\}$ we have the diffeomorphisms I_i and $\tilde{I}_i$, which in polar coordinates at $\vec{c}_i$ take the forms (32) and (33) respectively. These induce involutions on the space of functions, defined by

$$J_i(f) := \frac{a_i}{r_i} f \circ I_i \quad \text{and} \quad \tilde{J}_i(f) := \frac{a_i}{r_i} f \circ \tilde{I}_i \tag{37}$$

respectively, where f is any function. The crucial property of these maps is

$$\Delta_f \circ J_i = (a_i/r_i)^4 J_i \circ \Delta_f \quad \text{and} \quad \Delta_f \circ \tilde{J}_i = (a_i/r_i)^4 \tilde{J}_i \circ \Delta_f, \tag{38}$$

which in particular implies that the image of a harmonic function will again be harmonic, although with different singularity structure. The image of the constant function, $f \equiv 1$, under either of these maps is just $f' = a_i/r_i$, i.e., the pole of strength a_i at $\vec{c}_i$. Moreover, given the unit pole $f(\vec{x}) = 1/\|\vec{x} - \vec{d}\|$ at $\vec{d}$ outside S_i, then its image under J_i is

$$J_i(f) = \frac{a_i}{\|\vec{c}_i - \vec{d}\|} \frac{1}{\|\vec{x} - I_i(\vec{d})\|}, \tag{39}$$

and correspondingly for $\tilde{J}_i$. It represents a pole of strength $a_i/\|\vec{c}_i - \vec{d}\| < 1$ at the image point $I_i(\vec{d})$ (resp. $\tilde{I}_i(\vec{d})$).

Writing down the metric $h = \Phi^4 ds_f^2$ in polar coordinates centered at $\vec{c}_i$ one easily verifies that I_i ($\tilde{I}_i$) is an isometry of h if Φ is invariant under J_i ($\tilde{J}_i$). The construction of such an invariant Φ is by brute force: One averages the function $\Phi_0 \equiv 1$ over the free product of the groups generated by J_1, J_2 ($\tilde{J}_1, \tilde{J}_2$). The elements of this free-product-group are strings of alternating J_1's and J_2's, where for each string length $n \geq 1$ there are the two different elements $J_1 \circ J_2 \circ J_1 \cdots$ and $J_2 \circ J_1 \circ J_2 \cdots$. By definition, the string of length 0 is the identity element. Hence one sets

$$\Phi_N := 1 + \sum_{n=1}^{N} \sum J_{i_1} \circ \cdots \circ J_{i_n} (\Phi_0), \tag{40}$$

where the first sum is over the two different elements of length n. On $\mathbf{R}^3$ − {image points} the sequence Φ_N converges to a smooth function Φ for $N \to \infty$. Convergence follows because at level N the strengths of the new poles are suppressed by at least a factor of q^{N-1}, where $q = \sup_{i,j} a_i/(r_{ij} - a_j) < 1$. Note also that all image poles in S_i lie in fact in the interior of the concentric but smaller sphere of radius $a'_i := a_i^2/(r_{ij} - a_j)$. Cutting out the interiors of $S(a'_i, \vec{c}_i)$ $i = 1,2$ thus leaves the spheres S_i with small collar neighborhoods the two sides of which are isometrically mapped into each other by I_i (or $\tilde{I}_i$). Using two copies of the manifold so obtained we can pairwise identify these collar neighborhoods using these isometries so that an Einstein-Rosen manifold with two bridges results. Their topology is that of the twice punctured "handle" $S^1 \times S^2$ with each puncture corresponding to an asymptotically flat end. This construction generalizes to any number N of holes (or bridges), where as manifold one obtains the twice punctured connected sum of $N-1$ handles. (For the notion of connected sums see e.g. Giulini 1994.) For two holes of equal mass one may

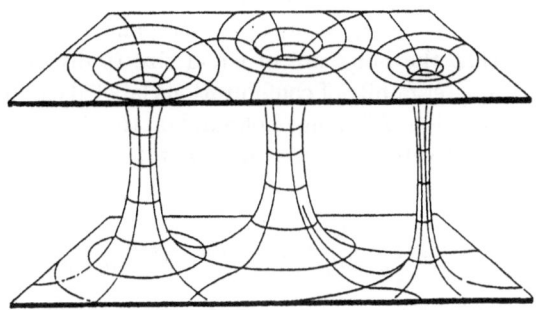

Fig. 4 Three Einstein-Rosen bridges

also just identify S_1 and S_2 and get Misner's wormhole (Misner 1960) if one uses inversions of the first kind, or its non-orientable counterpart if one uses inversions of the second kind (Giulini 1990). Both manifolds just have one end. In the second case one has the additional possibility to just close "close-off" the spheres S_i individually by identifying its antipodal points using $\tilde{I}_i$ (Giulini

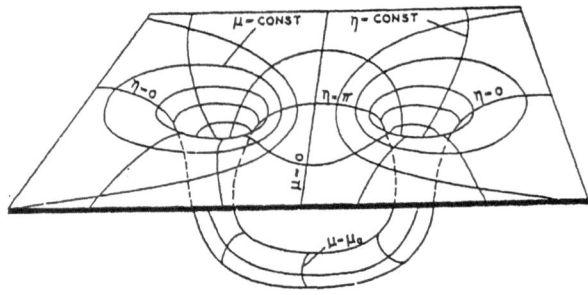

Fig. 5 The Misner Wormhole

1992). The manifold has the topology of the once punctured connected sum of two real projective spaces $\mathbf{R}P^3$. It is orientable and has only one asymptotically flat end. It can be seen as the generalization to two holes of the once punctured single $\mathbf{R}P^3$ obtained above. This construction also generalizes to any number N of holes and one obtains the once punctured connected sum of N $\mathbf{R}P^3$'s. These manifolds are doubly covered by the N-bridge manifolds discussed above.

3.3   Analytic Expressions

In the case of two holes there exists a geometrically adapted coordinate system – so called spherical bi-polar coordinates – which allows to write down explicit expressions. We take $a_1 = a_2 = a$, $\vec{c}_1 = d\,\vec{e}_z$ and $\vec{c}_2 = -d\,\vec{e}_z$. Taking the a_i's equal means that the holes are of equal size (individual mass). We thus consider a two parameter family of configurations labeled e.g. by mass (overall or individual) and separation. All image poles are on the z-axis whose strengths a_n and locations d_n (positively counted z coordinate) satisfy the coupled recursion relations

$$a_n = a_{n-1}\frac{a}{d + d_{n-1}}, \qquad d_n = d \mp \frac{a^2}{d + d_{n-1}}, \tag{41}$$

where the upper (lower) sign is valid for inversions of the first (second) kind. Using instead of a, d the parameters c, μ_0 defined by $a := c/\sinh\mu_0$, $d := c\coth\mu_0$ we can solve the recursion relations by

$$a_n = \frac{c}{\sinh n\mu_0}, \qquad d_n = c\coth n\mu_0, \tag{42}$$

for the upper sign, and for the lower sign

$$a_n = \frac{c}{\sinh n\mu_0}, \qquad d_n = c\coth n\mu_0 \qquad \text{for } n \text{ even}, \tag{43}$$

$$a_n = \frac{c}{\cosh n\mu_0}, \qquad d_n = c\tanh n\mu_0 \qquad \text{for } n \text{ odd}. \tag{44}$$

In the xz-plane we introduce bi-polar coordinates via $\exp(\mu - i\eta) = (\xi + c)/(\xi - c)$ with $\xi = z + ix$. By construction the lines of constant μ intersect those of constant η orthogonally. Both families consist of circles; those in the first family are centered on the z-axis with radius $c/\sinh\mu$ at $|z| = c\coth\mu$, and those in the second family on the x-axis with radius $c/\sin\eta$ at $|x| = c\cot\eta$. Rotating this

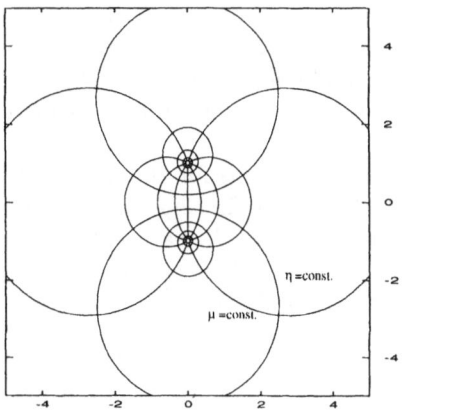

Fig. 6 The Coordinate System

system around the z-axis with azimuthal angle ϕ leads to the spherical bi-polar coordinates. Explicitly one obtains

$$x = c\frac{\sin\eta\cos\phi}{\cosh\mu - \cos\eta}, \quad y = c\frac{\sin\eta\sin\phi}{\cosh\mu - \cos\eta}, \quad z = c\frac{\sinh\mu}{\cosh\mu - \cos\eta}. \tag{45}$$

Together with (42-44) this gives

$$\frac{a_n}{\|\vec{x} \pm d_n\vec{e}_z\|} = \frac{[\cosh\mu - \cos\eta]^{1/2}}{[\cosh(\mu \pm 2n\mu_0) - \varepsilon\cos\eta]^{1/2}}, \tag{46}$$

where $\varepsilon = 1$ if one uses inversions of the first kind and $\varepsilon = -1$ if one uses those of the second kind. The final expression for the metric in (μ, η, ϕ)-coordinates can now be written down:

$$
h = \left[1 + \sum_{n=1}^{\infty} \left(\frac{a_n}{\|\vec{x} + d_n \vec{e}_z\|} + \frac{a_n}{\|\vec{x} - d_n \vec{e}_z\|} \right) \right]^4 d\vec{x} \cdot d\vec{x} \tag{47}
$$

$$
= \left[\sum_{n \in \mathbf{Z}} (\cosh(\mu + 2n\mu_0) - \varepsilon^n \cos\eta)^{-1/2} \right]^4 (d\mu^2 + d\eta^2 + \sin^2\eta \, d\phi^2). \tag{48}
$$

It nicely exhibits the isometries $(\mu, \eta, \phi) \mapsto (\mu + 2\mu_0, \eta, \phi)$ for $\varepsilon = 1$ and $(\mu, \eta, \phi) \mapsto (\mu + 2\mu_0, \pi - \eta, \phi)$ for $\varepsilon = -1$. The extrinsic curvature matrix for the surfaces of constant μ with respect to an orthonormal basis in η and ϕ direction is given by $2\Phi^{-3}\partial\Phi/\partial\mu$ times the unit matrix. Hence K has only a trace part (the surfaces of constant μ are totally umbillic) and vanishes iff $\mu = \pm\mu_0$. Hence in both cases, $\varepsilon = \pm 1$, the apparent horizons are also totally geodesic (this we already knew for $\varepsilon = 1$).

Next we turn to the expressions for the masses. We shall follow Lindquist (1963) and define the mass of the first hole by appropriately applying (35): We sum all the "bare masses" $2a_i$ in S_1, each enhanced by an interaction factor $1 + \chi_i'$ which includes the interactions of each pole in S_1 with any pole in S_2, but *not* with poles in S_1. This we write as

$$
m_1 = 2 \sum_{i \in S_1} a_i \left(1 + \sum_{j \in S_2} \frac{a_j}{r_{ij}} \right), \tag{49}
$$

with the obvious meaning of "$\in$". Since $m_1 = m_2$ we write m for the individual mass and M for the overall mass. The latter is just the sum of all $2a_i$. Using (42-44) one obtains (quantities referring to $\varepsilon = -1$ carry a tilde)

$$
m = 2c \sum_{n=1}^{\infty} \frac{n}{\sinh n\mu_0}, \qquad M = 4c \sum_{n=1}^{\infty} \frac{1}{\sinh n\mu_0}, \tag{50}
$$

for $\varepsilon = 1$, and for $\varepsilon = -1$

$$
\tilde{m} = 2c \sum_{n=1}^{\infty} \frac{2n}{\sinh 2n\mu_0} + 2c \sum_{n=0}^{\infty} \frac{2n+1}{\cosh(2n+1)n\mu_0}, \tag{51}
$$

$$
\tilde{M} = 4c \sum_{n=1}^{\infty} \frac{1}{\sinh 2n\mu_0} + 4c \sum_{n=0}^{\infty} \frac{1}{\cosh(2n+1)n\mu_0}. \tag{52}
$$

As mentioned above, we define the distance of the holes as the geodesic distance of the apparent horizons $\mu = \pm\mu_0$. The shortest geodesic connecting these two surfaces is $\eta = \pi$. For $\varepsilon = 1$ its length, l, may be expressed in closed form:

$$
l = 2c(1 + 2m\mu_0), \tag{53}
$$

with m from (50). I have not been able to find such a compact expression in the case $\varepsilon = -1$.

Like $\tilde{l}$, many quantities of interest cannot be evaluated in closed form. In these cases it may be useful to expand in powers of m/l. Numerical studies show that additional outer apparent horizons form (i.e. the holes merge) for values above $m/l \simeq 0.26$ (Smarr et al 1976), so that good convergence holds up to the merging ratio.

Comparing $\varepsilon = 1$ to $\varepsilon = -1$. We have seen that mathematically these two cases differ by allowing different topologies. But are there more physical aspects in which they differ? A natural question is how for fixed "physical" variables $m = \tilde{m}$ and $l = \tilde{l}$ the total energies M and $\tilde{M}$ differ (Giulini 1990). One finds

$$\frac{\tilde{M} - M}{M} = -\left(\frac{m}{2l}\right)^2 + \mathcal{O}(3), \tag{54}$$

showing that for $\varepsilon = -1$ the holes are slightly tighter bound (i.e. they attract stronger), although the additional energy gained until merge is only about $10^{-2}M$. This result is qualitatively unchanged if one uses Penrose's instead of Lindquist's definition of mass.

Another difference shows up in the deformation of the apparent horizons upon (adiabatic) approach of the two holes. One can define an intrinsic deformation parameter as follows: Regard (η, ϕ) as polar coordinates. The poles are the zeros of the Killing field ∂_ϕ. Define C_η as twice their geodesic distance. Among the orbits of ∂_ϕ is one of greatest length, C_ϕ. The deformation parameter is $D := (C_\eta - C_\phi)/C_\eta$. One obtains (Giulini 1990)

$$D = \frac{3}{2}\left(\frac{m}{2l}\right)^3 + \mathcal{O}(4), \tag{55}$$

$$\tilde{D} = \frac{3}{2}\left(\frac{m}{2l}\right)^2 + \mathcal{O}(3). \tag{56}$$

The power of 2 in (56) seems in conflict with the usual "tidal-force" interpretation. The shapes themselves are also different. Like eggs with the thick ends pointing towards each other in the first case, and prolonged symmetrically (with respect to reflections on the equator $\eta = \pi/2$) in the second.

Bibliography

[1] Araki H. (1959): On the time symmetric gravitational waves. Ann. Phys. (NY) **7**, 456–465

[2] Beig R. (1991): Time symmetric initial data and Penrose's quasilocal mass. Class. Quant. Grav. **8**, L205–L209

[3] Bishop N.T. (1982): The closed trapped region and the apparent horizon of two Schwarzschild black holes. Gen. Rel. Grav. **14** no. 9, 717–723

[4] Bowen J. (1984): Inversion symmetric initial data for N charged black holes. Ann. Phys. (NY) **165**, 17-37

[5] Bowen J., York J. (1980): Time-asymmetric initial data for black holes and black-hole collision. Phys. Rev. D **21**, 2047-2056

[6] Brandt S., Brügmann B. (1997): A simple construction of initial data for multiple black holes. Phys. Rev. Lett. **78**, 3606-3609

[7] Brill D (1959): On the positive definite mass of the Bondi-Weber-Wheeler time-symmetric gravitational waves. Ann. Phys. (NY) **7**, 466-483

[8] Brill D., Lindquist R. (1963): Interaction Energy in Geometrostatics. Phys. Rev. **131**, 471-476

[9] Čadež A. (1974): Apparent horizons in the two-black-hole problem. Ann. Phys. (NY) **83**, 449-457

[10] Choquet-Bruhat Y, Geroch R. (1969): Global aspects of the Cauchy problem in general relativity. Comm. Math. Phys. **3**, 334-357

[11] Choquet-Bruhat Y., York J. (1980): The Cauchy Problem. In: *General Relativity and Gravitation* Vol. 1, Ed. A. Held (Plenum Press, New York, London), 99-172

[12] Gibbons G.W. (1972): The time symmetric initial value problem for black holes. Comm. Math. Phys. **27**, 87-102

[13] Gibbons G.W. (1984): The isoperimetric and Bogomolny inequalities for black holes. In: *Global Riemannian Geometry*, Eds. T.J. Willmore and N. Hitchin (Ellis Horwood Limited, Chichester), 194-202

[14] Giulini D. (1990): Interaction energies for three-dimensional wormholes. Class. Quant. Grav. **7**, 1271-1290

[15] Giulini D. (1992): Two body interaction energies in classical general relativity. In: *Relativistic Astrophysics and Cosmology*, Eds. S. Gottlöber, J.P. Mücket, V. Müller (World Scientific, Singapore), 333-338

[16] Giulini D. (1994): 3-manifolds for relativists. Int. J. Theo. Phys. **33**, 913-930

[17] Hawking S.W., Ellis G.F.R. (1973): *The large scale structure of space-time* (Cambridge University Press)

[18] Huisken G., Ilmanen T. (1997): The Riemannian Penrose Inequality. Int. Math. Res. Not. **20**, 1045-1058

[19] Lindquist R. (1963): Initial value problem on Einstein-Rosen Manifolds. J. Math. Phys. **4**, 938-950

[20] Misner C. (1960): Wormehole initial conditions. Phys. Rev. **118**, 1110–1111

[21] Misner C. (1963): The method of images in geometrostatics. Ann. Phys. (NY) **24**, 102–117

[22] Ó Murchadha N., York J. (1974): Initial-value problem of general relativity. I. General formulation and physical interpretation. Phys. Rev. D **10**, 428–436

[23] Penrose R. (1982): Quasi-local mass and angular momentum in general relativity. Proc. Roy. Soc. Lond. A **381**, 53–63

[24] Penrose R. (1984): Mass in general relativity. In: *Global Riemannian Geometry*, Eds. T.J. Willmore and N. Hitchin (Ellis Horwood Limited, Chichester), 203–213

[25] Smarr L., Čadež A., DeWitt B., Eppley K. (1976): Collision of two black holes: Theoretical framework. Phys. Rev. D **14**, 2443–2452

[26] Spivak M. (1979): *Differential Geometry I*, 2nd ed., (Publish or Perish, Wilmington, Delaware), 462–463

[27] Tod K.P. (1983): Some examples of Penrose's quasi-local mass construction. Proc. Roy. Soc. Lond. A **388**, 457–477

[28] Witt D. (1986): Vacuum space-times that admit no maximal slice. Phys. Rev. Lett. **57**, 1386–1389

[29] York J. (1973): Conformally invariant orthogonal decomposition of symmetric tensors on Riemannian manifolds and the initial-value problem of general relativity. J. Math. Phys. **14**, 456–464

Numerical Approach to Black Holes

Edward Seidel

[1] Max-Planck-Institut für Gravitationsphysik, Schlaatzweg 1, D-14473 Potsdam, Germany
[2] University of Illinois, Urbana, IL 61801, USA

Abstract. I describe approaches to the study of black hole spacetimes via numerical relativity. After a brief review of the basic formalisms and techniques used in numerical black hole simulations, I discuss a series of calculations from axisymmetry to full 3D that can be seen as stepping stones to simulations of the full 3D coalescence of two black holes. In particular, I emphasize the interplay between perturbation theory and numerical simulation that build both confidence in present results and tools to interpret results of future simulations of coalescence.

1   Introduction

In these lectures I concentrate on recent results in 3D numerical studies of black holes, building towards extraction of waveforms in 3D. Numerical relativity is enjoying a surge of interest as computer power increases dramatically, allowing vastly larger simulations to be performed, and as gravitational wave detectors promise to actually see signals generated by events such as collisions of black holes in the Universe[1].

For these reasons, numerical evolutions of black hole data sets are becoming more and more common[2]. As black hole collisions are considered a most promising source of signals to be detected by these observatories, it is crucial to have a detailed theoretical understanding of the coalescence process that can only be achieved through numerical simulation. In particular, it is most important to be able to simulate accurately the excitation of the coalescing black holes, to follow the waves generated in the process, and to extract gravitational waveforms expected to be seen by detectors. I concentrate in these lectures on the foundations of this subject, on the present state of numerical black hole simulations, on the techniques used to extract waveforms, and on the prospects for obtaining accurate waveforms to be used in conjunction with data collected by gravitational wave observatories during the next five years.

Black holes present very difficult computational problems to overcome, as one must (a) deal with singularities inside the black holes, (b) follow the highly nonlinear regime in the coalescence process taking place near the horizons, and (c) calculate the linear regime in the radiation zone where the waves represent a very small perturbation on the background spacetime metric. In the next sections I discuss progress that has been made to handle each of these problems, and in particular I focus on a series of testbed calculations of increasing complexity that brings us closer to the goal of simulating true 3D coalescing black holes.

2   Formalisms, Initial Data, and Tools for Numerical Black Hole Studies

In this section I review briefly a number of important concepts and techniques needed for understanding numerical approaches to black hole evolutions and interpreting the results. This is not meant as a detailed review of these subjects, but rather as a brief description with pointers to sources of more information. I have borrowed heavily from the PhD thesis of Karen Camarda (with permission). The entire text of this thesis will be made available online from a link to the AEI web server (http://www.aei-potsdam.mpg.de) at some future date.

2.1   The 3+1 (ADM) formalism

This material is familiar to many, and developed in more detail in Giulini's lectures in this volume. Here I just sketch the main ideas mainly to introduce the notation used here. Einstein's equations treat the time and space variables on an equal footing. In order to solve these equations numerically, it is convenient to cast the equations in the form of a Cauchy problem. Once that is done, one must find an appropriate initial data set. Then, to evolve the initial data set, one must choose an evolution system and specify spatial boundary conditions and gauge conditions, as described below.

In the ADM splitting of the Einstein equations, one considers the spacetime to consist of a foliation of three dimensional spatial hypersurfaces, each the level surface of some time coordinate (see, e.g., [3]). The invariant distance between two infinitesimally separated events can then be written as

$$ds^2 = -(\alpha^2 - \beta^i \beta_i)dt^2 + 2\beta_i dx^i dt + \gamma_{ij} dx^i dx^j. \tag{1}$$

The so-called lapse function α determines the proper time $d\tau = \alpha dt$ measured by an observer falling normal to the hypersurface, or slice. The shift vector β^i determines the coordinate distance a constant coordinate point moves away from the normal vector to the slice as one advances from one slice to the next. Finally, γ_{ij} is the metric induced on the given 3D hypersurface specified by $t = const$.

Einstein's equations that govern the metric tensor γ_{ij} can be written in the deceptively simple form $G_{\mu\nu} = 8\pi T_{\mu\nu}$, where the Einstein tensor $G_{\mu\nu}$ is a nonlinear differential operator on the metric, and $T_{\mu\nu}$ is the stress energy tensor. In vacuum spacetimes $T_{\mu\nu}$ vanishes, and the Einstein equations reduce to setting each component of the 4-dimensional Ricci tensor, $^{(4)}R_{\mu\nu}$, to zero. We specialize to the vacuum case for the rest of this paper.

It is convenient to introduce a tensor known as the extrinsic curvature, K_{ij}, which describes the curvature of the 3D slice in the 4D Lorentzian space in which it is embedded. This quantity is defined via the Lie derivative of the 3-metric with respect to the future-pointing unit normal vector to the 3-surface, i.e.,

$$K_{ij} = -\frac{1}{2}\mathcal{L}_n \gamma_{ij}$$

$$= -\frac{1}{2\alpha}\left(\partial_t \gamma_{ij} - D_i \beta_j - D_j \beta_i\right). \tag{2}$$

In order to express the Einstein equations in terms of the 3+1 variables, one decomposes the Ricci tensor into its timelike and spacelike components. Setting the timelike components $^{(4)}R_{0\alpha}$ to zero leads to the Hamiltonian and momentum constraints for the 3-metric and the extrinsic curvature:

$$R + (trK)^2 - K^{ij}K_{ij} = 0 \tag{3}$$

$$D_j(K^{ij} - \gamma^{ij}trK) = 0. \tag{4}$$

As described in Giulini's lecture and below, these equations are used to provide initial data for evolution.

Setting the spacelike components $^{(4)}R_{ij}$ to zero, and using the definition of the extrinsic curvature, results in twelve evolution equations:

$$\partial_t \gamma_{ij} = -2\alpha K_{ij} + D_i\beta_j + D_j\beta_i \tag{5}$$

$$\begin{aligned} \partial_t K_{ij} = &-D_iD_j\alpha + \alpha[R_{ij} + (trK)K_{ij} - 2K_{ik}K^k{}_j] \\ &+\beta^k D_k K_{ij} + K_{ik}D_j\beta^k + K_{jk}D_i\beta^k. \end{aligned} \tag{6}$$

In these equations, R_{ij} is the Ricci tensor, R the scalar curvature, and D_i the covariant derivative associated with the metric of the spacelike slices, γ_{ij}.

Analytically, the evolution equations are guaranteed to preserve the constraint equations. Numerically, however, they may not. Therefore, it is important at least to monitor, if not to enforce, the constraint equations during the evolution, in order to gauge the size of numerical errors. There are numerous ideas on how to deal with the constraint equations during a numerical evolution, which I have discussed somewhat in Ref. [4].

This is the standard evolution system that is used in most calculations in numerical relativity, and for the numerical results presented in later sections in this paper. However, it is important to note that this evolution system has been completely overhauled in recent years into a form that is explicitly *hyperbolic*. The equations are recast into a completely first order form in space in time, and can be written in a special "flux conservative" form:

$$\partial_t \mathbf{u} + \partial_k F^k(\mathbf{u}) = S(\mathbf{u}) \tag{7}$$

where the vector $\mathbf{u}$ displays the set of variables and both "fluxes" F^k and "sources" S are vector valued functions. Furthermore, under a broad set of conditions the system can be diagonalized, with a complete set of eigenfields with real eigenvalues, and in that case the system is said to be hyperbolic. This technical property has a number of important consequences: *(a)* It is precisely the form of the equations known in hydrodynamics, for which many advanced numerical techniques have been developed. These techniques can now be applied to the Einstein equations for the first time. *(b)* The decomposition of the system into its characteristic fields and eigenvalues (speeds) allows one a better understanding of which quantities are propagating and at what speeds, which can be essential in identifying radiation, gauge information, and can be useful in applying boundary conditions for ingoing and outgoing quantities.

Building on earlier work by Choquet-Bruhat and Rugerri[5], Bona and Massó began to study this problem in the late 1980's, and by 1992 they had developed a fully hyperbolic system for the Einstein equations with harmonic lapse and zero shift[6]. This work was generalized recently to apply to a large family of slicing conditions and arbitrary shift[7]. Independently, another system was developed by Abrahams, Anderson, Choquet-Bruhat, and York[8]. This system contains an extra time derivative, so that a second order hyperbolic system for the extrinsic curvature K_{ab} is found. 1D and 3D codes based on these two new formulations are under development at present. These works have sparked considerable interest in the relativity community, and now I am aware of several more systems of hyperbolic equations for the Einstein system (see, e.g., [9, 10]and references therein). Although all results presented in this paper are based on the ADM formulation, a large scale computational effort at AEI is presently underway to compare various formulations of the equations, on problems such as those described below. This code, called Cactus, will be made available to the community sometime in 1998. An announcement will made on the AEI web page at http://www.aei-potsdam.mpg.de when it is ready.

2.2   Initial data

In this section, I again build on Giulini's lectures in this volume on initial data. For most of the results presented here, we will consider a family of distorted single black hole initial data. We generalize the Schwarzschild construction discussed in Giulini's section 3.1 to include a "Brill wave", which he also discussed in his section 2. In this construction the black hole has been distorted by the presence of an adjustable torus of nonlinear gravitational waves ("Brill waves") which surround it. The amplitude and shape of the torus can be specified by hand, as described below, and can create very highly distorted black holes. Such initial data sets, and their evolutions in axisymmetry, have been studied extensively, as described in Refs.[11, 12, 13]. For our purposes, we consider them as convenient initial data that create a distorted black hole that mimics the merger, just after coalescence, of two black holes colliding [14].

Following[12], we write the 3–metric as

$$dl^2 = \tilde{\psi}^4 \left(e^{2q} \left(d\eta^2 + d\theta^2 \right) + \sin^2 \theta d\phi^2 \right), \tag{8}$$

where η is a radial coordinate related to the Cartesian coordinates by

$$\sqrt{x^2 + y^2 + z^2} = e^{\eta}. \tag{9}$$

(We have set the scale parameter m in Giulini's section 3.1 to be 2.) Given a choice for the "Brill wave" function q, the Hamiltonian constraint leads to an elliptic equation for the conformal factor $\tilde{\psi}$. The function q represents the gravitational wave surrounding the black hole, and is chosen to be

$$q\left(\eta, \theta, \phi\right) = a \sin^n \theta \left(e^{-\left(\frac{\eta+b}{w}\right)^2} + e^{-\left(\frac{\eta-b}{w}\right)^2}\right)\left(1 + c \cos^2 \phi\right). \tag{10}$$

Thus, an initial data set is characterized by the parameters (a, b, w, n, c), where, roughly speaking, a is the amplitude of the Brill wave, b is its radial location, w its width, and n and c control its angular structure. Note that we have generalized the original axisymmetric construction of Ref. [12] to full 3D by the addition of the parameter c. If the amplitude a vanishes, the undistorted Schwarzschild solution results, leading to

$$\tilde{\psi} = 2 \cosh\left(\frac{\eta}{2}\right). \tag{11}$$

Thus, our Eq. (8) is exactly Giulini's Eq. (30) written in different coordinates.

We note that just as the Schwarzschild geometry has an isometry that leaves the metric unchanged under the operation $\eta \to -\eta$, our data sets also have this property, even in the presence of the Brill wave. As discussed in [15, 13] and below, this condition can also be applied during the evolution and in Cartesian coordinates as well.

2.3   Gauge conditions

The Einstein equations do not specify evolution equations for the lapse function α or the shift vector β^i. These quantities are gauge quantities and can be chosen at will. Thus, the first thing one needs to do before evolving the initial data is to decide on a gauge condition. We discuss here a few choices for a lapse function, also known as the slicing condition, because it determines which spatial slices will be used in the evolution. This subject is discussed in more detail in many places (see, e.g., [4]).

The simplest slicing condition is geodesic slicing, which amounts to setting $\alpha = 1$ and $\beta^i = 0$. In geodesic slicing, grid points correspond to freely falling observers. This slicing condition provides a good test of a numerical black hole code, since for a Schwarzschild black hole, an observer initially on the horizon will hit the singularity in a time of πM (see, e.g., [16]). Thus, a computer code should "crash" at this time. This has been an important test of many black hole codes, and has been discussed extensively in, e.g., [15].

Although geodesic slicing is simple, it does not allow one to cover enough of the spacetime because it allows slices to hit singularities very early in the evolution. One would like a "singularity avoiding slicing". Such a slicing results in a lapse function which is very small in regions close to a singularity, effectively stopping evolution there. One such slicing is maximal slicing. Maximal slicing is so-called because it produces the lapse that maximizes the volume of the spatial hypersurface. Because volume elements are small near a singularity, these slices avoid singularities. It can be shown that this property is equivalent to setting the trace of the extrinsic curvature equal to zero. Thus, to obtain the maximal slicing equation for the lapse, one takes the trace of the evolution equation for the extrinsic curvature (6) and sets $tr K$ equal to zero. This yields the following elliptic equation for the lapse:

$$D^i D_i \alpha = R\alpha. \tag{12}$$

Elliptic equations are time-consuming to solve numerically, especially in 3D, so a number of *algebraic* singularity-avoiding conditions on the lapse have been developed as well. These slicings are based on the fact that $det(\gamma_{ij})$ will shrink in the vicinity of a curvature singularity, and since $det(\gamma_{ij})$ is a simple algebraic combination of metric functions, it can be cheaply and easily computed. The algebraic slicings that have been tried generally have the form $\alpha = \alpha_0 f(\hat{\gamma})$, where $\hat{\gamma} = det(\hat{\gamma}_{ij})$. A particular favorite used by the groups at NCSA and AEI is obtained by setting $f(\hat{\gamma})$ to $1 + \ln \hat{\gamma}$, and was used for some of the 3D simulations discussed below. These algebraic slicings have various drawbacks as well. For one, they tend to be very "local"; if a problem develops in the three metric (say a large gradient) in one small region of the calculation, the lapse responds locally and immediately, which can exacerbate the problem[15]. For another problem, they have been associated with "slicing pathologies" or "coordinate shocks" as discussed by Alcubierre [17]. These problems aside, they have still been found to be very useful in numerical relativity.

2.4   Boundary conditions

When one is evolving a black hole spacetime numerically, one needs to specify boundary conditions at two places: the inner boundary, generally chosen at the black hole's "throat", or isometry surface, and the outer boundary, far away from the black hole. In this section I discuss present approaches to these two problems.

Inner Boundary

Throat Boundary Conditions The throat of a black hole is the inner-most surface which is locally an areal minimum. For a Schwarzschild black hole, the throat is the initial event horizon. The usual way to apply a boundary condition on the throat is to require the geometry, and thus the metric, to be identical under a certain mapping from the outside of the throat to the inside. In pictures, this makes "both sides of the wormhole" have the identical geometry. This mapping is known as an isometry condition, and was already discussed above and in Giulini's lectures.

The isometry condition takes the form of a map J which identifies two asymptotically flat sheets through the throat [18]:

$$\gamma_{ij}(\vec{x}) = \pm J_i^k J_j^\ell \gamma_{k\ell}(J(\vec{x})) \tag{13}$$

with

$$J\left(\vec{x}\right) = a^2 \frac{\vec{x} - \vec{c}}{\left|\vec{x} - \vec{c}\right|^2} + \vec{c} \tag{14}$$

and $J_i^j = \frac{\partial J^j}{\partial x^i}$, where a is the radius of the throat centered at $\vec{c}$. The mapping (13) is applicable to both the metric and the extrinsic curvature tensor components. The evolution equations themselves also respect this symmetry if the lapse and shift variables obey the isometry as well. Generally, the radial component of the shift must vanish on the throat, and the lapse and extrinsic curvature components must have the same sign (positive or negative).

The use of the isometry condition, coupled with singularity avoiding slicings, leads to time slices that advance very slowly inside the horizon and more rapidly outside. These slicings have been successful to varying degrees, enabling detailed studies of black hole spacetimes for time periods of up to order $t = 100M$, where M is the mass of the black hole. However, these conditions do not completely solve the problem; they merely serve to delay the breakdown of the numerical evolution. In the vicinity of the singularity, these slicings inevitably contain a region of abrupt change near the horizon. This behavior typically manifests itself in the form of sharply peaked profiles in the spatial metric functions [19], "grid stretching" [20], large coordinate shift [21] on the black hole throat, *etc.* These features are most pronounced where the time slices are sharply bent towards the past. Numerical simulations will eventually crash due to these pathological properties of the slicing. For discussions of these problems, see, e.g., [22, 23]. This is one of the fundamental problems of numerical black hole evolution.

Apparent Horizon Boundary Conditions Another approach to the inner boundary is under development by many groups at present. Cosmic censorship suggests that in physical situations, singularities are hidden inside black hole horizons. Because the region of spacetime inside the horizon cannot causally affect the region of interest outside the horizon, one is tempted to cut away the interior region containing the singularity and evolve only the singularity-free region outside. The procedure of cutting away the singular region will drastically reduce the dynamic range, making it easier to maintain accuracy and stability. With the singularity removed from the numerical spacetime, there is in principle no physical reason why black hole codes cannot be made to run indefinitely.

A number of recent papers [22, 23, 24, 25] have demonstrated that a horizon boundary condition can be realized. There are two basic ideas behind the implementation of the apparent horizon boundary condition: *(a)* A shift (or grid velocity in some cases [25]) is used to control the motion of the horizon, tying the spatial coordinates to the spatial geometry and causal structure. *(b)* It is important to use a finite differencing scheme which respects the causal structure of the spacetime. Since the horizon is a one-way membrane, quantities on the horizon can be affected only by quantities outside but not inside the horizon. Hence, in a finite differencing scheme which respects the causal structure, all quantities on the horizon can be updated solely in terms of known quantities residing on or outside the horizon.

These ideas are presently under development by various groups. Although I will not discuss them further here, as they are not used in the results presented below, they will most likely play an essential role in developing codes to study the full 3D coalescence of two black holes. Furthermore, any codes developed with these techniques will need to be tested on spacetimes such as those discussed below, to ensure that they are able to reproduce the waveforms from the same spacetimes.

Outer Boundary Appropriate conditions for the outer boundary have yet to be derived for 3D. In 1D and 2D codes, the outer boundary is simply placed far enough away that the spacetime is nearly flat there, and static or flat boundary conditions can be specified for the evolved functions. However, due to the constraints placed on us by limited computer memory, this is not currently possible in 3D. Most results to date have been computed with the evolved functions kept static at the outer boundary, even if the boundaries are too close for comfort in 3D!

There are several other approaches under development that promise to improve this situation greatly that I will not have time to explore here, but should be mentioned. First, by using perturbation theory, as described later in this paper, it is possible to identify quantities in the metric functions that obey wave equations. These can be used to provide boundary conditions on the metric and extrinsic curvature functions in an actual evolution, as described in a recent paper [26]. Secondly, one can use the hyperbolic formulations of the Einstein equations to find eigenfields, for which outgoing conditions can in principle be applied. Finally, "Cauchy-Characteristic matching" attempts to match spacelike slices to null slices at some finite radius, and the null slices can be carried out to scri. These methods have different strengths and weaknesses, but all promise to improve boundary treatments significantly, helping to enable longer evolutions than are presently possible.

One final point: In order to avoid unnecessary computations, when possible, a useful trick is to take advantage of symmetry inherent in initial data to evolve only one octant of the Cartesian grid. When this is done, boundary conditions must also be supplied for the planes $x = 0$, $y = 0$, and $z = 0$. The setting of boundary conditions on these planes is straightforward as they are determined by the symmetries of the problem.

3 Ladder of Credibility: Building Towards Waveforms for LIGO

In this section I present a series of ideas and calculations that lead to what I call a "Ladder of Credibility". Problems need to be studied in sequence from easier to more complicated, leading ultimately to the 3D spiraling coalescence problem. I will discuss the foundations of black hole perturbation theory, and show how it can be used both to test and to interpret results of full numerical simulations, beginning with axisymmetry, and building to full 3D simulations.

3.1 Perturbation theory

In this section I give a very brief overview of the theory of perturbations of the Schwarzschild spacetime. This is a topic which is very rich and has a long history. A more detailed discussion can be found in [27]. In the next section, I will discuss how we apply the theory to study black hole data sets.

One begins the analysis by writing the full metric as a sum of the Schwarzschild metric $\overset{o}{g}_{\alpha\beta}$ and a small perturbation $h_{\alpha\beta}$:

$$g_{\alpha\beta} = \overset{o}{g}_{\alpha\beta} + h_{\alpha\beta}. \tag{15}$$

One then plugs this expression into the vacuum Einstein equations to get an equation for the perturbation tensor $h_{\alpha\beta}$. One can separate off the angular part of the solution by expanding $h_{\alpha\beta}$ in spherical tensor harmonics, as was originally done by Regge and Wheeler [28]. For each ℓ, m mode, one gets separate equations for the perturbed metric functions, which we now denote by $h_{\alpha\beta}^{(\ell m)}$.

There are two independent expansions: one, known as even parity, which does not introduce any rotational motion to the hole, and one, known as odd parity, which does. We will concentrate here on even parity perturbations, as these are the ones which will be relevant for studying the datasets discussed here. The odd parity perturbations produce equations which are very similar, and both can be considered in the general case.

We also note here that this treatment is presently restricted to perturbations of Schwarzschild black holes. For the more general rotating case, one would like to use the Teukolsky formalism describing perturbations of Kerr. This is much more complicated, and has not yet been applied to numerical black hole simulations of the kind discussed in this paper. This is an important research topic that needs attention soon!

When dealing with perturbations in relativity, one must be careful about interpreting the various metric components $h_{\alpha\beta}$ in terms of physics. Under a coordinate transformation of the form $x^\mu \to x^\mu + \delta x^\mu$, the metric coefficients will transform as well. One can use this gauge freedom to eliminate certain metric functions to simplify the corresponding equations for the perturbations. Another, more powerful approach, developed first by Moncrief, is to consider linear combinations of the $h_{\alpha\beta}$ and their derivatives that are actually *invariant* under the gauge transformation above. In either case, the analysis leads to a single wave equation for the perturbations of the black hole:

$$\frac{\partial^2 \psi^{(\ell m)}}{\partial r^{*2}} - \frac{\partial^2 \psi^{(\ell m)}}{\partial t^2} + V^{(\ell)}(r)\,\psi^{(\ell m)} = 0, \tag{16}$$

where the potential function $V^{(\ell)}(r)$ is given by

$$V^{(\ell)}(r) = \left(1 - \frac{2M}{r}\right) \times \tag{17}$$

$$\left\{ \frac{1}{\Lambda^2} \left[\frac{72M^3}{r^5} - \frac{12M}{r^3} (\ell - 1)(\ell + 2) \left(1 - \frac{3M}{r}\right) \right] + \frac{\ell(\ell - 1)(\ell + 2)(\ell + 1)}{r^2 \Lambda} \right\},$$

where r is the standard Schwarzschild radial coordinate, and r^* is the so-called tortoise coordinate, given by $r^* = r + 2M \ln(r/2M - 1)$, and Λ is a function of r described below. For all ℓ, the potential function is positive and has a peak near $r = 3M$. This equation was first derived by Zerilli [29]. Regge and Wheeler found an analogous equation for the odd parity perturbations, which are much simpler than the even parity perturbations, 13 years earlier in 1957 [28]. Note that the potential depends on ℓ, but is independent of m. This remarkable equation (along with its odd-parity counterpart, known as the Regge-Wheeler equation) completely describes gravitational perturbations of a Schwarzschild black hole. It continues to be studied by many researchers now 40 years after this perturbation program was begun in 1957.

Using this equation, it has been shown that Schwarzschild is stable to perturbations, and has characteristic oscillation frequencies known as quasinormal modes. These modes are solutions to the Zerilli equation as given in Eq. (16) which are completely ingoing at the horizon ($r^* = -\infty$) and completely outgoing at infinity ($r^* = \infty$). For each ℓ-mode, independent of m, there is a fundamental frequency, and overtones. These are very important results! One expects that a black hole, when perturbed in an arbitrary way, will oscillate at these quasinormal frequencies. This will give definite signals to look for with gravitational wave observatories.

These quasinormal frequencies are complex, meaning they have an oscillatory and a damping part (not growing—black holes are stable!), so the oscillations die away as the waves carry energy away from the system. The frequencies depend only on the mass, spin, and charge of the black hole.

There are numerous ways in which this perturbation theory has become essential in numerical black hole simulations, and the rest of this paper will concentrate on this subject. First of all, the fact that perturbation theory reveals that black holes have quasinormal mode oscillations raises expectations about the evolution of distorted black holes: they should, at least in the linear regime, oscillate at these frequencies which should be seen in fully nonlinear numerical simulations. But are they still seen in highly nonlinear interactions, e.g., in the collision of two black holes? Secondly, as we will see, this perturbation theory provides a method by which to separate out the Schwarzschild background from the wave degrees of freedom, which can be used to find waves in numerical simulations. Finally, as the perturbations are governed by their own evolution equation, this equation should be useful to actually evolve some classes of black hole initial data, as long as they represent slightly perturbed black holes, and this can be used as an important check of fully nonlinear numerical codes. Certainly, during the late stages of black hole coalescence, the system will settle down to a slightly perturbed black hole, and numerical codes had better be able to accurately compute waves from such systems if they are to be used to help researchers find signals in actual data collected by gravitational wave observatories.

3.2   Waveform extraction

In this section I show how to take this perturbation theory and apply it in a practical way to numerical black hole simulations. One considers now the numerically generated metric $g_{\alpha\beta,num}$ to be the sum of a spherically symmetric part and a perturbation: $g_{\alpha\beta,num} = \overset{o}{g}_{\alpha\beta} + h_{\alpha\beta}$, where the perturbation $h_{\alpha\beta}$ is expanded in tensor spherical harmonics as before. To compute the elements of $h_{\alpha\beta}$ in a numerical simulation, one integrates the numerically evolved metric components $g_{\alpha\beta,num}$ against appropriate spherical harmonics over a coordinate 2–sphere surrounding the black hole. The orthogonality of the $Y_{\ell m}$'s allows one to "project" the contributions of the general wave signal into individual modes, as explained below. The resulting functions can then be combined in a gauge-invariant way, following the prescription given by Moncrief[30], leading directly to the Zerilli function. This procedure was originally developed by Abrahams[31] and developed further by various groups.

As mentioned above, we assume the general metric can be decomposed into its spherical and non-spherical parts. The spherical part $\overset{o}{g}_{\mu\nu}$ will of course be Schwarzschild, but we will in general not know the mass of this Schwarzschild background, or what coordinate system it will be in. However, in general, we know it can be written

$$\overset{o}{g}_{\mu\nu} = \begin{pmatrix} -N^2 & 0 & 0 & 0 \\ 0 & A^2 & 0 & 0 \\ 0 & 0 & R^2 & 0 \\ 0 & 0 & 0 & R^2\sin^2\theta \end{pmatrix} \tag{18}$$

where the functions N, A, and R are functions of our coordinate radius r and time t. Regge and Wheeler showed that $h_{\mu\nu}$ for even-parity perturbations can be written

$$h_{tt} = -N^2 H_0^{(\ell m)} Y_{\ell m} \tag{19}$$

$$h_{tr} = H_1^{(\ell m)} Y_{\ell m} \tag{20}$$

$$h_{t\theta} = h_0^{(\ell m)} Y_{\ell m,\theta} \tag{21}$$

$$h_{t\phi} = h_0^{(\ell m)} Y_{\ell m,\phi} \tag{22}$$

$$h_{rr} = A^2 H_2^{(\ell m)} Y_{\ell m} \tag{23}$$

$$h_{r\theta} = h_1^{(\ell m)} Y_{\ell m,\theta} \tag{24}$$

$$h_{r\phi} = h_1^{(\ell m)} Y_{\ell m,\phi} \tag{25}$$

$$h_{\theta\theta} = R^2 K^{(\ell m)} Y_{\ell m} + R^2 G^{(\ell m)} Y_{\ell m,\theta\theta} \tag{26}$$

$$h_{\theta\phi} = R^2 G^{(\ell m)} (Y_{\ell m,\theta\phi} - \cot\theta Y_{\ell m,\phi}) \tag{27}$$

$$h_{\phi\phi} = R^2 K^{(\ell m)} \sin^2\theta Y_{\ell m} + R^2 G^{(\ell m)} (Y_{\ell m,\phi\phi} + \sin\theta\cos\theta Y_{\ell m,\theta}) \tag{28}$$

The spherical part of the metric, given in the functions N, A, and R, can be obtained by projecting the full metric against Y_{00}, yielding the following expressions[1]:

$$N^2 = -\frac{1}{4\pi} \int g_{tt} d\Omega \tag{29}$$

$$A^2 = \frac{1}{4\pi} \int g_{rr} d\Omega \tag{30}$$

$$R^2 = \frac{1}{8\pi} \int \left(g_{\theta\theta} + \frac{g_{\phi\phi}}{\sin^2\theta} \right) d\Omega \tag{31}$$

Each ℓm-mode of $h_{\mu\nu}$ can then be obtained by projecting the full metric against the appropriate $Y_{\ell m}$:

$$h_1^{(\ell m)} = \frac{1}{\ell(\ell+1)} \int \left(g_{r\theta} Y_{\ell m,\theta}^* + \frac{g_{r\phi}}{\sin^2\theta} Y_{\ell m,\phi}^* \right) d\Omega \tag{32}$$

$$H_2^{(\ell m)} = \frac{1}{A^2} \int g_{rr} Y_{\ell m}^* d\Omega \tag{33}$$

$$G^{(\ell m)} = \frac{1}{R^2 \ell(\ell+1)(\ell-1)(\ell+2)} \int \left[\left(g_{\theta\theta} - \frac{g_{\phi\phi}}{\sin^2\theta} \right) \left(Y_{\ell m,\theta\theta}^* - \cot\theta Y_{\ell m,\theta}^* - \frac{1}{\sin^2\theta} Y_{\ell m,\phi\phi}^* \right) + \frac{4}{\sin^2\theta} g_{\theta\phi} (Y_{\ell m,\theta\phi}^* - \cot\theta Y_{\ell m,\phi}^*) \right] d\Omega \tag{34}$$

$$K^{(\ell m)} = \frac{\ell(\ell+1)}{2} G^{(\ell m)} + \frac{1}{2R^2} \int \left(g_{\theta\theta} + \frac{g_{\phi\phi}}{\sin^2\theta} \right) Y_{\ell m}^* d\Omega \tag{35}$$

In practice, we generally do not extract A to compute $H_2^{(\ell m)}$, but rather we assume it to have the form $1 - \frac{2M}{R}$, where we take M to be the ADM mass of

---

[1] We thank Gabrielle Allen for pointing out the error in the expression for R^2 in reference [11].

the spacetime. Thus, from this point forward we assume that the background spacetime metric is adequately represented by Schwarzschild *in Schwarzschild coordinates*. This may not be true in the general case, but in our simulations it has been adequate to assume this.

In the case where the background is Schwarzschild, Moncrief showed that the Zerilli function is gauge invariant in the sense discussed above, and can be constructed from the Regge-Wheeler variables as follows:

$$\psi^{(\ell m)} = \sqrt{\frac{2(\ell-1)(\ell+2)}{\ell(\ell+1)}} \frac{4RS^2 k_2^{(\ell m)} + \ell(\ell+1)R k_1^{(\ell m)}}{\Lambda}, \tag{36}$$

where

$$\Lambda \equiv \ell(\ell+1) - 2 + \frac{6M}{R} \tag{37}$$

$$k_1^{(\ell m)} \equiv K^{(\ell m)} + SRG_{,R}^{(\ell m)} - 2\frac{S}{R}h_1^{(\ell m)} \tag{38}$$

$$k_2^{(\ell m)} \equiv \frac{H_2^{(\ell m)}}{2S} - \frac{1}{2\sqrt{S}}\frac{\partial}{\partial R}\left(\frac{RK^{(\ell m)}}{\sqrt{S}}\right) \tag{39}$$

$$S \equiv 1 - \frac{2M}{R}. \tag{40}$$

In order to compute the Regge-Wheeler perturbation functions h_1, H_2, G, and K, one needs the spherical metric functions on some 2-sphere. We get these in 3D by interpolating the Cartesian metric functions onto a surface of constant coordinate radius, and computing the spherical metric functions from these using the standard transformation. Second order interpolation is used. In order to compute the needed radial derivatives of these functions, we first compute the derivative of the Cartesian metric functions with respect to the coordinate radius on the Cartesian grid. We interpolate these quantities onto the 2-sphere. Then, to get the derivatives with respect to the Schwarzschild radius R, we use the derivative of Eq. (31) with respect to the coordinate r, giving

$$\frac{\partial R}{\partial r} = \frac{1}{16\pi R}\int\left(g_{\theta\theta,r} + \frac{g_{\phi\phi,r}}{\sin^2\theta}\right)d\Omega \tag{41}$$

As in Ref. [11], it is convenient to normalize the Zerilli function so that the asymptotic energy flux in each mode is given by $\dot{E} = (1/32\pi)\dot{\psi}^2$. While previously only axisymmetric simulations have been studied, we can now study all non-trivial wave modes, including those with $m \neq 0$.

3.3   Applications

In the sections above, I showed how to extract the gauge invariant Zerilli function at a given radius on some time slice of the numerical spacetime. In the following sections I show several ways in which this information can be used, including

(a) *evolving* a numerically generated initial data set with perturbation theory, and (b) extracting waveforms from a fully nonlinear evolution, possibly from the same data set.

Axisymmetric Distorted Black Holes

Linear Evolution We first apply this method to axisymmetric initial data sets of the type discussed above. These are single black holes distorted by gravitational waves. If the amplitude is low enough, this should represent a small perturbation on a Schwarzschild black hole. The idea is to actually compute $\psi(r, t = 0)$, by extracting the Zerilli function at every radial grid point on the initial data slice, and use the Zerilli evolution equation to actually evolve the system as a perturbation. This will allow us to compare the waveform at some radius $\psi_{lin}(r_0, t)$ obtained in this way, to that obtained with a well-tested 2D axisymmetric code that performs full nonlinear evolutions. This will serve as a test of the initial Zerilli function being given to the linear code, and of the linear evolution code itself. It will also help us determine for which Brill wave amplitudes this procedure breaks down. Beyond a certain point, perturbation theory will fail and nonlinear effects will become important.

Let us first consider the data set $(a, b, w, n, c) = (0.05, 1, 1, 4, 0)$, in the notation above. In this case the Brill wave is initially far from the black hole, and will propagate in, hitting it and exciting the normal mode oscillations. In Figure 1 we show the $\ell = 2$ and $\ell = 4$ Zerilli functions as a function of time, at a radius of $r = 15M$. Data are shown from both the linear and 2D nonlinear codes. We see that for both functions, the linear and nonlinear results line up nicely until about $t = 50M$, when a phase shift starts to be significant. This phase shift and widening of the wave at late times is known from previous studies of numerical simulations of distorted black hole spacetimes in axisymmetry [11].

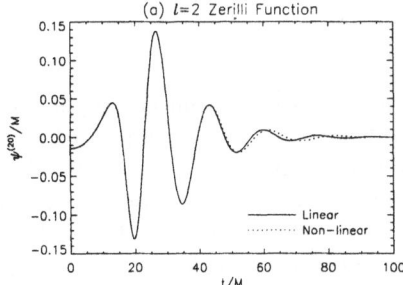 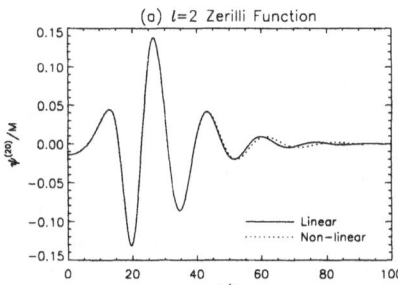

Fig. 1 We show the (a)$\ell = 2$ and (b)$\ell = 4$ Zerilli functions as a function of time, extracted during linear and 2D nonlinear evolutions of the data set $(a, b, w, n, c) = (0.05, 1, 1, 4, 0)$. The data were extracted at a radius of $r = 15M$.

As second example, let us look at $(a, b, w, n, c) = (0.05, 0, 1, 4, 0)$. In this case the Brill wave is initially right on the throat. In Figure 2 we show the $\ell = 2$ and the $\ell = 4$ waveforms as a function of time extracted at a radius of $r = 15M$. Again, data from both the linear and 2D nonlinear codes are shown. The data line up well until about $t = 50M$, when phase errors begin to be significant.

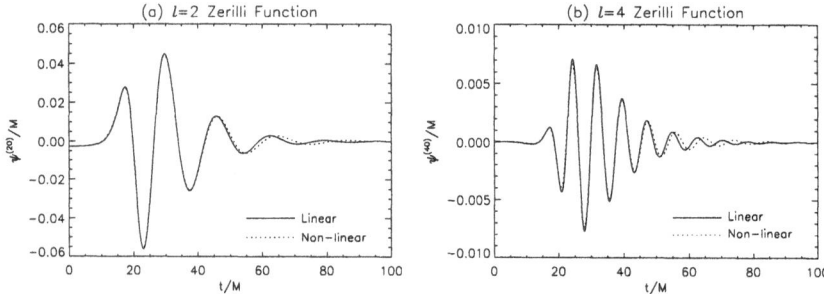

Fig. 2 We show the (a)$\ell = 2$ and (b)$\ell = 4$ Zerilli functions as a function of time, extracted during linear and 2D nonlinear evolutions of the data set $(a, b, w, n, c) = (0.05, 0, 1, 4, 0)$. The data were extracted at a radius of $r = 15M$.

This is the first step in our ladder of credibility. We have shown that these data sets provide an important testbed for a numerical black hole evolution. First they confirm that these data sets can be treated as linear perturbations on a Schwarzschild background, since both linear and fully nonlinear evolutions agree. Second, they remarkably confirm the results of a complex, nonlinear evolution code which evolves a black hole in maximal slicing. This gives great confidence in the ability of this code to treat black holes and extract waveforms, even in the more highly distorted cases where perturbation theory breaks down (but waveform extraction will not necessarily break down, at least far from the hole). We will use this technique in various ways below.

Axisymmetric Black Hole Collisions We now turn to another application of this basic idea of evolving dynamic black hole spacetimes with perturbation theory, but this time we consider two black holes colliding head on. This might seem to be impossible to treat perturbatively, but there are two limits in which perturbation theory has been shown to be incredibly successful. First, if the two holes are so close together initially that they have actually already merged into one, they might be considered as a single perturbed Schwarzschild hole (the so-called "close limit"). Using similar ideas to those discussed above, Price and Pullin and others [32, 33, 34, 35, 36] used this technique to produce waveforms for colliding black holes in the Misner and Brill and Lindquist data described by Giulini. In fact, the original paper of Price and Pullin [32] is what spurred on so much interest in these many applications of perturbation theory as a check on numerical relativity. Second, when the holes are very far apart, one can consider one black hole as a test particle falling into the other. Then one rescales the answer obtained by formally allowing the "test particle" to be a black hole with the same mass as the one it is falling into[37, 14, 33].

The details of this success has provided insights into the nature of collisions of holes, and should also apply to many systems of dynamical black holes. The waveforms and energies agree remarkably well with numerical simulations. Moreover, second order perturbation theory [36] spectacularly improved the agreement between the close limit and full numerical results for even larger distances between the holes, although ultimately beyond a certain limit the approximation is simply inappropriate and breaks down.

The success of these techniques suggests, among other things, that these are very powerful methods that can be used hand-in-hand with fully nonlinear numerical evolutions, and can be applied in a variety of black hole spacetimes where one might naively think they would not work. For these reasons, many researchers are continuing to apply these techniques to the axisymmetric case with more and more complicated black hole spacetimes (e.g., the collision of counter-rotating, spinning black holes, colloquially known as the "cosmic screw".) Furthermore, these techniques will become even more essential in 3D, where we cannot achieve resolution as high as we can with 2D codes.

Finally, this is yet another rung on the "ladder of credibility": we now have not only slightly perturbed Schwarzschild spacetimes to consider, but also a series of highly nontrivial colliding black hole spacetimes that are now well understood in axisymmetry due to the nice interplay between perturbative and fully numerical treatments of the same problems. These then provide excellent testbeds for 3D simulations, which we turn to next.

3D Testbeds Armed with robust and well understood axisymmetric black hole codes, we now consider the 3D evolution of axisymmetric distorted black hole initial data. These same axisymmetric initial data sets can be ported into a 3D code in cartesian coordinates, evolved in 3D, and the results can be compared to those obtained with the 2D, axisymmetric code discussed above.

In the first of these simulations, we study the evolution of the distorted single black hole initial data set $(a, b, w, n, c) = (0.5, 0, 1, 2, 0)$. As the azimuthal parameter c is zero, this is axisymmetric and can also be evolved in 2D. In Figure 3a we show the result of the 3D evolution, focusing on the $\ell = 2$ Zerilli function extracted at a radius $r = 8.7M$ as a function of time. Superimposed on this plot is the same function computed during the evolution of the same initial data set with a 2D code, based on the one described in detail in [11, 13]. The agreement of the two plots over the first peak is a strong affirmation of the 3D evolution code and extraction routine. It is important to note that the 2D results were computed with a different slicing (maximal), different coordinate system, and a *different spatial gauge*. Yet the physical results obtained by these two different numerical codes, as measured by the waveforms, are remarkably similar (as one would hope). A full evolution with the 2D code to $t = 100M$, by which time the hole has settled down to Schwarzschild, shows that the energy emitted in this mode at that time is about $4 \times 10^{-3} M$. This result shows that now it is possible in full 3D numerical relativity, in cartesian coordinates, to study the evolution and waveforms emitted from highly distorted black holes, even when the final waves leaving the system carry a small amount of energy.

In Fig. 3b we show the $\ell = 4$ Zerilli function extracted at the same radius, computed during evolutions with 2D and 3D codes. This waveform is more diffi-cult to extract, because it has a higher frequency in both its angular and radial dependence, and it has a much lower amplitude: the energy emitted in this mode is three orders of magnitude smaller than the energy emitted in the $\ell = 2$ mode, i.e., $10^{-6}M$, yet it can still be accurately evolved and extracted. This is quite a remarkable result, and bodes well for the ability of numerical relativity codes ultimately to compute accurate waveforms that will be of great use in inter-preting data collected by gravitational wave detectors. (However, as I point out below, there is a quite a long way to go before the general 3D coalescence can be studied!)

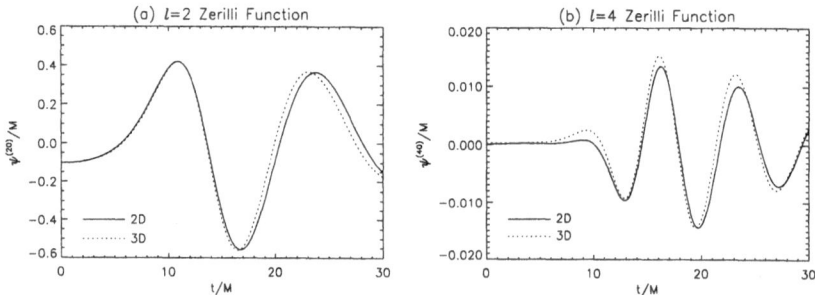

Fig. 3 We show the (a) $\ell = 2$ and (b) $\ell = 4$ Zerilli functions vs. time, extracted during 2D and 3D evolutions of the data set $(a, b, w, n, c) = (0.5, 0, 1, 2, 0)$. The functions were extracted at a radius of $8.7M$. The 2D data were obtained with 202×54 grid points, giving a resolution of $\Delta\eta = \Delta\theta = 0.03$. The 3D data were obtained using 300^3 grid points and a resolution of $\Delta x = 0.0816M$.

These results have been reported in much more detail in [38, 39].

True 3D Distorted Black Holes We now turn to radiation extraction in true 3D black hole evolutions. This is of major importance for the connection between numerical relativity and gravitational wave astronomy. Gravitational wave de-tectors such as LIGO, VIRGO, and GEO will measure these waves directly, and may depend on numerical relativity to provide templates to both extract the signals from the experimental data and to interpret the results. The preliminary results presented here represent the first time that non-axisymmetric modes of gravitational waves have been extracted in a numerical black hole spacetime.

In the sections above, I showed by careful comparison to 2D results that a 3D code is able to accurately simulate distorted black holes. Armed with these tests, we now consider evolutions of initial data sets which are not axisymmetric, *i.e.*, data sets which have non-vanishing azimuthal parameter c.

Black hole perturbation theory predicts that for each ℓ-mode, there is a series of quasinormal mode frequencies of the hole, as discussed above. However, we have previously dealt only with axisymmetric distorted black holes. For full 3D distorted holes, there are quasinormal mode frequencies for each m-mode as well. For rotating black holes, the m-modes are all distinct, and have different frequencies depending on the angular momentum of the hole. However, for non-rotating black holes, it turns out that for a given ℓ-value, all m-modes have the same quasinormal mode frequency (remember above we showed that the Zerilli potential $V(r)$, which governs the perturbation equations, is independent of m.) Hence, when we compare our results with the known quasinormal mode frequencies, we expect all m-modes to fit the $m = 0$ quasinormal mode.

In Figure 4 we show the $\ell = 2$ Zerilli functions computed during an evolution of the fully 3D initial data set $(a, b, w, n, c) = (0.5, 0, 1, 2, 0.5)$. This computation was done with 200^3 grid points and a resolution of $\Delta x = 0.106M$. The Zerilli functions were extracted at a radius of $r = 8.46M$. For each mode we also show the fit to the two lowest $\ell = 2$ quasinormal modes. The good agreement indicates that these waveforms have the correct frequency. We do not expect to be able to fit quasinormal modes to the waveforms initially, because here the waveform is highly dependent on the particular dynamics of the situation, and the black hole has not yet settled down to a perturbed Schwarzschild black hole.

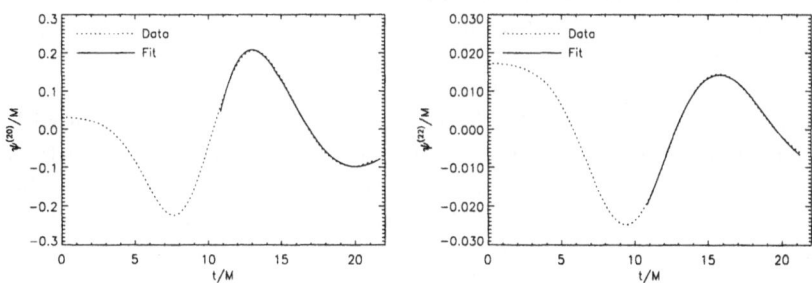

Fig. 4 We show the $\ell = 2$, $m = 0$ and $\ell = 2$, $m = 2$ Zerilli functions extracted at $r = 8.46M$ as a function of time for the evolution of the initial data set $(a, b, w, n, c) = (0.5, 0, 1, 2, 0.5)$. Also plotted are fits of the two lowest $\ell = 2$ quasinormal modes to each waveform. The calculation was done with 200^3 grid points and a resolution of $\Delta x = 0.106M$, placing the outer boundary along an axis at approximately $x = 21.2M$.

This $\ell = 2, m = 2$ mode is to my knowledge the first nonaxisymmetric wave mode extracted from a 3D black hole evolution. Such modes are extracted not just because they *can* be, but because they should be quite *important* for gravitational wave observatories. It turns out that the $\ell = 2, m = 2$ is considered to be one of the most promising black hole modes to be seen by gravitational wave detector. In realistic black hole coalescence, the final hole is expected to have a large amount of angular momentum, possibly near the Kerr limit $a = 1$. It turns out that this particular mode is one of the least damped (much less damped than for Schwarzschild, as seen here), and is also expected to be strongly excited[40]. Therefore, it is important to begin exhaustively testing the code's ability to generate and cleanly extract such nonaxisymmetric modes, even in the case studied here without rotation.

Many more modes can be extracted, including $\ell = 4, m$ modes, and details and analysis can be found in [38, 41]. Note that these nonspherical black hole evolutions have not yet been compared against perturbation theory evolutions. This is an important missing "rung" in the ladder of credibility that is in progress at present. These rungs will continue all the way to the "Holy Grail" of numerical relativity: a code that can truly simulate the full 3D coalescence of black holes in orbit about each other, complete with accurate waveforms for comparison with observations.

4   How Far Do We Have to Go?

Unfortunately, we still have a very long way to go! In these lectures I have tried to give an overview of the basic techniques needed in numerical relativity applied to black hole evolutions, along with an assessment of the state of the art in extracting waveforms. The progress has been good and exciting! However, although one can now do 3D evolutions of distorted black holes, and accurately extract waves emitted even at the $10^{-6}M$ level, the calculations one can presently do are actually very limited. With present techniques, the evolutions can only be carried out for a fraction of the time required to simulate the 3D orbiting coalescence. Most of what as been described here has been with certain symmetries, or with a single black hole in full 3D. At the present time, I am only aware of one attempt to study the collision of two black holes in 3D without any symmetries, which was recently reported by Brügmann [42]. However, this calculation is treated as a feasibility study, without detailed waveform extraction at this point, and again the evolution times are quite limited (less than those reported here.)

Many techniques to handle this more general case are under development, such as hyperbolic formulations of the Einstein equations and the advanced numerical methods they bring[25], adaptive mesh refinement that will enable placing the outer boundary farther away while resolving the strong field region where the waves are generated, and apparent horizon boundary conditions that excise the interiors of the black holes, thus avoiding the difficulties associated with singularity avoiding slicings.

All of these techniques, and others, may be needed to handle the more general, long term evolution of coalescing black holes. Each of these techniques may introduce numerical artifacts, even if at very low amplitude, to which the waveforms may be very sensitive. As new methods are developed and applied to numerical black hole simulations, they can now be tested on evolutions such as those presented here to ensure that the waveforms are accurately represented in the data.

5   Acknowledgments

This work has been supported by the Albert Einstein Institute (AEI) and NCSA, and much of the discussions are derived from Karen Camarda's PhD thesis. I am most thankful to her for allowing me to adapt parts of that work for use here, and for carrying out much of this work with me over the last few years. Among many colleagues who have contributed to this work on black holes, I also thank Andrew Abrahams, Gabrielle Allen, Pete Anninos, David Bernstein, Steve Brandt, David Hobill, Joan Massó, John Shalf, Larry Smarr, Wai-Mo Suen, John Towns, and Paul Walker.

I would like to thank K.V. Rao and the staff at NCSA for assistance with the computations. Calculations were performed at AEI and NCSA on an SGI/Cray Origin 2000 supercomputer.

Bibliography

[1] A. A. Abramovici *et al.*, Science **256**, 325 (1992).

[2] E. Seidel and W.-M. Suen, in *1995 Les Houches School on Gravitational Radiation*, edited by J.-P. Lasota and J.-A. Marck (1996), in press.

[3] J. York, in *Sources of Gravitational Radiation*, edited by L. Smarr (Cambridge University Press, Cambridge, England, 1979).

[4] E. Seidel, in *Relativity and Scientific Computing*, edited by F. Hehl (Springer-Verlag, Berlin, 1996).

[5] Y. Choquet-Bruhat and T. Ruggeri, Comm. Math. Phys **89**, 269 (1983).

[6] C. Bona and J. Massó, Phys. Rev. Lett. **68**, 1097 (1992).

[7] C. Bona, J. Massó, E. Seidel, and J. Stela, Phys. Rev. Lett. **75**, 600 (1995).

[8] A. Abrahams, A. Anderson, Y. Choquet-Bruhat, and J. York, Phys. Rev. Lett. (1995).

[9] S. Fritelli and O. Reula, Commun. Math. Phys. **166**, 221 (1994).

[10] H. Friedrich, (1996), aEI preprint 001.

[11] A. Abrahams, D. Bernstein, D. Hobill, E. Seidel, and L. Smarr, Phys. Rev. D **45**, 3544 (1992).

[12] D. Bernstein, D. Hobill, E. Seidel, and L. Smarr, Phys. Rev. D **50**, 3760 (1994).

[13] D. Bernstein, D. Hobill, E. Seidel, L. Smarr, and J. Towns, Phys. Rev. D **50**, 5000 (1994).

[14] P. Anninos, D. Hobill, E. Seidel, L. Smarr, and W.-M. Suen, Phys. Rev. D **52**, 2044 (1995).

[15] P. Anninos, K. Camarda, J. Massó, E. Seidel, W.-M. Suen, and J. Towns, Phys. Rev. D **52**, 2059 (1995).

[16] C. W. Misner, K. S. Thorne, and J. A. Wheeler, *Gravitation* (W. H. Freeman, San Francisco, 1973).

[17] M. Alcubierre, Phys. Rev. D **55**, 5981 (1997).

[18] G. Cook, Ph.D. thesis, University of North Carolina at Chapel Hill, Chapel Hill, North Carolina, 1990.

[19] L. Smarr and J. York, Phys. Rev. D **17**, 2529 (1978).

[20] S. L. Shapiro and S. A. Teukolsky, in *Dynamical Spacetimes and Numerical Relativity*, edited by J. M. Centrella (Cambridge University Press, Cambridge, England, 1986), pp. 74–100.

[21] D. Bernstein, D. Hobill, and L. Smarr, in *Frontiers in Numerical Relativity*, edited by C. Evans, L. Finn, and D. Hobill (Cambridge University Press, Cambridge, England, 1989), pp. 57–73.

[22] E. Seidel and W.-M. Suen, Phys. Rev. Lett. **69**, 1845 (1992).

[23] P. Anninos, G. Daues, J. Massó, E. Seidel, and W.-M. Suen, Phys. Rev. D **51**, 5562 (1995).

[24] M. A. Scheel, S. L. Shapiro, and S. A. Teukolsky, Phys. Rev. D **51**, 4208 (1995).

[25] C. Bona, J. Massó, and J. Stela, Phys. Rev. D **51**, 1639 (1995).

[26] A. M. Abrahams *et al.*, Phys. Rev. Lett. **80**, 1812 (1998).

[27] S. Chandrasekhar, *The Mathematical Theory of Black Holes* (Oxford University Press, Oxford, England, 1983).

[28] T. Regge and J. Wheeler, Phys. Rev. **108**, 1063 (1957).

[29] F. J. Zerilli, Phys. Rev. Lett. **24**, 737 (1970).

[30] V. Moncrief, Annals of Physics **88**, 323 (1974).

[31] A. Abrahams, Ph.D. thesis, University of Illinois, Urbana, Illinois, 1988.

[32] R. H. Price and J. Pullin, Phys. Rev. Lett. **72**, 3297 (1994).

[33] P. Anninos, R. H. Price, J. Pullin, E. Seidel, and W.-M. Suen, Phys. Rev. D **52**, 4462 (1995).

[34] A. Abrahams and R. Price, Phys. Rev. D **53**, 1972 (1996).

[35] J. Baker, A. Abrahams, P. Anninos, S. Brandt, R. Price, J. Pullin, and E. Seidel, Phys. Rev. D **55**, 829 (1997).

[36] R. J. Gleiser, C. O. Nicasio, R. H. Price, and J. Pullin, Physical Review Letters **77**, 4483 (1996).

[37] P. Anninos, D. Hobill, E. Seidel, L. Smarr, and W.-M. Suen, Phys. Rev. Lett. **71**, 2851 (1993).

[38] K. Camarda, Ph.D. thesis, University of Illinois at Urbana-Champaign, Urbana, Illinois, 1997.

[39] K. Camarda and E. Seidel, Phys Rev. D **57**, 3204 (1998).

[40] Éanna É. Flanagan and S. A. Hughes, Phys Rev. D **57**, 4535 (1998).

[41] K. Camarda and E. Seidel, in preparation.

[42] B. Brügmann, gr-qc/9708035.

Part IV

Beyond Classical General Relativity

Measurement Theory and General Relativity

Bahram Mashhoon

Department of Physics and Astronomy, University of Missouri-Columbia, Columbia, Missouri 65211, USA

Abstract. The theory of measurement is employed to elucidate the physical basis of general relativity. For measurements involving phenomena with intrinsic length or time scales, such scales must in general be negligible compared to the (translational and rotational) scales characteristic of the motion of the observer. Thus general relativity is a consistent theory of coincidences so long as these involve classical point particles and electromagnetic rays (geometric optics). Wave "optics" is discussed and the limitations of the standard theory in this regime are pointed out. A nonlocal theory of accelerated observers is briefly described that is consistent with observation and excludes the possibility of existence of a fundamental scalar field in nature.

1 Introduction

The quantum theory of measurement deals with observers and measuring devices that are all inertial. The universality of gravitational interaction implies, however, that gravitational fields cannot be ignored in general. Moreover, most measurements are performed in laboratories on the Earth, which — among other motions — rotates about its proper axis; in fact, measurements are generally performed by devices and observers that are accelerated. It is therefore necessary to investigate the assumptions that underlie the extension of physics to accelerated systems and gravitational fields. This amounts to a determination of the physical foundations of Einstein's theory of gravitation inasmuch as this theory is in agreement with all observational data available at present [1]. A critical examination of general relativity from the standpoint of measurement theory leads to certain basic limitations that are the main subject of this paper.

2 Physical Elements of General Relativity

The basic concepts of general relativity can be uniquely determined starting from the consideration of what observers would measure in physical experiments. This results in the four building blocks of general relativity that are described below.

(i) The fundamental laws of microphysics have been formulated with respect to inertial observers. The measurements of inertial observers in Minkowski spacetime are connected via inhomogeneous Lorentz transformations (i.e. Poincaré transformations). An inertial observer is an observer at rest in an inertial reference system; in fact, such an observer can be thought of as carrying a natural orthonormal tetrad frame $\lambda^{\mu}_{(\alpha)}$ along its worldline. Here $\lambda^{\mu}_{(0)} = dx^{\mu}/d\tau$ is the vector tangent to the worldline ("time axis") and $\lambda^\mu_{(i)}$, $i = 1, 2, 3$, are the natural spatial axes of the frame so that $\lambda^\mu_{(\alpha)} = \delta^\mu_\alpha$. Thus Maxwell's equations in this inertial frame refer to the fields actually measured by these standard observers, i.e. $F_{\mu\nu} \lambda^\mu_{(\alpha)} \lambda^\nu_{(\beta)} \rightarrow (\mathbf{E}, \mathbf{B})$. One can consider other inertial observers as being at rest in other inertial systems in uniform motion with respect to the original reference system described above. To express the measurements of the other observers, one could transform to their rest frames; alternatively, one could consider physics in the original inertial system and simply describe all measurements with respect to a single system of inertial coordinates $x^\alpha = (ct, \mathbf{x})$. In the latter case, which is adopted here for the sake of convenience, one can describe the determination of the electromagnetic field by a moving inertial observer as the projection of the field on the observer's frame,

$$\hat{F}_{(\alpha)(\beta)} = F_{\mu\nu} \hat{\lambda}^\mu_{(\alpha)} \hat{\lambda}^\nu_{(\beta)}. \tag{1}$$

Let us now suppose that inertial observers choose to employ arbitrary smooth spacetime coordinates $x'^\mu = x'^\mu(x^\alpha)$. It turns out that — so long as the observers remain inertial — this extension is purely mathematical in nature and can be accomplished without introducing any new physical assumption into the theory. Consider, for instance, the Lorentz force law for a particle of mass m and charge q,

$$m \frac{\mathrm{d}^2 x^\mu}{\mathrm{d}\tau^2} = q F^\mu{}_\nu \frac{\mathrm{d}x^\nu}{\mathrm{d}\tau}. \tag{2}$$

Here $\mathrm{d}\tau$ is the invariant spacetime interval measured along the path of the particle by the standard inertial observers, i.e. $\mathrm{d}\tau = c\,\mathrm{d}t/\gamma$ and γ is the Lorentz factor. Assuming the invariance of this interval under the change of coordinates, $\mathrm{d}\tau^2 = \eta_{\mu\nu}\mathrm{d}x^\mu\mathrm{d}x^\nu = g'_{\alpha\beta}\mathrm{d}x'^\alpha\mathrm{d}x'^\beta$ with

$$g'_{\alpha\beta} = \eta_{\mu\nu} \frac{\partial x^\mu}{\partial x'^\alpha} \frac{\partial x^\nu}{\partial x'^\beta}, \tag{3}$$

one can simply write equation (2) as

$$m \left[\frac{\mathrm{d}^2 x'^\rho}{\mathrm{d}\tau^2} + \Gamma'^\rho_{\alpha\beta}(x') \frac{\mathrm{d}x'^\alpha}{\mathrm{d}\tau} \frac{\mathrm{d}x'^\beta}{\mathrm{d}\tau} \right] = q F'^\rho{}_\sigma \frac{\mathrm{d}x'^\sigma}{\mathrm{d}\tau}, \tag{4}$$

with the Christoffel connection

$$\Gamma'^\rho_{\alpha\beta} = \frac{\partial^2 x^\mu}{\partial x'^\alpha \partial x'^\beta} \frac{\partial x'^\rho}{\partial x^\mu}, \tag{5}$$

and the auxiliary field variables

$$F'^{\rho\sigma}(x') = \frac{\partial x'^\rho}{\partial x^\mu} \frac{\partial x'^\sigma}{\partial x^\nu} F^{\mu\nu}(x). \tag{6}$$

In Euclidean space, one can always introduce curvilinear coordinates for the sake of convenience; similarly, one can introduce arbitrary (smooth and admissible) coordinates in Minkowski spacetime. In this way, tensors under the inhomogeneous Lorentz group become tensors under general coordinate transformations.

(ii) To extend measurements to accelerated observers in Minkowski spacetime, a physical hypothesis is required that would connect the measurement of accelerated and inertial observers. In the standard approach to the theory of relativity, the assumption is that an accelerated observer is at each instant physically equivalent to a hypothetical momentarily comoving inertial observer. Thus an accelerated observer passes through an infinite sequence of such hypothetical inertial observers. Mathematically, this basic assumption is equivalent to replacing a curve by its tangent vector at each point as illustrated in Figure 1. This assumption is clearly valid for Newtonian point particles, since at each

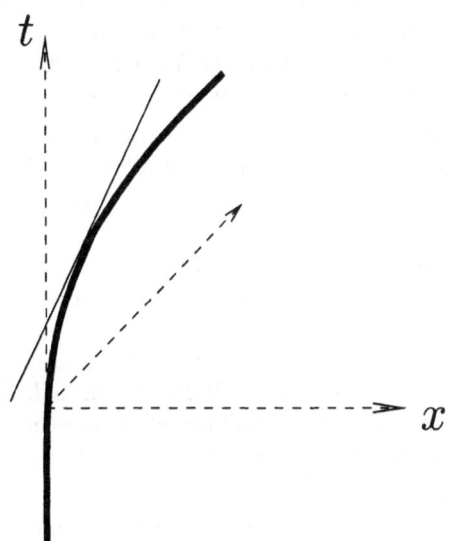

Fig. 1 The worldline of an accelerated observer in Minkowski spacetime is curved. The hypothesis of locality postulates that the observer is at each moment locally inertial.

instant the accelerated particle and the momentarily comoving inertial particle have the same state, i.e. the same position and velocity. Moreover, it can be naturally extended to all pointlike phenomena; that is, the assumption is also valid if all phenomena are thought of in terms of pointlike *coincidences* of Newtonian point particles and null rays. However, in more general cases involving intrinsic temporal and spatial scales the above assumption will be referred to as "the hypothesis of locality" [2]. Imagine, for instance, an accelerated measuring device;

clearly, it is affected by internal inertial effects. If these inertial effects integrate to a perceptible influence on the outcome of a measurement, the hypothesis of locality is violated. On the other hand, if the timescale of the measurement is so short that the influence of the inertial effects is negligible, then the device is "standard", i.e. its acceleration can be locally ignored. The hypothesis of locality applied to a clock implies that a standard clock will measure proper time τ along its path; therefore, the hypothesis of locality is the generalization of the "clock hypothesis" to all standard measuring devices [3, 4, 5, 6, 7]. Moreover, the local equivalence of an accelerated observer with an infinite sequence of comoving inertial observers endows the accelerated observer with the continuously varying tetrad system of the inertial observers. This variation can be characterized by a translational acceleration $\mathbf{g}(\tau)$ and a rotation of the spatial frame with frequency $\Omega(\tau)$; alternatively, one may associate acceleration scales (such as c^2/g and c/Ω) with the motion of the observer [8, 9].

The extension of measurements to all observers that can use arbitrary coordinates in Minkowski spacetime implies that one can formulate physical laws in a *generally covariant* form. To extend this covariance further to curved spacetime manifolds, Einstein's principle of equivalence is indispensable.

(iii) Einstein's principle of equivalence embodies the universality of the gravitational interaction and is the cornerstone of general relativity. This principle generalizes a result of Newtonian gravitation that is directly based upon the principle of equivalence of inertial and gravitational masses. Einstein postulated a certain equivalence between an observer in a gravitational field and an accelerated observer in Minkowski spacetime. This heuristic principle, when combined with the hypothesis of locality, implies that an observer in a gravitational field is locally inertial. Thus gravitation has to do with the way local inertial frames are connected to each other. The simplest possibility is through the pseudo-Riemannian curvature of the spacetime manifold; therefore, in general relativity the gravitational field is identified with the spacetime curvature.

(iv) The correspondence between general relativity and Newton's theory of gravitation is established via the gravitational field equation. That is, within the framework of Riemannian geometry the gravitational field equations are the simplest generalizations of Poisson's equation, $\nabla^2 \Phi_N = -4\pi G\rho$, for the Newtonian potential Φ_N. In general relativity, the Newtonian potential is generalized and replaced by the ten components of the metric tensor $g_{\mu\nu}$; similarly, the acceleration of gravity is replaced by the Christoffel connection $\Gamma^\mu_{\alpha\beta}$ and the tidal matrix $\partial^2 \Phi_N / \partial x^i \partial x^j$ is replaced by the Riemann curvature tensor $R_{\mu\nu\rho\sigma}$. In Newtonian gravitation, the trace of the tidal matrix is connected to the local density of matter ρ by Newton's constant of gravitation. Similarly, in general relativity the trace of the Riemann tensor is connected to the energy-momentum tensor of matter,

$$R_{\mu\nu} - \frac{1}{2}g_{\mu\nu}g^{\alpha\beta}R_{\alpha\beta} = \frac{8\pi G}{c^4}T_{\mu\nu}. \tag{7}$$

3   Measurements of Accelerated Observers

The primary measurements of an observer are those of duration and distance. In general relativity, the hypothesis of locality is indispensable for the interpretation of the results of measurements by accelerated observers. In particular, we define "standard" measuring devices to be those that are compatible with the locality assumption. Thus a standard clock measures proper time along its trajectory; similarly, a standard measuring rod is usually assumed to provide a proper measure of distance. At each instant of time, the accelerated observer is momentarily equivalent to a hypothetical comoving inertial observer; therefore, both observers have the instantaneous Euclidean space in common. It would appear then that placing standard measuring rods one next to the other and so on should lead to the proper measurement of spatial distances by accelerated observers.

An important issue is the extent to which such measurements of time and distance can lead to the establishment of an admissible coordinate system around the accelerated observer. Well-known investigations have led to the result that such coordinate systems have limited spatial extent given by the acceleration lengths (e.g. c^2/g and c/Ω), since these are the only length scales in the problem. The method of construction of accelerated coordinate systems could even be nonlocal; however, limitations would still exist as recently pointed out by Marzlin [10]. It might therefore appear that (local and nonlocal) coordinate systems could in general be constructed in a cylindrical region around the worldline of the accelerated observer. However, this conclusion is ultimately based upon the use of standard measuring rods whose existence turns out to be in conflict with the hypothesis of locality. A fundamental problem associated with length measurements is the following: a standard measuring rod, however small, has nevertheless a nonzero spatial extent whereas the hypothesis of locality is only pointwise valid. This implies a rather basic limitation on the measurement of length by accelerated observers and can be illustrated by the following thought experiment. Imagine two observers O_1 and O_2 at rest in an inertial frame. For $t \leq 0$, their coordinates $x^\alpha = (ct, \mathbf{x})$ are $(ct, 0, 0, 0)$ and $(ct, L, 0, 0)$, respectively. At $t = 0$, they are accelerated from rest along the x-direction *in exactly the same way* so that at time $t > 0$ each has a velocity $\mathbf{v} = v\hat{\mathbf{x}}$. The distance between O_1 and O_2 as measured by observers at rest in the inertial frame is always L, since

$$x_1(t) = \int_0^t v \, dt \qquad \text{and} \qquad x_2(t) = L + \int_0^t v \, dt \qquad (8)$$

for $t > 0$ and $x_2(t) - x_1(t) = L$. What is the distance between O_1 and O_2 as measured by comoving observers? It turns out that the hypothesis of locality provides a unique answer to this question only in the limit $L \to 0$. To show this, let us first note that at a given time $\hat{t} > 0$, O_1 and O_2 have the same speed $\hat{v} = c\beta$. The hypothesis of locality implies that the accelerated observers pass through an infinite sequence of momentarily comoving inertial observers. Thus imagine the Lorentz transformation between the inertial frame $x^\alpha = (ct, \mathbf{x})$ and the "instantaneous" inertial rest frame $x'^{\alpha} = (ct', \mathbf{x}')$ of the observers at $\hat{t}$ given by

$$c(t - \hat{t}) = \gamma(ct' + \beta x'), \quad x - \hat{x} = \gamma(x' + c\beta t'), \quad y = y', \quad z = z', \quad (9)$$

where $\gamma = (1 - \beta^2)^{-1/2}$ is the Lorentz factor at $\hat{t}$. The events with coordinates $O_1 : (c\hat{t}, x_1, 0, 0)$ and $O_2 : (c\hat{t}, x_2, 0, 0)$ in the original inertial frame have coordinates $O_1 : (ct'_1, x'_1, 0, 0)$ and $O_2 : (ct'_2, x'_2, 0, 0)$ in the instantaneous inertial frame. It follows from the Lorentz transformation (9) that $L' = x'_2 - x'_1 = \gamma L$. This has a simple physical interpretation: The Lorentz-FitzGerald contracted distance between O_1 and O_2 is always L, hence the "actual" distance between O_1 and O_2 must be larger by the Lorentz γ-factor. One can imagine that the distance between O_1 and O_2 is populated by a large number of hypothetical accelerated observers moving in exactly the same way as O_1 and O_2 and carrying infinitesimal measuring rods that are placed side by side to measure the distance under consideration.

It must be equally correct to replace the infinite sequence of inertial systems $x'^{\alpha} = (ct', \mathbf{x}')$ by a continuously moving frame. To this end, we must choose the worldline $\bar{x}^{\mu}(\tau)$ of one of the accelerated observers — such as O_1, O_2, or any of the hypothetical observers in between the two — and note that at any instant of proper time τ along the worldline, this fiducial observer is in a Euclidean space with Cartesian coordinates $\mathbf{X}$ in accordance with the hypothesis of locality. The connection between the coordinates x^{μ} in the original inertial frame and the new coordinates X^{μ} is given by $X^0 = \tau$ and

$$x^{\mu} = \bar{x}^{\mu}(X^0) + X^i \bar{\lambda}^{\mu}_{(i)}, \quad (10)$$

where $\bar{\lambda}^{\mu}_{(i)}$ is the natural tetrad frame along the worldline of the reference observer. Specifically, the fiducial observer is instantaneously inertial by the hypothesis of locality and hence assigns coordinates $X^0 = \tau$ and $X^i = \sigma \xi_{\mu} \bar{\lambda}^{\mu}_{(i)}$ to spacetime events. Here ξ^{μ} is a unit spacelike vector normal to $\bar{\lambda}^{\mu}_{(0)}$ at $\bar{x}^{\mu}(\tau)$ along a straight line that connects $\bar{x}^{\mu}(\tau)$ to an event with coordinates x^{μ} in the original background inertial frame, $\xi_{\mu} \bar{\lambda}^{\mu}_{(i)}$ are direction cosines and $\sigma = |\mathbf{X}|$ is the proper length of this spacelike line segment. To develop this approach further, it is necessary to specify the motion explicitly. Thus we assume that O_1 and O_2 are uniformly accelerated with acceleration g and we choose O_1 to be the fiducial observer. The natural orthonormal nonrotating tetrad frame along the worldline of O_1 is given by

$$\bar{\lambda}^{\mu}_{(0)} = (\gamma, \beta\gamma, 0, 0), \quad (11)$$
$$\bar{\lambda}^{\mu}_{(1)} = (\beta\gamma, \gamma, 0, 0), \quad (12)$$
$$\bar{\lambda}^{\mu}_{(2)} = (0, 0, 1, 0), \quad (13)$$
$$\bar{\lambda}^{\mu}_{(3)} = (0, 0, 0, 1), \quad (14)$$

just as for the Lorentz transformation (9). Then the inertial frame $x^\alpha = (ct, x, y, z)$ and the Fermi frame $X^\alpha = (cT, X, Y, Z)$ are connected by

$$ct = \left(X + \frac{c^2}{g}\right)\sinh\left(gT/c\right), \tag{15}$$

$$x = \left(X + \frac{c^2}{g}\right)\cosh\left(gT/c\right) - \frac{c^2}{g}, \tag{16}$$

$y = Y$ and $z = Z$. The spatial origin of the new coordinate system is occupied by O_1 such that $\bar{x}^\mu(\tau) = \mathcal{L}(\beta\gamma, \gamma - 1, 0, 0)$, where $\beta = \tanh(\tau_1/\mathcal{L})$, $\gamma = \cosh(\tau_1/\mathcal{L})$, $\mathcal{L} = c^2/g$ is the acceleration length and τ_1 is the proper time along O_1. As before, at any given time $\hat{t} > 0$ the events $O_1 : (c\hat{t}, x_1, 0, 0)$ and $O_2 : (c\hat{t}, x_2, 0, 0)$ now correspond to $O_1 : (\tau_1, X_1, 0, 0)$ and $O_2 : (\tau_2, X_2, 0, 0)$, where $x_2 - x_1 = L$ and $X_1 = 0$ by construction. The distance between O_1 and O_2 in this Fermi frame is then given by $L_F = X_2 - X_1 = X_2$. It follows from equations (15) and (16) that

$$c\hat{t} = \mathcal{L}\sinh(\tau_1/\mathcal{L}), \tag{17}$$

$$x_1 = \mathcal{L}\left[\cosh(\tau_1/\mathcal{L}) - 1\right], \tag{18}$$

$$c\hat{t} = (X_2 + \mathcal{L})\sinh(\tau_2/\mathcal{L}), \tag{19}$$

$$x_2 = (X_2 + \mathcal{L})\cosh(\tau_2/\mathcal{L}) - \mathcal{L}. \tag{20}$$

Equations (19) and (20) can be written as

$$(X_2 + \mathcal{L})^2 = (x_2 + \mathcal{L})^2 - c^2\hat{t}^2, \tag{21}$$

where $x_2 = x_1 + L$ and x_1 and $\hat{t}$ are given by equations (18) and (17), respectively. Thus one finds that

$$L_F = \mathcal{L}\left[(1 + 2\epsilon\gamma + \epsilon^2)^{1/2} - 1\right], \tag{22}$$

where $\epsilon = L/\mathcal{L} = gL/c^2$ and $\gamma = (1 + g^2\hat{t}^2/c^2)^{1/2}$. The length in the Fermi frame L_F must be compared with the corresponding result from the instantaneous Lorentz frame $L' = \gamma L$; indeed, the ratio L_F/L' approaches unity only in the limit $\epsilon \to 0$. This is a remarkable result that has far-reaching consequences. Let us note that for $\epsilon \ll 1$,

$$L_F/L' \approx 1 - \frac{1}{2}\beta^2\gamma\epsilon \tag{23}$$

to first order in ϵ; however, over a long time $\gg c/g$ the quantity $\beta^2\gamma\epsilon$ may not remain small compared to unity. Moreover, $L_F/L' \to 0$ as $g\hat{t}/c \to \infty$ and hence $\gamma \to \infty$. It follows from these considerations that consistency is achieved for $\gamma\epsilon \to 0$; hence, the acceleration length and time, i.e. c^2/g and c/g, respectively, place severe limitations on the domain of applicability of the hypothesis of locality. Furthermore, let us suppose that the Fermi frame is established along O_2 instead of O_1. Then the resulting distance would be different from L_F; however, all such lengths agree in the $\epsilon \to 0$ limit.

It is interesting to mention here another measure of distance from O_1 and O_2 using light signals. Let O_1 send a signal at τ_1^- that reaches O_2 at τ_2 and is immediately returned to O_1. The return signal reaches O_1 at τ_1^+, where $\tau_2 = (\tau_1^- + \tau_1^+)/2$. Observer O_1 would then determine the distance to O_2 via $L_{\mathrm{ph}} = c(\tau_1^+ - \tau_1^-)/2$, which works out to be

$$L_{\mathrm{ph}} = \mathcal{L} \ln\left(1 + L_{\mathrm{F}}/\mathcal{L}\right). \tag{24}$$

It is clear by symmetry that if O_2 initiates a light signal to O_1, etc., then the resulting light travel time would be different, since in equation (24) the Fermi length would be the one determined on the basis of O_2 as the fiducial observer. Nevertheless, for $\gamma\epsilon$ negligibly small all these length measurements agree with each other.

The simple example that has been worked out here can be generalized to arbitrary but identical velocity for O_1 and O_2. The comparison of the instantaneous local inertial frame with the continuously moving geodesic frame leads to the conclusion that the basic length and time scales under consideration must in general be negligible compared to the relevant acceleration scales. This has significant consequences for the comparison of theory and experiment in general relativity [2]; in particular, the physical significance of Fermi coordinates is in general further limited to the immediate neighborhood of the observer and wave equations are meaningful only within this domain.

It follows from these considerations that the physical dimensions of any standard measuring device must be negligible compared to the relevant acceleration length $\mathcal{L}$ and the duration of the measurement must in general be negligible compared to $\mathcal{L}/c$. These are not significant limitations for typical accelerations in the laboratory; for instance, for the Earth's acceleration of gravity $c^2/g \simeq 1\,\mathrm{lyr}$. Moreover, observers at rest on the Earth typically refer their measurements to rotating Earth-based coordinates; hence, this coordinate system is mathematically valid up to a "light cylinder" at a radius of $\mathcal{L} = c/\Omega \simeq 28\,\mathrm{AU}$. But physically valid length measurements can extend over a neighborhood of the observer with a radius much smaller than $\mathcal{L}$. In fact, this "light cylinder" has no bearing on astronomical observations, since observers simply take into account the absolute rotation of the Earth and reduce astronomical data by taking due account of aberration and Doppler effects.

The standard "classical" measuring device of mass μ has wave characteristics, given by its Compton wavelength $\hbar/\mu c$ and period $\hbar/\mu c^2$, that must be negligible in comparison with the scales of length and time that characterize the device as a consequence of the quasi-classical approximation. For instance, a clock of mass μ must have a resolution exceeding $\hbar/\mu c^2$; similarly, the mass of a clock with resolution θ must exceed $\hbar/\theta c^2$. These assertions follow from the application of the uncertainty principle to measurements performed by a standard device [11, 12]. When such quantum limitations are combined with the classical limitations discussed above, on finds that $\mathcal{L} \gg \hbar/\mu c$; therefore, the translational acceleration of a standard classical measuring device must be much less than $\mu c^3/\hbar$ and its rotational frequency must be much less than $\mu c^2/\hbar$. The idea of the existence of a maximal proper acceleration is due to Caianiello [13, 2, 14].

4   Measurements in Gravitational Fields

The physical results of the previous section can be extended to local measurements in a gravitational field via an interpretation of the Einstein principle of equivalence in terms of the gravitational Larmor theorem.

A century ago, Larmor [15] established a local equivalence between magnetism and rotation for all particles with the same charge to mass ratio (q/m). That is, charged particle phenomena in a magnetic field correspond to those in a frame rotating with the Larmor frequency $\Omega_{\mathrm{L}} = q\mathbf{B}/2mc$. This local relation is valid to first order in field strength for slowly varying fields and slowly moving charged particles. Such a correspondence also exists for electric fields and linearly accelerated frames. It turns out that Larmor's theorem can be generalized in a natural way to the case of gravitational fields.

The close analogy between Coulomb's law of electricity and Newton's law of gravitation leads to an interpretation of Newtonian gravity in terms of nonrelativistic theory of the gravitoelectric field. Moreover, any theory that combines Newtonian gravity with Lorentz invariance in a consistent manner is expected to contain a gravitomagnetic field as well. In fact, in general relativity the exterior spacetime metric for a rotating mass may be expressed in the linear approximation as

$$ds^2 = -c^2\left(1 - \frac{2}{c^2}\Phi_{\mathrm{N}}\right)dt^2 + \left(1 + \frac{2}{c^2}\Phi_{\mathrm{N}}\right)\delta_{ij}\,dx^i\,dx^j - \frac{4}{c}(\mathbf{A}_g \cdot d\mathbf{x})\,dt, \qquad (25)$$

where $\Phi_{\mathrm{N}} = GM/r$ is the Newtonian potential and $\mathbf{A}_g = G\mathbf{J} \times \mathbf{r}/cr^3$ is the gravitomagnetic vector potential. The gravitoelectric and gravitomagnetic fields are then given by $\mathbf{E}_g = -\nabla\Phi_{\mathrm{N}}$ and $\mathbf{B}_g = \nabla \times \mathbf{A}_g$, respectively.

It is possible to formulate a gravitational Larmor theorem [16] by postulating that the gravitoelectric and gravitomagnetic charges are given by $q_E = -m$ and $q_B = -2m$, respectively. In fact, $q_B/q_E = 2$ since gravitation is a spin-2 field. Thus $\Omega_{\mathrm{L}} = -\mathbf{B}_g/c$, which is consistent with the fact that an ideal gyroscope at a given position in space would precess in the gravitomagnetic field with a frequency $\Omega_{\mathrm{P}} = \mathbf{B}_g/c$. The general form of the gravitational Larmor theorem is then an interpretation of the Einstein principle of equivalence for linear gravitational fields in a finite neighborhood of an observer; for instance, in the gravitational field of the Earth an observer can be approximately inertial within the "Einstein elevator" if the "elevator" falls freely with acceleration $g \sim GM/r^2$ while rotating with frequency $\Omega \sim GJ/c^2r^3$. It follows that the relevant gravitoelectromagnetic acceleration lengths are given by c^2/g and c/Ω in this case and the restrictions discussed in the previous section would then apply to the measurements of an observer in a gravitational field as well. These limitations are generally expected to be important for the post-Newtonian corrections of high order in relativistic gravitational systems.

General relativity has found applications mostly in astronomical systems, where Newtonian results have been extended to the relativistic domain. In particular, small post-Newtonian corrections are usually included in the equations of motion. Suppose, for instance, that one is interested in the distance between the members of a relativistic binary system. It follows from our considerations that such a length — which corresponds in the Newtonian theory to the Euclidean distance — may not be well defined. However, the resulting discrepancy could be masked by other parameters; that is, this circumstance may be difficult to ascertain experimentally since the comparison of data with the theory generally involves parameters that are not independently available and whose particular values need to be determined from the data.

Let us next consider tidal accelerations within the "Einstein elevator". For a device of dimension $\hat{\delta}$, the tidal acceleration $\hat{g}$ is given by the Jacobi equation and can be estimated by $\hat{g} \sim K\hat{\delta}$, where K is a typical component of the tidal matrix $K_{ij} = c^2 R_{\mu\nu\rho\sigma} \lambda^{\mu}_{(0)} \lambda^{\nu}_{(i)} \lambda^{\rho}_{(0)} \lambda^{\sigma}_{(j)}$. According to the results of the previous section $\hat{g} \ll \mu c^3/\hbar$, where μ is the mass of the device. Imagine, for instance, such a device on a star of mass M and radius R that is undergoing "complete" spherical gravitational collapse. In this case , $K \sim GM/R^3$ and $\hat{\delta} \ll c^2/\hat{g}$ imply that $\hat{\delta}^2 \ll c^2 R^3/GM$. On the other hand, the requirements that $\mu \ll M$ and $\hat{\delta} \gg \hbar/\mu c$ result in

$$ R^3 \gg \frac{GM}{c^2} \left(\frac{\hbar}{Mc} \right)^2 = \frac{\hbar}{Mc} L_{\mathrm{P}}^2, \qquad (26) $$

where $L_{\mathrm{P}} = (\hbar G/c^3)^{1/2}$ is the Planck length ($\simeq 10^{-33}$ cm) that is the geometric mean of the gravitational radius GM/c^2 and the Compton wavelength $\hbar/Mc$ for any physical system [17]. Thus collapse to a classical point singularity is meaningless on the basis of these considerations.

5   Wave Phenomena

Classical waves have intrinsic scales and are thus expected to be in conflict with the hypothesis of locality; indeed, for an electromagnetic wave of (reduced) wavelength λ the expected deviation from the hypothesis of locality is expected to be of the form $\lambda/\mathcal{L}$. More specifically, let us consider the problem of determination of the period of an incident electromagnetic wave by an accelerated observer. The observer needs to measure at least a few oscillations of the wave before a reasonable determination of the period can be made; therefore, the curvature of the observer's worldline cannot be neglected unless $\lambda/\mathcal{L}$ is too small to be observationally significant. It follows that the instantaneous Doppler and aberration formulas are in general valid only in the eikonal limit $\lambda/\mathcal{L} \to 0$. The issues involved here can be illustrated by a simple thought experiment. Let us consider an observer rotating with uniform speed $c\beta$ and frequency Ω in the positive sense around the origin on a circle of radius $r = c\beta/\Omega$ in the (x, y)-plane. A plane electromagnetic wave of frequency ω is incident along the z-axis and the rotating observer measures its frequency. According to the hypothesis of locality, the observer is at each instant momentarily inertial and hence $\omega' = \gamma\omega$ according to the transverse Doppler effect. This is illustrated in Figure 2. On the other hand,

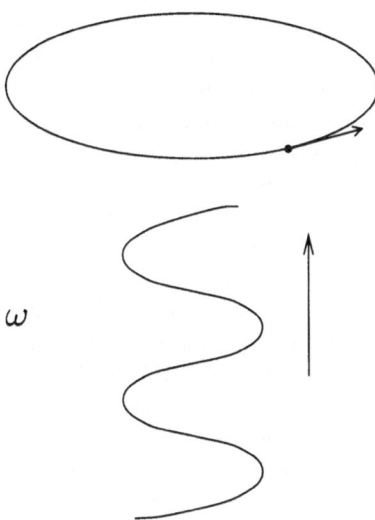

Fig. 2 A thought experiment involving the measurement of the frequency of a normally incident plane monochromatic electromagnetic wave of frequency ω by a uniformly rotating observer.

if we assume that the hypothesis of locality applies to the field measurement,

$$F_{(\alpha)(\beta)}(\tau) = F_{\mu\nu}\lambda^{\mu}_{(\alpha)}\lambda^{\nu}_{(\beta)}, \tag{27}$$

and the instantaneously determined electromagnetic field $F_{(\alpha)(\beta)}(\tau)$ is then Fourier analyzed over proper time — which is definitely a nonlocal procedure — to determine its frequency content, then we find that $\omega' = \gamma(\omega \mp \Omega)$. Thus $\omega' = \gamma\omega(1 \mp \lambda/\mathcal{L})$, where $\mathcal{L} = c/\Omega$; hence, the instantaneous Doppler result is recovered for $\lambda \to 0$. The upper (lower) sign here refers to right (left) circularly polarized incident wave. Apart from the Lorentz factor γ that refers to the time dilation involved here, the result for ω' has a simple physical interpretation: The electromagnetic field rotates with frequency ω $(-\omega)$ about the z-axis for an incident right (left) circularly polarized wave, so that the field rotates with respect to the observer with frequency $\omega - \Omega$ $(-\omega - \Omega)$. Thus the helicity of the radiation couples to the rotation of the observer, i.e. $\hbar\omega' = \gamma(\hbar\omega - \mathbf{s} \cdot \boldsymbol{\Omega})$; in fact, this is an example of the general phenomenon of spin-rotation coupling

[18, 19, 20, 21, 22, 23, 24]. For instance, for experiments on the Earth the "non-relativistic" Hamiltonian for a spin-$\frac{1}{2}$ particle should be supplemented by

$$\mathcal{H}_{SR} = -\mathbf{s} \cdot \Omega + \mathbf{s} \cdot \Omega_P, \qquad (28)$$

where Ω is the frequency of Earth's rotation and Ω_P is the gravitomagnetic precession frequency. The second term in equation (28) illustrates the gravitational Larmor theorem. It is interesting to note that $\hbar\Omega \sim 10^{-19}\,\mathrm{eV}$ and $\hbar\Omega_P \sim 10^{-29}\,\mathrm{eV}$; in fact, recent experiments [25, 26] have demonstrated the existence of the first term in (28). Moreover, the position dependence of the second term in (28) indicates the existence of a gravitomagnetic Stern-Gerlach force $-\nabla(\mathbf{s} \cdot \Omega_P)$ that is purely spin dependent and violates the universality of free fall. For instance, neutrons in different spin states in general fall differently in the gravitational field of a rotating mass; similarly, the gravitational deflection of polarized light is affected by the rotation of the mass. That is, in addition to, and about, the Einstein deflection angle $\Delta = 4GM/c^2D$, there is a splitting due to the helicity-rotation coupling by a much smaller angle $\delta = 4\lambda GJ/c^3D^3$, where D is the impact parameter for radiation propagating normal to the rotation axis and over a pole of the rotating mass [18, 16]. As $\lambda/\mathcal{L} \to 0$, $\delta \to 0$ and hence the standard result for a null geodesic is recovered.

To explain all of the experimental tests of general relativity, it is sufficient to consider all wave phenomena only in the JWKB limit. That is, geometric "optics" is all that is required; no gravitational effect involving wave "optics" has ever been detected thus far. An interesting opportunity for detecting such effects would come about if the quasinormal modes (QNMs) of black holes could be observed. The infinite set of QNMs corresponds to damped oscillations of a black hole that come about as the black hole divests itself of the energy of the external perturbation and returns to a stationary state; therefore, these ringing modes of black holes appear as $\mathcal{A}\exp(-i\omega t)$ at late times far from a black hole. Here $\mathcal{A}$ is the amplitude of the oscillation that depends on the strength of the perturbation as well as the black hole response, while $\omega = \omega_0 - i\Gamma$ with $\Gamma \geq 0$ is purely a function of mass M, angular momentum J and charge Q of the black hole, i.e. $\omega = \omega_{jmn}(M, J, Q)$, where j, m and n are parameters characterizing the total angular momentum of the radiation field, its component along the z-axis and the mode number, respectively [27]. The mode number $n = 0, 1, 2, \dots$, generally refers to the fundamental, first excited state, etc., of the perturbed black hole with j and m; in fact, Γ increases with n so that the higher excited states are more strongly damped. The fundamental least-damped gravitational mode with $j = 2$ and $n = 0$ for a Schwarzschild black hole is given by

$$\omega_0/2\pi \approx 10^4(M_\odot/M)\,\mathrm{Hz}, \qquad (29)$$

$$\Gamma^{-1} \approx 6 \times 10^{-5}(M/M_\odot)\,\mathrm{sec}, \qquad (30)$$

so that even this mode is rather highly damped and would therefore be very difficult to observe. The damping problem improves by an order of magnitude if the black hole rotates rapidly; however, the observational difficulties would still be considerable. The observation of such a mode would be very significant physically since, among other things, near an oscillating black hole $\mathcal{L}_g \sim GM/c^2$ and with $\lambda = c/\omega_0$, we have $\lambda/\mathcal{L}_g \sim 1$, so that wave "optics" can be explored in the gravitational field of a black hole.

It is necessary to examine the justification for the local field assumption (27), since it leads — in the thought experiment of Figure 2 — to the result that a normally incident right circularly polarized wave with $\omega = \Omega$ would stand completely still with respect to the observer. This circumstance is in contradiction with expectations based on elementary notions of relativity theory [28]. In fact, at $\omega = \Omega$ one has $\lambda/\mathcal{L} = 1$ and it is possible to argue that the hypothesis of locality must be violated. To this end, imagine an accelerated charged particle in the nonrelativistic approximation. The particle radiates electromagnetic waves with characteristic wavelength $\lambda \sim \mathcal{L}$; therefore, it is expected that such a particle would not be locally inertial and that (27) is violated. Indeed, the equation of motion of the particle is given by

$$m\frac{d^2\mathbf{x}}{dt^2} - \frac{2}{3}\frac{q^2}{c^3}\frac{d^3\mathbf{x}}{dt^3} + \cdots = \mathbf{f}. \tag{31}$$

The radiation reaction term — due originally to Abraham and Lorentz — ensures that the particle is not pointwise inertial, since its position and velocity are not sufficient to determine the state of the radiating particle.

These classical considerations must naturally extend to the quantum domain as well, since quantum theory is based on the notion of wave-particle duality. That is, we expect that the hypothesis of locality would be violated in the quantum regime. Consider, for instance, the determination of muon lifetime by Bailey *et al.* [29] involving muons (in a storage ring at CERN) undergoing centripetal acceleration of $g = \gamma^2 v^2/r \simeq 10^{21}$ cm sec^{-2}. If τ_μ^0 is the lifetime of the muon at rest, then the hypothesis of locality would imply that the lifetime in the storage ring would be $\tau_\mu = \gamma\tau_\mu^0$. In the experiment, $r \simeq 7$ m, $\gamma \simeq 29$ and time dilation is verified at the level of $\sim 10^{-3}$. On the other hand, the deviation from the hypothesis of locality is expected to be of the form $\lambda/\mathcal{L} \sim 10^{-13}$, where $\lambda = \hbar/mc$ is the Compton wavelength of the muon and $\mathcal{L} = c^2/g \simeq 1$ cm is the translational acceleration length. But the functional form of this deviation is not specified by our general intuitive considerations. In any case, $\lambda/\mathcal{L}$ is about ten orders of magnitude below the level of experimental accuracy [29]. In fact, the decay of the muon has been considered in this case by Straumann and Eisele by replacing the accelerated muon by the stationary state of a muon in a Landau level with very high quantum number [30]. It can be shown that the decay of such a state results in

$$\tau_\mu \simeq \gamma\tau_\mu^0 \left[1 + \frac{2}{3}(\lambda/\mathcal{L})^2\right], \tag{32}$$

so that the deviation from the hypothesis of locality is very small ($\sim 10^{-25}$) in this case but definitely nonzero.

6 Discussion

General relativity is a consistent theory of pointlike coincidences involving classical point particles and rays of radiation. The theory is robust and can be naturally extended to include wave phenomena ("minimal coupling"); however, general relativity is expected to have limited significance in this regime. From a basic standpoint, the main difficulty is the hypothesis of locality.

An accelerated observer passes through a continuous infinity of hypothetical inertial observers; therefore, the most general linear connection between the field measured by the accelerated observer $\mathcal{F}_{\alpha\beta}(\tau)$ and the locally measured field $F'_{\alpha\beta}(\tau) = F_{(\alpha)(\beta)}(\tau)$ that is consistent with causality is

$$\mathcal{F}_{\alpha\beta}(\tau) = F'_{\alpha\beta}(\tau) + \int_0^\tau \mathcal{K}_{\alpha\beta}{}^{\gamma\delta}(\tau,\tau')\, F'_{\gamma\delta}(\tau')\, d\tau'. \tag{33}$$

Here the observer is inertial for $\tau \leq 0$ and the absence of the kernel $\mathcal{K}$ would be equivalent to the hypothesis of locality; moreover, if $\mathcal{K}$ is directly connected with acceleration, then the deviation from the hypothesis of locality is generally of order $\lambda/\mathcal{L}$. Assuming that $\mathcal{K}$ is a convolution-type kernel (i.e. it depends only on $\tau - \tau'$), it is possible to determine $\mathcal{K}$ uniquely based on the assumption that no observer can ever stay at rest with respect to a basic radiation field. This is simply a generalization of the well-known result of Lorentz invariance, so that the *motion* of an electromagnetic wave would then become independent of the observer. We extend the observer independence of wave notion to all basic radiation fields and elevate this notion to the status of a fundamental physical principle [28]. Writing equation (27) as $F' = \Lambda F$, our basic assumption implies that the *resolvent* kernel $\mathcal{R}$ is given by [31]

$$\mathcal{R} = \frac{d\Lambda(\tau)}{d\tau} \Lambda^{-1}(0). \tag{34}$$

It follows that for a scalar field ($\Lambda = 1$), $\mathcal{R} = 0$ and hence $\mathcal{K} = 0$; therefore, an observer can in principle stay at rest with respect to a scalar field. This is contrary to our basic assumption, which then excludes fundamental scalar fields. In this way, a nonlocal theory of accelerated observers has been developed that is in agreement with all available observational data [31]. Moreover, novel inertial effects are predicted by the nonlocal theory. For instance, let us recall the thought experiment (cf. Figure 2) involving plane electromagnetic radiation of frequency ω normally incident on an observer rotating counterclockwise with $\Omega \ll \omega$; the nonlocal theory predicts that the field amplitude measured by the observer is larger by a factor of $1 + \Omega/\omega$ for positive helicity radiation and smaller by a factor of $1 - \Omega/\omega$ for negative helicity radiation. For radio waves with $\lambda \simeq 1$ cm and an observer rotating at a frequency of 50 Hz, we have $\Omega/\omega = \lambda/\mathcal{L} \simeq 10^{-8}$.

Finally, it should be mentioned that no thermal ambience is encountered for an accelerated observer on the basis of the approach adopted in this paper. This is consistent with the absence of any experimental evidence for such a thermal ambience at present [20]. That is, either (27) or (33) can be used to determine the quantum radiation field according to an accelerated observer once the quantum field in the inertial frame is given. Indeed, the nonlocal theory has been developed based on the assumption that no quanta are created or destroyed merely because an observer accelerates ("quantum invariance condition").

Bibliography

[1] A. Einstein, *The Meaning of Relativity* (Princeton University Press, Princeton, 1955).

[2] B. Mashhoon, *Found. Phys.* (Wheeler Festschrift) **16** (1986) 619; *Phys. Lett. A* **122** (1987) 67, 299; *Phys. Lett. A* **126** (1988) 393; *Phys. Lett. A* **143** (1990) 176; *Phys. Lett. A* **145** (1990) 147.

[3] H. Heintzmann and P. Mittelstaedt, *Springer Tracts Mod. Phys.* **47** (1968) 185.

[4] W.A. Rodrigues Jr. and E.C. De Oliveira, *Phys. Lett. A* **140** (1989) 479.

[5] S.R. Mainwaring and G.E. Stedman, *Phys. Rev. A* **47** (1993) 3611.

[6] R. Neutze and W. Moreau, *Phys. Lett. A* **179** (1993) 389; *Phys. Lett. A* **183** (1993) 141.

[7] F.V. Kowalski, *Phys. Rev. A* **53** (1996) 3761.

[8] B. Mashhoon, in: *Quantum Gravity and Beyond*, edited by F. Mansouri and J. Scanio (World Scientific, Singapore, 1993), p. 257.

[9] W. Rindler and L. Mishra, *Phys. Lett. A* **173** (1993) 105.

[10] K.-P. Marzlin, *Phys. Rev. D* **50** (1994) 888; *Gen. Rel. Grav.* **26** (1994) 619; *Phys. Lett. A* **215** (1996) 1.

[11] E. Schrödinger, *Preuss. Akad. Wiss. Berlin Ber.* **12** (1931) 238.

[12] H. Salecker and E.P. Wigner, *Phys. Rev.* **109** (1958) 571.

[13] E.R. Caianiello, *Riv. Nuovo Cimento* **15** (1992) no.4.

[14] A. Feoli, G. Lambiase, G. Papini and G. Scarpetta, *Nuovo Cimento B* **112** (1997) 913; G. Lambiase, G. Papini and G. Scarpetta, *Nuovo Cimento B* **112** (1997) 1003.

[15] J. Larmor, *Phil. Mag.* **44** (1897) 503.

[16] B. Mashhoon, *Phys. Lett. A* **173** (1993) 347.

[17] B. Mashhoon and H. Quevedo, *Nuovo Cimento B* **110** (1995) 291.

[18] B. Mashhoon, *Nature* **250** (1974) 316; *Phys. Rev. D* **11** (1975) 2679; *Phys. Rev. Lett.* **61** (1988) 2639; *Phys. Lett. A* **139** (1989) 103; *Phys. Rev. Lett.* **68** (1992) 3812; *Phys. Lett. A* **198** (1995) 9.

[19] F.W. Hehl and W.T. Ni, *Phys. Rev. D* **42** (1990) 2045; F.W. Hehl, J. Lemke and E.W. Mielke, in: *Geometry and Theoretical Physics*, edited by J. Debrus and A.C. Hirshfeld (Springer, Berlin, 1991), p. 56; J. Audretsch, F.W. Hehl and C. Lämmerzahl, in: *Relativistic Gravity Research*, edited by J. Ehlers and G. Schäfer (Springer, Berlin, 1992), p. 368.

[20] Y.Q. Cai and G. Papini, *Phys. Rev. Lett.* **66** (1991) 1259; *Phys. Rev. Lett.* **68** (1992) 3811; Y.Q. Cai, D.G. Lloyd and G. Papini, *Phys. Lett. A* **178** (1993) 225.

[21] M.P. Silverman, *Phys. Lett. A* **152** (1991) 133; *Nuovo Cimento D* **14** (1992) 857.

[22] J. Huang, *Ann. Physik* **3** (1994) 53.

[23] I.D. Soares and J. Tiomno, *Phys. Rev. D* **54** (1996) 2808.

[24] L.H. Ryder, *J. Phys. A: Math. Gen.* **31** (1998) 2465.

[25] D.J. Wineland *et al.*, *Phys. Rev. Lett.* **67** (1991) 1735.

[26] B.J. Venema *et al.*, *Phys. Rev. Lett.* **68** (1992) 135.

[27] H. Liu and B. Mashhoon, *Class. Quantum Grav.* **13** (1996) 233.

[28] B. Mashhoon, *Found. Phys. Lett.* **6** (1993) 545.

[29] J. Bailey *et al.*, *Nature* **268** (1977) 301; *Nucl. Phys. B* **150** (1979) 1.

[30] A.M. Eisele, *Helv. Phys. Acta* **60** (1987) 1024.

[31] B. Mashhoon, *Phys. Rev. A* **47** (1993) 4498; in: *Proc. VII Brazilian School of Cosmology and Gravitation*, edited by M. Novello (Editions Frontières, Gif-sur-Yvette, 1994), p. 245.

Boson Stars in the Centre of Galaxies?

Franz E. Schunck and Andrew R. Liddle

Astronomy Centre, University of Sussex, Falmer, Brighton BN1 9QJ, UK

Abstract. We investigate the possible gravitational redshift values for boson stars with a self-interaction, studying a wide range of possible masses. We find a limiting value of $z_{\mathrm{lim}} \simeq 0.687$ for stable boson star configurations. We can rule out the direct observability of boson stars. X-ray spectroscopy is perhaps the most interesting possibility.

1 Introduction

The idea of the boson star goes back to Kaup [6]. A boson star is a gravitationally bound collection of bosonic particles, arising as a solution of the Klein–Gordon equation coupled to general relativity. Many investigations of the possible configurations have been carried out; for reviews see [8, 5, 9]. For non-self-interacting bosons of mass m, the mass of a typical configuration is of order m_{Pl}^2/m, to be compared with a typical neutron star mass of $m_{\mathrm{Pl}}^3/m_{\mathrm{neutron}}^2$ which is about a solar mass. Here m_{Pl} is the Planck mass.

The situation is very different if the boson stars have even a very weak self-interaction. [1] showed that the maximum mass of stable configurations is then of order $\lambda^{1/2} m_{\mathrm{Pl}}^3/m^2$, where λ is the scalar field self-coupling, normally assumed to be of order unity. Then boson star configurations exist with mass (and radius) similar to that of neutron stars, if the bosons, like neutrons, have a mass around 1 GeV. They can also be much heavier, should the bosons be lighter. We allow ourselves to consider a very wide range of possibilities for the boson star mass and radius. If boson stars exist, they provide an alternative explanation for stellar systems in which an object is inferred to have a high mass; conventionally, a 'star' with mass greater than a few solar masses is assumed to be a black hole.

We investigate the implications of assuming that the material from which boson stars are made interacts with neighbouring baryonic material and photons just gravitationally, as the relation between a visible galaxy and its dark matter halo. An example already existing in the literature is the *boson–fermion star* ([2, 3]), which is made up of bosons and neutrons interacting only gravitationally. However, while a galaxy halo can be described using Newtonian theory, boson stars close to the maximum allowed mass are general relativistic objects. This gives such objects a new characteristic, a gravitational redshift ([10]).

The boson star model is described by the Lagrange density of a massive complex self-gravitating scalar field $\mathcal{L} = \sqrt{|\,g\,|}\left[m_{\mathrm{Pl}}^2 R/8\pi + \partial_\mu \Phi^* \partial^\mu \Phi - U(|\Phi|^2)\right]/2$, where R is the curvature scalar, g the determinant of the metric $g_{\mu\nu}$, and Φ is a *complex* scalar field with a potential U. We take $\hbar = c = 1$. Since want to have an additional global $U(1)$ symmetry (conserved particle number), we can take the following potential $U = m^2|\Phi|^2 + \lambda|\Phi|^4/2$, where m is the scalar mass and λ a dimensionless constant measuring the self-interaction strength. For spherically symmetric solutions we use the static line element $ds^2 = \exp(\nu(r))dt^2 - \exp(\mu(r))dr^2 - r^2(d\vartheta^2 + \sin^2\vartheta \, d\varphi^2)$. The most general scalar field ansatz consistent with this metric is $\Phi(r,t) = P(r)\exp(-i\omega t)$, where ω is the frequency.

2   Gravitational redshift and detectability of boson stars

The maximum possible redshift for a given configuration is obtained if the emitter is exactly at the center $R_{int} = 0$. The receiver is always practically at infinity. For all other redshifts in between, we define the redshift function $1 + z_g(x) \equiv \exp\left(-\nu(x)/2\right)$, where $x = \omega r$. A boson star with the maximum mass gives the highest value one can obtain from stable configurations (unstable configurations can yield very high redshift values) ([7]). We find that with increasing self-interaction values $\Lambda := \lambda m_{Pl}^2/4\pi m^2$ also the maximal redshift value grows [10]. For $\Lambda \to \infty$, we find the asymptotic value $z_{lim} \simeq 0.687$.

The mass M of a boson star composed of non-self-interacting particles is inversely proportional to m, while the mass of a self-interacting boson star is proportional to $\sqrt{\lambda}/m^2$; see [1]. Taking $\lambda \sim 1$, then for small m (to be precise, provided $\Lambda \gg 1$) the self-interacting star is much more massive. For example, if we want to get a boson star with a mass of order 10^{33}g (a solar mass), then we need $m \sim 10^{-10}$ eV for $\lambda = 0$, or $m \propto \lambda^{1/4}$ GeV if $\lambda \gg 10^{-38}$ (we see that the self-coupling has to be extraordinarily tiny to be negligible). In this example, the scalar particle has a mass comparable to a neutron, leading to a boson star with the dimensions of a neutron star. If we reduce the scalar mass further, to $m \sim 1$ MeV, then we find $M \sim 10^{39}\sqrt{\lambda}$ g and $R \sim 10^6\sqrt{\lambda}$ km; this radius is comparable to that of the sun, but encloses 10^6 solar masses. These parameters are reminiscent of supermassive black holes, for example as in Active Galactic Nuclei;; the mass–radius relation is effectively fixed just by the objects being relativistic. In all cases, the density of the boson stars makes their direct detection as difficult as in the case of black holes; in particular, they cannot be resolved in any waveband, cf. [10].

However, even if boson stars cannot be directly resolved, their influence might still be visible if material in their vicinity is sufficiently luminous. It is necessary to find a certain amount of luminous matter within the gravitational potential of the boson star. This could, for example, be HI gas clouds as seen in galaxies. One might also expect accretion disks about boson stars, though there the luminosity could be dominated by regions outside the gravitational potential and the boson star would be indistinguishable from a black hole.

The most promising technique for observing supermassive boson stars is to consider a wave-band where they might be extremely luminous, e.g. X-rays. A very massive boson star, say $10^6 M_\odot$ is likely to form an accretion disk, and since its exterior solution is Schwarzschild it is likely to look very similar to an AGN with a black hole at the center. In X-rays, it has been claimed by [4] that using ASCA data they have probed to within 1.5 Schwarzschild radii. A boson star configuration provides a non-singular solution where emission can occur from arbitrarily close to the center. The signature they use is a redshifted wing of the Iron K-line. If such techniques have their validity confirmed, it may ultimately be possible to use X-ray spectroscopy to map out the shape of the gravitational potential close to the event horizon or boson star.

The rotation curves about a boson star ([10]) show an increase up to a maximum with more than one-third of the velocity of light followed by a Keplerian decrease. If boson stars exist, then such enormous rotation velocities are not necessarily signatures of black holes. In [11], a model with massless bosonic particles was applied to fit rotation curve data of spiral and dwarf galaxies.

Acknowledgments

FES is supported by a Marie Curie research fellowship (European Union TMR programme) and ARL is supported by the Royal Society. We would like to thank John Barrow and Eckehard W. Mielke for helpful discussions and comments.

Bibliography

[1] Colpi, M., Shapiro, S.L., Wasserman, I. (1986): Boson stars: Gravitational equilibria of self-interacting scalar fields. Phys. Rev. Lett. **57**, 2485–2488

[2] Henriques, A. B., Liddle, A. R., Moorhouse, R. G. (1989): Combined boson-fermion stars. Phys. Lett. B **233**, 99–106

[3] Henriques, A. B., Liddle, A. R., Moorhouse, R. G. (1990): Combined boson–fermion stars: Configurations and stability. Nucl. Phys. B **337**, 737–761

[4] Iwasawa, K., et al. (1996): The variable iron K emission line in MCG–6-30-15. Mon. Not. R. Astron. Soc. **282**, 1038–1048

[5] Jetzer, Ph. (1992): Boson stars. Phys. Rep. **220**, 163–227

[6] Kaup, D. J. (1968): Klein–Gordon geon. Phys. Rev. **172**, 1331–1342

[7] Kusmartsev, F. V., Mielke, E. W., Schunck, F. E. (1991): Gravitational stability of boson stars. Phys. Rev. D **43**, 3895–3901

[8] Lee, T. D., Pang, Y. (1992): Nontopological solitons. Phys. Rep. **221**, 251–350

[9] Liddle, A. R., Madsen, M. S. (1992): The structure and formation of boson stars. Int. J. Mod. Phys. D **1**, 101–143

[10] Schunck, F. E., Liddle, A. R. (1997): The gravitational redshift of bosons stars. Phys. Lett. B **404**, 25–32

[11] Schunck, F. E. (1997): Massless scalar field models rotation curves of galaxies. In Klapdor-Kleingrothaus, H. V., Ramachers Y. (eds.) *Aspects of dark matter in astro- and particle physics*, World Scientific Press, Singapore, p. 403–408.

Black Holes in Two Dimensions

Yuri N. Obukhov[1] and Friedrich W. Hehl[2]

[1] Department of Theoretical Physics, Moscow State University, 117234 Moscow, Russia
[2] Institute for Theoretical Physics, University of Cologne, D-50923 Köln, Germany

Abstract. Models of black holes in $(1 + 1)$-dimensions provide a theoretical labora-tory for the study of semi-classical effects of realistic black holes in Einstein's theory. Important examples of two-dimensional models are given by string theory motivated *dilaton gravity*, by ordinary general relativity in the case of *spherical symmetry*, and by *Poincaré gauge gravity* in two spacetime dimensions. In this paper, we present an in-troductory overview of the exact solutions of two-dimensional classical Poincaré gauge gravity (PGG). A general method is described with the help of which the gravita-tional field equations are solved for an arbitrary Lagrangian. The specific choice of a torsion-related coframe plays a central role in this approach. Complete integrability of the *general* PGG model is demonstrated in vacuum, and the structure of the black hole type solutions of the *quadratic* models with and without matter is analyzed in detail. Finally, the integrability of the general dilaton gravity model is established by recasting it into an effective PGG model.

1 Introduction

Standard General Relativity (GR) is trivial in two dimensions. Nevertheless, two–dimensional (2D) models of gravity which differ from GR have recently received considerable attention [2]-[35]. The interest in 2D gravity is strongly supported by the fact that usual four-dimensional GR, in the case of spherical symmetry, is described by an effective 2D gravitational model of the dilaton type. Such a dimensional reduction provides a technical tool for the study of long standing problems in black hole physics, including an understanding of the final state of a black hole with an account of the back reaction of the quan-tum evaporation process, see [9, 5, 35]. On the other hand, lower–dimensional black hole physics is discussed in the context of string theory motivated dila-ton gravity (see, e.g., [9, 15, 18, 19, 20, 28, 30, 27, 34]) and in the framework of PGG ("Poincaré gauge gravity") [1]-[8],[29, 31, 33],[37]-[41]. The approaches [1]-[8] are attempts to construct string theories with a dynamical geometry [10]-[17],[33]. In this paper we present an overview of the black hole solutions in classical two–dimensional PGG. At the same time, 2D gravity is of interest in itself as a theoretical laboratory which offers a simple way to study difficult non-perturbative quantization problems [13].

In the studies of both, classical and quantized 2D models, it is of crucial importance to find exact solutions of the field equations. Here we describe an elegant method developed in [29]-[41] with the help of which one can explicitly integrate the field equations of classical PGG with and without matter sources. The central point is to use a specific coframe built up from the one–form of the torsion trace and its Hodge dual. The early proofs of the integrability of the *quadratic* PGG models in vacuum were based on the component approach and relied on specific gauge choices, like the conformal or the light-cone gauge [10]-[26]. The coupling to gauge, scalar, and spinor matter fields was shown to destroy the integrability in general. A peculiar but common feature of the *standard* matter sources (gauge, scalar, and spinor fields) in two dimensions is that for all of them the spin current vanishes. Thus, quite generally, the Lorentz connection is explicitly decoupled from 2D matter. Hence the material energy–momentum current is symmetric and covariantly conserved with respect to the Riemannian connection. The absence of the spin–connection coupling considerably facilitates the integration of the field equations.

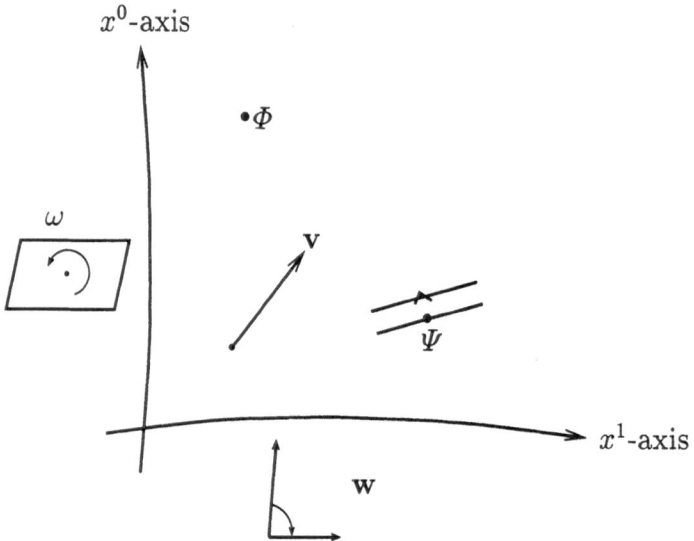

Fig. 1 *Two-dimensional spacetime*: A 0-form (scalar) Φ has one component, a 1-form Ψ two components, and a 2-form ω one component. A vector $\mathbf{v}$ has two components, a bivector $\mathbf{w}$ one component.

The structure of the paper is as follows: Sec. 2 contains an introduction to 2D Riemann-Cartan geometry. In Secs. 3 and 4, we demonstrate the integrability of PGG with an arbitrary gravitational Lagrangian and prove the consistency of our method in general. As a particular application, a *quadratic* model with an action containing squares of torsion and curvature is discussed in vacuum (Sec. 5) and in the presence of conformally invariant matter (Sec. 6). The properties of the exact solutions of black hole type are described in detail. Finally, in Sec. 7, we apply

x^0-axis

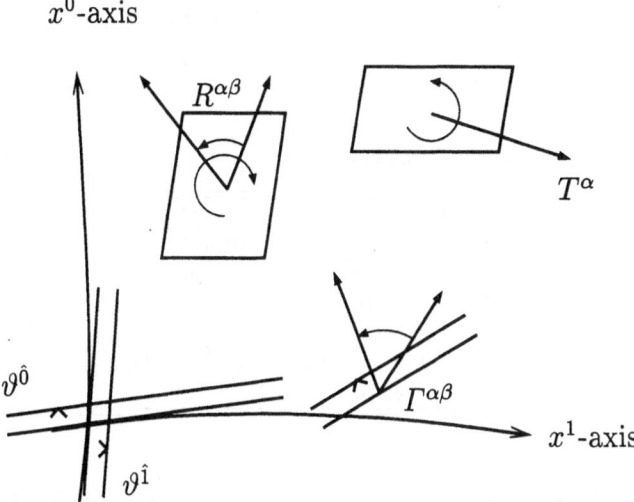

Fig. 2 *Two-dimensional Riemann-Cartan spacetime*: Coframe ϑ^α (2 components), connection $\Gamma^{\alpha\beta} = -\Gamma^{\beta\alpha}$ (2 components), torsion T^α (2 components), curvature $R^{\alpha\beta} = -R^{\beta\alpha}$ (1 component). Here, $\vartheta^\alpha = \{\vartheta^{\hat{0}}, \vartheta^{\hat{1}}\}$ is a natural coframe, i.e., $\vartheta^{\hat{0}} = dx^0$, $\vartheta^{\hat{1}} = dx^1$.

the general method to the (purely *Riemannian*) string theory motivated dilaton gravity models by rewriting them in form of an effective *Poincaré-Brans-Dicke* theory. The general solution of an arbitrary two-dimensional dilation gravity model is obtained in explicit form.

2   Two-dimensional Riemann-Cartan spacetime

The Riemann–Cartan geometry has rather remarkable properties in two dimensions, see Fig.1.

In the PGG approach, the *orthonormal* coframe one–form ϑ^α and the linear connection one–form $\Gamma^{\alpha\beta}$ are considered to be the translational and the Lorentz gauge potentials of the gravitational field, respectively. The corresponding field strengths are given by the torsion two-form $T^\alpha := D\vartheta^\alpha$ and the curvature two-form $R^{\alpha\beta} := d\Gamma^{\alpha\beta} - \Gamma^{\alpha\gamma} \wedge \Gamma_\gamma{}^\beta$, see Fig.2. The frame $e_\alpha = e^i{}_\alpha \, \partial_i$ is dual to the coframe $\vartheta^\beta = e_j{}^\beta \, dx^j$, i.e., $e_\alpha \rfloor \vartheta^\beta = e^i{}_\alpha \, e_i{}^\beta = \delta^\beta_\alpha$. The spacetime manifold M is equipped with a metric

$$g = g_{ij} \, dx^i \otimes dx^j \, . \tag{2.1}$$

Thus its coframe components satisfy

$$o_{\alpha\beta} = e^i{}_\alpha \, e^j{}_\beta \, g_{ij} \, , \qquad (o_{\alpha\beta}) = \mathrm{diag}(-1, +1). \tag{2.2}$$

In an orthonormal frame, the curvature, like the connection, is antisymmetric in α and β.

Table 1 Gauge field strengths, matter currents, and η–basis

	Type		Components	
Object	Valuedness	p-form	n-dimensions	n = 2
T^α	vector	2	$n^2(n-1)/2$	2
$R^{\alpha\beta}$	bivector	2	$n^2(n-1)^2/4$	1
Σ^α	vector	$n-1$	n^2	4
$\tau^{\alpha\beta}$	bivector	$n-1$	$n^2(n-1)/2$	2
η^α	vector	$n-1$	n^2	4

For a 2D Riemann–Cartan space, as we can take from Table 1, we have two translation generators and one rotation generator. This allows us to introduce a Lie (or right) duality operation, that is, a duality with respect to the Lie–algebra indices, which maps a vector into a covector, and vice versa:

$$\psi_\alpha^\star := \eta_{\alpha\beta}\,\psi^\beta\,, \qquad \psi^\alpha = \eta^{\alpha\beta}\,\psi_\beta^\star. \qquad (2.3)$$

Here the completely antisymmetric tensor is defined by $\eta_{\alpha\beta} := \sqrt{|\det o_{\mu\nu}|}\,\epsilon_{\alpha\beta}$, where $\epsilon_{\alpha\beta}$ is the Levi–Civita symbol normalized to $\epsilon_{\hat0\hat1} = +1$ (a circumflex on top of a number identifies the number as an anholonomic or frame index). For $\psi^\beta = \vartheta^\beta$ we get $\eta_\alpha := \ast\vartheta_\alpha = \vartheta_\alpha^\star$, where $\ast$ denotes the Hodge (or left) dual. Using the Lie (or right) duality in two dimensions, we can appreciably compactify the notation, see Table 2.

Local Lorentz transformations are defined by the 2×2 matrices $\Lambda_\beta{}^\alpha(x) \in SO(1,1)$,

$$\Lambda_\alpha{}^\beta = \delta_\alpha^\beta \cosh\omega + \eta_\alpha{}^\beta \sinh\omega. \qquad (2.4)$$

Table 2 2D geometrical objects

n=2	Valuedness	p-form	Components
$\Gamma^\star := (1/2)\,\eta_{\alpha\beta}\,\Gamma^{\alpha\beta}$	scalar	1	2
$t^\alpha := \ast T^\alpha$	vector	0	2
$T := e_\alpha \rfloor T^\alpha$	scalar	1	2
$t^2 := o_{\alpha\beta}\,t^\alpha\,t^\beta$	scalar	0	1
$R^\star = d\Gamma^\star$	scalar	2	1
$R := e_\alpha \rfloor e_\beta \rfloor R^{\alpha\beta}$	scalar	0	1

The gauge transformations of the basic gravitational field variables read

$$\vartheta'^\alpha = (\Lambda^{-1})_\beta{}^\alpha\, \vartheta^\beta = \vartheta^\alpha \cosh\omega - \eta^\alpha \sinh\omega, \tag{2.5}$$

$$\Gamma'_\alpha{}^\beta = \Lambda_\alpha{}^\gamma \Gamma_\gamma{}^\delta (\Lambda^{-1})_\delta{}^\beta - \Lambda_\alpha{}^\gamma d(\Lambda^{-1})_\gamma{}^\beta = \Gamma_\alpha{}^\beta + \eta_\alpha{}^\beta d\omega, \tag{2.6}$$

or $\qquad \Gamma^{*\prime} = \Gamma^* - d\omega. \tag{2.7}$

The curvature 2–form has only one irreducible component, and it can be expressed in terms of the curvature scalar $R := e_\alpha\rfloor e_\beta\rfloor R^{\alpha\beta}$:

$$R^{\alpha\beta} = -\frac{1}{2} R\, \vartheta^\alpha \wedge \vartheta^\beta. \tag{2.8}$$

In two dimensions torsion is irreducible and reduces to its vector piece

$$T^\alpha = -t^\alpha \eta, \tag{2.9}$$

where the vector–valued torsion zero–form t^α is defined via the Hodge dual

$$t^\alpha := *T^\alpha. \tag{2.10}$$

When the torsion square is not identically zero, i.e. $t^2 := t_\alpha t^\alpha \neq 0$, we call the corresponding manifold M a *non-degenerate Riemann–Cartan spacetime*. In this case, using the scalar-valued torsion one–form $T := e_\alpha\rfloor T^\alpha$, we can write a coframe as

$$\vartheta^\alpha = -\frac{1}{t^2}\left(T\,\eta^{\alpha\beta}\,t_\beta + *T\,t^\alpha\right) = -\frac{1}{t^2}\left(T\,t^{*\alpha} + *T\,t^\alpha\right). \tag{2.11}$$

Thus, the torsion one–form T and its dual $*T$ specify a coframe with respect to which one can expand all the 2D geometrical objects. When $t^2 \neq 0$, this coframe is non-degenerate, hence the terminology of a non-degenerate Riemann–Cartan space. In this case, the volume two–form can be calculated, in the non-degenerate case, as an exterior square of the torsion one–form T:

$$\eta := \frac{1}{2}\,\eta_{\alpha\beta}\,\vartheta^\alpha \wedge \vartheta^\beta = \frac{1}{t^2}\,*T \wedge T. \tag{2.12}$$

Defining a coframe of a 2D Riemann-Cartan spacetime in terms of the torsion one–form turns out to be extremely convenient, and, in fact, underlies the integrability of the 2D gravity models with and without matter.

3   The field equations of PGG: invariant formulation

The total action of the interacting matter field Ψ and the PGG fields in two dimensions reads

$$W = \int\left[L(\vartheta^\alpha, \Psi, D\Psi) + V(\vartheta^\alpha, T^\alpha, R^{\alpha\beta})\right], \tag{3.1}$$

where the matter Lagrangian two–form L will be specified later.

One can prove that torsion and curvature can enter a general gravitational Lagrangian V only in form of the scalars $t^2 := o_{\alpha\beta}\, t^\alpha t^\beta$ and R. The gravitational Lagrangian density is denoted by $\mathcal{V} := *V$. Then the general gauge invariant PGG Lagrangian reads

$$V(\vartheta^\alpha, T^\alpha, R^{\alpha\beta}) = V(\vartheta^\alpha, t^2, R) = -\mathcal{V}(t^2, R)\, \eta. \tag{3.2}$$

The partial derivatives

$$P := -2\left(\frac{\partial \mathcal{V}}{\partial t^2}\right), \quad \kappa := 2\left(\frac{\partial \mathcal{V}}{\partial R}\right), \tag{3.3}$$

i.e. the generalized gravitational field momenta ('excitations'), define two functions $P = P(t^2, R)$ and $\kappa = \kappa(t^2, R)$ which are assumed to be smooth and *nontrivial:* $P \neq 0$, $\kappa \neq 0$.

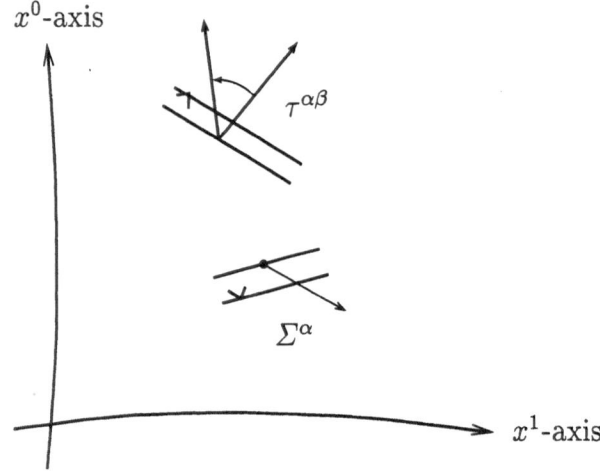

Fig. 3 Energy–momentum current Σ^α (4 components) and spin current $\tau^{\alpha\beta}$ (2 components) in two–dimensional spacetime.

The variational derivatives

$$\Sigma_\alpha := \frac{\delta L}{\delta \vartheta^\alpha}, \quad \tau_{\alpha\beta} := \frac{\delta L}{\delta \Gamma^{\alpha\beta}}, \tag{3.4}$$

yield the energy–momentum and the spin one–forms of matter, respectively, see Fig.3. Using the right duality, one straightforwardly replaces the bivector-valued spin by the scalar–valued one–form $\tau^\star := \frac{1}{2}\eta_{\alpha\beta}\, \tau^{\alpha\beta}$. Similarly, instead of using the vector-valued energy-momentum one-form Σ_α, it turns out to be more convenient to introduce two scalar-valued one-forms

$$S := t^\alpha \Sigma_\alpha, \qquad S^* := t_\alpha \, \eta^{\alpha\beta} \Sigma_\beta = t^{*\alpha} \Sigma_\alpha. \tag{3.5}$$

Analogously to (2.11), which expresses the coframe in terms of T and $*T$, one can rewrite the energy–momentum current in terms of S and S^*:

$$\Sigma_\alpha = \frac{1}{t^2} \left(t_\alpha \, S + \eta_{\alpha\beta} t^\beta \, S^* \right) = \frac{1}{t^2} \left(t_\alpha \, S + t_\alpha^* \, S^* \right). \tag{3.6}$$

If we use the Hodge star, we find by straightforward algebra

$$S^* + *S = *(\vartheta^\alpha \wedge \Sigma_\alpha) *T - *(\eta^\alpha \wedge \Sigma_\alpha) T. \tag{3.7}$$

Let us recall that $\vartheta^\alpha \wedge \Sigma_\alpha$ describes the trace of the energy-momentum whereas $\eta^\alpha \wedge \Sigma_\alpha$ represents its antisymmetric part. The latter is related to the spin via the second Noether identity:

$$2 \, d\tau^* = \eta^\alpha \wedge \Sigma_\alpha. \tag{3.8}$$

The general field equations of PGG arise from independently varying (3.1) with respect to the coframe ϑ^α and the connection $\Gamma^{\alpha\beta}$. Remarkably, these equations can be rewritten in a completely coordinate and gauge invariant form

$$d(P^2 \, t^2) = 2P(\tilde{\mathcal{V}} T + S), \tag{3.9}$$

$$d(P *T) = (P \, t^2 - 2 \, \tilde{\mathcal{V}}) \, \eta + \vartheta^\alpha \wedge \Sigma_\alpha, \tag{3.10}$$

$$d\kappa = -P \, T + 2\tau^*, \tag{3.11}$$

$$P t^2 (\Gamma^* + du) = \tilde{\mathcal{V}} \, *T + S^*, \tag{3.12}$$

where

$$\tilde{\mathcal{V}} := \mathcal{V} + P \, t^2 - \frac{1}{2} \kappa \, R \tag{3.13}$$

is the so–called modified Lagrangian function and

$$t^2 \, du := \eta_{\alpha\beta} \, t^\alpha dt^\beta. \tag{3.14}$$

The term du in (3.12) is physically irrelevant, since a 2D Lorentz transformation can create such an Abelian shift, see (2.7). In (3.9)-(3.12), the source terms of the matter field Ψ are represented by S and S^*, by the energy–momentum trace $\vartheta^\alpha \wedge \Sigma_\alpha$, and by the spin τ^*. We marked them in the field equations by letters in boldface. Besides the gravitational field equations, we have the matter field equation. For matter described by a p–form field Ψ it reads

$$\frac{\delta L}{\delta \Psi} = \frac{\partial L}{\partial \Psi} - (-1)^p D \frac{\partial L}{\partial D \Psi} = 0. \tag{3.15}$$

As we have seen, the system (3.9)-(3.12) involves the energy–momentum one–forms S and S^* as sources. They satisfy the following equations which can be derived from the Noether identities:

$$d(PS) = T \wedge P(S - R\tau^*) + \frac{1}{2}\tilde{\mathcal{V}}\,d\tau^*, \tag{3.16}$$

$$d(PS^*) = T \wedge P(S^* + R*\tau^*) + \frac{2}{t^2}S \wedge S^* + \left(\tilde{\mathcal{V}} - Pt^2\right)\vartheta^\alpha \wedge \Sigma_\alpha. \tag{3.17}$$

In 2D, the specific feature of standard matter (scalar, spinor, Abelian and non–Abelian gauge fields) is that the spin current is zero:

$$\tau_{\alpha\beta} = 0 \qquad \text{(standard matter)}. \tag{3.18}$$

Thus only the canonical energy–momentum one–form Σ_α enters the gravitational field equations as a source. Moreover, in view of (3.8), it becomes symmetric: $\eta^\alpha \wedge \Sigma_\alpha = 0$. In this paper, we limit ourselves to the discussion of massless, conformally invariant matter models. Then we have

$$\vartheta^\alpha \wedge \Sigma_\alpha = 0 \qquad \text{(massless, conformally invariant matter)}, \tag{3.19}$$

and the corresponding terms drop out from the field equation (3.10) and the Noether identity (3.17). Consequently, only S is left as a source. From now on, *we will specialize* to this physically most interesting case obeying (3.18) and (3.19). Under these conditions, (3.7) simply reduces to

$$S^* + *S = 0. \tag{3.20}$$

Accordingly, the complete system (3.9)-(3.12) and (3.16)-(3.17), with (3.18) and (3.19), should be jointly solved with the matter field equation (3.15).

Consistency check of the invariant formulation

As it is clearly suggested by the field equation (3.11), the function κ of the Riemann-Cartan curvature R (and, in general, of t^2) can be conveniently treated as one of the local coordinates on a two–dimensional manifold M. However, one has always to check the consistency of the scheme by explicitly calculating the curvature from the connection which itself is obtained from the field equations. This was done for the vacuum solutions of the general PGG model in [29] and for non-vacuum solutions of the quadratic models in [31]. Here we will demonstrate consistency in general, for arbitrary matter sources and arbitrary gravitational Lagrangian. We consider the non-trivial non-degenerate case with $t^2 \neq 0$.

Eq.(3.12) yields the general solution for the Lorentz connection. Starting from the definitions (3.3) and using (3.11), it is straightforward to compute the differential of the modified Lagrangian:

$$d\tilde{\mathcal{V}} = \frac{1}{2}\left(\frac{1}{P}d(P^2t^2) + RPT\right) - R\tau^*. \tag{3.21}$$

With the help of this relation and eqs.(3.9), (3.10), (3.17), (3.7), one finds

$$d[P(\widetilde{\mathcal{V}} *T + S^{\star})] = \frac{1}{P^2 t^2} \, d(P^2 t^2) \wedge [P(\widetilde{\mathcal{V}} *T + S^{\star})] + \frac{1}{2} R P^2 T \wedge *T. \quad (3.22)$$

The consistency proof is completed by taking the exterior differential of the left- and right-hand sides of equation (3.12). With the help of (3.21) and (3.22), this yields

$$d\Gamma^{\star} = -\frac{1}{2} R \eta.$$

4   Exact solutions of PGG with arbitrary gravitational Lagrangian

The two cases of two–dimensional PGG should be treated separately, the degenerate case with $t^2 = 0$ and the non–degenerate one with $t^2 \neq 0$. We will formulate our answers for matter sources obeying (3.18) and (3.19).

4.1   Degenerate torsion solutions

If $t^2 = 0$, the torsion one–form is either self- or anti-self-dual,

$$T = \pm *T. \quad (4.1)$$

Then (3.10) and (3.11) yield $\widetilde{\mathcal{V}} = 0$. This in turn, with (3.13) and (3.3), yields

$$f(R) := \mathcal{V} - R \frac{\partial \mathcal{V}}{\partial R} = 0. \quad (4.2)$$

For a given Lagrangian $\mathcal{V} = \mathcal{V}(R)$, the t^2-dependence drops out because of the degeneracy, the solutions of $f(R) = 0$ determine some $R = R_1$, $R = R_2, \ldots$ Therefore the curvature is constant, $R = const$ and, by implication, also the R-dependent Lorentz field momentum, $\kappa = const$. Then, from (3.11), one finds $T = 0$. Finally, an analysis of the matter field equation and of the Noether identities shows that only trivial matter configurations are allowed: A constant field in the case of a zero–form Ψ, e.g..

Summarizing, we see that the degenerate solutions of PGG reduce to the torsionless de Sitter geometry,

$$T^{\alpha} = 0, \quad R = const, \quad \Psi = const, \quad (4.3)$$

where the constant values of the curvature are roots of equation (4.2). Incidentally, the same turns out also to be true for some conformally non-invariant matter, for a massive scalar field with arbitrary self–interaction, e.g.. In the rest of the paper we will mainly consider the non-degenerate case with $t^2 \neq 0$.

4.2   Non-degenerate vacuum solutions

Let us now specialize to the *vacuum* field equations. Accordingly, in (3.9)-(3.11) we have to put $S = 0$, $\vartheta^\alpha \wedge \Sigma_\alpha = 0$, and $\tau^* = 0$. The formal general solution is obtained as follows: Let us introduce a coordinate system (κ, λ) which is related to the torsion 1-form basis $(T, *T)$ via

$$P T = -d\kappa, \qquad\qquad P *T = B d\lambda, \qquad\qquad (4.4)$$

with some function $B(\kappa, \lambda)$. Consequently, the volume 2–form is given by

$$\eta = \frac{B}{P^2 t^2}\, d\kappa \wedge d\lambda, \qquad\qquad (4.5)$$

cf. (2.12). The first equation in (4.4) is simply the field equations (3.11).

Substitution of the ansatz (4.4) into (3.9) and (3.10) results in

$$\frac{\partial}{\partial\kappa}(P^2 t^2) = -2\tilde{\nu}, \qquad \frac{\partial}{\partial\lambda}(P^2 t^2) = 0, \qquad\qquad (4.6)$$

$$\frac{\partial}{\partial\kappa}\ln B = \frac{\partial}{\partial\kappa}\ln(P^2 t^2) + \frac{1}{P}. \qquad\qquad (4.7)$$

Formal integration of (4.7) yields the solution

$$B = B_0(\lambda)\, P^2 t^2 \exp\left(\int \frac{d\kappa}{P}\right), \qquad\qquad (4.8)$$

where $B_0(\lambda)$ is an arbitrary function of λ only.

Provided the gravitational Lagrangian V, and hence P, is smooth, there always exists a solution of the first order ordinary differential equations (4.6). This describes $P^2 t^2$ as a function of κ and λ, thus completing our formal demonstration of the integrability of the general two-dimensional vacuum PGG. The complete non-degenerate vacuum solution is evidently of the black hole type with the metric

$$g = -\frac{d\kappa^2}{P^2 t^2} + P^2 t^2 \exp\left(2\int \frac{d\kappa}{P}\right) d\lambda^2. \qquad\qquad (4.9)$$

Here, without restricting generality, we put $B_0 = 1$. *Torsion* and *curvature* for our solution are obtained by inverting the relations $P = P(t^2, R)$, $\kappa = \kappa(t^2, R) \rightarrow t^2 = t^2(P, \kappa)$, $R = R(P, \kappa)$. For the solution to be unique, one must assume the relevant Hessian $(\frac{\partial^2 V}{\partial t^2 \partial t^2}, \frac{\partial^2 V}{\partial R \partial R})$ to be non-degenerate. It is straightforward to derive from (3.12) the *curvature* scalar of the general solution:

$$R = \frac{P^2 t^2}{B} \frac{\partial}{\partial\kappa}\left(\frac{B}{P^2 t^2} \frac{\partial}{\partial\kappa}(P^2 t^2)\right). \qquad\qquad (4.10)$$

The position of the horizon(s) is evidently determined by the zeros of the metric coefficient $g_{\lambda\lambda} = P^2 t^2 \exp\left(2 \int d\kappa/P\right)$. It is impossible to say more without explicitly specializing the gravitational Lagrangian. An important particular case is represented by quadratic PGG which will be discussed in the next two sections.

5 Exact vacuum solutions of PGG with quadratic gravitational Lagrangian

Let us now analyze two-dimensional PGG with a gravitational Lagrangian *quadratic* in torsion and curvature,

$$V = - \left(\frac{a}{2} T_\alpha * T^\alpha + \frac{1}{2} R^{\alpha\beta} \eta_{\alpha\beta} + \frac{b}{2} R_{\alpha\beta} * R^{\alpha\beta} \right) - \Lambda \eta. \tag{5.1}$$

Here a, b, and Λ are the coupling constants. Using (5.1) in (3.3), we find:

$$P = a, \qquad \kappa = b R - 1. \tag{5.2}$$

The modified Lagrangian function reads

$$\tilde{V} = \frac{a}{2} t^2 - \frac{b}{4} R^2 + \Lambda. \tag{5.3}$$

We will concentrate here on *vacuum solutions*. The general scheme for an arbitrary Lagrangian V was given in the previous section. The degenerate solutions (4.3) are de Sitter spacetimes:

$$T^\alpha = 0, \qquad R = \pm R_{\mathrm{dS}}, \qquad R_{\mathrm{dS}} := 2\sqrt{\frac{\Lambda}{b}}. \tag{5.4}$$

Also the non-degenerate solutions can be easily obtained. Substituting (5.2)-(5.3) into (4.6), we explicitly find for the scalar *torsion* square

$$-t^2 = 2M_0 \, e^{-bR/a} - \frac{b}{2a} R^2 + R + \frac{2\Lambda}{a} - \frac{a}{b}. \tag{5.5}$$

The integration constant M_0 has the physical meaning of the mass of a point-like source. This can be derived from the existence of the timelike Killing vector field $\zeta = \partial_\lambda$ which yields the conserved energy–momentum 1–form

$$\varepsilon_{\mathrm{RC}} := \zeta^\alpha \Sigma_\alpha = -e^{bR/a} \left(\frac{1}{2} d(t^2) + \frac{1}{a^2} \tilde{V} \, d\kappa \right). \tag{5.6}$$

This 1–form is *strongly* conserved, i.e. $d\varepsilon_{\mathrm{RC}} = 0$ even when the field equations are not fulfilled. One can verify that

$$\varepsilon_{\mathrm{RC}} = dM, \qquad M := -\frac{e^{bR/a}}{2} \left(t^2 - \frac{b}{2a} R^2 + R + \frac{2\Lambda}{a} - \frac{a}{b} \right). \tag{5.7}$$

When the *vacuum* field equations are satisfied, $\varepsilon_{\mathrm{RC}} = 0$, and thus $M = M_0$.

Black hole structure of non-degenerate spacetimes

The degenerate solutions are torsionless de Sitter spacetimes with constant Riemannian curvature R_{dS}, see (5.4). The properties of the non-degenerate solutions are obviously defined by the values of the coupling constants (a, b, Λ) and by the value of the first integral M_0. Let us denote

$$M_\pm := \frac{e^{\mp b R_{dS}/a}}{2} \left(\frac{a}{b} \pm R_{dS} \right). \tag{5.8}$$

We recognize that always $M_- \leq M_+$, equality is achieved only for vanishing cosmological constant: $M_- = M_+ = a/(2b)$ for $R_{dS} = \Lambda = 0$. For sufficiently large Λ, namely when $\Lambda > a^2/(4b)$ (or, equivalently, $R_{dS} > a/b$) one finds negative M_-, otherwise $M_\pm \geq 0$. The special case $\Lambda = a^2/(4b)$ in a de Sitter gauge gravity model was discussed in [36, 38] (then $R_{dS} = a/b$ and $M_- = 0$).

As we already know, the spacetime metric is given by the line element (4.9) with (5.2) and (5.5) inserted. Clearly, instead of κ, we can use the scalar curvature R as a spatial coordinate. Another convenient choice is a "radial" coordinate

$$r := \exp \left(\frac{b}{a} R \right). \tag{5.9}$$

The meaning of the quantities (5.8) becomes clear when we analyze the metric coefficient

$$g_{\lambda\lambda} = -2M_0 \, e^{bR/a} + e^{2bR/a} \left(\frac{b}{2a} R^2 - R - \frac{2\Lambda}{a} + \frac{a}{b} \right). \tag{5.10}$$

Its zeros define the positions of the horizons R_h,

$$g_{\lambda\lambda}(R_h) = 0. \tag{5.11}$$

At $R = -\infty$, we have $g_{\lambda\lambda} = 0$. However, this point is not a horizon but a true singularity with infinite curvature. This corresponds to $r = 0$ which one can consider as the position of a central point-like source mass. Such a singularity is, in general, hidden by the horizons.

The position, the number, and the type of horizons are completely determined by the total mass M_0. We can distinguish five qualitatively different configurations:

(i) $M_0 < M_-$: one horizon at

$$R_h < -R_{dS}. \tag{5.12}$$

(ii) $M_0 = M_-$: two horizons at

$$R_{h_1} < -R_{dS}, \qquad R_{h_2} = R_{dS}. \tag{5.13}$$

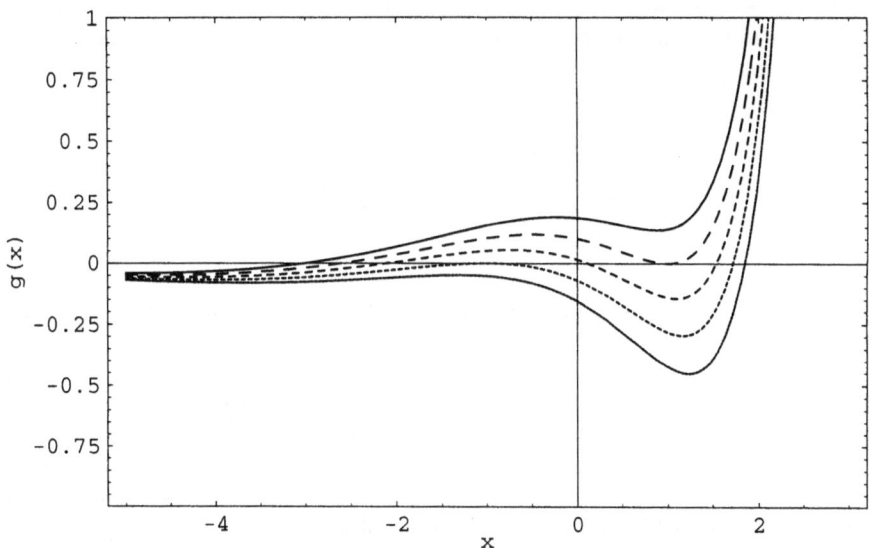

Fig. 4 Metric coefficient $g(x) := g_{\lambda\lambda}/R_{dS}$ as a function of $x := R/R_{dS}$: Moving from upper to lower curves corresponds to the cases (i)-(v), respectively.

(iii) $M_- < M_0 < M_+$: three regular horizons at

$$R_{h_1} < -R_{dS} < R_{h_2} < R_{dS}, \qquad R_{h_c} > R_{dS}. \qquad (5.14)$$

(iv) $M_0 = M_+$: two horizons at

$$R_{h_1} = -R_{dS}, \qquad R_{h_2} > R_{dS}. \qquad (5.15)$$

(v) $M_0 > M_+$: one regular horizon at

$$R_h > R_{dS}. \qquad (5.16)$$

Here we have assumed $M_0 > 0$ which is physically reasonable. For zero or negative values of M_0, the smallest horizons disappear for the cases (i), (ii), and (iii). In Fig. 4, the function (5.10) is depicted for $\Lambda = \frac{a^2}{16b}$ and positive values of M_0.

The geometry of these solutions is obtained by an appropriate gluing of a charged black hole to a de Sitter spacetime. In (iii), the largest R, namely R_{h_c}, describes the cosmological event horizon which hides de Sitter singularity at $R = \infty$ ($r = \infty$) from an observer at $R < R_{h_c}$. Other horizons, with $R_{h_{1,2}}$, describe the charged black hole. The cases (ii) and (iv) correspond to the extremal Reissner-Nordström black hole.

The Hawking temperature of the black holes is related to their surface gravity. The latter can be straightforwardly calculated from the knowledge of the timelike Killing vector $\zeta = \partial_\lambda$:

$$\sqrt{-\frac{1}{2}(\nabla_i\zeta_j)(\nabla^i\zeta^j)}\Bigg|_{R=R_h} = \frac{b}{2a^2}\left(R_h{}^2 - R_{dS}{}^2\right). \tag{5.17}$$

The temperature vanishes for the extremal black hole configurations (ii) and (iv).

It seems worthwhile to mention that $g_{\lambda\lambda}$ reaches a local maximum at R_{h_1} [case (iv)], and a local minimum at R_{h_2} [case (ii)], when $M_- \neq M_+$. For $M_- = M_+$, the three cases (ii), (iii), and (iv) degenerate to a configuration with a horizon at $R_{h_1} = R_{h_2} = 0$, which is an inflexion point of $g_{\lambda\lambda}$. The qualitative behavior is given in Fig. 5 for positive values of M_0.

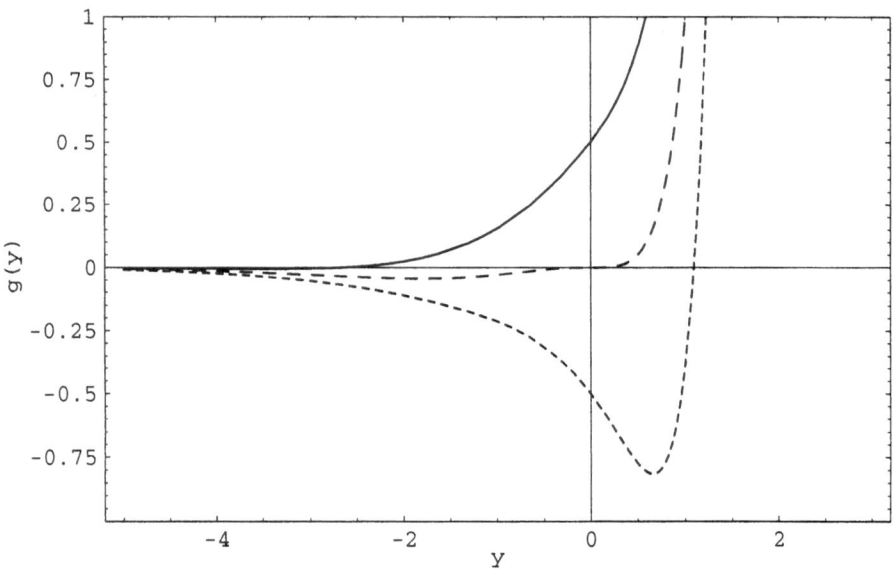

Fig. 5 The case $R_{dS} = \Lambda = 0$: The metric coefficient $g(y) := g_{\lambda\lambda}b/a$ is plotted as a function of $y := Rb/a$. Upper, central, and lower curves describe the cases (i), (ii)-(iv), and (v), respectively.

6 Quadratic PGG with massless matter

6.1 Massless fermions

Dirac spinors in two dimensions have two (complex) components,

$$\psi = \begin{pmatrix} \psi_1 \\ \psi_2 \end{pmatrix}. \tag{6.1}$$

The spinor space at each point of the spacetime manifold is related to the tangent space at this point via the spin-tensor objects. In the approach with Clifford-algebra valued forms, the central object is the Dirac one–form

$$\gamma = \gamma_\alpha \vartheta^\alpha, \tag{6.2}$$

which satisfies

$$\gamma \otimes \gamma = g, \qquad \gamma \wedge \gamma = -2\gamma_5 \eta. \tag{6.3}$$

The γ_5 matrix is implicitly defined by

$$*\gamma = \gamma_5 \gamma. \tag{6.4}$$

We are using the following explicit realization of the Dirac one–form:

$$\gamma := \begin{pmatrix} 0 & -(\vartheta^{\hat{0}} - \vartheta^{\hat{1}}) \\ (\vartheta^{\hat{0}} + \vartheta^{\hat{1}}) & 0 \end{pmatrix}. \tag{6.5}$$

The Dirac matrices γ^α satisfy the usual identity $\gamma^\alpha \gamma^\beta + \gamma^\beta \gamma^\alpha = 2g^{\alpha\beta}$.

The gauge and coordinate invariant Lagrangian two–form for the massless Dirac spinor field can be written in the form

$$L = \frac{i}{2}(\bar{\psi}\gamma \wedge d\psi + d\bar{\psi} \wedge \gamma\psi). \tag{6.6}$$

It is well known that in two dimensions there is no interaction of spinors with the Lorentz connection, and the above Lagrangian contains ordinary exterior differentials and not the covariant ones. Nevertheless, the theory is explicitly invariant under local Lorentz rotations.

The (Dirac) field equation is obtained from the variation of L with respect to $\bar{\psi}$ and reads

$$\gamma \wedge d\psi - \frac{1}{2}(d\gamma)\psi = 0. \tag{6.7}$$

The degenerate case was described in Sec. 4.1. Here we assume that $t^2 \neq 0$. Thus the one–forms T and $*T$ can be treated as the coframe basis in a two–dimensional

Riemann-Cartan spacetime. In the explicitly gauge invariant approach, it is convenient to define, instead of the original spinor (6.1), two complex (four real) Lorentz invariant functions:

$$\varphi_1 = u_1 + iu_2 := \sqrt{(t^{\hat{0}} + t^{\hat{1}})}\,\psi_1, \quad \varphi_2 = v_1 + iv_2 := \sqrt{(t^{\hat{0}} - t^{\hat{1}})}\,\psi_2. \qquad (6.8)$$

The Dirac equation (6.7) yields for the real variables u_A, v_A, $A = 1, 2$ (using (2.11))

$$d\left[\frac{u_A u_B}{t^2}(T - *T)\right] = 0, \qquad d\left[\frac{v_A v_B}{t^2}(T + *T)\right] = 0. \qquad (6.9)$$

This can be immediately integrated. In particular, the Poincaré lemma (locally) guarantees the existence of two real functions. We denote them by x and y, such that

$$\frac{|\varphi_1|^2}{t^2}(T - *T) = dx, \qquad \frac{|\varphi_2|^2}{t^2}(T + *T) = dy. \qquad (6.10)$$

This evidently provides local null coordinates (x, y) for the spacetime manifold. Introducing the phases of the complex spinor components explicitly,

$$\varphi_1 = |\varphi_1|e^{i\alpha}, \quad \varphi_2 = |\varphi_2|e^{i\beta}, \qquad (6.11)$$

we find, using (6.9), that these phases depend only on one of the above variables,

$$\alpha = \alpha(x), \quad \beta = \beta(y). \qquad (6.12)$$

This construction gives the general exact solution of the massless Dirac equation in an arbitrary two–dimensional Riemann–Cartan spacetime.

The energy–momentum one–form is straightforwardly obtained,

$$S = A_1(T - *T) + A_2(T + *T), \qquad (6.13)$$

where we denoted

$$A_1 := -\frac{d\alpha}{dx}\frac{|\varphi_1|^4}{t^2}, \quad A_2 := -\frac{d\beta}{dy}\frac{|\varphi_2|^4}{t^2}. \qquad (6.14)$$

In (6.13), we have $S^* = - * S$, well in accordance with (3.20).

6.2   Massless bosons

Let us now turn to a gravitationally coupled massless scalar field ϕ with the Lagrangian two–form

$$L = -\frac{1}{2}\,d\phi \wedge * d\phi. \qquad (6.15)$$

Variation with respect to ϕ yields the Klein-Gordon equation:

$$*d*d\phi = 0. \tag{6.16}$$

In non-degenerate spacetimes with $t^2 \neq 0$, we can use the torsion coframe basis and can write, in the most general case,

$$d\phi = \Phi_1(T - *T) + \Phi_2(T + *T), \tag{6.17}$$

with some functions $\Phi_{1,2}$. We substitute (6.17) into the Klein-Gordon equation (6.16). Then it turns out that locally there exists such a scalar function z that

$$\Phi_1(T - *T) - \Phi_2(T + *T) = dz. \tag{6.18}$$

This describes a general solution of the Klein–Gordon equation.

For the energy-momentum one–form S we get, similarly to (6.13),

$$S = A_1(T - *T) + A_2(T + *T), \tag{6.19}$$

where now

$$A_1 = -t^2\Phi_1^2, \quad A_2 = -t^2\Phi_2^2. \tag{6.20}$$

6.3   Chiral solutions

Both, the massless Dirac equation and the massless Klein–Gordon equation, admit chiral solutions. For fermions chirality means that only one component of the spinor field is nontrivial. For bosons chirality can be formulated in terms of self- or anti-self-duality of the "velocity" one–form $d\phi$. In both cases, the field equations describe right- or left-moving configurations. In this section we describe the corresponding gravitational field for the quadratic PGG model (5.1).

Let us assume that $\varphi_2 = \psi_2 = 0$ for the fermion and $\Phi_2 = 0$ for the boson field. Then $A_2 = 0$ in (6.14) as well as in (6.20). Hence the energy–momentum one–form S is anti-self-dual $S^* = S$. These are chiral configurations.

The integrals (6.10) and (6.18),

$$T - *T = \frac{t^2}{|\varphi_1|^2}\,dx \quad \text{(spinor)}, \quad T - *T = \frac{1}{\Phi_1}\,dz \quad \text{(scalar)}, \tag{6.21}$$

together with the equations (3.11), (5.2), suggest a natural interpretation of the variables R and x (or R and z, respectively,) as two local spacetime coordinates. Clearly, x and z are different in each case, but we can unify the two problems without risk of confusion. For the torsion coframe the equations (6.21) and (3.11) explicitly yield

$$*T = -\left(\frac{b}{a}dR + Bdx\right), \quad T = -\frac{b}{a}dR, \tag{6.22}$$

for the vacuum case, see (4.4). Hence the volume two–form reads

$$\eta = \frac{1}{t^2} *T \wedge T = \frac{bB}{at^2}\, dx \wedge dR, \tag{6.23}$$

and the spacetime metric is given by

$$g = \frac{1}{t^2}\left[\left(B\,dx + \frac{b}{a}\,dR\right)^2 - \frac{b^2}{a^2}\,dR^2\right]. \tag{6.24}$$

Here we use the unifying notation

$$B = \frac{t^2}{|\varphi_1|^2}. \tag{6.25}$$

for fermions, while for bosons this function relates the two coordinate systems via $dz = \Phi_1 B\, dx$.

The spacetime geometry is completely described when one solves the field equations (3.9)-(3.11), (3.16)-(3.17), thus finding the functions t^2 and B explicitly. By means of (6.13) and (6.19), the energy–momentum one–form turns out to be

$$S = A\, dx, \tag{6.26}$$

where

$$A := -|\varphi_1|^2 \frac{d\alpha}{dx} \quad \text{(spinor)}, \quad A := -t^2 B \Phi_1^2 \quad \text{(scalar)}. \tag{6.27}$$

Substituting (6.26), (6.22) into (3.9)-(3.11) and (3.16), one finds

$$\frac{\partial t^2}{\partial R} = -\frac{2b}{a^2}\tilde{V}, \qquad \frac{\partial t^2}{\partial x} = \frac{2}{a}A, \tag{6.28}$$

$$\frac{1}{b}\frac{\partial \ln B}{\partial R} + \frac{1}{a} - \frac{2}{a^2 t^2}\tilde{V} = 0, \tag{6.29}$$

$$\frac{\partial A}{\partial R} + \frac{b}{a}A = 0. \tag{6.30}$$

The equation (3.17) is redundant. This can be compared with the vacuum case (4.6),(4.7).

The system (6.28)-(6.30) is solved by

$$A = f(x)\, e^{-bR/a}, \tag{6.31}$$

$$B = B_0(x)\, t^2 e^{bR/a}, \tag{6.32}$$

$$-t^2 = 2\, M(x)\, e^{-bR/a} - \frac{b}{2a}\,R^2 + R + \frac{2\Lambda}{a} - \frac{a}{b}, \tag{6.33}$$

where

$$M(x) := -\frac{1}{a} \int f(x)dx, \qquad (6.34)$$

with the arbitrary functions $f(x)$ and $B_0(x)$. Without loss of generality we can put $B_0 = 1$ since a redefinition of x is always possible. For completeness, let us write down the Lorentz connection. Inserting (6.31)-(6.33) into (3.12), we find

$$\Gamma^* = d\tilde{u} + \frac{e^{bR/a}}{2} \left(R - \frac{a}{b} \right) dx, \qquad (6.35)$$

where $\tilde{u}$ is a pure gauge contribution.

The gravitational field defined by (6.31)-(6.33) has the same form for chiral fermionic and bosonic sources. However, the function $f(x)$ is different for each particular physical source.

For *fermions* combining (6.32), (6.25), (6.31), and (6.27), we find

$$|\varphi_1|^2 = e^{-bR/a}, \qquad f(x) = -\frac{d\alpha}{dx}. \qquad (6.36)$$

Hence the solution for the chiral fermion field, in terms of its invariant complex component, reads

$$\varphi_1 = \exp\left(-\frac{b}{2a}R + i\alpha(x) \right), \qquad (6.37)$$

whereas the metric is described by (6.24) with B as specified in (6.32) and

$$M(x) = \frac{\alpha(x)}{a} + M_0, \qquad (6.38)$$

where M_0 is an arbitrary integration constant.

For *bosons*, combining (6.31)-(6.32) with (6.27) and (6.17), one finds

$$f(x) = -(\Phi_1 B)^2 = -\left(\frac{d\phi}{dx} \right)^2, \qquad (6.39)$$

and the scalar field $\phi(x)$ remains an arbitrary function of x.

The physical meaning of the solutions obtained is clear. In Sec. 5 it was shown that the structure of a static black hole in vacuum is determined by the value of the total mass M_0 entering the torsion square (5.5). The massless chiral (fermionic and bosonic) matter contributes a variable "mass" $M(x)$ to the torsion square (6.33). As a result, the black hole in general becomes non-static (6.24).

One can illustrate this process of a restructuring of a black hole by matter falling into it [9, 41]. Let us consider $f(x) = -m\,\delta\left(\frac{x-x_0}{a}\right)$. In view of (6.26) and (6.31), this function describes a point-like "impulse" of matter: the field energy is zero everywhere except for a single moving point (recall that x is a null coordinate). Then, for (6.34), one obtains

$$M(x) = M_0 + m\,\theta(x - x_0). \tag{6.40}$$

In the region $x < x_0$ we have a static black hole with mass M_0, whereas for $x > x_0$ its mass increases to $M_0 + m$.

6.4   Conformally non-invariant matter

The above results are restricted to the chiral case and the conformally invariant massless matter sources. Some remarks are necessary for the more general cases. The non–chiral solutions were obtained in [31] for fermionic and in [33] for bosonic matter. In general, the resulting system cannot be integrated analytically, and a numeric analysis is needed. It is possible, though, to obtain exact analytic solutions for certain models with a complicated matter content: a nonlinear spinor field interacting with scalars, e.g., see the discussion of the instanton type solutions in [2]-[4].

The lack of conformal invariance does not always lead to serious difficulties. Let us consider Yang-Mills theory, for example, with an arbitrary in general non-Abelian gauge group. The dynamical variable is the gauge potential or, equivalently, the Lie algebra-valued connection one–form A^B. The Yang-Mills Lagrangian is constructed from the corresponding gauge field strength two–form F^B:

$$L_{YM} = -\frac{1}{2}\,F^B \wedge *F_B. \tag{6.41}$$

The energy–momentum current attached to (6.41) reads

$$\Sigma_\alpha := e_\alpha \rfloor L_{YM} + (e_\alpha \rfloor F^B) \wedge *F_B = -\frac{1}{2}f^2\,\eta_\alpha\,, \tag{6.42}$$

where $f^2 := f^B f_B$, $f_B = *F_B$. As in the previous cases, the spin current vanishes, $\tau_{\alpha\beta} = 0$, since the Lorentz connection does not couple to the Yang–Mills potential.

Observe that the energy–momentum trace, in contrast to four dimensions, does *not* vanish:

$$\vartheta^\alpha \wedge \Sigma_\alpha = -f^2\eta \neq 0. \tag{6.43}$$

The results described in Sec. 3 refer only to the conformally invariant case. Thus one needs a proper generalization for the case (6.43). Quite fortunately, the situation is greatly simplified due to the constancy of f^2. Although, of course, the Lie algebra–valued scalar field f_A is not constant in view of the nonlinear nature of the Yang-Mills equations

$$D * F_A = d\,f_A + c_{ABC}A^B\,f^C = 0, \tag{6.44}$$

obviously its square is conserved: $f^2 = const.$

As a result, the gravitational field equations (3.9)-(3.11) are modified by a following simple shift of the cosmological constant:

$$\Lambda \to \tilde{\Lambda} = \Lambda + \frac{1}{2} f^2. \tag{6.45}$$

In particular, the complete integrability of the vacuum system is not disturbed by Yang–Mills matter, and one again recovers the static black hole solutions described in Sec. 5.

7 Black hole solution for general dilaton gravity

In this section we demonstrate that our method also successfully works in the string motivated dilaton models, cf. [9, 5, 28, 30, 27, 18, 19, 34, 35].

Let us denote the purely Riemannian curvature scalar by a tilde: $\tilde{R}$. In general, the same notation is used for all geometrical objects and operations which are defined by the torsion-free Riemannian connection $\tilde{\Gamma}^{\alpha\beta}$ (Christoffel symbols). Let be given, in two dimensions, the gravitational potential ϑ^{α} on one side and a scalar field Φ and a Yang–Mills potential $A^B = A_i^B dx^i$ on the other, the matter side. These fields are interacting with each other. A corresponding general Lagrangian two–form reads

$$V_{\text{dil}} = \eta \left(\mathcal{F}(\Phi)\tilde{R} + \mathcal{G}(\Phi)(\partial_\alpha\Phi)^2 + \mathcal{U}(\Phi) + \mathcal{J}(\Phi)(F_{ij}^B)^2 \right). \tag{7.1}$$

Here the kinetic terms are constructed from $\partial_\alpha\Phi := e_\alpha\rfloor d\Phi$ and $F^B = \frac{1}{2}F_{ij}^B dx^i \wedge dx^j$, respectively. For the string motivated dilaton models, the coefficient functions read:

$$\mathcal{F}(\Phi) = e^{-2\Phi}, \quad \mathcal{G}(\Phi) = \gamma e^{-2\Phi}, \quad \mathcal{U}(\Phi) = e^{-2\Phi}U(\Phi), \quad \mathcal{J}(\Phi) = \frac{-e^{(\epsilon-2)\Phi}}{4}, \tag{7.2}$$

where $\gamma = 4$ and $U(\Phi) = c$ in the tree approximation of string theory. A number of physically interesting models correspond to different values of γ, ϵ, and $U(\Phi)$.

7.1 Main result

Locally one can always treat the scalar function Φ as a coordinate on a two-dimensional spacetime manifold. Denote the second coordinate by λ.

In terms of the local coordinates (λ, Φ), the metric of the general solution of the gravitational field equation of the model (7.1) reads

$$ds^2 = -4\,h(\Phi)\,e^{-2\nu(\Phi)}\,d\lambda^2 + \frac{(\mathcal{F}')^2}{h(\Phi)}\,d\Phi^2, \tag{7.3}$$

where

$$h(\Phi) = e^{\nu(\Phi)} \left(\frac{M_0}{2} + \int^{\Phi} d\varphi \left(\mathcal{U}(\varphi) + \frac{Q_0^2}{2\mathcal{J}(\varphi)} \right) \mathcal{F}'(\varphi) \, e^{-\nu(\varphi)} \right), \qquad (7.4)$$

$$\nu(\Phi) = \int^{\Phi} d\varphi \, \frac{\mathcal{G}(\varphi)}{\mathcal{F}'(\varphi)}. \qquad (7.5)$$

Here M_0 and Q_0^2 are the integration constants which are related to the total mass and the (squared) charge of a solution. For completeness, let us give the solution for the Yang–Mills field:

$$F^B = f^B \, \eta, \qquad f^B f_B = \left(\frac{Q_0}{\mathcal{J}(\Phi)} \right)^2. \qquad (7.6)$$

The proof of this result, see below, is obtained by the method developed for two-dimensional PGG.

7.2 Dilaton models and PGG

In two dimensions, torsion is represented by its trace one-form $T := e_\alpha \rfloor T^\alpha$, see Table 2. Then the Riemann-Cartan connection decomposes into the Riemannian and post-Riemannian pieces as follows:

$$\Gamma_{\alpha\beta} = \tilde{\Gamma}_{\alpha\beta} - \vartheta_\alpha \, e_\beta \rfloor T + \vartheta_\beta \, e_\alpha \rfloor T = \tilde{\Gamma}_{\alpha\beta} - \eta_{\alpha\beta} *T. \qquad (7.7)$$

By differentiation we find a corresponding decomposition of the Riemann-Cartan curvature. For the Hilbert-Einstein two-form this yields

$$R^{\alpha\beta} \eta_{\alpha\beta} = \tilde{R}^{\alpha\beta} \eta_{\alpha\beta} + 2 \, d *T. \qquad (7.8)$$

Let us consider the "scalar-tensor" type PGG model with the Lagrangian

$$V_0 = -\frac{1}{2} \left[\xi(\Phi) \, T_\alpha \wedge *T^\alpha + \omega(\Phi) \, R^{\alpha\beta} \eta_{\alpha\beta} \right]. \qquad (7.9)$$

Here $\xi(\Phi)$ and $\omega(\Phi)$ are the scalar functions which describe a variable gravitational "constant" à la Jordan and Brans-Dicke. Variation of (7.9) with respect to the connection yields the field equation

$$\xi(\Phi) \, T = \omega'(\Phi) \, d\Phi. \qquad (7.10)$$

Hereafter the prime denotes derivative with respect to Φ.

As we see, the torsion trace turns out to be an exact one-form, and the scalar dilaton field plays a role of its generalized potential. Substituting (7.10) into (7.8) and (7.9), one finds (dropping an exact form):

$$V_0 = -\frac{1}{2}\left(\omega \tilde{R}^{\alpha\beta}\eta_{\alpha\beta} - \frac{(\omega')^2}{\xi}\,d\Phi \wedge *d\Phi\right) \qquad (7.11)$$

$$= \eta\left(\frac{\omega}{2}\tilde{R} + \frac{(\omega')^2}{2\xi}(\partial_\alpha\Phi)^2\right). \qquad (7.12)$$

This is evidently equivalent to the dilaton model (7.1) with

$$\mathcal{F}(\Phi) = \frac{\omega(\Phi)}{2}, \qquad \mathcal{G}(\Phi) = \frac{(\omega'(\Phi))^2}{2\xi(\Phi)}. \qquad (7.13)$$

The equivalence of (7.9) and (7.12) can be also verified by comparing the relevant sets of field equations.

The scalar-tensor PGG model (7.9) can be straightforwardly generalized in order to include the potential for the dilation field Φ and a possible interaction with the matter field Ψ:

$$L_{\text{tot}} = V + L_{\text{mat}}(\Psi, D\Psi, \Phi), \qquad (7.14)$$

$$V = -\frac{1}{2}\,\xi(\Phi)T_\alpha \wedge *T^\alpha - \frac{1}{2}\,\omega(\Phi)R^{\alpha\beta}\eta_{\alpha\beta} + \mathcal{U}(\Phi)\,\eta. \qquad (7.15)$$

As we know, for standard matter in two dimensions (scalar, spinor, Abelian and non–Abelian gauge fields), the spin current is zero, $\tau_{\alpha\beta} = 0$, and hence the "second" field equation (7.10), which results from varying the connection, remains the same for the generalized model (7.15). This fact is the basis of the equivalence of the scalar-tensor model (7.15) and the general dilaton type model (7.1) with the same identifications of the coefficient functions (7.13). The dilaton field potential $\mathcal{U}(\Phi)$ is taken from (7.1), whereas

$$L_{\text{mat}} = 2\mathcal{J}(\Phi)\,F_B \wedge *F^B = \mathcal{J}(\Phi)\,(F_{ij}^B)^2\,\eta \qquad (7.16)$$

represents a specific matter field Ψ, with $\Psi = A^B$. In general, we may have a larger set of matter fields.

7.3 Proof

The explicit construction of the general solution for dilaton gravity, (7.3), (7.4), and (7.5), is obtained by the same machinery as that developed for two-dimensional PGG.

Again, we have a Lagrangian which depends on the torsion square t^2 and the curvature scalar R. The gravitational field momenta (3.3) read

$$P = \xi(\Phi), \qquad \kappa = -\omega(\Phi). \qquad (7.17)$$

The Lagrangian two–form (7.15) can be transformed into the corresponding Lagrangian density in (3.2): $\mathcal{V} = -\frac{1}{2}(\xi t^2 + \omega R) - \mathcal{U}$. Hence (3.13) yields $\tilde{\mathcal{V}} = \frac{1}{2}\xi t^2 - \mathcal{U}$.

In the absence of matter, which can be enforced in (7.16) by putting $\mathcal{J}(\Phi) = 0$, we immediately obtain the gravitational field equations (3.9)-(3.11) in coordinate and gauge invariant form:

$$d(\xi^2 t^2) = 2\left(\mathcal{U}(\Phi) - \frac{1}{2}\xi(\Phi)\,t^2\right) d\kappa, \qquad (7.18)$$

$$d(\xi{*}T) = 2\mathcal{U}(\Phi)\,\eta, \qquad (7.19)$$

$$d\kappa = -\xi\,T. \qquad (7.20)$$

If a Yang–Mills field is present, its contribution manifests itself in a simple "deformation" of the potential function $\mathcal{U}(\Phi)$, similar to the shift of the cosmological constant in (6.45). Indeed, taking into account the Yang-Mills field equation for (7.16),

$$D(\mathcal{J}(\Phi) * F_A) = d(\mathcal{J}(\Phi)f_A) + c_{ABC}A^B\,\mathcal{J}(\Phi)f^C = 0, \qquad (7.21)$$

where $f_A := *F_A$, we straightforwardly obtain the first integral

$$f^B f_B\,(\mathcal{J}(\Phi))^2 =: Q_0^2. \qquad (7.22)$$

With the help of this result, one can prove that the field equations (7.18),(7.19) are formally the same, if the potential $\mathcal{U}(\Phi)$ is replaced by

$$\mathcal{U}(\Phi) \longrightarrow \mathcal{U}(\Phi) + \frac{1}{2}Q_0^2/\mathcal{J}(\Phi). \qquad (7.23)$$

In order to simplify the notation, we will treat both cases simultaneously, considering the system (7.18)-(7.20) with $\mathcal{U}(\Phi)$ properly defined.

The integration of (7.18)-(7.20) is straightforward. At first, after substituting (7.17), we immediately obtain a linear equation for $\xi^2 t^2$,

$$(\xi^2 t^2)' = \frac{\omega'}{\xi}\,\xi^2 t^2 - 2\mathcal{U}\omega', \qquad (7.24)$$

where the prime denotes a derivative with respect to Φ. Formal integration yields

$$-\xi^2 t^2 = 2\,e^{\nu(\Phi)}\left(M_0 + \int^\Phi d\varphi\,\mathcal{U}(\varphi)\,\omega'(\varphi)\,e^{-\nu(\varphi)}\right), \qquad (7.25)$$

$$\nu(\Phi) = \int^\Phi d\varphi\,\frac{\omega'(\varphi)}{\xi(\varphi)}. \qquad (7.26)$$

The construction of the metric can be completed along the lines of our general method. Namely, since $t^2 \neq 0$, one can construct a zweibein from the torsion one-form T and its dual $*T$. According to (7.20), we may consider either $\kappa = -\omega(\Phi)$

or the the field Φ itself as a first local coordinate. The second ("time") coordinate, say λ, is then naturally associated with another leg of the zweibein,

$$\xi * T := B \, d\lambda, \tag{7.27}$$

with some function $B = B(\lambda, \kappa(\Phi))$. Substituting the latter into (7.19) and taking into account the volume 2–form $\eta = \frac{B}{\xi^2 t^2} \, d\kappa \wedge d\lambda$, see (2.12), we obtain an equation for B:

$$\frac{\partial \ln B}{\partial \kappa} = \frac{2\mathcal{U}}{\xi^2 t^2}. \tag{7.28}$$

On integration

$$B = B_0(\lambda) \, \xi^2 t^2 \exp\left(\int \frac{d\kappa}{\xi}\right) = B_0(\lambda)\xi^2 t^2 \, e^{-\nu(\Phi)}, \tag{7.29}$$

and, again without loss of generality, one can put the function $B_0(\lambda) = 1$.

Accordingly, the metric finally reads:

$$g = -\frac{d\kappa^2}{\xi^2 t^2} + \xi^2 t^2 \exp\left(2\int \frac{d\kappa}{\xi}\right) d\lambda^2 = \xi^2 t^2 \, e^{-2\nu(\Phi)} d\lambda^2 - \frac{(\omega')^2}{\xi^2 t^2} \, d\Phi^2. \tag{7.30}$$

Recalling the identifications (7.13), which establish the relation between the dilaton and PGG, we arrive at the result (7.3) by putting $h(\Phi) := -\xi^2 t^2/4$. Then the equations (7.25) and (7.26) reduce to (7.4) and (7.5), respectively.

7.4   Concluding remarks

The general solution described in Sec. 7.1 contains all the exact black hole configurations reported earlier in the literature as particular cases which correspond to specific choices of the coefficient functions $\mathcal{F}(\Phi), \mathcal{G}(\Phi), \mathcal{U}(\Phi), \mathcal{J}(\Phi)$, cf. [9, 28, 30, 27, 34], e.g..

The new results are most useful for the investigation of the dynamical picture of a gravitational collapse for nontrivial matter sources. Of particular interest is the case of a non-minimally coupled scalar field which describes the semi-classical correction to the Lagrangian due to Hawking radiation.

Acknowledgments

We are grateful to Marc Toussaint for useful remarks and for reading the manuscript and to Ralph Metzler for help in drawing the figures.

Bibliography

[1] A. Achucarro, *Lineal gravity from planar gravity*, Phys. Rev. Lett. **70** (1993) 1037-1040.

[2] K.G. Akdeniz, A. Kizilersü, and E. Rizaoğlu, *Instanton and eigen-modes in a two-dimensional theory of gravity with torsion*, Phys. Lett. **B215** (1988) 81-83.

[3] K.G. Akdeniz, A. Kizilersü, and E. Rizaoğlu, *Fermions in two-dimensional theory of gravity with dynamical metric and torsion*, Lett. Math. Phys. **17** (1989) 315-320.

[4] K.G. Akdeniz, O.F. Dayi, and A. Kizilersü, *Canonical description of a two-dimensional gravity*, Mod. Phys. Lett. **A7** (1992) 1757-1764.

[5] T. Banks, A. Dabholkar, M.R. Douglas, and M. O'Loughlin, *Are horned particles the end point of Hawking evaporation?*, Phys. Rev. **D45** (1992) 3607-3616.

[6] D. Cagnemi, *One formulation for both lineal gravities through a dimensional reduction*, Phys. Lett. **B297** (1992) 261-265.

[7] D. Cagnemi and R. Jackiw, *Geometric gravitational force on particles moving in a line*, Phys. Lett. **299** (1993) 24-29.

[8] D. Cagnemi and R. Jackiw, *Poincaré gauge theory for gravitational forces in (1+1) dimensions*, Ann. Phys. (N.Y.) **225** (1993) 229-263.

[9] C.G. Callan, S.B. Giddings, J.A. Harvey, and A. Strominger, *Evanescent black holes*, Phys. Rev. **D45** (1992) R1005-R1009.

[10] M.O. Katanaev, *Complete integrability of two dimensional gravity with dynamical torsion*, J. Math. Phys. **31** (1990) 882-891.

[11] M.O. Katanaev, *Conformal invariance, extremals, and geodesics in two-dimensional gravitation with torsion*, J. Math. Phys. **32** (1991) 2483–2496.

[12] M.O. Katanaev, *All universal coverings of two-dimensional gravity with torsion*, J. Math. Phys. **34** (1993) 700-736.

[13] M.O. Katanaev, *Canonical quantization of the string with dynamical geometry and anomaly free nontrivial string in two dimensions*, Nucl. Phys. **B416** (1994) 563-605.

[14] M.O. Katanaev, *Euclidean two–dimensional gravity with torsion*, J. Math. Phys. **38** (1997) 946-980.

[15] M.O. Katanaev, W. Kummer, and H. Liebl, *Geometric interpretation and classification of global solutions in generalized dilaton gravity*, Phys. Rev. **D53** (1996) 5609-5618.

[16] M.O. Katanaev and I.V. Volovich, *Two-dimensional gravity with dynamical torsion and strings*, Ann. Phys. (N.Y.) **197** (1990) 1-32.

[17] M.O. Katanaev and I.V. Volovich, *Theory of defects in solids and three-dimensional gravity*, Ann. Phys. *(N.Y.)* **216** (1992) 1-28.

[18] T. Klösch and T. Strobl, *Classical and quantum gravity in* (1 + 1)-*dimensions. Part 1: A unifying approach*, Class. Quantum Grav. **13** (1996) 965-984.

[19] T. Klösch and T. Strobl, *Classical and quantum gravity in* (1 + 1)-*dimensions. Part 2: The universal coverings*, Class. Quantum Grav. **13** (1996) 2395-2422.

[20] W. Kummer, *Deformed iso*(2, 1) *symmetry and non-Einsteinian 2d-gravity with matter*, in: Proc. of the Conf. "Hadron Structure 1992", Stará Lesná *(Slovakia), 6-11 Sept. 1992*, D. Brunsko and J. Urban, eds. (Košice University Publications, Košice 1992) 48-56.

[21] W. Kummer and D.J. Schwarz, *Renormalization of R^2 gravity with dynamical torsion in d=2*, Nucl. Phys. **B382** (1992) 171-186.

[22] W. Kummer and D.J. Schwarz, *General analytic solution of R^2 gravity with dynamical torsion in two dimensions*, Phys. Rev. **D45** (1992) 3628-3635.

[23] W. Kummer and D.J. Schwarz, *Two-dimensional R^2-gravity with torsion*, Class. Quantum Grav. Suppl. **10** (1993) S235-S238.

[24] W. Kummer and D.J. Schwarz, *Comment on "Canonical description of a two–dimensional gravity"*, Mod. Phys. Lett. **A8** (1993) 2903.

[25] W. Kummer and P. Widerin, *Non-Einsteinian gravity in d = 2: symmetry and current algebra*, Mod. Phys. Lett. **A9** (1994) 1407-1414.

[26] W. Kummer and P. Widerin, *Conserved quasilocal quantities and general covariant theories in two dimensions*, Phys. Rev. **D52** (1995) 6965-6975.

[27] O. Lechtenfeld and C.R. Nappi, *Dilaton gravity and no-hair theorem in two dimensions*, Phys. Lett. **B288** (1992) 72-76.

[28] M. D. McGuigan, C.R. Nappi, and S.A. Yost, *Charged black holes in two-dimensional string theory*, Nucl. Phys. **B375** (1992) 421-450.

[29] E.W. Mielke, F. Gronwald, Yu.N. Obukhov, R. Tresguerres, and F.W. Hehl, *Towards complete integrability of two-dimensional Poincaré gauge gravity*, Phys. Rev. **D48** (1993) 3648-3662.

[30] C.R. Nappi and A. Pasquinucci, *Thermodynamics of two-dimensional black-holes*, Mod. Phys. Lett. **A7** (1992) 3337-3346.

[31] Yu.N. Obukhov, *Two-dimensional Poincaré gauge gravity with matter*, Phys. Rev. **D50** (1994) 5072-5086.

[32] Yu.N. Obukhov and S.N. Solodukhin, *Dynamical gravity and conformal and Lorentz anomalies in two dimensions*, Class. Quantum Grav. **7** (1990) 2045-2054.

[33] Yu.N. Obukhov, S.N. Solodukhin, and E.W. Mielke, *Coupling of lineal Poincaré gauge gravity to scalar fields*, Class. Quantum Grav. **11** (1994) 3069-3079.

[34] J.G. Russo and A.A. Tseytlin, *Scalar-tensor quantum gravity in two dimensions*, Nucl. Phys. **B382** (1992) 259-275.

[35] J.G. Russo, L. Susskind, and L. Thorlacius, *End point of Hawking radiation*, Phys. Rev. **D46** (1992) 3444-3449.

[36] S.N. Solodukhin, *Topological 2D Riemann–Cartan–Weyl gravity*, Class. Quantum Grav. **10** (1993) 1011-1021.

[37] S.N. Solodukhin, *Black-hole solution in 2D gravity with torsion*, Pis'ma ZhETF **57** (1993) 317-322; JETP Lett. **57** (1993) 329-334.

[38] S.N. Solodukhin, *Two-dimensional black hole with torsion*, Phys. Lett. **B319** (1993) 87-95.

[39] S.N. Solodukhin, *Cosmological solutions in 2D Poincaré gravity*, Int. J. Mod. Phys. **D3** (1994) 269-272.

[40] S.N. Solodukhin, *On exact integrability of 2-D gravity*, Mod. Phys. Lett. **A9** (1994) 2817-2823.

[41] S.N. Solodukhin, *Exact solution of two–dimensional Poincaré gravity coupled to fermion matter*, Phys. Rev. **D51** (1995) 603-608.

[42] T. Strobl, *All symmetries of non-Einsteinian gravity in d=2*, Int. J. Mod. Phys. **A8** (1993) 1383-1397.

Part V

Thermodynamics

Black Hole Thermodynamics

Gernot Neugebauer

Theoretisch-Physikalisches Institut, Friedrich-Schiller-Universität Jena,
Max-Wien-Platz 1, D-07743 Jena, Germany

Abstract. The aim of this lecture is to build a bridge between ordinary thermodynamics and black hole thermodynamics. To this end, we review the principles of general relativistic irreversible thermodynamics and derive a universal parameter thermodynamics for rotating fluids ("rotating bodies"). To extend the procedure to black holes, the black hole limit of the rigidly rotating disk of dust solution is discussed. It then turns out that the first law of black hole mechanics is just an extension of the Gibbs equation for the disk. Problems with the second law of black hole thermodynamics are discussed.

1 Introduction

Thermodynamics and Einstein's Theory of Gravitation are closely connected. The first law of thermodynamics asserts the validity of the energy principle for every thermodynamic system. On the other hand, energy and momentum form the source term T_{ik} of the Einstein equations

$$G_{ik} \equiv R_{ik} - \frac{1}{2} R g_{ik} = 8\pi T_{ik} \tag{1}$$

where the left hand side consists of terms containing the gravitational fields g_{ik}. Hence, every thermodynamic system is self-gravitating, and fundamental thermodynamic investigations have to involve the phenomenon of gravitation. The consideration of astrophysical objects raises very interesting questions such as: What happens to the entropy of a star falling in a black hole? Does the entropy disappear into the black hole, lowering the entropy of the universe? Is that a violation of the second law of thermodynamics which states that entropy can never be annihilated? To answer such questions we have to examine the phenomenological nature of entropy as a measure of irreversibility of physical processes. Hence, I will first try to find a phenomenological description of irreversibility in general relativity. (For the sake of brevity, many wonderful concepts in statistical thermodynamics like phase space, ergodic behavior, coarse graining, statistical operator and information, must unfortunately be omitted.)

2 How can irreversibility be described?

In order to get an idea for the general relativistic description of the phenomenon of irreversibility let us have a look at classical mechanics in the Lagrangian formulation.

The one-dimensional motion of a mass point (mass m) in a potential $V(x)$ follows from a variational principle for the action W,

$$W := \int_{t_1}^{t_2} dt\, L(x, \dot{x}), \qquad L := \frac{m}{2}\dot{x}^2 - U(x), \tag{2}$$

where the Lagrangian L depends on the coordinate x and the velocity $\dot{x}$. A variation δx vanishing at $t = t_1$, $t = t_2$ induces a variation δW of the action,

$$x \to x + \delta x, \quad \delta x|_{t_1, t_2} = 0: \quad \delta W = \int_{t_1}^{t_2} dt\, \frac{\delta L}{\delta x}\, \delta x, \tag{3}$$

where $\delta L / \delta x$ is the variational derivative,

$$\frac{\delta L}{\delta x} = \frac{\partial L}{\partial x} - \frac{d}{dt}\frac{\partial L}{\partial \dot{x}}. \tag{4}$$

Reversible motion of the mass point obeys the variational principle

$$\delta x \text{ arbitrary}: \quad \delta W = 0 \tag{5}$$

with the consequence

$$\frac{\delta L}{\delta x} = 0: \quad m\ddot{x} = -\frac{\partial U}{\partial x}. \tag{6}$$

There is no restriction on the variation δx (except the second relation in eq. (3)). Obviously, this cannot be true for irreversible motions (e.g. under the influence of friction), since irreversibility creates a distinction between the future and the past directions in time. We define virtual future states $x + \delta x$ by the special variation $\delta x = \dot{x}\delta\omega$, $\delta\omega \geq 0$ and satisfy the second equation in (3) by $\delta\omega|_{t_1, t_2} = 0$. With these definitions, *irreversible* motions can be characterized by

$$\delta x = \dot{x}\,\delta\omega, \quad \delta\omega \geq 0: \quad \delta W = \int_{t_1}^{t_2} dt\, \frac{\delta L}{\delta x}\,\dot{x}\,\delta\omega \geq 0 \tag{7}$$

with the consequence

$$\sigma := \frac{\delta L}{\delta x}\dot{x} \geq 0: \quad -\left(\frac{\partial U}{\partial x} + m\ddot{x}\right)\dot{x} \geq 0. \tag{8}$$

On account of the time translation invariance of the Lagrangian L one obtains the Noether identity

$$\sigma = -\frac{dH}{dt}, \quad H := \dot{x}\frac{\partial L}{\partial \dot{x}} - L, \tag{9}$$

where H is the Hamiltonian. Obviously,

$$\frac{dH}{dt} \leq 0 \tag{10}$$

must hold for all irreversible processes and justifies the variational principle (7) *a posteriori*. The most simple way to ensure the inequality $\sigma \geq 0$ is with the linear ansatz

$$-\frac{\partial U}{\partial x} - m\ddot{x} = l\dot{x}, \quad l \geq 0, \tag{11}$$

which is frequently used to describe friction phenomena in particle mechanics. Indeed, $l \geq 0$ guarantees $\sigma \geq 0$.

3 Relativistic thermodynamics

It turns out [1] that the procedure outlined in section 2 can be applied to Einstein's gravitational theory step by step. Again, we choose a particular system and consider a one-component fluid with two state variables (a,b). According to (2) we may discuss an action integral W

$$W := \int_{\mathcal{D}} d^4x \sqrt{-g} L, \quad L = L(g_{ik}, g_{ik,l}, g_{ik,l,m}; a, b) \tag{12}$$

where $\mathcal{D}$ is an arbitrary compact domain of space time. g is the determinant of the metric tensor g_{ik} $(i, k, l = 1 - 4)$. Thus the Lagrangian L depends on the state variables V_A,

$$(V_A) = (g_{ik}, a, b) . \tag{13}$$

The variations corresponding to (3) may be written in the form[1]

$$V_A \rightarrow V_A + \delta V_A, \quad \delta V_A|_{\partial \mathcal{D}} = 0: \quad \delta W = \int_{\mathcal{D}} d^4x \frac{\delta L \sqrt{-g}}{\delta V_A} \delta V_A, \tag{14}$$

where

$$\frac{\delta L \sqrt{-g}}{\delta V_A} := \frac{\partial L \sqrt{-g}}{\partial V_A} - \left(\frac{\partial L \sqrt{-g}}{\partial V_{A,i}} \right)_{,i} + \left(\frac{\partial L \sqrt{-g}}{\partial V_{A,i,k}} \right)_{,i,k} . \tag{15}$$

To describe irreversibility in the sense of (7) we need a definition of the virtual future states $\bar{V}_A = V_A + \delta V_A$. The following ansatz has proved successful [1], [2]:

$$\delta V_A = \delta \omega \mathcal{L}_{\vartheta} V_A, \quad \delta \omega \geq 0, \quad \delta \omega|_{\partial \mathcal{D}} = 0, \tag{16}$$

---

[1] Summation convention. $\partial \mathcal{D}$ is the boundary of $\mathcal{D}$.

where $\mathcal{L}_{\vartheta}$ is the Lie-derivative with respect to a future pointing, time-like vector field ϑ^i.

Introducing the unit vector u^i, $u^i u_i = -1$, we may represent ϑ^i in the form

$$\vartheta^i = \frac{u^i}{T}, \quad T > 0 . \tag{17}$$

We call T the invariant temperature and ϑ^i the temperature vector. Analogously to eq. (7) we assume the inequality

$$\delta W = \int_D d^4 x \, \frac{\delta L \sqrt{-g}}{\delta V_A} \, \delta V_A \geq 0 \tag{18}$$

with δV_A as in eq. (16). From this equation it follows that

$$\sigma := \frac{1}{\sqrt{-g}} \frac{\delta L \sqrt{-g}}{\delta V_A} \, \mathcal{L}_{\vartheta} V_A \geq 0 . \tag{19}$$

Because of the diffeomorphism invariance of L, the infinitesimal diffeomorphism generated by an arbitrary vector field is a local symmetry of the theory. Choosing ϑ^i as the generator of the diffeomorphism and applying Noether's theorem, one can show that σ is the divergence of a well-defined four current S^i (cf. [1] [2]),

$$\sigma = S^i{}_{;i} , \tag{20}$$

cf. eq. (9).

We let S^i denote the entropy current, and consequently σ denotes the entropy production density. Thus

$$S^i{}_{;i} = \sigma \geq 0 \tag{21}$$

is the second law of thermodynamics in a field theoretical formulation.

In order to demonstrate the consequences of these definitions let us apply them to the particular system mentioned above. To describe a one-component fluid, we choose

$$(V_A) = (g_{ik}, \, a_1 = T, a_2 = \varrho) \tag{22}$$

and

$$L = -\left(\frac{R}{16\pi} + f\right), \tag{23}$$

where f is the density of the free energy depending on temperature T and baryonic mass density ϱ. The local thermodynamics of the volume elements of the fluid may be derived from the Gibbs equation

$$df = \tilde{\mu} \, d\varrho - s \, dT , \tag{24}$$

where $\bar{\mu}$ and s are the chemical potential (free enthalpy) and the entropy density s, respectively. The hydrostatic pressure p and the energy density ε can be introduced via Legendre transformations,

$$\varepsilon = f + Ts, \quad p = \bar{\mu}\varrho - f . \tag{25}$$

We are now able to calculate the entropy production density (19) for fluids. We get

$$\sigma = \frac{1}{16\pi}\left(G^{ik} - 8\pi \overset{0}{T}{}^{ik}\right) \mathcal{L}_\vartheta g_{ik} + (\varrho u^i)_{;i} \frac{\bar{\mu}}{T} \geq 0 \tag{26}$$

with

$$\overset{0}{T}{}^{ik} = -\left[(\varepsilon + p)u^i u^k + p\varrho^{ik}\right] . \tag{27}$$

Again, a linear ansatz (cf. (11))

$$G^{ik} - 8\pi \overset{0}{T}{}^{ik} = l^{iklm}\mathcal{L}_\vartheta g_{lm} \tag{28}$$

with the phenomenological coefficients

$$l^{iklm} = l^{iklm}(V_A) \tag{29}$$

and the local conservation of baryonic matter,

$$(\varrho u^i)_{;i} = 0 \tag{30}$$

guarantee the semi-definiteness of the entropy production density,

$$\sigma = \frac{1}{16\pi} l^{iklm}\mathcal{L}_\vartheta g_{ik}\mathcal{L}_\vartheta g_{lm} \geq 0 , \tag{31}$$

provided the coefficients l_{iklm} of the quadratic form (31) obey certain positivity criteria. Eq. (29) indicates that the phenomenological coefficients may depend on the state variables. Let us now discuss eqs. (28) which are the Einstein equations for fluids with heat conduction and viscosity. They appear, in our context, as the linear phenomenological relations of the Onsager theory, cf. [3]. Obviously, the energy momentum tensor of the fluid consists of the "reversible" perfect fluid part $\overset{0}{T}{}^{ik}$ and the "irreversible" part $(1/8\pi)l^{iklm}\mathcal{L}_\vartheta g_{lm}$ representing the influence of the irreversible processes. l^{iklm} can be constructed from the coefficients of heat conductivity and viscosity, from the metric g^{ik} and from the four velocity u^i of the medium. The Lie derivative of the metric tensor,

$$\mathcal{L}_\vartheta g_{ik} \equiv \vartheta_{i;k} + \vartheta_{k;i} \tag{32}$$

consists of the gradients of temperature and velocity which are the "driving" forces of the irreversible processes of heat conduction and viscosity. A more detailed discussion may be found in [2]. Summarizing our results, we can now answer the question posed in the heading of section 2: *Irreversible dynamics implies $\delta W \geq 0$* (at least at the level of linear phenomenological relations, cf. eqs. (11) and (28)). Moreover, $\delta W \geq 0$ together with the "Onsager philosophy" of linear phenomenological relations which is a basic idea of the so-called phenomenological thermodynamics of irreversible processes, cf. [3], provides us with the basic equations (11) and (28) of irreversible dynamics.

The theory as outlined here, can be used to formulate the equilibrium conditions of thermodynamic systems. However, it is not hyperbolic, and thus predicts acausalities in the the context of propagation phenomena. There is a "second order" theory developed by ISRAEL and STEWART [4]–[7] in which perturbations are known to propagate causally via hyperbolic differential equations. A detailed analysis may be found in HISCOCK and LINDBLOM [8], [9]. It should be possible to find a Lagrange formulation of the Israel-Stewart theory by replacing the matter term, $f(T, \varrho)$, in eq. (23) by a more general expression (which would also depend on the gradients of the thermodynamic state variables).

4   Parameter thermodynamics

We turn, now, to the consideration of the thermodynamic equilibrium of the fluid. The general equilibrium condition $\sigma = 0$ together with eq. (31) tells us that the Lie derivative of the metric has to vanish (in the sense that the "driving" forces no longer act, $\mathcal{L}_\vartheta g_{ik} \to 0$), and the integrability condition $\overset{0}{T}{}^{ik}{}_{;k} = 0$ of the field equations (28) contributes a further condition. Thus we arrive at the particular equilibrium conditions

$$\mathcal{L}_\vartheta g_{ik} = \vartheta_{i;k} + \vartheta_{k;i} = 0, \quad \left(\frac{\tilde{\mu}}{T}\right)_{,i} = 0 \tag{33}$$

for one component fluids. We now want to apply these conditions to rotating bodies. The Killing equations for the temperature vector ensure that rotating bodies in a state of thermodynamic equilibrium must be stationary. In addition, we may assume that theses bodies are axially symmetric. Though physical intuition tells us that a steadily rotating fluid ball should take the shape of a body of revolution (otherwise it would emit gravitational radiation and thus couldn't be stationary), a rigorous proof of the axisymmetry is not available.

Any axisymmetric and stationary metric admits a 2-dimensional Abelian group of motions G_2,

$$\xi_{i;k} + \xi_{k;i} = 0, \quad \eta_{i;k} + \eta_{k;i} = 0, \quad \xi^i{}_{,k}\eta^k - \eta^i{}_{,k}\xi^k = 0$$
$$\xi_i \xi^i < 0, \quad \eta^i \eta_i > 0 \tag{34}$$

with Killing vector fields ξ^i and η^i. The space-like vector η^i generating axial symmetry has closed (compact) trajectories and vanishes on the axis of rotation.

From these equations we conclude that we can choose a coordinate system in which the Killing vectors take the form $\xi^i = \delta_4^i$, $\eta^i = \delta_3^i$, and the metric tensor $g_{ik}(x)$ is independent of the time coordinate $x^4 = t$ and the space (azimuthal) coordinate $x^3 = \varphi$. Using eqs. (34), the Killing vector ϑ^i takes the form

$$\vartheta^i = \frac{u^i}{T} = \frac{1}{T_0} \left(\xi^i + \Omega_0 \eta^i \right) \tag{35}$$

(T_0 and Ω_0 are constants). Then, by $u^i u_i = -1$, we obtain

$$\left(\frac{T_0}{T} \right)^2 = - \left(\xi_i + \Omega_0 \eta_i \right) \left(\xi^i + \Omega_0 \eta^i \right) \equiv e^{2V} . \tag{36}$$

V is the generalization of the "corotating" Newtonian gravitational potential. Eq. (36) can be used to write the second equilibrium condition (33) in the symmetric form of the Tolman conditions [10],

$$T e^V = T_0, \quad \tilde{\mu} e^V = \tilde{\mu}_0 , \tag{37}$$

where $\tilde{\mu}_0$ is a constant. With the help of eqs. (35) and (37) we have "integrated" the equilibrium conditions (33). Since the angular velocity Ω_0 must be a constant, the rotational motion u^i of equilibrium configurations (bodies in their equilibrium state) is rigid. For compact fluid configurations, the conditions (34) and (35) together with the remaining field equations

$$G_{ik} = 8\pi \overset{0}{T}_{ik} = -8\pi \left[(\varepsilon + p) u^i u^k + p \varrho^{ik} \right] \tag{38}$$

imply [11]

$$\varepsilon_{iklm} \eta^i \xi^k \xi^{l;m} = 0 = \varepsilon_{iklm} \xi^i \eta^k \eta^{l;m} , \tag{39}$$

i.e. the space-time of rotating bodies admits 2-spaces orthogonal to the 2-dimensional group orbits formed by the Killing trajectories (ε_{iklm} is the Levi-Civita tensor). Condition (39) excludes "coils", where the source has a toroidal topology.

We now have to discuss the Einstein equations (38) under the conditions (35), (37). The most elegant way to do so is to go back to the action functional (12), with the Lagrangian (23), and to plug in the equilibrium conditions (35), (37). Since the variable pair $(T, \tilde{\mu})$ enters the equilibrium conditions (37), it is advisable to make use of the corresponding thermodynamic potential $p = p(T, \tilde{\mu})$ defined by

$$dp = \varrho \, d\tilde{\mu} + s \, dT . \tag{40}$$

This can be done by performing the Legendre transformation $f \to p$ (25) in eqs. (23), (12). One then obtains

$$W \to W_p = \int_{\mathcal{D}} d^4x \sqrt{-g} \left(\frac{R}{16\pi} - p \right) . \tag{41}$$

Furthermore, one can (using a coordinate system with $\xi^i = \delta_4^i$, $\eta^i = \delta_3^i$) specify the domain $\mathcal{D}$ as the region between two asymptotically flat space-like hypersurfaces $t_0 =$ constant, $t_1 =$ constant, which intersect the world lines of the body. Let Σ ($t =$ constant) be an intermediate asymptotically flat space-like hypersurface. It follows immediately from eq. (40) that

$$W_p = \int_{t_0}^{t_1} dt\, G_p = (t_1 - t_0)G_p, \quad G_p = \int_\Sigma d^3x\, \sqrt{-g}\left(\frac{R}{16\pi} - p\right). \quad (42)$$

Thus it is sufficient to restrict our attention from W_p to G_p. By eq. (37) we obtain

$$p = p(T, \tilde{\mu}) = p\left(T_0\, e^{-V}, \tilde{\mu}_0\, e^{-V}\right). \quad (43)$$

On the surface of the body the pressure must vanish. Let us assume that the material of the body is able to form a closed surface (note that there are potentials $p = p\left(T_0\, e^{-}V, \tilde{\mu}_0\, e^{-}V\right)$ which have no zeros). Then as a consequence of eq. (43), V must be constant on the surface,

$$V|_{\text{surface}} = V_0, \quad p\left(T_0\, e^{-V_0}, \tilde{\mu}_0\, e^{-V_0}\right) = 0. \quad (44)$$

Hence, one of the constants $(T_0, \tilde{\mu}_0)$ may be replaced by the zero V_0 and the pressure function can be written in the form

$$p = h\left(\frac{T_0}{\tilde{\mu}_0}, e^{V_0 - V}\right), \quad h\left(\frac{T_0}{\tilde{\mu}_0}, 1\right) = 0. \quad (45)$$

In the adapted coordinate system ($\xi^i = \delta_4^i$, $\eta^i = \delta_3^i$) V is a function of the metric and the angular velocity Ω_0, cf. eq. (36). Thus, p depends on the metric and the (constant) parameters (V_0, Ω_0). (p also depends on the ratio $\tilde{\mu}_0/T_0$ and could depend on further material parameters. However, we will focus our attention on these two "kinematic" parameters.). Finally, the equilibrium action G_p is a functional of the metric and a function of the parameters (Ω_0, V_0),

$$G_p = G_p[g_{ik}; V_0, \Omega_0]. \quad (46)$$

To derive a (global) Gibbs equation for G_p (comparable with a fundamental equation like $T\,\delta S = \delta U + p\,\delta V$) let us compare two infinitesimally close states $[g_{ik}; V_0, \Omega_0]$ and $[g_{ik} + \delta g_{ik}; V_0 + \delta V_0, \Omega_0 + \delta \Omega_0]$. Then, by eqs. (40) and (45) and allowing variations δg_{ik} vanishing at the boundary of the domain $\mathcal{D}$, we obtain [12]

$$\delta G_p = -\mathcal{J}\,\delta\Omega_0 - \mathcal{N}\,\delta V_0 - \frac{1}{2}\delta M - \frac{1}{16\pi}\int_\Sigma d^3x\, \sqrt{-g}\,\delta g_{ik}\left(G^{ik} - 8\pi\overset{0}{T}{}^{ik}\right),$$

$$(47)$$

where

$$J := \int_{\Sigma} \mathrm{d}^3 x \sqrt{-g} \left([\varepsilon + p]\, u_i \eta^i\, \mathrm{e}^{-V} \right) , \qquad (48)$$

$$\mathcal{N} := \int_{\Sigma} \mathrm{d}^3 x \sqrt{-g} (\varepsilon + p) . \qquad (49)$$

J is the angular momentum of the system and δM denotes the variation of a surface term M (which will turn out to be the total (energy-) mass of the rotating body). Eq. (47) may be compared with a similar expression in the Ginzburg-Landau theory of super-conductivity. According to that theory, the gravitational field $g_{ik}(x)$ is an "order parameter" associated with different equilibrium configurations of a rotating body. From eq. (47) we may conclude

$$(i)\ \delta \left(G_p + \frac{M}{2} \right)\Bigg|_{V_0, \Omega_0} = 0 \quad \Longleftrightarrow \quad G^{ik} = 8\pi \overset{0}{T}{}^{ik} \qquad (50)$$

$$(ii)\, \text{If}\ \ G^{ik} = 8\pi \overset{0}{T}{}^{ik}\ \ \text{then}\ \ \delta \left(G_p + \frac{M}{2} \right) = -J\, \delta\Omega_0 - \mathcal{N}\, \delta V_0 . \qquad (51)$$

The first statement implies that, for fixed parameters (V_0, Ω_0), the Einstein equations for a rigidly rotating body in its state of equilibrium can be obtained from a variational principle. Related principles have already been discussed by HARTLE and SHARP [14], BARDEEN [15], [16], CARTER [17], and BARDEEN, CARTER and HAWKING [18]. Our procedure is comparable with the derivation of the equilibrium conditions from the thermodynamic potentials. The second equation states that the rotating body (matter and field) is an equilibrium system characterized by the thermodynamic potential

$$G = G_p + \frac{M}{2} \qquad (52)$$

and the fundamental equation

$$\delta G = -J\, \delta\Omega_0 - \mathcal{N}\, \delta V_0 . \qquad (53)$$

"On shell" – i.e., when the field equations hold – we have

$$G = \frac{1}{2} \int_{\Sigma} \mathrm{d}^3 x \sqrt{-g}(p - \varepsilon) + \frac{M}{2} , \qquad (54)$$

where

$$M = 2\,\Omega J + \int_{\Sigma} \mathrm{d}^3 x \sqrt{-g}(\varepsilon + 3p) \qquad (55)$$

is the total mass. Obviously,

$$\mathcal{N} = M - \Omega \mathcal{J} - G .$$ (56)

It can be useful to perform a Legendre transformation

$$\mathcal{S} = e^{-V_0} \mathcal{N}$$ (57)

in eq. (53). We obtain

$$\delta \mathcal{S} = e^{-V_0} \left(\delta M - \Omega_0 \, \delta \mathcal{J} \right)$$ (58)

and have thus found a potential in the additive quantities M and $\mathcal{J}$. Eqs. (53) and (58) establish a parameter thermodynamics [12] which is useful e.g. for calculating the far field parameters M and $\mathcal{J}$ from the source parameters V_0 and Ω_0 with the aid of a single potential G,

$$M = M(V_0, \, \Omega_0) = \Omega_0 \, \mathcal{J} + G - \frac{\partial G}{\partial V_0}$$ (59)

$$\mathcal{J} = \mathcal{J}(V_0, \, \Omega_0) = -\frac{\partial G}{\partial \Omega_0} .$$ (60)

Black hole physicists would call an expression like (58) a "first law of (parameter) thermodynamics".

Strictly speaking, the functional G_p depends also on the parameter combination $T_0/\tilde{\mu}_0$. If we had wanted to take this additional parameters into account, we would have had to replace the fundamental equation (58) by the more general expression [13]

$$T_0 \, \delta S = \mathrm{d}M - \Omega_0 \, \delta \mathcal{J} - \tilde{\mu}_0 \, \delta \mathcal{M} ,$$

where $\mathcal{M}$ is the baryonic mass of the fluid, cf. eq. (69). However, the parameter $T_0/\tilde{\mu}_0$ is not relevant if one starts with an equation of state $\varepsilon = \varepsilon(p)$. Making the aforementioned generalisation has no affect in the following chapters.

5   'Thermodynamics' of the disk of dust

Encouraged by our success in deriving eq. (58) we may ask if a parameter thermodynamics could also be derived for black holes. Starting with a 'normal' rotating body we could try to find a *parametric transition* to the black hole state and to extend equations like (58) to apply to black holes. This proposition sounds a little unlikely to succeed for two reasons:

(i) It is known from studies of static fluid spheres that there is a gap between the last "normal" matter configuration and the spherical static Schwarzschild black hole.

(ii) Little is known about rotating bodies and their black hole limits.

It is the aim of this section to show that a recently found solution for the rigidly rotating *disk of dust* [19]–[22] (an approximate solution of the problem was given by BARDEEN and WAGONER [23]) could help to overcome these difficulties, since it has the desired black hole limit.

Dust matter consists of particles (mass elements) which interact exclusively via gravitational forces. In the language of hydrodynamics, dust is a perfect fluid without pressure ($p = 0$). As a consequence, its energy density ε coincides with the baryonic mass density (particle number density) ϱ,

$$\varepsilon = \varrho \, , \tag{61}$$

the motion is geodesic,

$$\dot{u}_i = u_{i;k}\, u^k = 0 \, , \tag{62}$$

and the baryonic mass is a conserved quantity,

$$\left(\varrho u^i\right)_{;i} = 0 \, . \tag{63}$$

At first glance, such a matter model would not seem like a suitable candidate for a thermodynamic treatment: From eqs. (24) and (25) one obtains

$$T\,\mathrm{d}\frac{s}{\varrho} = \mathrm{d}\frac{\varepsilon}{\varrho} + p\,\mathrm{d}\frac{1}{\varrho} \, , \tag{64}$$

and, because of $p = 0$, $\varepsilon = \varrho$

$$T\,\mathrm{d}\frac{s}{\varrho} = 0 \, . \tag{65}$$

However, the remarkable fact emerges that, despite its oversimplified ordinary thermodynamics, the parameter thermodynamics of the model seems quite normal.

The details of the disk of dust solution may be found in [22]. In the context of the present discussions, it suffices to point out that dust is a hydrodynamic model for a many particle system, where the particles (mass elements) move on the geodesic lines (62) of their own gravitational field. In our case, they form a rigidly rotating disk (a flat "galaxy"). The distribution (surface density) of the particles cannot be prescribed *a priori*. Instead it must be calculated from the Einstein equations. Fig. 1 shows typical densities in the Newtonian region ($\mu \ll 1$) as well as in the ultra-relativistic region ($\mu \lesssim \mu_0 = 4.62966184\ldots$; the "centrifugal" parameter μ is defined by $\mu = 2\varrho_0^2 \Omega_0^2 e^{-2V_0}$, where ϱ_0 is the coordinate radius of the disk).

We turn, now, to the discussion of the "first law of parameter thermodynamics" (58). For dust ($p = 0$, $\varepsilon = \varrho$) eqs. (49), (55), (57), and (58) yield

$$\mathcal{N} = M - 2\Omega_0 \mathcal{J} \, , \tag{66}$$

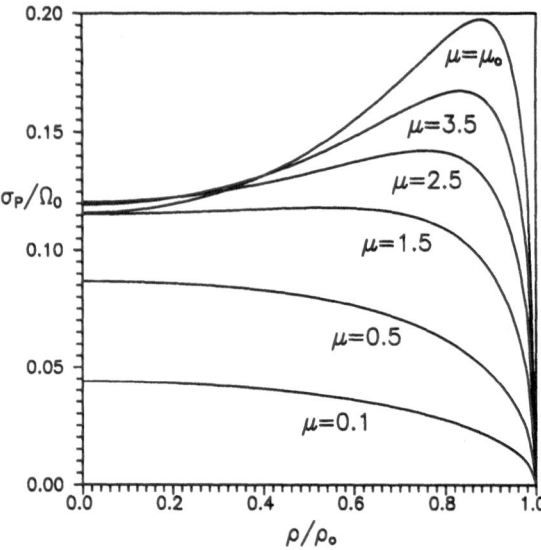

Fig. 1 The normalized surface mass-density σ_p/Ω_0 of the rigidly rotating disk of dust as a function of the normalized radial coordinate ϱ/ϱ_0. ϱ_0 is the coordinate radius of the disk.

and

$$\delta Y^{\mathrm{d}} = \frac{1}{M - 2\Omega_0 J}(\delta M - \Omega_0 \delta J) \,, \tag{67}$$

where

$$Y^{\mathrm{d}} = \log \mathcal{S} \,. \tag{68}$$

For dust, $\mathcal{S}$ has a simple meaning. Because of eq. (30), the baryonic mass $\mathcal{M}$ of the disk is given by

$$\mathcal{M} = \int_{\Sigma} \mathrm{d}^3 x \sqrt{-g}\varrho\, \mathrm{e}^{-V} \,. \tag{69}$$

As a consequence of eqs. (44), (49) and (57) we obtain

$$\mathcal{M} = \mathrm{e}^{-V_0} \int_{\Sigma} \mathrm{d}^3 x \sqrt{-g}\varrho = \mathrm{e}^{-V_0}\mathcal{N} = \mathcal{S} \,, \tag{70}$$

so, $\mathcal{S}$ is simply the baryonic mass of the rigidly rotating disk of dust. After a simple rearrangement of eq. (67) we have

$$\delta(Y^{\mathrm{d}} - \log M) = \frac{\Omega_0 M}{M^4/J^2 - (\Omega_0 M)(M^2/J)}\,\delta(M^2/J) \,. \tag{71}$$

Obviously, $\Omega_0 M$ and $S/M = M^{-1}\exp Y^{\mathrm{d}}$ must be functions of the similarity variable M^2/J,

$$2\Omega_0 M = F(M^2/J), \qquad S = M H(M^2/J), \qquad (72)$$

where

$$H'(x) = \frac{F(x)H(x)}{x[2x - F(x)]} \ . \qquad (73)$$

Hence, $F(x)$ is the only unknown function. To calculate it one needs details of the disk solution which cannot be discussed here. Fig. 2 shows the result of an earlier discussion [19], [22] and indicates that a solution exists for the parameter interval $0 \le M^2/J < 1$. The Maclaurin disk as the Newtonian limit corresponds to $M^2/J \to 0$, whereas the solution approaches the extreme Kerr solution in the ultrarelativistic limit $M^2 \to J$.

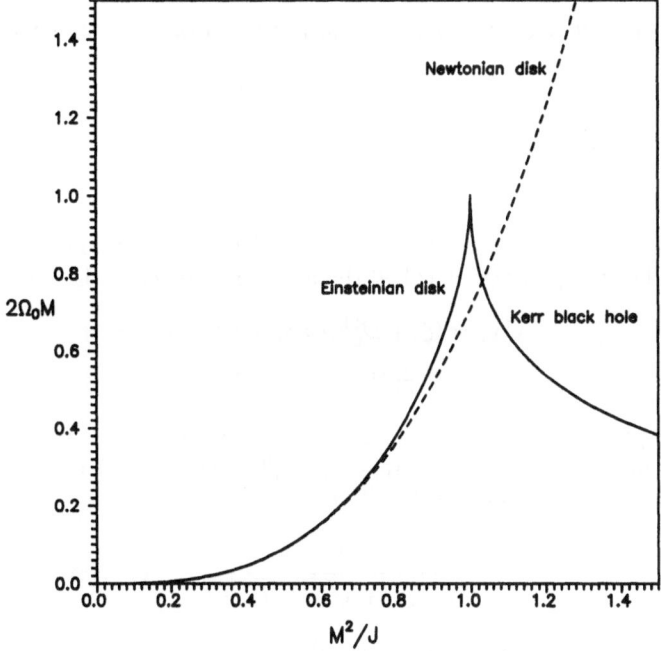

Fig. 2 The normalized angular velocity $2\Omega_0 M$ as a function of the mass M and the angular velocity J for the classical Maclaurin disk (dashed line), the general relativistic dust disk and the Kerr black hole.

On the other hand, for a Kerr black hole (which is stationary and axisymmetric) it can be shown that there exists a linear combination, χ^i, of the Killing

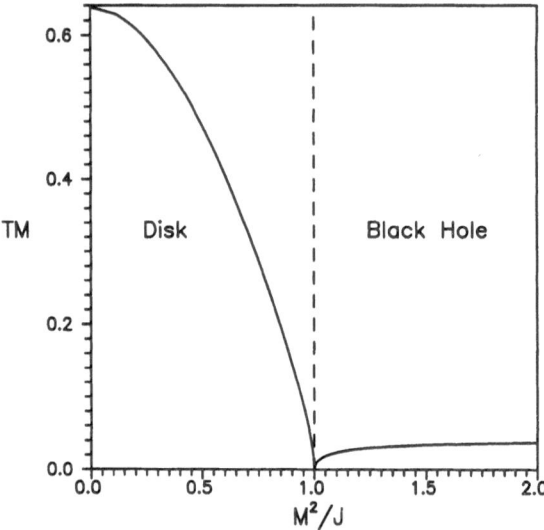

Fig. 3 TM vs. M^2/J for the dust disk and the Kerr black hole.

vectors ξ^i and η^i,

$$\chi^i = \xi^i + \Omega_0^{bh}\eta^i \tag{74}$$

which is normal to the event horizon and satisfies the relation $\chi^i\chi_i = 0$ on the horizon. The constant Ω_0^{bh} is called the angular velocity of the horizon. From

$$\chi^i\chi_i = (\xi^i + \Omega_0^{bh}\eta^i)(\xi_i + \Omega_0^{bh}\eta_i) = 0$$
$$\xi^i = \delta_4^i \qquad \eta^i = \delta_3^i, \tag{75}$$

and the explicit knowledge of the Kerr solution, see e.g. R. WALD's textbook [24], one obtains the parameter relation $\Omega_0^{bh} = \Omega_0^{bh}(M, J)$ for the Kerr black hole. Surprisingly, the product $2\Omega_0^{bh} M$,

$$2\Omega_0^{bh}M = G\left(\frac{M^2}{J}\right) = \frac{1}{(M^2/J) + \sqrt{M^4/J^2 - 1}} \qquad (1 \le M^2/J < \infty) \tag{76}$$

is again a function of the similarity variable M^2/J and coincides with the corresponding disk function F at $M^2 = J$, cf. Fig. 2. Hence, the Kerr angular velocity Ω_0^{bh} can be considered to be the extension of the angular velocity Ω_0 of the disk.

6   Parameter thermodynamics of black holes

Unfortunately, our theorem (51) does not apply to black holes directly. However, we can extend the parameter thermodynamics of dust into the black hole region,

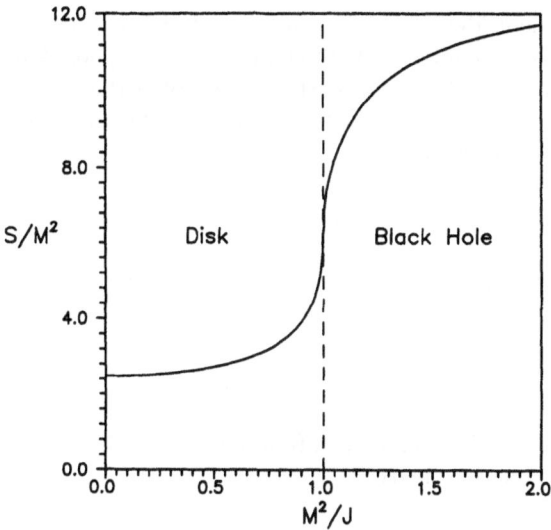

Fig. 4 "Entropy" S for the disk of dust and the Kerr black hole

since the black hole function $G(M^2/\mathcal{J})$ is an extension of the dust function $F(M^2/\mathcal{J})$. Consequently, eqs. (67) or (71) may be used to find an entropy-like potential Y^{bh} for Kerr black holes

$$\delta Y^{\mathrm{bh}} = \frac{1}{M - 2\Omega_0 \mathcal{J}}(\delta M - \Omega_0^{\mathrm{bh}}\delta \mathcal{J}) , \qquad (77)$$

extending the disk of dust potential (68). Using eqs. (76) and (77), it is not difficult to show that

$$Y^{\mathrm{bh}} = \frac{1}{2}\log \mathcal{A} , \qquad (78)$$

where

$$\mathcal{A} = 8\pi \left(M^2 + \sqrt{M^4 - \mathcal{J}^2}\right) \qquad (79)$$

is the area of the event horizon, see [24], p. 327. Introducing the surface gravity κ, of the Kerr black hole, see [24], p. 331,

$$\kappa = \frac{\sqrt{M^4 - \mathcal{J}^2}}{2M(M^2 + \sqrt{M^4 - \mathcal{J}^2})} , \qquad (80)$$

we may write eq. (67) in the form

$$\frac{\kappa}{8\pi}\delta\mathcal{A} = \delta M - \Omega_0^{\mathrm{bh}}\delta\mathcal{J} , \qquad (81)$$

which is the famous Bekenstein formulation of the first law of black hole mechanics [25] – [28]. Finally, we may connect $Y^{\mathrm{bh}} = \frac{1}{2} \log \mathcal{A}$ and $Y^{\mathrm{d}} = \log \mathcal{M}$ at the transition point $M^2 = \mathcal{J}$. It should be noted that Y^{d} and Y^{bh} are only defined up to additive constants. This fact permits the introduction of a common potential Y for both regions,

$$
Y = \begin{cases} \log \mathcal{M} + \log \gamma & \left(0 \le \dfrac{M^2}{\mathcal{J}} < 1 , \quad \text{disk region}\right) \\[2mm] \dfrac{1}{2} \log \mathcal{A} & \left(1 \le \dfrac{M^2}{\mathcal{J}} < \infty , \quad \text{black hole region}\right) \end{cases}
\tag{82}
$$

and to fix the constant γ by

$$
\mathcal{A}\big|_{M^2=\mathcal{J}} = \gamma^2 \, M^2\big|_{M^2=\mathcal{J}} .
\tag{83}
$$

It can be shown [22] that $\mathcal{M}/M$ is a function of $M^2/\mathcal{J}$, cf. Fig. 5. The detailed analysis of this function yields

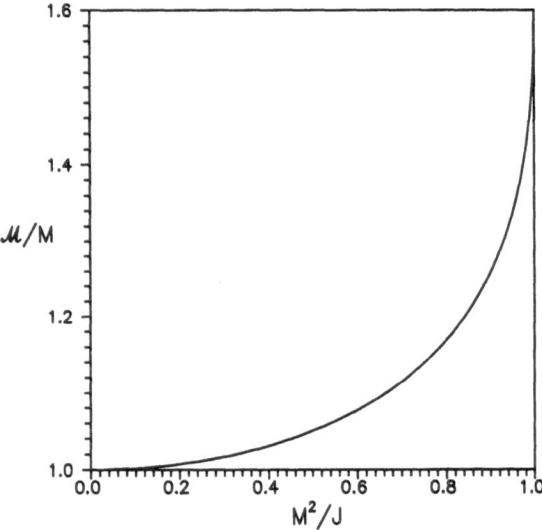

Fig. 5 Ratio of baryonic mass $\mathcal{M}$ and total mass M for the rigidly rotating disk of dust

$$
\frac{\mathcal{M}}{M}\bigg|_{M^2=\mathcal{J}} = 1.59562\ldots
\tag{84}
$$

and together with $\mathcal{A}\big|_{M^2=\mathcal{J}} = 8\pi M^2$

$$
\gamma = \sqrt{8\pi}\,\frac{M}{\mathcal{M}_0}\bigg|_{M^2=\mathcal{J}} = 3.13965\ldots
\tag{85}
$$

Aided by the first law of black hole mechanics we can introduce a common "entropy" S for the disk as well as for the black hole region by

$$S = \frac{1}{4} e^{2Y} = \begin{cases} \dfrac{\gamma^2 \mathcal{M}^2}{4} & \text{(disk region)} \\[2ex] \dfrac{\mathcal{A}}{4} & \text{(black hole region)} \end{cases} \tag{86}$$

satisfying the first law of parameter thermodynamics for disks and black holes

$$T \delta S = dM - \Omega_0 \, d\mathcal{J} \tag{87}$$

with

$$T = \begin{cases} \dfrac{2(M - 2\Omega_0 \mathcal{J})}{\gamma \mathcal{M}^2} = \dfrac{2e^{V_0}}{\gamma \mathcal{M}} & \text{(disk region)} \\[3ex] \dfrac{2(M - 2\Omega_0^{\text{bh}} \mathcal{J})}{\mathcal{A}} = \dfrac{\kappa}{2\pi} & \text{(black hole region)} \end{cases} \tag{88}$$

Fig. 3 illustrates the equation of state $T = T(M, \mathcal{J})$.

7 Discussion

We have seen that the action integral for phenomenological matter "on shell", i.e. when the Einstein equations hold, becomes a "thermodynamic" potential satisfying a first law of parameter thermodynamics (cf. eq. (51)) which, applied to the rigidly rotating disk of dust, leads to the first law of black hole mechanics (81). In this context it should be mentioned, that generalized first laws of black hole mechanics can be derived in very general field theories, see HEUSLER and STRAUMANN [29], [30], IYER and WALD [31], [32] and Wald [33] (Interested readers could also consult the lecture by C. KIEFER in this volume.)

The deciding step in formulating black hole thermodynamics was made by BEKENSTEIN [25], who pointed out that Hawking's area theorem [27] might be closely analogous to the second law of thermodynamics, which states that the entropy S of a closed system never decreases[2],

$$\Delta S \geq 0 .$$

Since the area $\mathcal{A}$, of the black hole horizon, can also never decrease,

$$\Delta \mathcal{A} \geq 0 ,$$

and $\mathcal{A}$ according to equation (81), is a potential in the extensive quantities mass M and angular momentum $\mathcal{J}$, one is prompted to interpret $\mathcal{A}$ (times a constant) as the entropy S^{bh}, of the black hole,

$$S^{\text{bh}} = \frac{1}{4} \mathcal{A} ,$$

---

[2] ΔS is the entropy difference between two equilibrium states.

and κ as the temperature T^{bh}, of the black hole,

$$T^{bh} = \frac{\kappa}{2\pi} \; .$$

However, the identifications made here would have remained formal without HAWKING's analysis of quantum particle creation effects in the field of a black hole and his result [34] that a black hole radiates particles with a black body spectrum exactly at the temperature $\kappa/2\pi$. (In physical units, $T^{bh} = (\hbar\kappa)/(2\pi ck)$ and $S^{bh} = (c^3 k \mathcal{A}/(4G\hbar))$, where k and G are the Boltzmann constant and the gravitational constant, respectively.) Hence, quantum theory is essential for the interpretation of black hole thermodynamics. (The interested reader is referred to the lectures of C. KIEFER and A. WIPF in this volume.)

The quantum physical interpretation of the geometrical quantities $\mathcal{A}$ and κ is only a first step towards a "unified" thermodynamics of black holes and "normal" matter. Certainly, BEKENSTEIN's proposal, [25], [26], to introduce a total generalized entropy S as the sum of the "normal" matter entropy S^m and the black hole entropy S^{bh},

$$S = S^m + S^{bh}$$

and to postulate

$$\Delta S \geq 0 \; ,$$

is a very interesting attempt to generalize the second law (for equilibrium states).

However, physical processes are irreversible and therefore one wants to describe phenomena such as the local production of entropy (cf. eq. (21)) or the entropy transport between subsystems one or more of which could be a black hole. It would be a fascinating task to try to develop the underlying theoretical concepts.

Encouraged by the success of parameter thermodynamics (which is a special case of ordinary thermodynamics) in yielding the first law of black hole mechanics (see eq. (87)) and Figs. 2, 3, 4), one would be tempted to speculate that the Noether entropy current S^i of eq. (20) could likewise have a meaning for black holes. According to [1], [2], S^i has the structure

$$S^i = \vartheta_k \, R^{ik} + \text{"matter terms"} \; ,$$

where ϑ_k is a temperature vector (17). An equation like this could provide a "geometrical" representation of the role played by the entropy, and the second law (20) could be a consequence of this equation and additional energy and causality conditions.

In order to solve problems like these it would seem to be necessary to study more examples for the transition of "normal" thermodynamic systems into their black hole state, and to discuss equilibrium systems consisting of "normal" matter and black holes[3]. One example can be found in W. ISRAEL's lecture where he discusses the thermodynamics of a collapsing shell.

The answer to the question of whether eqs. (86), (87), (88) permit the interpretation of $S = \gamma^2 \mathcal{M}^2/4$ (or a functional of this quantity) as the entropy of the disk of dust, is not easy to find. Interestingly, $\gamma^2 \mathcal{M}^2/4$ is a conserved quantity for a disk consisting of a fixed number of particles (fixed baryonic mass). This quantity cannot change during a transition from one equilibrium state to another. The situation would change dramatically, if the disk reached a black hole state (see Fig. 2). After such a phase transition the conservation of baryonic mass is no longer required, and S (which is now the black hole area) can increase in time. Note that S as defined in eq. (86) is differentiable (C^1) at the transition point $M^2 = \mathcal{J}$, see Fig. 4.

I would like to thank R. Meinel and G. Schäfer for valuable discussions and D. Ruder, A. Kleinwächter, and D. Moran for the great assistance during the preparation of the manuscript.

Bibliography

[1] Neugebauer, G., Int. J. Theor. Phys. **16**, 241 (1977)

[2] Neugebauer, G., Relativistische Thermodynamik (Akademie-Verlag Berlin 1980) WTB-Reihe Vol. 142

[3] Meixner, J., H.G. Reik in: S. Flügge (Ed.), Handbuch der Physik, Vol. III/2 (Springer-Verlag Berlin 1959)

[4] Israel, W., Ann. Phys. (N.Y.) **100**, 310 (1976)

[5] Stewart, J.M., Proc. Roy. Soc. Lond. **A357**, 59 (1977)

[6] Israel, W., J.M. Stewart, Proc. Roy. Soc. Lond. **A365**, 43 (1979)

[7] Israel, W., J.M. Stewart, Ann. Phys. (N.Y.) **118**, 341 (1979)

[8] Hiscock, W.A., L. Lindblom, Ann. Phys. (N.Y.) **151**, 466 (1983)

[9] Hiscock, W.A., L. Lindblom, Phys. Rev. **D35**, 3723 (1987)

[10] Tolman, R.C., Relativity, Thermodynamics, and Cosmology (Oxford Univ. Press 1934)

[11] Kundt, W., M. Trümper, Z. Phys. **192**, 419 (1966)

[12] Neugebauer, G., H. Herold in: J. Ehlers, G. Schäfer (Eds.), Lecture Notes in Physics **410**, 305 (Springer-Verlag Berlin 1992), p. 305

---

[3] E.g. what is the entropy of a black hole surrounded by a dust ring?

[13] Neugebauer, G. in: Z. Perjes (Ed.) Relativity Today (World Scientific Singapore 1988), p. 134

[14] Hartle, J.B., D.H. Sharp, Astrophys. J. **147**, 317 (1967)

[15] Bardeen, J.M., Astrophys. J. **162**, 71 (1970)

[16] Bardeen, J.M. in: C. DeWitt and B.S. DeWitt (Eds.), Black Holes (Gordon and Breach New York 1973) p. 241

[17] Carter, B. in: C. DeWitt and B.S. DeWitt (Eds.), Black Holes (Gordon and Breach New York 1973) p. 57

[18] Bardeen, J.M., B. Carter and S.W. Hawking, Commun. Math. Phys. **31**, 161 (1973)

[19] Neugebauer, G., R. Meinel, Astrophys. J. **414**, L97 (1993)

[20] Neugebauer, G., R. Meinel, Phys. Rev. Lett. **73**, 2166 (1994)

[21] Neugebauer, G., R. Meinel, Phys. Rev. Lett. **75**, 3046 (1995)

[22] Neugebauer, G., A. Kleinwächter, R. Meinel, Helv. Phys. Acta **69**, 472 (1996)

[23] Bardeen, J.M., R.V. Wagoner, Astrophys. J. **158**, L65 (1969); **167**, 359 (1971)

[24] Wald, R., General Relativity (The University of Chicago Press Chicago and London 1984)

[25] Bekenstein, J.D., Phys. Rev. **D7**, 2333 (1973)

[26] Bekenstein, J.D., Phys. Rev. **D9**, 3292 (1974)

[27] Hawking, S.W., Phys. Rev. **D13**, 191 (1976)

[28] Hawking, S.W., Phys. Rev. **D14**, 2460 (1976)

[29] Heusler, M., N. Straumann, Class. Quantum Grav. **10**, 1299 (1993)

[30] Heusler, M., N. Straumann, Phys. Lett. **B315**, 55 (1993)

[31] Iyer, V., R.M. Wald, Phys. Rev. **D50**, 846 (1994)

[32] Iyer, V., R.M. Wald, Phys. Rev. **D52**, 4430 (1995)

[33] Wald, R.M., Report gr-qc/9702022 (1997) to appear in the Proceedings of the Symposium on Black Holes and Relativistic Stars (in honor of S. Chandrasekhar), December 14–15, 1996

[34] Hawking, S.W., Commun. Math. Phys. **43**, 199 (1975)

Gedanken Experiments in Black Hole Thermodynamics

Werner Israel

Canadian Institute for Advanced Research Cosmology Program, Department of Physics and Astronomy, University of Victoria, P.O. Box 3055, Victoria B.C., V8W 3P6 Canada

Abstract. Quantum effects at high accelerations and in strong gravitational fields give rise to counter-intuitive effects, such as "acceleration radiation", and mysterious connections, such as the relation between black hole area and entropy. Idealized experiments, using accelerated mirrors, descending boxes and massive shells, can help to provide intuitive understanding and operational definitions of these concepts.

1 Introduction

In the mid-1970s, just 50 years after Heisenberg, Schrödinger and Dirac deciphered the mechanism of the quantum world, the quantum revolution impacted upon general relativity in profound ways whose full implications we still struggle to understand. The roots of these developments are traceable to the influence of John Wheeler, Bryce De Witt and Yakov Zel'dovich who stressed already in the 1960s the fundamental importance of quantum effects in black holes, strong gravitational fields and the microstructure of spacetime at the Planck level.

Jean Pierre Luminet, in his contribution to this volume (Chap. 1), describes how the striking parallels between the laws of thermodynamics and black hole dynamics uncovered by Christodoulou (1970) and Bardeen, Carter and Hawking (1973) emboldened Jacob Bekenstein in 1972 to advance the then heretical suggestion that this might be more than a formal similarity — that the area of a black hole might actually be a physical measure of its entropy.

About the same time, Steven Fulling (1973) undertook the first study of quantum field theory in uniformly accelerated ("Rindler") coordinates. He noted that the natural ground state for a uniformly accelerated observer is different from the Minkowski vacuum. There is a characteristic temperature

$$T_{\mathrm{acc}} = \hbar a/2\pi \tag{1}$$

(a is the acceleration and I have set Boltzmann's constant $k_B = 1$) associated with this difference, as first pointed out by Paul Davies (1975), and operationally clarified by William Unruh (1976).

Meanwhile, in 1974, Hawking had published his celebrated discovery that black holes evaporate, and that the process is thermal, with characteristic temperature:

$$T_{\mathrm{BH}} = \hbar \kappa/2\pi \tag{2}$$

(where κ is the so-called surface gravity of the horizon), thereby giving precise expression to Bekenstein's hypothesis.

One of the most surprising predictions of quantum field theory in flat space was reported by Davies and Fulling in 1976. A mirror, moving through empty Minkowski space with non-uniform acceleration, should emit a flux of negative energy in the "forward" direction (the direction in which its acceleration increases) and (to conserve energy) an equal flux of positive energy backwards. This arises from the interaction of the mirror with zero-point fluctuations of a quantum field.

The late 1970s saw extensive development of all these topics. Yet the deepest questions remain with us today. What, at bottom, is this entity which we have become accustomed to call "black hole entropy", and where (if its localization has a meaning) does it hide — near the surface or deep inside the hole? How does it contrive to have a universal geometrical form, irrespective of the nature and fate of the materials which collapsed to form the hole and presumably linger on inside it? What becomes of the quantum correlations between exterior and interior when the hole finally evaporates?

The field has been dredged for twenty years but is still capable of yielding up new surprises. There has, for instance, been a recent suggestion [1], based on indirect and formal arguments, that for extremal (zero surface gravity) black holes, entropy is not related to area but is actually zero. This would mean that black holes obey the third law of thermodynamics in its strongest (Planck) form.

Some of these strange phenomena refer specifically to curved spacetimes, others are already present in Minkowski space. Yet one has an impression of deep underlying connections.

My purpose here is to show that a good deal of the mystery can be dispelled by considering simple idealized experiments, i.e., by attempting to understand black hole thermodynamics directly in the spirit of Carnot. The exposition may thus serve as a supplement or introduction to the systematic treatment by my colleagues elsewhere in this volume.

2   Acceleration in Minkowski space

The work of Davies and Unruh is often summed up in the phrase: a uniformly accelerated observer "perceives" the Minkowski vacuum as apparently filled with thermal radiation at the acceleration temperature T_{acc} given by (1). What does this really mean? A better way to say it: the ground state for a uniformly accelerated observer has *negative* energy. So he views the Minkowski vacuum as elevated (thermally) above his ground state.

But what is the origin of this negative energy? Can one understand it operationally? Let us turn to the result of Davies and Fulling: a non-uniformly accelerated mirror in Minkowski space radiates negative energy "forwards" and positive energy backwards. Davies and Fulling calculated the flux F explicitly for a mirror in $(1+1)$ spacetime dimensions interacting with zero-point fluctuations of a massless scalar field. (The reader is invited to recover their result in a simpler way in one of the accompanying exercises.) They found for the two-dimensional stress-energy:

$$T_{\mu\nu} = F \, l_\mu l_\nu \,, \quad F = -\frac{\hbar}{12\pi} \frac{\mathrm{d}a}{\mathrm{d}\tau} \,. \tag{3}$$

Here, $a(\tau)$ is the mirror's acceleration at (retarded) proper time τ; l_μ is a lightlike propagation vector pointing in the direction of positive a and normalized by $l \cdot u = -1$ at the retarded point; $u^\mu = \mathrm{d}z^\mu/\mathrm{d}\tau$ is the unit tangent to the mirror's world-line $x^\mu = z^\mu(\tau)$.

Now let us see how the Davies-Unruh and Davies-Fulling results are related.

We consider a one-dimensional "box" with perfectly reflecting walls, initially empty and at rest. The box is gently set into motion and ends up in a state of uniform acceleration. En route, it necessarily passes through a phase of non-uniform acceleration during which the rear wall emits negative energy forwards, the front wall positive energy backwards. These two fluxes are not quite equal in magnitude, since the rear wall must change acceleration a little faster in order to reproduce the Lorentz contraction of the box observed in the laboratory. Thus there is a net preponderance of negative energy, which accumulates inside the box.

An observer inside the box sees the negative process as adiabatic. It cannot affect the internal quantum state, which remains the ground state. But this state now has an energy density lower than the Minkowski vacuum outside the accelerating box.

We shall verify presently that this mechanism even provides a *quantitatively* accurate explanation of the negative energy density inside the box, at least in (1+1)-dimensions.

3   Black hole entropy, descending boxes and the Boulware state

The idea that black hole entropy and area might be proportional had to contend from the outset with a gedanken experiment proposed by Robert Geroch in 1971. This was an idealized process for depositing arbitrary amounts of entropy into a spherical black hole without any change in its mass or area.

Fill up a small, weightless box far from the hole with mass-energy in the form of radiation. Slowly lower the box toward the horizon, making it do work by hauling up a counterweight. When it reaches the horizon, its total (material plus potential) energy is zero: all of its initial mass-energy has been extracted as work. Now open a trapdoor in the box's floor, and allow the radiation to drain into the hole. You thus deposit all of its entropy but none of its mass.

The net result is that the entropy of the black hole has been increased but its mass and area are unchanged. It is therefore impossible to maintain proportionality between black hole area and entropy.

It took ten years to spot the loophole in this argument. In 1982 Unruh and Wald [2] called attention to the fact that the black hole appears surrounded by a bath of quantum (acceleration) radiation to a local *stationary* (hence accelerated) observer. The Archimedean buoyancy of this bath opposes free descent of the box, and it grows without bound near the horizon where the local acceleration temperature becomes infinite.

Alternatively, we may adopt the viewpoint of a local observer in free fall, for whom the quasi-stationary box is accelerated upwards. Thus negative energy enters the box and builds up as the box is lowered further and its acceleration increases. This acts as a debit on the net mass-energy content of the box, which eventually becomes zero and then negative. The box has now reached a floating point. It is a bubble of negative energy in a sea of zero energy (or nearly zero energy: the effects of vacuum polarization and Hawking radiation are everywhere bounded and very small for astrophysical-size black holes, and for simplicity we here ignore them). To make it descend further one would have to press downwards, and therefore no further mass-energy can be extracted. Unruh and Wald were able to show that radiation dropped from the floating point retains enough energy (along with its entropy) to maintain the correct proportionality between black hole area and entropy.

In describing such gedanken experiments it is essential to distinguish between the viewpoints of a local stationary observer (who "sees" a bath of acceleration radiation) and an inertial observer (who does not). Acceleration radiation has no more objective reality than centrifugal force. For examples it does not gravitate, i.e., its stress-energy does not appear on the right-hand side of the Einstein equations. Yet it provides a perfectly valid description of certain experiments if applied consistently. It is the failure to observe a consistent distinction between the stationary and inertial viewpoints which has led repeatedly to confusion in the literature.

A useful re-analysis of the Geroch experiment which emphasizes these points was published by W. J. Anderson [3] in 1994. In the course of his analysis, Anderson noted a further interesting point: the ground state for the inside of the slowly descending box is what is known as the "Boulware state".

I must digress for a moment to explain what this is and place is in context. We are concerned with a quantum field — for definiteness let us say a massless scalar field — propagating on a stationary, asymptotically flat gravitational background.

The Boulware state is the one which appears empty to observers at infinity. Further down, vacuum polarization effects associated with the increasing gravity will induce a (real) non-vanishing effective stress-energy with a negative energy density of order $-T_{\text{acc}}^4/\hbar^3 \sim -\hbar a^4$ in (3+1)-dimensions, where a is the local upward force required to hold a unit mass stationary.

The Boulware state is the natural ground state of a quantum field propagating through the spacetime of a stationary star. But it would be unstable in a black hole spacetime, since its stress-energy diverges at horizons.

We may think of the Boulware energy density as a superposition of two negative-energy fluxes — an outflux, infinitely blueshifted and diverging along the future horizon $\mathcal{H}^+$, and an influx which (in the case of an *eternal* black hole with a past horizon $\mathcal{H}^-$) diverges along $\mathcal{H}^-$.

These divergences can be cured by altering the boundary conditions at infinity, i.e., changing the quantum state. The negative-energy fluxes and their divergences can be neutralized by injecting two steady streams of positive energy, an influx from past lightlike infinity $\mathcal{I}^-$, and an outflux to future lightlike infinity $\mathcal{I}^+$. These new fluxes can be maintained by enclosing the black hole in a hot, perfectly reflecting sphere of large radius kept at the Hawking temperature. The new state represents a black hole in thermal equilibrium with its own radiation inside a large spherical cavity. It is known as the *Hartle-Hawking state*.

If the black hole was formed by the gravitational collapse of an initially stationary star, there is no past horizon needing regularization by an influx from infinity, and $\mathcal{I}^-$ can be left empty. We require only the positive-energy outflux to $\mathcal{I}^+$. The state defined by these boundary conditions agrees with the Boulware state in the remote past, and in the future (after the collapse) represents a situation in which positive energy flows out to infinity, with a corresponding influx of *negative* energy through the future horizon, i.e., it represents a black hole formed by collapse which radiates freely into space and absorbs negative energy (which would gradually eat away its mass if we took back-radiation into account). This is the *Unruh state*, appropriate for an evaporating black hole.

In (1+1)-dimensions, the quantum stress tensors for the various states take a simple explicit form, and it is then easy to make these remarks more concrete. This is the subject of one of the accompanying exercises (see also the lectures of Dr. Wipf), so I will confine myself to summarizing the results for the two states of most interest to us here, the Hartle-Hawking and Boulware states.

It is sufficiently general to consider 2-metrics of the form

$$ds^2 = \frac{dr^2}{f(r)} - f(r)dt^2 . \tag{4}$$

We denote by $\kappa(r)$ the redshifted gravitational force, i.e. the upward acceleration $a(r)$ of a stationary test particle reduced by the redshift factor $f^{1/2}(r)$, so that

$$\kappa(r) = \frac{1}{2}f'(r) . \tag{5}$$

A horizon is characterized by $r = r_0$, $f(r_0) = 0$ and its surface gravity defined by $\kappa_0 = \kappa(r_0)$.

Quantum effects induce an effective quantum stress-energy T_{ab} $(a, b, \ldots = r, t)$ in the geometry (4). If we assume no net energy flux $(T_t^r = 0)$ — thus excluding the Unruh state — T_{ab} is completely specified by a quantum energy density $\rho = -T_t^t$ and pressure $P = T_r^r$.

In (1 + 1)-dimensions the only result we need from quantum field theory is the so-called conformal or trace anomaly T_a^a. Classically, the trace would vanish for a conformally invariant field. In the quantum theory, the anomaly is most easily understood formally as arising from dimensional regularization of the stress tensor: the conformal invariance of the field in 2 dimensions is broken in $(2+\epsilon)$ dimensions, and this leaves a finite residue in the limit $\epsilon \to 0$. The trace anomaly is a c-number, i.e., it is independent of the state. In $(1+1)$ dimensions it is proportional to the curvature scalar R. Its explicit form for a massless scalar field is

$$T_a^a = \frac{\hbar}{24\pi} R, \tag{6}$$

with $R = -f''(r)$ for the metric (4).

It is now simple to integrate the conservation law $T_{a;\,b}^b = 0$ and obtain expressions for the individual components ρ and P. This gives

$$fP = -\frac{\hbar}{24\pi} \left(\kappa^2 + \text{const.} \right) \tag{7}$$

Different choices of the constant of integration correspond to different boundary conditions, i.e., different quantum states.

For the Hartle-Hawking state, we require ρ and P to be regular on the horizon $r = r_0$, giving

$$P_{\mathrm{HH}} = \frac{\hbar}{24\pi} \frac{\kappa_0^2 - \kappa^2}{f}, \quad \rho_{\mathrm{HH}} = P_{\mathrm{HH}} + \frac{\hbar}{12\pi} \frac{d\kappa}{dr} \tag{8}$$

When $r \to \infty$ this reduces to (setting $f \to 1$, $k \to 0$)

$$\rho_{\mathrm{HH}} \approx P_{\mathrm{HH}} \approx \frac{\pi}{6\hbar} T_{\mathrm{BH}}^2, \quad T_{\mathrm{BH}} = \frac{\hbar}{2\pi} \kappa_0, \tag{9}$$

which is appropriate for one-dimensional scalar radiation at the Hawking temperature T_{BH}.

For the Boulware state the boundary condition is $\rho = P = 0$ when $r = \infty$. The integration constant in (7) must vanish, and we find

$$P_{\mathrm{B}} = -\frac{\hbar}{24\pi} \frac{\kappa^2}{f}, \quad \rho_{\mathrm{B}} = P_{\mathrm{B}} + \frac{\hbar}{12\pi} \frac{d\kappa}{dr}. \tag{10}$$

If a horizon were present, the Boulware stress-energy would diverge there like

$$\rho_{\mathrm{B}} \approx P_{\mathrm{B}} \approx -\frac{\pi}{6\hbar} T_{\mathrm{acc}}^2, \quad T_{\mathrm{acc}} = \frac{\hbar}{2\pi} a, \tag{11}$$

where $a = \kappa/f^{1/2}$ is the local acceleration of a stationary observer.

4   Descending box in $(1+1)$ dimensions: energy density of the internal ground state

Following Anderson [3], we now attempt to calculate the energy density inside an initially empty one-dimensional box which slowly descends from infinity in the 2-geometry with metric (4). We shall then compare the answer with the Boulware energy density (10).

As the box descends, its gradually increasing upward acceleration causes negative and positive energy fluxes to flow into it from the floor (F) and ceiling (C) respectively. I shall assume these to be given by Davies-Fulling formula (3). Thus,

$$\left(\frac{dE}{dt}\right)_F = -\frac{\hbar}{12\pi}\left(\frac{da}{dt}\right)_F . \tag{12}$$

There is no a priori justification for this assumption in a curved space: the Davies-Fulling formula was derived for a mirror moving through Minkowski space. By ignoring the possible appearance of curvature-dependent terms in (12), we are assuming a "quantum equivalence principle" for the local rate of emission from the mirror.

The downflux of positive energy from the ceiling has a form similar to (12), but we must take into account that this energy is boosted by a factor $(f_C/f_F)^{1/2}$ in dropping to the floor. The net rate at which the energy in the box is changing is therefore

$$\frac{dE}{dt} = \frac{\hbar}{12\pi}f_F^{-1/2}\left\{\left(f^{1/2}\frac{da}{dt}\right)_C - \left(f^{1/2}\frac{da}{dt}\right)_F\right\} . \tag{13}$$

Introducing the proper vertical height $z = \int f^{-1/2}dr$, we can write, for floor and ceiling $(i = F, C)$

$$f_i^{1/2}\frac{da_i}{dt} = f_i^{1/2}\frac{da_i}{dz_i}\frac{dz_i}{dt} = f\frac{da_i}{dr_i}\frac{dz_i}{dt} ,$$

in which the proper velocities dz_i/dt are the same at both ends for a box of fixed proper length $l = z_C - z_F$. If l is small, (13) reduces to

$$\frac{dE}{dt} = \frac{\hbar}{12\pi}f^{-1/2}\frac{d}{dr}\left(f\frac{da}{dr}\right)\left(f^{1/2}l\right)\left(f^{-1/2}\frac{dr}{dt}\right) .$$

The energy accumulated in a box which descends from infinity to radius r is obtained by integrating this expression:

$$E = l\frac{\hbar}{12\pi}\int_\infty^r f^{-1/2}\frac{d}{dr}\left(f\frac{da}{dr}\right) .$$

Recalling that $a = (1/2)f^{-1/2}f'$ and integrating by parts yields

$$E/l = \frac{\hbar}{24\pi}\left(f'' - a^2\right) \tag{14}$$

for the energy per unit length inside the one-dimensional box.

In particular, if $f(r) = r$, we have the case of a box driven from rest into uniform acceleration on Minkowski space. Then (14) shows (as promised in Sec.2), that the internal ("Rindler") ground state has an energy depressed below the Minkowski vacuum by just the amount that allows an internal observer to assert that the Minkowski vacuum is "hot" relative to his ground state, with a relative energy density which is thermal and equal to $\frac{1}{6}\pi T^2_{\mathrm{acc}}/\hbar$.

In general, (14) agrees precisely with the Boulware energy density (10), even to the inclusion of the curvature-dependent contribution arising from the conformal anomaly.

In one sense, the conclusion we have reached — that the ground state inside the box is energetically indistinguishable from the Boulware state — is not surprising. The box started out empty at infinity. (Emptiness at infinity is characteristic of the Boulware state.) In its slow descent — an adiabatic process — its quantum state cannot change. The modes whose absence defines the ground state inside the box are always Boulware modes.

What *is* surprising is that the simplistic approach we have used to calculate the energy density of this state should yield the "correct" answer (10) *exactly*. We assumed that the Davies-Fulling expression (3) for the fluxes from the walls would transfer without change from flat to curved space. (That this seems to work exactly may well be an accident of two-dimensionality, perhaps related to the circumstance that every 2-space is conformally flat, and our fields are conformally invariant.) We further assumed that the two walls radiate independently of each other and incoherently, so that we could simply add their fluxes. But in actual fact the boundary conditions which have to be imposed at the reflecting walls exclude all but a denumerable set of Boulware modes from the interior. This affects the internal quantum state, as signalled by appearance of Casimir stresses. Neglect of these effects is a priori justified (as an approximation) only if the separation l of the walls is large compared with the characteristic Boulware wave-length ($l \gg \hbar/T_{\mathrm{acc}} \sim a^{-1}$), i.e., only at large accelerations. Fortunately, it is just in this regime that these quantum effects are of special interest.

5   Moving mirrors and boxes in $(3+1)$ dimensions

The sort of hands-on reasoning illustrated in previous sections should retain its heuristic value in three-dimensional problems. What is less clear at this time, because of the paucity of exact three-dimensional results, is whether it remains a reasonably accurate quantitative description.

No-one has yet found a general expression for the flux emitted by a plane mirror moving in $(3+1)$-dimensional Minkowski space, i.e. the three-dimensional analogue of the Davies-Fulling formula (10). (Even for uniform acceleration, the expression for the stress-energy is extremely formidable, though it reduces to a simple and intuitive form far from the mirror.) If we repeat the argument of Sec.4 to compare the energy density inside a three-dimensional box brought into uniform acceleration in Minkowski space, using the simple ansatz

$$F = \hbar \left(c_1 a^2 \frac{\mathrm{d}a}{\mathrm{d}\tau} + c_2 a \frac{\mathrm{d}^2 a}{\mathrm{d}\tau^2} \right) \tag{15}$$

for the three-dimensional flux (which is at least dimensionally correct), we shall obtain the correct Rindler value for a massless scalar field,

$$\rho_R = -\frac{\pi^2}{30\hbar^3}T_{\text{acc}}^4 \, ,$$

provided c_1 is chosen to be $(-1/360\pi^2)$. (The value of c_2 is immaterial to this argument because $d^2 a/d\tau^2$ is negligible if the acceleration changes slowly.)

However, it would be unrealistic to expect the true flux to have a form as simple as (15), except perhaps in some "effective" or asymptotic sense. There is, in fact, no reason to expect it to have any purely local form (depending only on the acceleration and its derivatives at the retarded point of the mirror nearest to the observer) — it could be a functional depending on the mirror's entire past history. A hint that this problem is really difficult comes from tracing back in time the path of any photon which has left the mirror obliquely. "After" its "first" bounce the photon recedes from the mirror with a normal component of velocity which is a fixed proper fraction of the speed of light. But the accelerating mirror comes arbitrarily close to the speed of light, and must overtake the receding photon again, and then again. Thus, any photon now reaching us obliquely from the mirror must have ricocheted off the mirror an infinite number of times in the past.

Things are easier for a *spherical* mirror. (In particular, ricocheting does not occur.) In 1979, Frolov and Serebriany [5] gave the complete solution of this problem for a spherical mirror moving with uniform acceleration a in Minkowski spacetime. Its history is given by

$$x^2 + y^2 + z^2 - t^2 = 1/a^2 \, .$$

This has a higher degree of 4-dimensional symmetry than the history of a plane mirror. All Green's functions (and hence also the quantum stress-energy) can be expressed in simple closed form using the method of images. (This merely involves transcribing to the Minkowskian sector the potential of a unit charge in the presence of a conducting 3-sphere in Euclidean 4-space.)

In particular, Frolov and Serebriany found that the stress-energy actually vanishes, both inside and outside the mirror, for a conformal scalar field in its ground state. Since it does not vanish for a *static* spherical mirror this is evidence of a rather remarkable cancellation between Casimir forces and the effects of acceleration in the spherical case.

Encouraged by the simplicity of these results for uniform acceleration, Warren Anderson and I have undertaken a perturbation analysis to determine the flux emitted by a spherical mirror which is changing its acceleration slowly. (This would suffice — modulo wall effects — to derive the energy accumulated inside a "box" in the form of a spherical shell slowly descending in a spherically symmetric gravitational field.) This calculation is straightforward, though tedious, and leads, remarkably, to purely local expressions for the perturbated (Wightman) Green's function and stress-energy. (The effect of the perturbation on individual modes involves the entire past history of the mirror, but in the effects on the Wightman function, particle and anti-particle modes interfere destructively.) In the limit $a \to 0$, our results agree with expressions derived by Ford and Vilenkin [6] for a *plane* mirror with small (variable) acceleration.

The spherical flux is a complicated expression, including terms of the general form (15) (with variable dimensionless coefficients c_1, c_2) but also others involving lower time-derivatives. Disentangling truly "radiative" and Casimir-like terms is a more difficult task here than in $(1 + 1)$ dimensions. At the time of writing, the proper physical interpretation still falls under the heading of "work in progress" and I shall not discuss it further here [7].

6   Bekenstein-Hawking entropy and black holes

In Section 2.5 of his lecture in this volume, Dr. Luminet introduced us to the laws of classical black hole dynamics. These follow directly from the Einstein field equations, and they relate geometrical properties of the horizon (area and surface gravity) to the purely mechanical quantities mass, charge and angular momentum. Thus, for a spherical black hole of mass m, charge e, horizon area $A_H = 4\pi r_+^2$ and surface gravity κ, the first law reads

$$\frac{\kappa}{8\pi} \mathrm{d}A_H = \mathrm{d}m - \frac{e}{r_+} \mathrm{d}e \ . \tag{16}$$

The second law (Hawking's area theorem) states that A_H is non-decreasing in any (classical) interaction of a black hole with its environment involving matter with non-negative energy density.

At this stage we may admit quantum theory into the picture. We then learn that black holes evaporate thermally, with a characteristic (Hawking) temperature

$$T_H = \hbar\kappa/2\pi \ . \tag{17}$$

All of this is nowadays well understood and generally accepted.

It is at this point that things become less straightforward. Let us (following Bekenstein) make the very appealing hypothesis that black holes may be considered on the quantum level to be simply hot objects satisfying the laws of ordinary thermodynamics. A formal comparison of (16) and (17) with the first law of thermodynamics then leads us to the cabalistic formula

$$S_{\mathrm{BH}} = \frac{1}{4} A_H/\hbar \tag{18}$$

in relativistic units ($G = c = 1$). For the purposes of the present discussion it is perhaps better not to beg the question by calling S_{BH} the black hole entropy. I shall refer to it as the Bekenstein-Hawking entropy.

The Bekenstein-Hawking entropy undoubtedly tells us something about the black hole. But what? Is it real or subjective? What exactly does it measure?

Where does it appear — on or near the horizon, or deep inside the hole? At what stage in the hole's evolution is it created — immediately upon formation by gravitational collapse, or only gradually over the long course of evaporation, like steam from a simmering kettle? What is the dynamical mechanism that makes S_{BH} a universal function, independent of the hole's past history or detailed internal condition?

These questions are embarrassing, because we do not know how to answer them. Creation of entropy is a non-equilibrium process. We have at our disposal an entire arsenal of sophisticated techniques for evaluating partition functions and treating the thermofield statistics of quantum fields propagating in $(3 + 1)$-dimensional black hole spacetimes, in particular the thermofield dynamics of Takahashi and Umezawa, the imaginary-time formalism of Matsubara and topological analysis of the black hole Euclidean sector ("Gibbons-Hawking instanton"). But all of these are geared to stationary backgrounds and thermal equilibrium. Dynamical models have been constructed for the evaporation of $(1 + 1)$-dimensional black holes (for a review, see [8]), but these run into difficulties and have brought us no nearer to answering the basic questions.

Nevertheless, we can discuss and weigh various alternatives. To mention first the subtlest possibility: it is conceivable that no quantum entropy is irreversibly created by the hole, no information is lost, and S_{BH} is merely a measure of our temporary loss of access (during the lifetime of the black hole) to correlations beneath the horizon. When the black hole finally evaporates completely, these correlations will be fully restored to us. A black hole formed by collapse of matter in a pure state will evaporate into a radiation field whose distribution appears thermal on a coarse-grained level but will be recognized to be in a pure state once the last photons have left the hole. A scenario of this general type is the one which tends to be favored by particle theorists.

Alternatively, it is possible that black hole formation and evaporation is accompanied by an irreversible increase of entropy, and we come back to the questions how, where and when?

The original (pre-Hawking 1974) motivation for assigning an entropy to a black hole was to keep account of the thermal entropy of objects thrown into the hole (Wheeler's teacup experiment). The generalized second law of black hole thermodynamics (sum of Bekenstein-Hawking entropy and external entropy non-decreasing) lends support to this view of BH entropy, though current proofs [9]–[11] are hedged with qualifications (e.g. quasi-stationarity).

Nevertheless, it is not possible (as emphasized more than twenty years ago by Wolfgang Kundt [12]) simply to identify S_{BH} with the thermal entropy of all the matter which collapsed to form and feed the hole. For a solar-mass black hole, S_{BH} is 10^{20} times as large as the entropy of its stellar progenitor. Since it is an issue of principle we are concerned with, we could even consider an idealized Oppenheimer-Snyder collapse of cold, pressureless, viscous-free dust in which no material entropy ever develops. It is possible to slice spacetime at arbitrarily late times by a complete spacelike hypersurface Σ which enters the hole and extends inwards to the (still nonsingular) centre of the collapsing cloud, yet encounters no material entropy anywhere.

The view that entropy is somehow created in the process of evaporation also meets with difficulties. Black hole evaporation is very nearly — and can be made exactly — reversible. We simply enclose the hole in a container, so that it comes into equilibrium with its own radiation, and then allow radiation to leak out of the container a little bit at a time. This process is reversible and cannot create entropy.

Zurek and Thorne [9] have suggested that the Bekenstein-Hawking entropy should be interpreted as the logarithm of "the number of quantum-mechanically distinct ways that the hole could have been made". (This requires some touching up for an evaporating black hole, since the number of possible past histories cannot diminish with time.) Formally, this statement is unexceptionable, assuming that the technical difficulties involved in making it precise can be overcome. Physically, it demotes S_{BH} to the status of a "class badge", which can tell us nothing about the state of an individual black hole formed in a specific way. To be sure, this is already implicit in the universality of S_{BH}, but the Zurek-Thorne interpretation simply accepts this universality without offering any dynamical explanation of it.

The most concretely dynamical and visualizable proposal stems from Frolov and Novikov [13]. This links S_{BH} with modes (produced by vacuum fluctuations) propagating "outwards" just inside and alongside the horizon. These modes have positive frequency but their energy is negative as calibrated for an observer at infinity (i.e., including the contribution of gravitational potential energy). Their spectrum is thermal with characteristic temperature T_H. Detailed implementation of this picture is so far still plagued with divergences and ambiguities. And we are left in the dark about the origins of the irreversibility (if any) associated with S_{BH}.

If the endpoint of evaporation is indeed a thermal state, then, as we have seen, we cannot ascribe the entropy created to the evaporation process itself or to thermodynamical entropy of the material which collapsed to form the hole in the first place. Which leaves us in a quandary.

Let us retreat to something simpler — an example from elementary thermodynamics. Gas inside a long cylinder confined to a volume V_1 by a piston. Now the piston is raised, suddenly or with supersonic speed, to a higher position. The gas molecules drift upward to fill up the larger volume, say V_2, and there is an increase of entropy. What is the origin of this new entropy? No heat enters the system, no work is done, there is no viscosity to generate heat inside the system. This is a non-equilibrium process, and thermodynamics simply does not have the vocabulary to describe the mechanism by which the entropy changes.

Which does not mean, of course, that thermodynamical methods cannot be used to calculate the *amount* of the change. We merely have to devise a second idealized reversible process which connects the initial and final equilibrium states, and apply the first law of thermodynamics. We allow the gas to push up the piston slowly and reversibly, meanwhile injecting heat (at the proper temperature) to compensate for the work done and keep the internal energy (and hence temperature) unchanged. The first law then gives the change of entropy as

$$\Delta S = k_B \ln (V_2/V_1)^N$$

for a gas of N particles.

The temperature does not appear in this expression. We see that the change of entropy in the first (non-ideal) process has to be described in statistical-geometrical terms as a (macroscopically) irreversible increase of the phase-volume accessible to and occupied by the system as a result of a change of its geometry.

It is perhaps not unreasonable to speculate by analogy that S_{BH} is some-how linked to the geometrical changes attendant upon the formation of an event horizon in gravitational collapse. Between the inner and outer horizons of a non-extremal hole, the collapsing system gains access to a new domain of configuration space marked by dynamical irreversibility (everything is forced to descend) and infinitely large in extent (the radial coordinate r is now timelike and the spatial 3-cylinders r=const. extend infinitely in the t-direction). The close link of S_{BH} to geometry and topology is well-known in the black hole Euclidean sector [1, 12]. It surely has a Lorentzian counterpart.

7   Reversible black hole creation and black hole entropy

No direct insight into the statistical origins of "black hole entropy" can come from thermodynamics. But if entropy really is a meaningful state function for black hole equilibrium states, then thermodynamics can tell us its value and provide at least an idealized operational definition of it. While we thus learn little of the real nature of black hole entropy, in compensation we are allowed free access to all the paraphemalia which are the stock-in-trade of the thermodynamicist — idealized engines employing unreal working substances, thin ("adiabatic") walls impermeable to the passage of heat and the like. Reversibility and the first law of thermodynamics are the only inviolable rules.

To find the entropy of any thermodynamical equilibrium state, one invents a reversible process which arrives at that state from a state of known entropy (e.g., a state of infinite dispersion and zero entropy), and computes how the entropy changes in that process using the first law of thermodynamics. In this and following section I shall report on some work done recently with Frans Pretorius and Dan Vollick [15] on the reversible creation of a spherical black hole. Our process involves the quasi-static contraction of a massive thin spherical shell to its gravitational radius, taking into account that (to maintain reversibility) the shell must remain in thermal equilibrium with its own acceleration radiation at every stage. This, incidentally, also provides an operational approach to questions raised recently [1] concerning the entropy of external black holes.

For the static spherical geometries outside and inside the shell it will be general enough for our purposes to adopt expressions of the form

$$ds^2 = \frac{dr^2}{f(r)} + r^2 d\Omega^2 - f(r)dt^2 .\tag{19}$$

This covers as special cases Minkowski, Schwarzschild, Reissner-Nordström and de Sitter backgrounds.

Of course the classical stress-energy associated (via the Einstein equations) with this metric is not the stress-energy of the ground state for the quantum fields which live in the spacetime. We know that this is the Boulware state, whose stress-energy $(T_{\mu\nu})_B$ depends on the types and number of fields and is unknown. For an ordinary star, $(T_{\mu\nu})_B$ is utterly negligible, but for a shell pressed close to its gravitational radius it generally grows without bound, and its back-reaction can certainly not be ignored. We cannot compute this back-reaction, but we can compensate for it. Drawing from an energy reservoir at infinity (in a realistic setting this could be the cosmic microwave background) we fill up the shell's surroundings with material whose stress-energy $\Delta T_{\mu\nu}$ tops up $(T_{\mu\nu})_B$ to form a thermal bath which shares the shell's local acceleration temperature $T_{\rm acc}(R)$ where they touch at the shell radius $r = R$. This "topped-up Boulware state" (TUB) is constructed in thermal quantum field theory by periodically identifying the coordinate t in the Euclidean sector with period equal to the reciprocal of the shell's redshifted acceleration temperature $T_\infty = f^{1/2}(R)T_{\rm acc}(R)$. Then the TUB's local temperature will vary with depth in accordance with Tolman's law

$$T(r)(-g_{tt})^{\frac{1}{2}} = T_\infty = \text{const.}\tag{20}$$

To local stationary observers, whose ground state is the Boulware state, the TUB's stress-energy

$$(T_{\mu\nu})_{\rm TUB} = (T_{\mu\nu})_B + \Delta T_{\mu\nu}\tag{21}$$

appears as a thermal bath at local temperature $T(r)$. As an example, in $(1+1)$-dimensions it is easy to check from (7) and (10) that the thermal top-up $\Delta T_a^b = \text{diag}(\Delta P, -\Delta\rho)$ takes the form

$$\Delta P = \Delta\rho = \frac{\pi}{6\hbar}T^2(r)$$

for a massless scalar field, which looks exactly like the flat space expression. This even extends to $(3+1)$-dimensions, at least to a good approximation. In the Page approximation [17], one obtains the isotropic distribution

$$\Delta P = \frac{1}{3}\Delta\rho = \frac{\pi^2}{30\hbar^3}T^4(r)\tag{22}$$

for a conformal scalar field, again exactly as in flat space, cf. [16].

It would be equally apt to call the TUB a "generalized Hartle-Hawking stae". Indeed, it tends smoothly to the Hartle-Hawking state in the limit when the shell approaches its gravitational radius. The stress-energy of both states is bounded everywhere (including the horizon), and is small (of order T_∞^4) for large black holes. But it does not vanish asymptotically at large radii.

Therefore, in order to keep effects of back-reaction under control, we imagine the TUB to be encased in a large spherical container of radius R_{big}. Back-reaction will be negligible if the TUB's total energy ($\sim T_\infty^4 R_{\text{big}}^3$ in Planck units) is small compared to the shell's mass M, i.e. (in conventional unit)

$$R_{\text{big}}/(2GM/c^2) \ll (M/m_{\text{pl}})^{\frac{2}{3}} \approx 10^{25} (M/M_\odot)^{2/3} ,$$

where m_{pl} is the Planck mass and $M_\odot$ the solar mass. I shall suppose this condition satisfied, and accordingly neglect the stress-energy and entropic contributions of the TUB.

Our phenomenological vantage point allows a dualistic approach. The thermal equilibrium condition $T_{\text{TUB}} = T_{\text{shell}}$ for $r = R$ cooresponds to the viewpoint of a local *stationary* observer. On the other hand, for the TUB's stress-energy we adopt the "objective" (gravitating) value which appears on the right-hand side of the Einstein equations; this would correspond to what is measured by a local *inertial* observer. We are taking this stress-energy (and the associated TUB entropy) to be negligible for a large black hole.

Both of these quantities would appear far from negligible to a stationary observer near the horizon. It is *his* point of view that we would be forced to adopt in a statistical analysis, since no technique is currently available for doing statistical thermodynamics in anything other than the system's rest-frame. Thus, the many current papers on black hole thermofield statistics have to resort to various devices (e.g., "brick-wall" cut-offs or renormalization of Newton's gravitational constant [18]) to deal with the resulting divergence at the horizon. A further, related question is whether and how to make allowance for gravitational back-reaction in such calculations: the decompositions (21), (22) assume a *fixed* geometrical background (19).

8   Dynamics of a spherical shell

Before proceeding with the thermodynamics of the shell, I shall pause here to assemble the basic mechanical formulae.

Quite generally for an arbitrary surface layer, there is a distributional equivalent of the Einstein field equations which relates the surface energy tensor S^{ab} to the jump (denoted by [...]) of the extrinsic curvature K_{ab} as one crosses the 3-surface in the direction of its unit normal n^μ. For a timelike shell history n is a spacelike vector, but it is sometimes of interest (and it costs nothing extra to allow for this possibility in our formulae) to consider spacelike "transition layers", which might, for instance, represent schematically rapid phase transitions in cosmology or the deep interior of black holes [19]. Accordingly, we take

$$n \cdot n \equiv n^\mu n_\mu = \eta \equiv \pm 1 .$$

The extrinsic curvature for a given embedding is given by

$$K_{ab} = e_{(a)} \cdot \delta n / \delta \xi^b \,,$$

where $e_{(a)}$ are the three holonomic base vectors associated with the intrinsic coordinates ξ^a, and $\delta n / \delta \xi^b$ is a 4-dimensional absolute derivative.

The Lanczos distributional field equations and conservation laws then read

$$-8\pi\eta S_{ab} = [K_{ab} - g_{ab}K], \quad \eta S^b_{a;b} = -[e^\alpha_{(a)} T^\beta_\alpha n_\beta] \,, \tag{23}$$

where T^β_α is the 4-dimensional stress-energy of the ambient medium. The intrinsic 3-metric $g_{ab} = e_{(a)} \cdot e_{(b)}$ is continuous across the layer. Specializing now to spherical metric as

$$ds^2 = R^2(\tau)d\Omega^2 - \eta d\tau^2 \,, \tag{24}$$

so that (for a timelike history) $R(\tau)$ is the radius of the shell at proper time τ.

Since $K^\tau_\tau - K = -2K^\theta_\theta$, (23) implies

$$4\pi\eta S^\tau_\tau = [K^\theta_\theta] \,. \tag{25}$$

All essential information about the shell's dynamics is contained in (25) and the conservation law.

The latter is simplified if we now assume that the interior and exterior geometries are both described by metrics of the form (19) — of course, with different functions $f_1(r)$ and $f_2(r)$ respectively. The special form (19) implies that $T^r_r = T^t_t$, and hence that $e^\alpha_{(a)} T_\alpha{}^\beta n_\beta = 0$ on both sides, i.e. that the ambient medium does no work on the shell, and the conservation law reduces to

$$\frac{\mathrm{d}}{\mathrm{d}\tau}(S^\tau_\tau A) + P\frac{\mathrm{d}A}{\mathrm{d}\tau} = 0, \quad P = S^\theta_\theta \,, \tag{26}$$

where $A = 4\pi R^2$ is the shell's area. This has an obvious intuitive meaning.

It is equally straightforward to obtain the explicit form of (25). If $u^\alpha = \mathrm{d}x^\alpha/\mathrm{d}\tau \equiv \dot{x}^\alpha$ is the shell "4-velocity" (spacelike for a transition layer!) we find (enumerating the four-dimensional coordinates as r, θ, φ, t)

$$u^\alpha = (\dot{R}, 0, 0, \dot{t}), \quad n_\alpha = (\dot{t}, 0, 0, -\dot{R}) \,.$$

The normalization $n \cot n = -u \cdot u = \eta$ gives

$$n^\alpha \partial_\alpha r = f\dot{t} = \epsilon(\eta f + \dot{R}^2)^{1/2} \,,$$

where $\epsilon = \mathrm{sign}(n^\alpha \partial_\alpha r) = \pm 1$ is another sign factor which tells us whether the radial coordinate r increases outwards. (In a closed universe or inside a black hole it need not!) Finally,

$$K^\theta_\theta = \frac{1}{2}n^\mu \partial_\mu \ln (r^2) = fn_r/R$$

and (25) has now been reduced to

$$4\pi\eta R S_\tau^\tau = [\epsilon(\eta f + \dot{R}^2)^{1/2}] . \tag{27}$$

Introducing the self-explanatory notation $\sigma = -S_\tau^\tau$, $M = 4\pi R^2\sigma$, (26) and (27) become

$$\eta\frac{M}{R} = -\left[\epsilon(\eta f + \dot{R}^2)^{1/2}\right] , \qquad \frac{\mathrm{d}M}{\mathrm{d}\tau} + P\frac{\mathrm{d}A}{\mathrm{d}\tau} = 0 . \tag{28}$$

The acceleration $\ddot{R}$ can be found by differentiating the first of these equations, eliminating $\dot{M}$ by using the second equation, then dividing through by $\dot{R}$. For a *static* shell we have $\dot{R} = \ddot{R} = 0$, and this yields an expression for the surface pressure P:

$$\eta\frac{M}{R} = -\left[\epsilon(\eta f)^{1/2}\right] , \qquad 16\pi P = \left[\frac{\epsilon}{(\eta f)^{1/2}}\left(f' + \frac{2f}{R}\right)\right] , \tag{29}$$

in which it is understood that $f_i = f_i(R)$ $(i = 1, 2)$.

It is instructive to see the explicit form of these expressions, say for a shell having gravitational (Schwarzschild) mass m and electric charge e, with a flat interior. We set $f_1(r) = 1$, $f_2(r) = 1 - 2m/r + e^2/r^2$, $\epsilon = \eta = 1$ in (29) and find

$$m = M - \frac{1}{2}\frac{M^2 - e^2}{R} , \qquad P = \frac{M^2 - e^2}{16\pi R^2(R - M)} . \tag{30}$$

These expressions have obvious Newtonian counterparts and simple intuitive meanings. We can immediately derive the differential relation (for fixed charge e)

$$\mathrm{d}m = V(\mathrm{d}M + P\mathrm{d}A) \tag{31}$$

where

$$V = f_2^{\frac{1}{2}} = (1 - M/R) \tag{32}$$

is the gravitational potential or redshift factor which recalibrates locally measured mass and work to energies available to an observer at infinity.

9   Shell thermodynamics

We can move on now to thermodynamical considerations. Since gravitational "force" is discontinuous across the shell, observers at the inner and outer faces have different accelerations. This means the inner and outer TUBs in which the shell is immersed are at different temperatures. To maintain reversibility, an "adiabatic" diaphragm (impermeable to heat) must be interposed between the faces. Other strange properties of the shell will emerge as we proceed. Their unreality is irrelevant. The shell's *raison d'être* is purely functional, to serve as a reversible link which enables us to keep track of how the entropy changes between the initial and final equilibrium states — an infinitely dispersed mass and a black hole.

We can picture the shell as a pair of concentric spherical plates (inner and outer masses M_1 and M_2) bonded together by a massless and thermally inert interstitional layer of negligible thickness. How we distribute the total shell mass $M = M_1 + M_2$ between the plates is arbitrary. Formally, the simplest expedient is to choose M_1 so that spacetime is flat between the plates. This generally makes M_1 negative, but we are not bound by positive-energy conventions. (In the simplest case, and the one of most immediate interest, the spherical cavity interior to the shell is empty, hence flat, and $M_1 = 0$)

The two plates thus separate three concentric spherical zones: an inner zone where the metric function $f(r)$ in (19) as $f_1(r)$, a very thin intermediate layer where $f = 1$, and an outer zone where $f = f_2(r)$. Applying (29) in them to the inner and the outer pair of adjoining zones, we find the masses M_i and surface pressures P_i ($i = 1, 2$) of the two plates:

$$\frac{M_i}{R} = \zeta(1 - V_i), \quad 16\pi P_i = \zeta\left(\frac{f'_i}{V_i} - \frac{2M_i}{R^2}\right). \tag{33}$$

Here, f_i and $V_i = f_i^{\frac{1}{2}}$ are understood to be evaluated at $r = R$, the common radius of the two plates, and $\zeta = (-1)^i$ is a sign factor, needed because flat space is on the outside of the inner plate. The temperatures T_i of the plates are given by the accelerations at the inner and outer shell faces:

$$T_i = \hbar a_i/2\pi, \quad a_i = \frac{1}{2}f'_i(R)/V_i(R). \tag{34}$$

This formula gives the intensive thermodynamical variable T_i as a function of the two parameters M_i and r, though not in the intensive combination $\sigma_i = M_i/4\pi R^2$. The thermodynamical equation of state for the material making up each plate should be flexible enough to allow us to choose the plate's mass and area independently. This requires that there is a second density parameter n_i, independent of σ_i, and also a function of M_i and R. Its functional form is strongly constrained (though not determined completely, see below) by the Gibbs-Duhem relations, together with the requirement that the equation of state $T_i = T_i(\sigma_i, n_i)$ for the intensive variables be equivalent to (34). Since the plates are merely abstract entropy-carrying devices, the physical significance one attaches to n_i is quite immaterial. To fix ideas, one might think of it as "molecular density".

Written in terms of densities, the Gibbs-Duhem relations are (temporarily dropping the indices i):

$$dS = \beta d\sigma - \alpha dn, \quad S = \beta(\sigma + P) - \alpha n. \tag{35}$$

Here, S is entropy per unit area, $\beta = T^{-1}$, $\alpha = \mu\beta$ and μ is the chemical potential associated with n. These relations look more familiar in their extensive form. Setting

$$S_{\text{tot}} = SA, \quad M = \sigma A, \quad N = nA,$$

we find that (35) are equivalent to

$$dS_{\text{tot}} = dM + P dA - \mu dN, \quad T S_{\text{tot}} = M + PA - \mu N. \tag{36}$$

Now, in the equation

$$n d\alpha = \beta \, dP + (\sigma + P) \, d\beta \tag{37}$$

the right-hand side is completely known, since we have explicit formulae for σ, P and T in terms of M and R. Thus, (37) can be reduced to

$$n d\alpha = \frac{1}{2} \frac{\sigma}{\gamma} \, d\left(\frac{1}{\zeta \sigma \gamma} \right), \tag{38}$$

where it is convenient to define

$$\gamma^2 = f'(R)/(8\pi \zeta \sigma V) \tag{39}$$

The functions n and α can be chosen arbitrarily subject only to the restriction (38), imposed by the Gibbs-Duhem relations. The simplest option is to select plate materials having the "canonical" equation of state (denoted by an asterisk, and restoring index i)

$$n_i^* = \sigma_i/\gamma_i, \quad \alpha_i^* = (2\zeta \sigma_i \gamma_i)^{-1}. \tag{40}$$

Now, from (34) and (39) we have the identity

$$T_i = 2\hbar \zeta \sigma_i \gamma_i^2. \tag{41}$$

Hence the canonical chemical potential $\mu_i^* = T_i \alpha_i^*$ obeys the simple relation

$$\mu_i^* n_i^* = \sigma_i. \tag{42}$$

Substituting in (35), and noting from (33) and (39) that the surface pressures can be expressed as

$$P_i = \frac{1}{2} \sigma_i \left(\gamma_i^2 - 1 \right), \tag{43}$$

we obtain the entropy densities of the plates as

$$S_i^* = \beta_i P_i = \frac{1}{4} \hbar^{-1} \zeta (1 - \gamma_i^{-2}). \tag{44}$$

At large radii in an asymptotically flat outer space, $f_2(r) \approx 1 - 2m_2/r$. Hence from (39) or (43), at wide shell dispersion $M_2 \to m_2$ and

$$\gamma_2^2 \approx 1 - \frac{1}{2} M_2/R \quad (R \to \infty). \tag{45}$$

If we want to think of n_i as "molecular density", we see from (40) that we should assign unit positive rest-mass to each "molecule" in the outer plate, and from (42) that canonical molecules have unit chemical potential when the plate material is cold.

When its slow contraction terminates, the shell is hovering just outside the horizon $r = r_0$ of the external geometry — $f_2(r_0) = 0$ — whose surface gravity $\kappa_2 = \frac{1}{2}f_2(r_0)$ is assumed to be non-zero. (The extremal case is dealt with in the following section.) Then (39) shows that γ_2^2 diverges according to

$$\gamma_2^2 \approx \frac{\kappa_2}{(M_2/R^2)}V_2^{-1}, \quad V_2^2 \approx 2\kappa_2(R - r_0) \quad (R \to r_0). \tag{46}$$

Hence, from (44),

$$\lim_{R \to r_0} S_2^* = \frac{1}{4}\hbar^{-1}, \tag{47}$$

i.e., the entropy of the outer plate is one quarter of its area in Planck units, in the black hole limit.

In the simplest situation, the spherical cavity inside the shell is empty and flat, so that, for the inner plate,

$$f_1(r) = 1, \quad M_1 = P_1 = S_1 = T_1 = 0, \tag{48}$$

and the outer plate contributes all of the mass and entropy of the shell. We can then read off the entropy of the final (black hole) state from (47) as equal to the Bekenstein-Hawking entropy.

It should be noted that this conclusion goes significantly beyond its verification of the generalized second law [9, 10, 11] which show that one quarter of the change in area (when the black hole slowly ingests material) is equal to the entropy absorbed. Here we have derived the entropy-area relation in integral form, eliminating the possibility of an additive constant.

The reversible shell-contraction described above suggests a fairly obvious operational definition for S_{BH}. But would such a definition be additive? Imagine that, in the field of a pre-existing black hole with Bekenstein-Hawking entropy S_{old} (or of any object, e.g., a star, having this thermodynamical entropy), we lower a shell of entropy $S_{\mathrm{shell}} = S_2 + s_1$ to the radius where an outer black hole, of area A_{new}, is about to form, so that $S_2 = \frac{1}{4}A_{\mathrm{new}}/\hbar$ according to (48). Is the new Bekenstein-Hawking entropy obtainable by simple addition as $S_{\mathrm{old}} + S_{\mathrm{shell}}$? At this stage of the operation, definitely not. The upper plate by itself already accounts for the full Bekenstein-Hawking entropy of the new configuration, so it would be necessary for the negative entropy S_1 of the lower plate to cancel exactly the entropy of whatever was inside the cavity initially, i.e., $S_1 + S_{\mathrm{old}}$ would need to vanish. In general, this is not true. However, we are still free to carry out a further reversible procedure: we can sweep all the material inside the cavity outwards and onto the new shell. (If this material includes an inner black hole, this will involve dilating the shell representing it until it merges with the lower plate of the new shell — in effect, a (reversible) "evaporation" of the inner black hole.) This "flattens" the cavity inside the new shell and annihilates the lower plate. The total entropy at the end, now carried entirely by the upper plate, is $S'_{\text{shell}} = \frac{1}{4} A_{\text{new}}/\hbar$, which is S_{BH} for the final black hole. In this specific sense, S_{BH} may be called "additive", but it is perhaps more correct to say it is "forgetful": S_{BH} for the final configuration betrays no clue about the entropy originally contained in the space now occupied by the hole.

These considerations suggest the following operational definition: S_{BH} is the maximum thermodynamical entropy that could be stored in the material which gathered and collapsed to form the black hole, if we imagine all of this material compressed into a thin layer near its gravitational radius.

Of course, this imagined process bears no resemblance to any real scenario of black hole formation. But it actually describes rather well the time-reversed process of evaporation since Hawking's mechanism of virtual pair-creation is a skin-effect confined to a thin layer near the horizon. In the real evaporation process the escaping particles are temporarily detained near the horizon by redshift effects on their way out of the hole; the shell model collects all of them at one time. Kundt's description of S_{BH} as "evaporation entropy" sums up the situation rather well, with the previse that the evaporation process itself (being virtually reversible) cannot be the *source* of S_{BH}; it acts only as its conduit.

The key result (47) was established for a special, "canonical" form of plate material. But since physical significance attaches only to the initial (infinitely dispersed mass) and final (black hole) states, and not on the (reversible) route by which we get from one to the other, one would expect (47) to be insensitive to the properties of its plate materials. Let us examine how far this is true.

The most general functions n and α satisfying the Gibbs-Duhem relations (with T_i equal to the acceleration temperatures at the two plates) are obtained by replacing α_i^* in (32) by an arbitrary function of itself, $g_i(\alpha_i^*)$, and respecting the invariance of $n_i d\alpha_i$ in (38). This yields the general formulas

$$\alpha_i = g_i(\alpha_i^*), \quad n_i = n_i^*/g_i'(\alpha_i^*) \tag{49}$$

$$\mu_i n_i/\sigma_i = g_i/\alpha_i^* g_i'. \tag{50}$$

Thus the most general expression for the entropy densities of the plates is

$$4\hbar\zeta S_i = 1 + \gamma_i^{-2}(1 - 2\mu_i n_i/(\sigma_i)). \tag{51}$$

We see that the conclusion (47), i.e.,

$$\lim_{R \to r_0} S_2 = \frac{1}{4}\hbar^{-1}, \tag{52}$$

is unchanged under arbitrary transformations (49) which leaves $\mu_i n_i$ bounded in the high-temperature limit. Indeed, we can allow transformations which are singular in this limit, provided only that

$$\lim_{T_i \to \infty} \mu_i n_i/(\sigma_i T_i) = 0, \tag{53}$$

recalling (41).

We cannot expect (52) to be invariant under arbitrary singular transformations. At the root of this problem is the fact that the black hole endstate is a *singular* state for the plate material (P_2 and T_2 become infinite). In these circumstances there is no a priori justification for excluding or constraining asymptotically singular behaviour of thermodynamical quantities. But it is reassuring that the loose constraint (53) guarantees that our conclusions are independent of the plate material for a very broad class of equations of state.

In passing, let us note that the freedom contained in (49) can be gainfully used to "improve" the behaviour of the plate material at low temperatures. For canonical material $S_i^*/T_i \sim R$ ($R \to \infty$). By suitable choice of g_i in (49) in the limit $\alpha_i^* \to \infty$, we can arrange $S_i/T_i \to 0$.

10   The third law

There are essentially two distinct versions of the third law of thermodynamics [20]. The first, proposed by Ernst in 1906, states that isothermal processes become isentropic in the limit of zero temperature. An essential form says that the temperature of a system cannot be reduced to zero in a finite number of operations.

In the third edition of his "Thermodynamik" published in 1911, Planck enunciated a stronger form of the third law: "The entropy of any system tends, as $T \to 0$, to an absolute constant, which may be taken as zero." (There are difficulties in making this formulation precise [21], involving enumeration of degenerate ground states in the statistical theory, which are not immediately relevant to us here.)

In their 1973 paper on "The four laws of (classical) black hole dynamics", Bardeen, Carter and Hawking proposed a form of the third law patterned after Nernst's unattainability principle: "It is impossible by any process, no matter how idealized, to reduce the surface gravity to zero in a finite sequence of operations."

A more specific form [22], which makes precise the meaning of "finite sequence of operations" states: "A non-extremal black hole cannot become extremal at a finite advanced time in any continuous process in which the stress-energy of accreted matter stays bounded and satisfies the weak energy condition." (It becomes clear from this formulation that certain quantum processes which involve absorbtion of negative energy can violate the third law — but not Hawking evaporation, since the influx of negative energy falls to zero near extremality like $-T_H^4$ in $(3 + 1)$-dimensions.)

For a long time it was believed that there is no black hole analogue to Planck's version of the third law. Recently, however, this has become a matter of controversy. Arguments based on black hole instanton topology and pair creation [1] suggest that the entropy of extremal black holes is zero. On the other hand, the remarkable indirect calculations of "black hole entropy" by counting states of strings on D-branes recover the old value $S_{BH} = \frac{1}{4}A/\hbar$ for extremal black holes [23].

We can examine this question from our observational point of view. As the simplest illustration, let us consider the quasi-static contraction of an extremally charged spherical shell with empty interior. Setting $|e| = m$ in (30) we find $M = |e|$, $P = 0$. If the shell is made of "canonical" material, (44) yields $S^* = 0$ at all stages of the contraction, leading to

$$S_{\text{extreme BH}} = 0. \tag{54}$$

The universality of the standard area formula thus breaks down in the extremal case.

The result (54) is, however, quite sensitive to the equation of state of the plate material. For arbitrary material, (51) gives for the shell's entropy density

$$S = \frac{1}{2}\hbar^{-1}(1 - \mu n/\sigma), \tag{55}$$

whose value can be made arbitrary by choice of the function $g(\alpha^*)$ in (49), (50). (In the extremal case, $\gamma^2 = 1$ and $\alpha^* = \gamma^{-1}$ according to (44) and (40).)

Thus, it would appear that an extremal black hole differs from a generic one in that its entropy is not independent of its mode of formation and past history. One point which seems to emerge rather clearly from the shell model is that the distinction between the extremal and non-extremal cases may be less a matter of zero versus non-zero temperature than of finite versus infinite temperatures measured by local stationary observers. There are many precedents for a situation where a simplicity and universality found at high temperatures breaks down at ordinary temperatures.

11   Concluding remarks

Our gedanken-experiments suggest an operational definition of Bekenstein-Hawking entropy and also a specific form of the generalized second law: In interactions of a black hole with its environment involving absorption or evaporation of material with prescribed mass, charge and angular momentum, S_{BH} changes by precisely the *maximum* thermodynamical entropy which could be stored in this material when it is compressed into a thin shell just outside the horizon.

This formulation in its turn suggests that S_{BH} should be considered as *effective entropy* for black holes, in the same sense that 6000 K is an effective temperature for the sun. As far as interactions with their environments are concerned both objects are indistinguishable from shells (of the same size and mass) whose entropy or temperature have the effective values. (The analogy is imperfect, since we are ignoring solar "hair"–sunspots, etc.)

This point of view provides a rationale for the numerous current attempts to understand black hole entropy and information statistically by focusing on the neighbourhood of the horizon. At the same time, it offers no direct encouragement for the hope that such efforts will lead to insight into the deeper properties of black holes.

A diametrically opposing view is suggested by the recent, uncannily successful calculations of Bekenstein-Hawking entropy by counting states of strings attached to D-branes [23]. This is not a very specific test of superstring theory: indeed, one would expect any plausible microtheory to reproduce the results of thermodynamics in the semi-classical limit. But it is welcome evidence that superstring theory passes a minimal consistency check in its claim that it offers a glimpse of the microstructure underlying space, time and matter, and further that the Bekenstein-Hartle formula actually tells us something about the inner depths of a black hole.

The sharply contrasting viewpoints expressed in the last two paragraphs are by no means necessarily incompatible. We can readily appreciate this if we recall by analogy that there are two very different but concordant ways of deriving the luminosity of the sun: either from its surface temperature and radius or from the nuclear reaction rates in its core. The first simply meters the radiation on its way out; the second tracks it down at the source.

12   Acknowledgments

Most of what I have learned about this subject emerged in numerous stimulating discussions with my collaborators Warren Anderson, Franz Pretorius and Dan Vollick, and I should like to take this opportunity to express my indebtedness to them.

The work was supported by NSERC of Canada and by the Canadian Institute for Advanced Research.

Bibliography

[1] S. W. Hawking, G. T. Horowitz and S. F. Ross, *Phys. Rev.* **D51**, 4302 (1995). G. W. Gibbons and R. E. Kallosh, *Phys. Rev.* **D51**, 2839 (1995).

[2] W. G. Unruh and R. Wald, *Phys. Rev.* **D25**, 942 (1982).

[3] W. J. Anderson, *Phys. Rev.* **D50**, 4786 (1994).

[4] P. Candelas and D. Deutsch, *Proc. Roy. Soc.* **A 354**, 79 (1977).

[5] V. P. Frolov and E. M. Serebriany, *J. Phys.* **A12**, 2415 (1979).

[6] L. H. Ford and A. Vilenkin, *Phys. Rev.* **D25**, 2569 (1982).

[7] W. J. Anderson, Ph.D. Thesis, Univ. of Alberta, Edmonton (1997); W. J. Anderson and W. Israel, paper in preparation.

[8] E. Martinec, *Class. Quantum Grav.* **13**. 1 (1996).

[9] W. H. Zurek and K. S. Thorne, *Phys. Rev. Letters* **54**, 2171 (1985).

[10] R. M. Wald, in *Black Hole Physics* (ed. V. De Sabbata and Z. Zhang), Kluver, Boston 1992.

[11] V. P. Frolov and D. N. Page, *Phys. Rev. Letters* **71**, 3902 (1993).

[12] W. Kundt, *Nature* **259**, 30 (1976);
D. Wilkins, *J. Gen. Rev. Grav.* **11**, 45 (1979).

[13] V. Frolov and I. Novikov, *Phys. Rev.* **D48**, 4545 (1993).
V. Frolov, in *Relativistic Astrophysics* (ed. B. J. T. Jones and D. Marković), Cambridge Univ. Press, Cambridge 1997, p.183.

[14] S. Liberati, *Nuov. Cim* **112B**, 405 (1997).

[15] F. Pretorius, D. Vollick and W. Israel, Phys.Rev. **D57**, 6311 (1998); S. Mukohyama and W. Israel, Los-Alamos e-print gr-qc/9806012.

[16] B.P.Jensen, J.G.McLaughlin and A.C.Ottewill, Phys.Rev. **D45**, 3002 (1992).

[17] D.N. Page, *Phys. Rev.* **D25**, 1499 (1982).

[18] G. 't Hooft, *Nucl. Phys.* **B256**, 727 (1985); L. Susskind and J. Uglum, *Phys. Rev.* **D50**, 2700 (1994); F. Larsen and F. Wilczek, *Nucl. Phys.* **B458**, 249 (1996); G. 't Hooft, Los-Alamos e-print gr-qc/9706058.

[19] V. P. Frolov, M. A. Markov and V. F. Mukhanov, *Phys. Rev.* **D41**, 383 (1990).

[20] H.B. Callen, *Thermodynamics*, Wiley, New York, 1960, Chap. 10; J. Kestin, *A Course in Thermodynamics*, Blaisdell, Waltham, Mass. 1968, Vol II, Chap. 23.

[21] R.M. Wald, *Phys. Rev.* **D56**, 6467 (1997).

[22] W. Israel, *Phys. Rev. Letters* **57**, 397 (1986).

[23] G.T. Horowitz, in *Black Holes and Relativistic Stars* (ed. R.M. Wald) University of Chicago Press 1998, Chap. 12, pp. 241–266 (gr-qc/970472).

Internal Structure of Black Holes

Werner Israel

Canadian Institute for Advanced Research Cosmology Program, Department of
Physics and Astronomy, University of Victoria, P.O. Box 3055, Victoria B.C., V8W
3P6 Canada

Abstract. Gravitational collapse to a black hole leaves a decaying wake of gravitational waves. Some of those waves are absorbed into the hole and have dramatic effects
on the geometry near the inner (Cauchy) horizon because of a diverging blueshift.
Determining the inner structure of the hole is really an evolutionary problem with precisely known initial data. The evolution can in principle be followed to within Planck
distances of the singularity at the inner horizon, using only well–established physical
laws (the Einstein field equations). This lecture reviews recent progress in this area,
highlighting outstanding problems and gaps.

1 Introduction

The sector $r \leq 2m$ of the Schwarzschild geometry was long considered outside
the realm of legitimate scientific enquiry. For more than 40 years, $r = 2m$ was
confusingly termed the "Schwarzschild singularity", a surface which, even for
those who understood that it was not actually singular in a mathematical sense,
was still a border where Einstein's theory parted company with sensible physics.

After 1960, as people gradually got accustomed to the Kruskal extension and
to the reality of quasars and gravitational collapse, there was a general fall–back
(with some notable exceptions: John Wheeler, Igor Novikov) to what was essentially the astrophysicist's position: that $r = 2m$ demarcates a region which can
never influence the outside world, and is therefore of no operational significance
or physical interest. (One may perhaps discern vestiges of this attitude in the
motivation underlying some of the recent attempts to burden this impalpable
boundary surface with all of the hole's entropy and information.)

But the justification for believing in a *permanent* demarcation of this kind
collapsed with the discovery of black hole evaporation in 1974. It is becoming
more and more obvious that a full understanding of black hole evolution will not
be achievable so long as a particular (and eventually the most crucial) sector of
the spacetime is left out of consideration.

A fairly recent first incursion into this area is a program to explore the early,
classical phase of the hole's internal evolution. The word "evolution" is used
advisedly. We are really concerned, not with a structural, but an evolutionary
problem. Inside the hole the radial coordinate r is timelike; deeper layers correspond to later times. Causality therefore allows us to take a piecemeal approach,
exploring first the outer, classical layers (which will in fact take us nearly all the

way down to the outermost classical singularity), and deferring the more difficult questions of quantum gravity which arise when curvatures begin to approach
Planck levels at greater depths.

I can confine myself here to a very brief elementary survey, since there is now
a considerable literature on this topic, including reviews both introductory [1, 2]
and more advanced [3]. A Workshop on black hole interiors was recently held in
Haifa, and the Proceedings [4] may be consulted for further details.

2   Internal evolution: classical formulation

The classical phase of the hole's internal evolution presents us with a problem
which is mathematically quite definite and, in principle, straightforward. It is a
hyperbolic initial-value problem of Cauchy's type. The evolution equations are
the classical Einstein field equations. The initial data are set on or near the event
horizon. The task is to evolve these data forward in time up to the point where a
singularity is imminent. (At this stage the classical evolution equations fail and
the quantum regime takes over.)

Thanks to the enormous simplification arising from the no-hair property (see
the lectures of Luminet and Heusler elsewhere in this volume), and to the work
of Richard Price, the initial data are known with some precision (in marked
contrast to the situation in cosmology). Subject to one caveat (possible failure
of the cosmic censorship hypothesis), it is generally expected that the unchecked
collapse of an arbitrarily deformed spinning mass produces a black hole whose
exterior field will converge asymptotically toward a stationary configuration, a
member of the Kerr (or, in the presence of charge, Kerr-Newman) family.

The settling process is accompanied by emission and absorption of gravitational waves. Richard Price found in 1972 that these wave-tails decay according
to an inverse power of exterior time. For a general wavelike perturbation $\delta\Phi$ of
multipole order l (e.g., $l = 2$ for a quadrupole gravitational wave $\delta g_{\mu\nu}$) propagating downwards into the hole,

$$\delta\Phi \sim v^{-\frac{1}{2}p+1} \quad (v \to \infty) \tag{1}$$

near the event horizon. Here, v is exterior advanced time and $p = 4l + 4$.

Thus, the initial conditions for the hole's internal evolution, which are set
near the event horizon, consist of the Kerr-Newman geometry perturbed by an
externally decaying wave-tail of the form (1).

It is found that the exterior decay of the perturbation and the convergence
of the exterior field to its asymptotic Kerr-Newman form extend inside the hole.
But here the convergence is non-uniform and breaks down entirely along the
ingoing sheet of the inner horizon, the so-called Cauchy horizon (CH), which
appears in the Penrose conformal map as an inward extension of future lightlike
infinity $\mathcal{I}^+$. (See Figure 1; also Section 2.7 of Dr Luminet's introductory lectures
in this volume.)

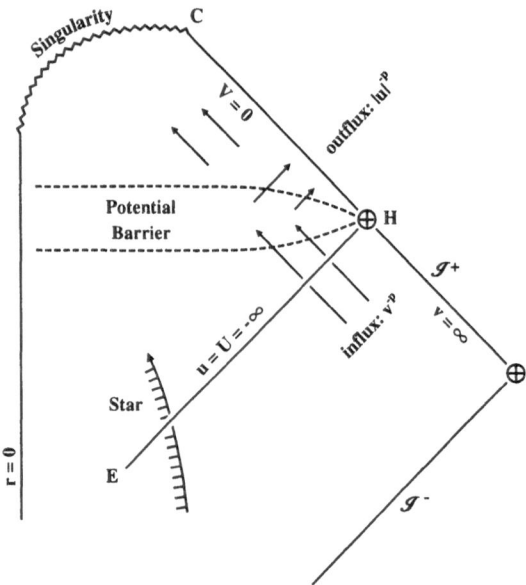

Fig. 1 History of spherical charged black hole, with infalling radiative tail. EH is the event horizon, CH the Cauchy horizon. The two angular coordinates are suppressed. H is an ideal point of this Penrose compactified map. In reality, CH and EH are 3cylinders $S^2 \times R^1$ of different radii which never intersect. See Figure 2 for a more faithful representation of the geometry.

The Cauchy horizon's instability and its source were first noticed by Penrose in 1968. Let us introduce the lightlike coordinate

$$V = -e^{-\kappa_0 v} \ . \tag{2}$$

This is a Kruskal advanced time, in terms of which the asymptotic Kerr-Newman metric is transformed into a manifestly regular form at its inner horizon. The ingoing sheet CH corresponds to $v = \infty$, $V = 0$. (The constant κ_0 is the surface gravity of this horizon. These matters are discussed in detail, e.g., in [5] and in Ray d'Inverno's introductory text [6].) An ingoing wave $\sim \sin \omega v$ appears speeded up near CH in terms of the locally regular coordinate V, and its flux of energy is received with a diverging blueshift $e^{2\kappa_0 v}$ by interior observers falling through CH. In the case of the wave-tail (1), its external power-law decay is overwhelmed by the exponential blueshift. The resulting divergent stress-energy has back-reaction effects which we shall learn about in the following sections; they are dramatic and surprising.

The blueshift instability of CH was confirmed in the decade 1973–82 by a number of investigations which studied the propagation of test fields on a *fixed* (usually Reissner–Nordström) black hole background. Near CH, amplitudes of wave tail perturbations typically decay like

$$\delta\Phi \sim (\ln|V|)^{-\frac{1}{2}p+1} \quad (V \to 0^-), \tag{3}$$

as one would anticipate from (1) and (2). Thus, field amplitudes Φ (in particular, the metric $g_{\mu\nu}$ for gravitational–wave perturbations) stay regular on CH to first order, but their derivatives $\partial_V\Phi$ diverge.

Since the Einstein equations are quadratic in first–order gradients $\nabla\Phi$, $\nabla g_{\mu\nu}$, it was widely expected that taking back-reaction into account would make the amplitudes themselves blow up on CH at higher orders of perturbation theory. However, up to now, detailed calculations have not borne out these expectations (though a final verdict on this question is still open). This is basically because the divergences are lightlike: roughly speaking, terms quadratic in $\partial_V\Phi$ are suppressed because the associated metric coefficient g^{VV} is zero. It thus appears that the Cauchy horizon of a black hole formed in a generic (nonspherical) collapse is indeed singular, but this singularity may have an unexpectedly mild and orderly structure.

3 Spherical models

In its most general form the problem confronting us is formidable indeed, and for a first reconnaissance it is advisable to make whatever simplifications we can, while (hopefully) not losing the essence of the physics.

Setting the angular momentum equal to zero while retaining a nonzero charge makes the black hole spherical without radically changing its horizon structure. Spherical scalar waves are a reasonable simulation of quadrupole gravitational waves insofar as their gravitational back-reaction is concerned. The large blueshift near the Cauchy horizon suggests further use of an optical approximation, which replaces the infalling waves by a stream of radially moving lightlike particles.

The earliest models [5, 7, 8] accordingly considered the effect of radial streams of "photons" on the interior geometry of a spherical hole with a (fixed) charge e.

The formulation of this spherically symmetric problem is elementary. No single chart covers the domain of interest to us, and it is therefore convenient to cast the equations in two-dimensionally covariant form, with the usual Schwarzschild coordinates r, t replaced by an arbitrary pair $x^A (A, B, \ldots = 0, 1)$ which label the 2-spheres. The 4-metric is then

$$ds^2 = g_{AB}dx^A dx^B + r^2 d\Omega^2, \tag{4}$$

with all coefficients functions of x^0, x^1 only.

The Schwarzschild mass function $M(x^A)$ is defined in terms of $\nabla r(x^A)$ by

$$g^{AB}(\partial_A r)(\partial_B r) = f \equiv 1 - 2M/r. \tag{5}$$

(In standard Schwarzschild coordinates this reduces to the familiar form $g^{rr} = 1 - 2M/r$.)

In the presence of a point charge e, the electric field energy excluded from a sphere of radius r is (in Gaussian units) $\frac{1}{2}e^2/r$. It proves convenient to separate off the (uninteresting for our purposes) electrostatic terms, and to define a non-electric mass function $m(x^A)$ by

$$M = m - \frac{1}{2}\frac{e^2}{r} . \tag{6}$$

One can now recast the Einstein field equations $G_{\alpha\beta} = 8\pi T_{\alpha\beta}$ as two-dimensionally covariant equations for the scalar fields $r(x^A), m(x^A)$. One finds [5] for m,

$$\partial_A m = 4\pi r^2 \left(T_A{}^B - \delta_A{}^B T_D{}^D\right) \partial_B r , \tag{7}$$

where again the Maxwellian contributions have been separated from the stress-energy components. (In Schwarzschild coordinates, (7) takes a form familiar from elementary texts:

$$\partial m(r,t)/\partial r = 4\pi r^2(-T_t^t), \quad \partial m/\partial t = 4\pi r^2 T_t^r .)$$

The equation for r is

$$\nabla_A \nabla_B r = -4\pi r T_{AB} + \kappa g_{AB} . \tag{8}$$

It is straightforward to check from (7) and (8) that m satisfies the $(1+1)$-dimensional wave equation [5]

$$\Box m = -16\pi^2 r^3 T_{AB} T^{AB} + 8\pi r f P_\perp , \tag{9}$$

which elegantly brings out the nonlinearity normally hidden in the Einstein field equations.

All covariant derivatives refer to the 2-metric g_{AB}; I have defined the "local surface gravity" $\kappa = (m - e^2/r)r^{-2}$; and $P_\perp = T_\theta^\theta$ is the (non-electric) transverse pressure. In (8) and (9), to simplify the appearance, I have set $T_A{}^A = 0$, a condition which indeed holds for the matter sources of interest to us (massless scalar fields, cross-flowing radial streams of photons).

For the $(1+1)$-dimensional wave operator the causal Green's function is simply a product of retarded and advanced lightlike step functions. Hence the solution of $\Box m = \sigma$ is

$$m(A) = -\frac{1}{2}\int_{\diamondsuit} \sigma dS + m(C) + m(D) - m(B), \tag{10}$$

where $dS = (-^2g)^{1/2}d^2x$ is the invariant element of area and the integration is over a diamond–shaped region $ACBD$ with lightlike sides. By aligning side CA

with CH, a considerable amount can be gleaned from (9) and (10) about the behavior of the mass function near the Cauchy horizon.

For an influx of radiation blueshifted near CH according to

$$T_{VV} \sim (\partial_V \Phi)^2 \sim (\ln |V|)^{-p} V^{-2} \quad (V \to 0^-), \tag{11}$$

(cf. 3) the right–hand side of (9) acts as a diverging source for m, provided some outflow is present too. (For a pure inflow [7], T_{AB} is light-like, $T_{AB} T^{AB} = 0$ and m undergoes little change near CH.) One finds [5] that the mass function diverges at CH roughly like

$$m \sim (\ln |V|)^{-p} |V|^{-1}, \tag{12}$$

an effect sometimes called "mass inflation". More specifically [9], if the outflow is caused by backscatter of the inflow,

$$m \sim |uv|^{-p} e^{\kappa_0 (u+v)} \quad (v \to \infty, u \to -\infty), \tag{13}$$

where u is interior retarded time, calibrated to decrease toward $-\infty$ at the event horizon and toward the rear of the (past-endless) 3-cylinder CH (see Figure 1).

Since the neighborhood of CH inside the hole cannot causally influence the outside, the drastic change (12) in the internal geometry remains undetected outside the hole. Outside observers continue to register a mass close to that of the progenitor star.

The straightforward interpretation of (12) is that the Cauchy horizon increasingly appears to observers approaching it as a sphere of infinite mass, a formidable singularity indeed! However, it is spread over the surface of the inner horizon, and hence is pancake-like and locally weak in a sense which Amos Ori [8] has made precise. The Weyl curvature and tidal forces ($\sim m/r^3$) do become infinite, but the cumulative tidal *deformation* of bodies falling toward CH remains bounded, and indeed small, up to the very moment of crossing. Stated in a more coordinate-dependent way, the curvature and affine connection blow up at CH but the metric (in suitable coordinates) does not.

(Parenthetically, it is entertaining and instructive to note that the gravitational effects near CH can be simulated rather well by a simple (though exotic) Newtonian model [10]. Imagine a "planet" whose radius r_0 is the same as CH, whose mass is infinite and whose external gravity is screened by an "atmosphere" of equal and opposite mass. Approaching the planetary surface the negative atmospheric density $\rho(r)$ is assumed to grow like

$$\rho(r) \sim -(\ln \Delta r)^{-p} (\Delta r)^{-2}$$

(compare (11)) as $\Delta r \equiv (r - r_0) \to 0^+$. It is then easy to check, for example, that the exterior Newtonian potential remains bounded for all $r \geq r_0$, and that falling meteorites and spacecraft are not tidally disrupted before they actually reach the surface.)

Not only are cumulative tidal effects weak near CH; the intrinsic 3-geometry of this lightlike hypersurface is also only weakly affected. For a spherical horizon, the 3-geometry is completely defined by the evolution of its area, and this is determined by the transverse (i.e. out-) flux focussing its generators. Since the outflux is not blueshifted, CH contracts only slowly, finally tapering to a strong spacelike singularity when its area becomes zero (Figure 1).

4   Mass inflation: a simple mechanical model

The phenomena near CH are not familiar from everyday laboratory practice, and it is useful to have at hand the simplest example that conveys the essence of what is happening.

Our example will schematically model the infalling and outgoing radiation near CH as a concentric pair of thin lightlike shells. But first a general remark on the dynamics of a (timelike) spherical shell.

Consider a shell of radius $r = R(\tau)$, a function of proper time τ, moving in the field of an internal distribution of gravitational (i.e. Schwarzschild) mass m_-. Both m_- and the exterior Schwarzschild mass m_+ are constant. The shell's proper mass $M(\tau)$ satisfies

$$dM + Pd(4\pi R^2) = 0 .$$

(It is conserved if the surface pressure $P = 0$.) The relation

$$m_+ - m_- = M \left(1 - \frac{2m_-}{R} + \dot{R}^2\right)^{1/2} - \frac{1}{2}\frac{M^2}{R} , \quad \dot{R} \equiv \frac{dR}{d\tau} ,$$

(whose derivation is sketched in Chap. 17, Section 8), expresses the total conserved, gravitating mass $m_+ - m_-$ of the shell as a sum of four terms (when the square root is expanded to first order): the rest-mass M, the kinetic energy $\frac{1}{2}M\dot{R}^2$, the mutual potential energy $-Mm_-/R$ and a self–potential energy $-\frac{1}{2}M^2/R$.

The one essential point to note is that (at least in this way of formulating the dynamics) the "potential energy" (in general a mutual, unlocalized property shared by a pair of bodies) here contributes to the gravitating mass of the *outer* body: it is a binding energy.

There are circumstances in which the outer body can be released from its gravitational binding, with a consequent increase of its gravitating mass. This provides an intuitive understanding of mass inflation. Consider a pair of massive thin spherical shells, one moving inwards, the other outwards, in a Reissner–Nordström!spacetime with central charge e. Let us assume that the shells are transparent and simply pass through each other when they cross at radius r_0. At this moment, their mutual potential energy, of order $-m_{in}m_{out}/r_0$, is suddenly transferred from the contracting to the expanding shell, resulting in a redistribution of their masses and an increase of the ambient Schwarzschild mass in the space between them.

This is illustrated most simply and graphically if both shells are made of light. Their histories divide the spacetime into four sectors $A, \ldots, D$ according to the scheme

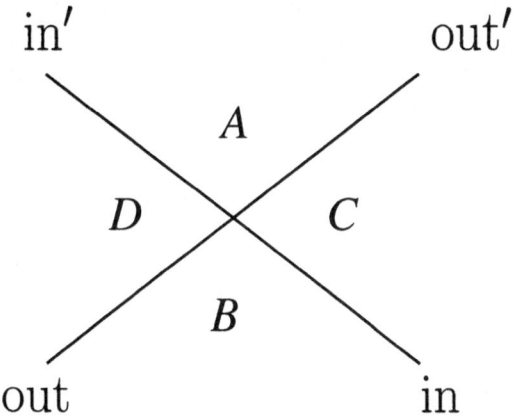

The metric in each of these sectors has the Reissner–Nordström form

$$ds^2 = \frac{dr^2}{f(r)} + r^2 d\Omega^2 - f(r)dt^2,$$
$$f(r) = 1 - 2m/r + e^2/r^2, \tag{14}$$

but with different masses $m_A, \ldots, m_D$. Conservation of energy at the moment of crossing is expressed by

$$f_A(r_0)f_B(r_0) = f_C(r_0)f_D(r_0), \tag{15}$$

first noted in 1985 by Dray and 't Hooft and by Redmount, and called the DTR relation. It can be considerably generalized [11]. (A simple derivation of (15) is appended as an exercise to this Chapter.)

Equation (15) can be re-expressed in a number of equivalent forms; for example,

$$\frac{m_A - m_D}{m_C - m_B} = \frac{f_D}{f_B}, \quad \frac{m_C - m_A}{m_B - m_D} = \frac{f_C}{f_B},$$

which translates to

$$m'_{\text{in}} = (f_D/f_B)m_{\text{in}}, \quad m'_{\text{out}} = (f_C/f_B)m_{\text{out}}. \tag{16}$$

Conservation of the total energy follows directly from (16) and (14):

$$m'_{\text{in}} + m'_{\text{out}} = m_{\text{in}} + m_{\text{out}}.$$

As already indicated, the shells may be considered to represent schematically the infalling and outgoing fluxes inside a spherical charged hole. If crossing occurs just outside the Cauchy horizon of sector B, so that $f_B(r_0)$ is negative and numerically very small, $f_D(r_0)$ negative and $f_C(r_0)$ positive (since $m_C = m_B - m_{\text{in}}$ is significantly smaller than m_B), it follows from (16) that m'_{in} and m_A are very much larger than m_{in} and m_B. (The new mass m'_{out} is then correspondingly negative, signifying that the "outgoing" shell is now burdened with so much binding energy that it can no longer reach infinity, i.e., it is trapped inside the hole and is now actually contracting.)

This represents a wholesale conversion of gravitational energy into material (kinetic) energy of infall. The inner horizon of a black hole (like a closed universe in cosmological inflation) is a bottomless well of gravitational energy.

5   Is the spherical picture generic?

The spherical analysis described in Sec. 3 has led us to the tentative picture of a black hole's interior summarized in Figure 1: the final strong, spacelike singularity has a weaker lightlike precursor extending backwards along the Cauchy horizon and characterized by mass inflation.

It is now time to take leave of toy models and confront the question of how far any spherical model can be trusted to provide a representative picture of the real conditions inside a generic rotating black hole. From the beginning this question aroused understandable skepticism, and the issue remains unsettled. But some of the early doubts have been clarified and laid to rest.

Even in the spherical case, an early numerical study of scalar wave tails absorbed by a charged hole suggested that CH is destroyed and that the $r = 0$ spacelike singularity is all-enveloping. But analytical work [9] and more refined numerics [12] have not confirmed this suggestion.

More serious questions arise when the restriction of spherical symmetry is given up. It has been widely felt that lightlike singularities cannot be generic. The argument runs something like this. We already know a class of singularities which are functionally generic in the sense that the solutions depend on 8 physically arbitrary functions of 3 variables (6 components of intrinsic metric and 6 components of extrinsic curvature for an initial spatial hypersurface, less 4 coordinate degrees of freedom. Imposing definite (e.g. vacuum) field equations would subject these 8 functions to 4 initial-value constraints.) These are the BKL (Belinskii–Khalatnikov–Lifshitz) or "mixmaster" chaotic oscillatory singularities [13]. Now, the BKL singularities are *spacelike*, and — presumably — they exhaust the set of generic singularities. If that is so, then lightlike singularities must be less generic, and could not develop from generic initial data. A generic perturbation should drive any such lightlike singularity into a spacelike one.

There is a flaw in this argument, and it was pinpointed by Ori and Flanagan [14]. They showed that it was actually possible to construct a generic class of *weak* lightlike singularities. And one should indeed expect intuitively that the Einstein equations (which are quasi-linear) should propagate generic weak discontinuities and mild singularities along characteristic (i.e., lightlike) hypersurfaces.

It is thus not possible to rule out the presence of weak lightlike singularities inside black holes *merely* on the grounds that they are non-generic. Of course, this should not be construed as a proof that such singularities actually do occur in generic black holes!

6   Covariant double-null dynamics

To enter into technical details concerning the latest non-spherical analyses [15] of black hole interiors would be beyond the scope of these introductory lectures. But I shall outline the general formalism [16] employed in these studies, because it is a versatile technique, useful in a variety of situations where the physics singles out particular lightlike surfaces.

This is a $(2+2)$-embedding formalism, adapted to a double foliation of space-time by a net of two intersecting families of lightlike hypersurfaces. It yields a simple and geometrically transparent decomposition of the Einstein field equations.

A number of such formalism (referenced in [16]) have been developed since 1980. The version I shall present here has the feature that it is two-dimensionally covariant and thus very compact.

To accustom ourselves to the notation, let us begin with a résumé of the basics of the familiar ADM $(3 + 1)$-dimensional decomposition. This involves a foliation by spacelike hypersurfaces $t = $ const., with parametric equations

$$x^\alpha = x^\alpha(\xi^a, t) \quad (\alpha, \beta \ldots = 0, \ldots, 3; \ a, b \ldots = 1, 2, 3) \ .$$

The basis vectors $e_{(a)}$ tangent to a hypersurface and associated with the intrinsic coordinates ξ^a have 4-dimensional components $e_{(a)}^\alpha = \partial x^\alpha / \partial \xi^a$.

The basic geometrical entities associated with a hypersurface are its intrinsic metric and extrinsic curvature, defined by

$$g_{ab} = e_{(a)} \cdot e_{(b)} \equiv g_{\alpha\beta} e_{(a)}^\alpha e_{(b)}^\beta, \quad K_{ab} = (\nabla_\alpha n_\beta) e_{(a)}^\alpha e_{(b)}^\beta \ ,$$

where n^α is the unit timelike normal: $n \cdot n = -1$.

The "shift" s^a, defined by

$$s^a = e_\alpha^{(a)} \partial x^\alpha / \partial t = -n^\alpha \partial \xi^a / \partial x^\alpha, \quad e^{(a)} \equiv g^{ab} e_{(b)} \ ,$$

is a measure of how much one has to veer off the normal in order to join two points on neighboring hypersurfaces having the same intrinsic coordinates ξ^a, according to the schemes illustrated in the sketch.

(Traditionally, s^a is denoted N^a.)

An arbitrary 4-dimensional displacement dx^α can be decomposed as

$$dx^\alpha = e_{(a)}^\alpha (d\xi^a + s^a dt) + N n^\alpha dt \ , \tag{17}$$

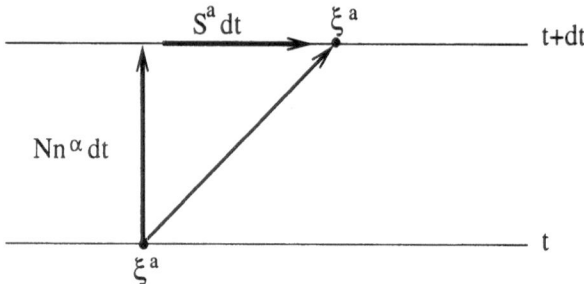

where the "lapse" $N = -n_\alpha \partial x^\alpha / \partial t$. From (17) follows at once the standard ADM decomposition of the 4-metric:

$$ds^2 = g_{\alpha\beta} dx^\alpha dx^\beta = g_{ab}(d\xi^a + s^a dt)(d\xi^b + s^b dt) - N^2 dt^2 .$$

However, most of this becomes useless if a hypersurface $t = $ const. is lightlike. The intrinsic metric g_{ab} is now degenerate (so there is no inverse g^{ab}), and K_{ab} provides no extrinsic information because n^α is now tangent to the hypersurface.

For a non-degenerate treatment of the lightlike case, one is forced to fall back on a $(2 + 2)$-decomposition. We shall suppose that we are given a double foliation of spacetime by a net of two intersecting families of lightlike hypersurfaces Σ^0 ($u^0 = $ const.) and Σ^1 ($u^1 = $ const.). The condition that the gradients ∇u^0, ∇u^1 are non-parallel future lightlike vectors is expressed by

$$\nabla u^A \cdot \nabla u^B = e^{-\lambda} \eta^{AB} \quad (A, B \ldots = 0, 1)$$

for some scalar function $\lambda(x^\alpha)$, where the matrix

$$\eta^{AB} = \begin{pmatrix} 0 & -1 \\ -1 & 0 \end{pmatrix} = \eta_{AB}$$

is used to raise and lower upper-case indices. The normal generator $l_\alpha^{(A)}$ of Σ^A is parallel to ∇u^A, and it is convenient to normalize it as $l^{(A)} = e^\lambda \nabla u^A$.

Each pair of hypersurfaces Σ^0 and Σ' intersect in a 2-surface S. We assign intrinsic coordinates θ^a ($a, b \ldots = 2, 3$) to these surfaces such that their parametric equations are smooth functions

$$x^\alpha = x^\alpha(u^A, \theta^a) . \tag{18}$$

The basis vectors $e_{(a)}$ tangent to S and its intrinsic 2-metric are then defined by

$$e_{(a)}^\alpha = \partial x^\alpha / \partial \theta^a, \quad g_{ab} = e_{(a)} \cdot e_{(b)} .$$

Associated with the two normals $l_{(A)}$ to S there are two shift vectors

$$s_A^a = e_\alpha^{(a)} \partial x^\alpha / \partial u^A = -l_{(A)}^\alpha \partial \theta^a / \partial x^\alpha .$$

An arbitrary small displacement dx^α in spacetime can now be resolved into its normal and tangential parts:

$$dx^\alpha = l^\alpha_{(A)}du^A + e^\alpha_{(a)}(d\theta^a + s^a_A du^A) \, .$$

It follows that the spacetime metric is decomposable as

$$ds^2 = g_{\alpha\beta}dx^\alpha dx^\beta = e^\lambda \eta_{AB}du^A du^B + g_{ab}(d\theta^a + s^a_A du^A)(d\theta^b + s^b_B du^B) \, .$$

The two normals also define two extrinsic curvatures

$$K_{Aab} = \left(\nabla_\beta l_{(A)\alpha}\right) e^\alpha_{(a)} e^\beta_{(b)} \, ,$$

which are easily shown to be symmetric in a, b. (A certain scale-arbitrariness is inherent in this definition, since we are free to rescale the null vectors $l_{(A)}$.)

The Lie bracket of $l_{(0)}$ and $l_{(1)}$ contains further geometrical information about the double foliation. One finds

$$[l_{(B)}, l_{(A)}] = \epsilon_{AB}\omega^a e_{(a)} \, , \tag{19}$$

where

$$\omega^a = \epsilon^{AB}(\partial_B s^a_A - s^b_B s^a_{A;b}) \, .$$

The semi-colon indicates 2-dimensional covariant differentiation associated with the metric g_{ab}, and ϵ_{AB} is the 2-dimensional permutation symbol.

From (19) we can read off the geometrical significance of the "twist" ω^a. The curves tangent to the generators $l_{(0)}, l_{(1)}$ mesh together to form 2-surfaces (orthogonal to the surface S) if and only if $\omega^a = 0$. If this were true, it would be consistent to allow the coordinates θ^a to be dragged along both sets of generators, and thus to gauge both shift vectors to zero.

The 2-dimensionally invariant operator associated with differentiation along the normal direction $l_{(A)}$ is denoted by D_A. Acting on any 2–dimensional geometrical object $X^a_{b...}$, D_A acts as follows:

$$D_A X^a_{b...} = \left(\partial_A - \mathcal{L}_{s^d_A}\right) X^a_{b...} \, .$$

Here, ∂_A is the partial derivative with respect to u^A and $\mathcal{L}_{s^d_A}$ the 2-dimensional Lie derivative with respect to the 2–vector s^d_A. For example,

$$D_A g_{ab} = \partial_A g_{ab} - 2s_{A(a;b)} = 2K_{Aab} \, .$$

Geometrically, $D_A X^a_{b...}$ is the projection onto S of the 4-dimensional Lie derivative with respect to $l^\alpha_{(A)}$ of the equivalent tangential 4-tensor

$$X^\alpha_{\beta...} \equiv X^a_{b...} e^\alpha_{(a)} e^{(b)}_\beta \, \cdots \, .$$

The objects K_{Aab}, ω^a and D_A are all simple projections onto S of 4-dimensional geometrical objects. Consequently they transform very simply under 2-dimensional coordinate transformations. Under the transformation

$$\theta^a \to \theta^{a'} = f^a(\theta^b, u^A) \tag{20}$$

(which leaves u^A and hence the surfaces Σ^A and S unchanged), g_{ab}, ω_a and K_{Aab} transform cogrediently with

$$e_{(a)} \to e'_{(a)} = e_{(b)} \partial \theta^b / \partial \theta^{a'} .$$

On the other hand, the shift vectors s^a_A will undergo a more complicated gauge-like transformation, arising from the u-dependence in (20).

(To avoid any possible misunderstanding, I should perhaps stress that this formalism is covariant under two completely independent groups of transformations. In the parametric equations (18), the 4-dimensional coordinates x^α and the 2-dimensional coordinates θ^a can be transformed independently of each other. In this sense, g_{ab} is at the same time a "2-tensor" and a "4-scalar".)

We now already have sufficient geometrical groundwork to display the simple form the Ricci components take in the double-null formalism (tetrad components are denoted by, e.g. $R_{aA} = R_{\alpha\beta} e^\alpha_{(a)} l^\beta_{(A)}$):

$$^{(4)}R_{ab} = \frac{1}{2} {}^{(2)}R g_{ab} - e^{-\lambda}(D_A + K_A)K^A_{ab}$$

$$+ 2e^{-\lambda} K^d_{A(a} K^A_{b)d} - \frac{1}{2} e^{-2\lambda} \omega_a \omega_b - \lambda_{;ab} - \frac{1}{2}\lambda_{,a}\lambda_{,b}$$

$$R_{AB} = -D_{(A} K_{B)} - K_{Aab}K^{ab}_B + K_{(A} D_{B)}\lambda$$

$$- \frac{1}{2}\eta_{AB}\left[(D^E + K^E)D_E\lambda - e^{-\lambda}\omega^a\omega_a + (e^\lambda)^{;a}_{\ a}\right]$$

$$R_{Aa} = K^b_{Aa;b} - \partial_a K_A - \frac{1}{2}\partial_a D_A \lambda + \frac{1}{2}K_A\partial_a\lambda$$

$$+ \frac{1}{2}\epsilon_{AB}e^{-\lambda}\left[(D^B + K^B)\omega_a - \omega_a D^B\lambda\right] ,$$

where $^{(2)}R$ is the curvature scalar associated with the 2-metric g_{ab}, and $K_A \equiv K^a_{Aa}$.

The economy and geometrical transparency of these formulae are self-evident. In particular, the shift vectors, which are largely an artifact of the choice of coordinates θ^a, make no explicit appearance.

Double-null formalism are of particular value in contexts where lightlike hypersurfaces are prominent: the characteristic initial-value problem (analytical and numerical), the dynamics of horizons, gravitational radiation, Planck-energy collisions and light-cone quantization.

7 Interior of a generic rotating black hole: general remarks

The task of piecing together a comprehensive picture of the generic black hole interior poses, even in its classical phases, a formidable challenge on which work is still in progress [3, 4, 16, 17, 18]. All I can offer here are some brief general comments.

One expects the essential physics to be captured by the following prototypical example. Given an initially stationary (i.e., Kerr) black hole, introduce near the event horizon $r = r_+$ an initially small vacuum perturbation $\delta g_{\alpha\beta}$ representing an ingoing packet of gravitational waves decaying with advanced time v according to Price's power law:

$$\delta g \sim h(\theta, \varphi_+) v^{-\frac{1}{2}p+1} \quad (r \approx r_+, \; v \to \infty) . \tag{21}$$

The challenge is to trace the classical evolution of this perturbation — including all nonlinear (i.e, back-reaction) effects — forward in time, and downward into the hole, all the way to the Cauchy horizon CH, using Einstein's vacuum equations.

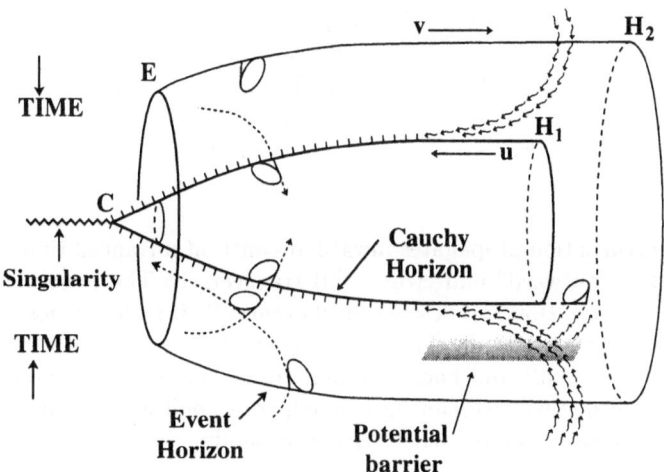

Fig. 2 View of black hole interior (not compactified, one angular variable suppressed). The figure shows future light-cones, and a stream of initially infalling radiation, partially scattered off the potential barrier between the event and Cauchy horizons, with the unscattered portion accumulating along the Cauchy horizon. This representation reveals more clearly than Figure 1 that most of the scattering takes place in a layer well separated from the belt of large blueshift near CH.

In (21), it is important to note that the initial amplitude h is regular only when expressed in terms of the Eddington-Kerr advanced angular coordinate φ_+ [19], not the Boyer-Lindquist coordinate φ — the two coordinates are related by an r-dependent winding translation which becomes singular at horizons. The advanced angle φ_+ is constant along the ingoing principal null rays of the Kerr geometry and along generators of the event and Cauchy horizons.

On a fixed Kerr background, a test wave packet sent in at a late advanced time v_1 will propagate inwards with constant angular momentum but angular velocity increasing because of frame-dragging. After scattering from the potential barrier between the horizons (see Figure 1), its transmitted part will edge up to CH as highly blueshifted "gravitons" moving parallel to its generators (since that is the only non-spacelike direction tangent to a lightlike surface). The flux is blueshifted by a factor $e^{2\kappa_0 v_1}$, where κ_0, the surface gravity of the Kerr inner horizon, is a *constant* (independent of latitude), even though CH is nonspherical.

The back-reaction is to be determined by formulating a characteristic initial-value problem. Initial data are set on an intersecting pair of lightlike hypersurfaces Σ^u (u=const.) and Σ^v (v=const.), with Σ^v pointing into the hole (Figure 3). Data on Σ^v can be trivially fixed by placing it wholly in the initially stationary (pure Kerr) sector of the geometry.

The correct formulation has Σ^u just above or coincident with the event horizon EH (Figure 3), and initial data on it specified as the Kerr geometry perturbed by an infalling wave tail like (21).

These initial data now evolve in accordance with the vacuum field equations. A broadbrush picture of the evolution can be gleaned by dividing it schematically into four stages, (a)–(d), as indicated in Figure 3. We follow a small-amplitude wave packet which originates on a segment of the event horizon at some late advanced time $v = v_1$:

(a) The perturbation propagates inward at constant advanced time. Part of it is scattered "outward" and crosses CH transversely. The remainder continues in free propagation and descent until eventually it sidles in close to CH. This phase, (a), ends when the wave has penetrated so deep into the environs of CH that blueshift and back–reaction are becoming appreciable. Up to this point, the disturbance can be treated, to a good approximation, as a test field propagating on a fixed Kerr background.

(b) With further descent at constant advanced time and deeper immersion into this high-blueshift layer, there is a transition from incipient to extreme blueshift, and to a third regime, (c), of fully developed back-reaction.

(c) This appears to be the most extended part of the evolution. Experience with the spherical and planar [20] cases, as well as the general analyses [16, 17, 18], suggest that this strong blueshift regime is relatively stable, changing with retarded time on the timescale associated with the slow shrinkage and deformation of CH due to the (weak and unblueshifted) transverse flux arising from backscatter.

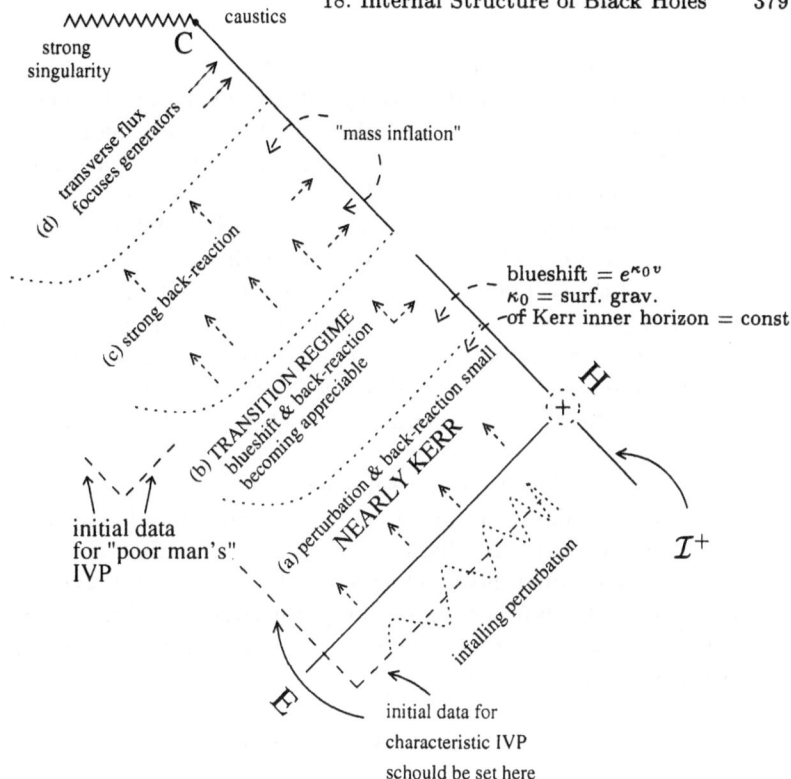

Fig. 3 From stages, (a)-(d), in the evolution of a wave-packet entering a black hole at the event horizon EH. This is a Penrose conformal map similar to Figure 1.

On CH itself — if one may loosely speak of this (mild) singularity as part of the manifold — regime (c) extends backwards, though steadily weakening, for the entire (endless) past history, as Figure 3 shows more clearly.

(d) Ultimately the generators of CH, focused by the transverse flux, form caustics and the Cauchy horizon breaks up, terminating in a strong singularity, presumably spacelike and (classically) of BKL mixmaster type.

Phases (b) and (d) are the most difficult to analyze and least understood. But even (a) presents problems, because the wave equation on a Kerr background is separable only for a special (harmonic or exponential) time-dependence, not well suited for power-law tails.

The detailed studies to date [16, 17, 18, 19] have been largely restricted to a simplified formulation which allows one to bypass the transition regime (b). In this "poor man's version", the initial surface Σ^u, whose proper placing is along or above EH, is taken to lie inside the hole (Figure 3). It now intersects CH, and its future end is therefore immersed in a layer of strong blueshift and back-reaction where initial data are not known a priori.

One must then resort to the strategy of making a plausible guess about the initial conditions in this layer, and then check that this guess is at least self-consistent by showing that the general character of the assumed conditions near CH is preserved by the evolution. In this way, it has been possible to put together a fairly plausible description of phase (c) as a self-sustaining, mildly singular, large-blueshift regime. But the evidence that this is what *must* emerge from phases (a) and (b) still falls well short of being compelling.

The most complete studies along these lines — by Brady and Chambers [17] and Brady, Droz and Morsink [16], building in part on earlier work by Ori [18] — employ the double-null formalism described in the previous section (or an earlier variant thereof due to Sean Hayward) to analyze the characteristic initial-value problem.

These investigations lead to a self-consistent description of phase (c) which is qualitatively similar to the spherical (Sec.3) and planar [20] pictures: a locally weak singularity along CH, characterized by mass inflation now overlaid by an imploding gravitational shock. The metric of two-dimensional sections of CH is not affected by blueshift and varies slowly; the transverse 2-metric orthogonal to these sections is exponentially compressed, i.e. the exponent λ in (21) decreases linearly with advanced time:

$$\lambda = -\kappa_0 v + \text{(less divergent terms)} \quad (v \to \infty) \ .$$

It is a consequence of the Einstein equations and the assumed initial regularity of the intrinsic geometry of CH in phase (c) that κ_0 must be a *constant* — a condition which already appears as a constraint that one has to impose on the initial data along Σ^u in the "poor man's" formulation. The constancy of the unperturbed inner horizon's surface gravity provides a hint that this condition does indeed emerge from phase (a). But the actual outcome will depend on the unexplored question of how the geometry evolves through the transition phase (b).

8   Summary and conclusion

Analytical and numerical studies have provided us with a fairly complete description of the effect of wave tails on the interior of a spherical (charged) black hole. In the nonspherical (rotating) case, partial analyses and general arguments suggest a qualitatively similar picture, but the evidence for this is still fragmentary.

Nonspinning, uncharged collapse produces an all-enveloping, crushing, spacelike singularity inside the hole. In the presence of spin or charge this picture appears to be modified and softened. Only a finite segment of the final spacelike singularity survives; it is joined to an infinitely long and milder lightlike precursor.

Knowledge of the classical geometry at late advanced times near the Cauchy horizon is a launchpad which provides initial conditions for the subsequent quantum phase of evolution. Preliminary explorations of this quantum phase, using spherical models [10, 21], lead to indeterminacies which will not be resolved until we have a manageable quantum theory of gravity. We have only charted a coastline; exploration of the hinterland is a task for the next century.

Acknowledgments
I should like to thank Warren Anderson, Roberto Balbinot, Claude Barrabès, Alfio Bonanno, Patrick Brady, Serge Droz, Sharon Morsink and Eric Poisson, whose enthusiasm and insights have sustained my interest and deepened my understanding. Also Friedrich Hehl, Claus Kiefer and Ralph Metzler for the hospitable and stimulating atmosphere they provided at Bad Honnef.

The work was supported by NSERC of Canada and the Canadian Institute for Advanced Research.

Bibliography

[1] I. D. Novikov, "Physics of black holes", in *Proceedings of Symposium Nonlinear Phenomena in Accretion Disks around Black Holes* (Island, June 1997), preprint Theoretical Astrophysics Centre, Copenhagen-Aarhus ref. 1997-036.

[2] S. Droz, W. Israel and S. M. Morsink, Physics World **9**, 34 (1996).

[3] V. P. Frolov and I. D. Novikov, *Black Hole Physics*, Kluwer, Boston 1998, Chap.14;

A. Ori, J. Gen. Rel. Grav. **29**, 881 (1997);

W. Israel, in *Black Holes and Relativistic Stars (Proc. Symp. in honor of S. Chandrasekhar)* (ed. R. Wald), Univ. of Chicago Press 1998, Chap.7.

[4] L. M. Burko and A. Ori (eds.), *Internal Structure of Black Holes and Spacetime Singularities* (IOP Publishing 1998).

[5] E. Poisson and W. Israel, Phys. Rev. **D41**, 1796 (1990).

[6] R. d'Inverno, Introducing Einstein's Relativity, Oxford U.P. 1992;

(German transl.: R. d'Inverno, Einführung in die Relativitätstheorie, VCH, Weinheim 1994.)

[7] W. A. Hiscock, Physics Letters **83A**, 110 (1981).

[8] A. Ori, Phys. Rev. Letters **67**, 789 (1991).

[9] A. Bonanno, S. Droz, W. Israel and S. M. Morsink, Phys. Rev. **D50**, 755 (1994); Proc. Roy. Soc. **A450**, 553 (1995); L. M. Burko and A. Ori, gr-qc/9711032.

[10] R. Balbinot, P. R. Brady, W. Israel and E. Poisson, Physics Letters **A161**, 223 (1991).

[11] C. Barrabès and W. Israel, Phys. Rev. **D43**, 1129 (1991).

[12] P. R. Brady and J. D. Smith, Phys. Rev. Letters **75**, 1256 (1995); L. M. Burko, Phys. Rev. Letters **79**, 4958 (1997).

[13] C. W. Misner, Phys. Rev. Letters **22**, 1071 (1969); V. A. Belinski, E. M. Lifshitz and I. M. Khalatnikov, Advances in Physics **19**, 525 (1970).

[14] A. Ori and E. M. Flanagan, Phys. Rev. **D53**, R1754 (1996).

[15] P. R. Brady, S. Droz and S. M. Morsink, "The late-time singularity inside non-spherical black holes", gr-qc/9805008.

[16] P. R. Brady, S. Droz, W. Israel and S. M. Morsink, Class. Quantum Grav. **13**, 221 (1996).

[17] P. R. Brady and C. M. Chambers, Phys. Rev. **D51**, 4177 (1995).

[18] A. Ori, Phys. Rev. Letters **68**, 2117 (1992).

[19] W. Krivan, P. Laguna and P. Papadopoulos, Phys. Rev **D54**, 4728 (1996); Phys. Rev. **D56**, 3395 (1997).

[20] A. Bonanno, S. Droz, W. Israel and S. M. Morsink, Can. J. Phys. **72**, 755 (1995) [erratum: **73**, 251 (1996)].

[21] R. Balbinot and E. Poisson, Phys. Rev. Letters **70**, 13 (1993);
W. G. Anderson, P. R. Brady, W. Israel and
S. M. Morsink, Phys. Rev. Letters **70**, 1041 (1993).

Part VI

Quantum Theory

Quantum Fields near Black Holes

Andreas Wipf

Theoretisch-Physikalisches Institut, Friedrich-Schiller-Universität Jena,
Max-Wien-Platz 1, D-07743 Jena, Germany

1 Introduction

In the theory of quantum fields on curved space-times one considers gravity
as a classical background and investigates quantum fields propagating on this
background. The structure of spacetime is described by a manifold $\mathcal{M}$ with
metric $g_{\mu\nu}$. Because of the large difference between the Planck scale (10^{-33}cm)
and scales relevant for the present standard model ($\geq 10^{-17}$cm) the range of
validity of this approximation should include a wide variety of interesting phenomena, such as particle creation near a black hole with Schwarzschild radius
much greater than the Planck length.

The difficulties in the transition from flat to curved spacetime lie in the absence of the notion of global inertial observers or of Poincaré transformations
which underlie the concept of particles in Minkowski spacetime. In flat spacetime,
Poincaré symmetry is used to pick out a preferred irreducible representation of
the canonical commutation relations. This is achieved by selecting an invariant vacuum state and hence a particle notion. In a general curved spacetime
there does not appear to be any preferred concept of particles. If one accepts
that quantum field theory on general curved spacetime is a quantum theory of
fields, not particles, then the existence of global inertial observers is irrelevant
for the formulation of the theory. For linear fields a satisfactory theory can be
constructed. Recently Brunelli and Fredenhagen [1] extended the Epstein-Glaser
scheme to curved space-times (generalising an earlier attempt by Bunch [2]) and
proved perturbative renormalizability of $\lambda\phi^4$.

The framework and structure of Quantum field theory in curved space-times
emerged from Parker's analysis of particle creation in the very early universe [3].
The theory received enormous impetus from Hawking's discovery that black holes
radiate as black bodies due to particle creation [4]. A comprehensive summary
of the work can be found in the books [5].

2 Quantum Fields in Curved Spacetime

In a general spacetime no analogue of a 'positive frequency subspace' is available and as a consequence the states of the quantum field will not possess a
physically meaningful particle interpretation. In addition, there are spacetimes,
e.g. those with time-like singularities, in which solutions of the wave equation
cannot be characterised by their initial values. The conditions of *global hyper-
bolicity* of $(\mathcal{M}, g_{\mu\nu})$ excludes such 'pathological' spacetimes and ensures that

the field equations have a well posed initial value formulation. Let $\Sigma \subset \mathcal{M}$ be a hypersurface whose points cannot be joined by time-like curves. We define the *domain of dependence* of Σ by

$$D(\Sigma) = \{p \in \mathcal{M} | \text{every inextendible causal curve through } p \text{ intersects } \Sigma\}.$$

If $D(\Sigma) = \mathcal{M}$, Σ is called a *Cauchy surface* for the spacetime and $\mathcal{M}$ is called *globally hyperbolic*. Globally hyperbolic spacetimes can be *foliated* by a one-parameter family of smooth Cauchy surfaces Σ_t, i.e. a smooth 'time coordinate' t can be chosen on $\mathcal{M}$ such that each surface of constant t is a Cauchy surface [6]. There is a *well posed initial value problem* for linear wave equations [7]. For example, given smooth initial data $\phi_0, \dot{\phi}_0$, then there exists a unique solution ϕ of the *Klein-Gordon equation*

$$\Box_g \phi + m^2 \phi = 0, \qquad \Box_g = \frac{1}{\sqrt{-g}} \partial_\mu (\sqrt{-g} g^{\mu\nu} \partial_\nu) \tag{1}$$

which is smooth on all of $\mathcal{M}$, such that on Σ we have $\phi = \phi_0$ and $n^\mu \nabla_\mu \phi = \dot{\phi}_0$, where n^μ is the unit future-directed normal to Σ. In addition, ϕ varies continuously with the initial data.

For the phase-space formulation we slice $\mathcal{M}$ by space-like Cauchy surfaces Σ_t and introduce unit normal vector fields n^μ to Σ_t. The spacetime metric $g_{\mu\nu}$ induces a spatial metric $h_{\mu\nu}$ on each Σ_t by the formula

$$g_{\mu\nu} = n_\mu n_\nu - h_{\mu\nu}.$$

Let t^μ be a 'time evolution' vector field on $\mathcal{M}$ satisfying $t^\mu \nabla_\mu t = 1$. We decompose it into its parts normal and tangential to Σ_t,

$$t^\mu = N n^\mu + N^\mu,$$

where we have defined the *lapse function* N and the *shift vector* N^μ tangential to the Σ_t. Now we introduce adapted coordinates $x^\mu = (t, x^i), i = 1, 2, 3$ with $t^\mu \nabla_\mu x^i = 0$, so that $t^\mu \nabla_\mu = \partial_t$ and $N^\mu \partial_\mu = N^i \partial_i$. The metric coefficients in this coordinate system are

$$g_{00} = g(\partial_t, \partial_t) = N^2 - N^i N_i \quad \text{and} \quad g_{0i} = g(\partial_t, \partial_i) = -N_i,$$

where $N_i = h_{ij} N^j$, so that

$$ds^2 = (N dt)^2 - h_{ij}(N^i dt + dx^i)(N^j dt + dx^j)$$
$$(\partial \phi)^2 = \frac{1}{N^2}(\partial_0 \phi - N^i \partial_i \phi)^2 - h^{ij} \partial_i \phi \partial_j \phi.$$

The determinant g of the 4-metric is related to the determinant h of the 3-metric as $g = -N^2 h$. Inserting these results into the Klein-Gordon action

$$S = \int L dt = \frac{1}{2} \int \eta \left(g^{\mu\nu} \partial_\mu \phi \partial_\nu \phi - m^2 \phi^2 \right), \qquad \eta = \sqrt{|g|} d^4 x,$$

one obtains for the momentum density, π, conjugate to the configuration variable ϕ on Σ_t

$$\pi = \frac{\partial L}{\partial \dot{\phi}} = \frac{\sqrt{h}}{N}\big(\dot{\phi} - N^i\partial_i\phi\big) = \sqrt{h}\big(n^\mu\partial_\mu\phi\big).$$

A point in classical phase space consists of the specification of functions (ϕ, π) on a Cauchy surface. By the result of Hawking and Ellis[7], smooth (ϕ, π) give rise to a unique solution to (1). The space of solutions is independent on the choice of the Cauchy surface.

For two (complex) solutions of the Klein-Gordon equation the inner product

$$(u_1, u_2) \equiv i \int_\Sigma \Big(\bar{u}_1 n^\mu\nabla_\mu u_2 - (n^\mu\nabla_\mu\bar{u}_1)u_2\Big)\sqrt{h}\,d^3x = i\int \big(\bar{u}_1\pi_2 - \bar{\pi}_1 u_2\big)d^3x$$

defines a natural symplectic structure. Natural means, that (u_1, u_2) is independent of the choice of Σ. This inner product is not positive definite. Let us introduce a complete set of conjugate pairs of solutions $(u_k, \bar{u}_k)$ of the Klein-Gordon equation[1] satisfying the following ortho-normality conditions

$$(u_k, u_{k'}) = \delta(k, k') \Rightarrow (\bar{u}_k, \bar{u}_{k'}) = -\delta(k, k') \quad\text{and}\quad (u_k, \bar{u}_{k'}) = 0.$$

There will be an infinity of such sets. Now we expand the field operator in terms of these modes:

$$\phi = \int d\mu(k)\Big(a_k u_k + a_k^\dagger\bar{u}_k\Big) \quad\text{and}\quad \pi = \int d\mu(k)\Big(a_k\pi_k + a_k^\dagger\bar{\pi}_k\Big),$$

so that

$$(u_k, \phi) = a_k \quad\text{and}\quad (\bar{u}_k, \phi) = -a_k^\dagger.$$

By using the completeness of the u_k and the canonical commutation relations one can show that the operator-valued coefficients $(a_k, a_k^\dagger)$ satisfy the usual commutation relations

$$[a_k, a_{k'}] = [a_k^\dagger, a_{k'}^\dagger] = 0 \quad\text{and}\quad [a_k, a_{k'}^\dagger] = \delta(k, k'). \tag{2}$$

We choose the Hilbert space $\mathcal{H}$ to be the Fock space built from a 'vacuum' state Ω_u satisfying

$$a_k\Omega_u = 0 \quad\text{for all}\quad k, \qquad (\Omega_u, \Omega_u)_{\mathcal{H}} = 1. \tag{3}$$

The 'vectors' $\Omega_u, a_k^\dagger\Omega_u, \ldots$ comprise a basis of $\mathcal{H}$. The scalar product given by (2,3) is positive-definite.

---

[1] the k are any labels, not necessarily the momentum

If $(v_p, \bar{v}_p)$ is a second set of basis functions, we may as well expand the field operator in terms of this set

$$\phi = \int d\mu(p) \left(b_p v_p + b_p^\dagger \bar{v}_p \right).$$

The second set will be linearly related to the first one by

$$v_p = \int d\mu(k) \left((u_k, v_p) u_k - (\bar{u}_k, v_p) \bar{u}_k \right) \equiv \int d\mu(k) \left(\alpha(p,k) u_k + \beta(p,k) \bar{u}_k \right).$$

The inverse transformation reads

$$u_k = \int d\mu(p) \left(v_p \bar{\alpha}(p,k) - \bar{v}_p \beta(p,k) \right).$$

As a consequence, the Bogolubov-coefficients are related by

$$\alpha \alpha^\dagger - \beta \beta^\dagger = 1 \quad \text{and} \quad \alpha \beta^t - \beta \alpha^t = 0. \tag{4}$$

If the $\beta(k,p)$ vanish, then the 'vacuum' is left unchanged, but if they do not, we have a nontrivial *Bogolubov transformation*

$$\left(a \ a^\dagger \right) = \left(b \ b^\dagger \right) \begin{pmatrix} \alpha & \beta \\ \bar{\beta} & \bar{\alpha} \end{pmatrix} \quad \text{and} \quad \begin{pmatrix} b \\ b^\dagger \end{pmatrix} = \begin{pmatrix} \bar{\alpha} & -\bar{\beta} \\ -\beta & \alpha \end{pmatrix} \begin{pmatrix} a \\ a^\dagger \end{pmatrix} \tag{5}$$

which mixes the annihilation and creations operators. If one defines a Fock space and a 'vacuum' corresponding to the first mode expansion, $a_k \Omega_u = 0$, then the expectation of the number operator $b_p^\dagger b_p$ defined with respect to the second mode expansion is

$$\left(\Omega_u, b_p^\dagger b_p \Omega_u \right) = \int d\mu(k) |\beta(p,k)|^2.$$

That is, the old vacuum contains new particles. It may even contain an infinite number of new particles, in which case the two Fock spaces cannot be related by a unitary transformation.

Stationary and static spacetimes. A spacetime is *stationary* if there exist coordinates for which the metric is time-independent. This property holds iff spacetime admits a time-like Killing field $K = K^\mu \partial_\mu$ and hence a natural choice for the mode functions u_k: We may scale K such that the Killing time t is the proper time measured by at least one comoving clock. Now we may choose as basis functions u_k the eigenfunctions of the Lie derivative,

$$iL_K u_k = \omega(k) u_k \quad \text{and} \quad iL_K \bar{u}_k = -\omega(k) \bar{u}_k,$$

where the $\omega(k) > 0$ are constant. The $\omega(k)$ are the frequencies relative to the particular comoving clock and the u_k and $\bar{u}_k$ are the positive and negative frequency solutions, respectively. Now the construction of the vacuum and Fock space is done as described above.

In a *static spacetime*, K is everywhere orthogonal to a family of hyper-surfaces and hence satisfies the Frobenius condition $\tilde{K} \wedge d\tilde{K} = 0$, $\quad \tilde{K} = K_\mu dx^\mu$. We may introduce adapted coordinates: t along the congruence ($K = \partial_t$) and x^i in one hypersurface such that the metric is time-independent and the shift vector N_i vanishes,

$$(g_{\mu\nu}) = \begin{pmatrix} N^2(x^i) & 0 \\ 0 & -h_{ij}(x^i) \end{pmatrix}.$$

As modes we use

$$u_k = \frac{1}{\sqrt{2\omega(k)}} e^{-i\omega(k)t} \phi_k(x^i)$$

which diagonalise L_K and for which the Klein-Gordon equation simplifies to

$$\mathcal{K}\phi_k \equiv \left(-\frac{N}{\sqrt{h}}\partial_i\left(N\sqrt{h}h^{ij}\partial_j\right) + N^2m^2\right)\phi_k = \omega_k^2\phi_k.$$

Since $n^\mu\partial_\mu = N^{-1}\partial_t$, the inner product of two mode functions is

$$(u_1, u_2) = \frac{\omega_1 + \omega_2}{2\sqrt{\omega_1\omega_2}} \, e^{i(\omega_1-\omega_2)t} \underbrace{\int \bar{\phi}_1\phi_2 \, N^{-1}\sqrt{h}\,d^3x}_{(\phi_1,\phi_2)_2}.$$

The elliptic operator $\mathcal{K}$ is symmetric with respect to the L_2 scalar product $(.,.)_2$ and may be diagonalised. Its positive eigenvalues are the $\omega^2(k)$ and its eigenfunctions form a complete 'orthonormal' set on Σ, $(\phi_k, \phi_{k'})_2 = \delta(k, k')$. It follows then that the u_k form a complete set with the properties discussed earlier.

Ashtekar and Magnon [8] and Kay [9] gave a rigorous construction of the Hilbert space and Hamiltonian in a stationary spacetime. They started with a *conserved positive scalar product* $(.,.)_E$

$$(\phi_1, \phi_2)_E = \int_\Sigma T_{\mu\nu}(\phi_1, \phi_2)K^\nu n^\mu \sqrt{h}d^3x,$$

where the bilinear-form on the space of complex solutions is defined by the metric 'stress tensor':

$$T_{\mu\nu}(\phi, \psi) = \frac{1}{2}\left(\phi^\dagger_{,\mu}\psi_{,\nu} + \phi^\dagger_{,\nu}\psi_{,\mu} - g_{\mu\nu}\left(\nabla\phi^\dagger\nabla\psi - m^2\phi^\dagger\psi\right)\right).$$

This 'stress tensor' is symmetric and conserved and hence $\nabla_\mu(T^{\mu\nu}K_\nu) = 0$. It follows that the norm is invariant under the time-translation map

$$\alpha_t^*(\phi) = \phi \circ \alpha_t \quad \text{or} \quad \left(\alpha_t^*(\phi)\right)(x) = \phi\left(\alpha_t(x)\right),$$

generated by the Killing field K. When completing the space of complex solutions in the 'energy-norm' one gets a complex (auxiliary) Hilbert space $\tilde{\mathcal{H}}$. The time translation map extends to $\tilde{\mathcal{H}}$ and defines a one-parameter unitary group

$$\alpha_t^* = e^{i\tilde{h}t}, \qquad \tilde{h} \quad \text{self-adjoint.}$$

Note, that from the definition of the Lie derivative,

$$\frac{d}{dt}(\alpha_t^*\phi)|_{t=0} = -L_K\phi = i\tilde{h}\phi.$$

The conserved inner product (ϕ_1, ϕ_2) can be bounded by the energy norm and hence extends to a quadratic form on $\tilde{\mathcal{H}}$. Let $\tilde{\mathcal{H}}^+ \subset \tilde{\mathcal{H}}$ be the positive spectral subspace in the spectral decomposition of $\tilde{h}$ and let P be the projection map $P : \tilde{\mathcal{H}} \to \tilde{\mathcal{H}}^+$. For all real solutions we may now define the *scalar product* as the inner product of the projected solutions, which are complex. The one-particle Hilbert space $\mathcal{H}$ is just the completion of the space $\tilde{\mathcal{H}}^+$ of 'positive frequency solutions' in the Klein-Gordon inner product.

Hadamard states. For a black hole the global Killing field is not everywhere time-like. One may exclude the non-time-like region from spacetime which corresponds to the imposition of boundary conditions. One may also try to retain this region but attempt to define a meaningful vacuum by invoking physical arguments. In general spacetimes there is no Killing vector at all. One probably has to give up the particle picture in this generic situation.

In (globally hyperbolic) spacetimes without any symmetry one can still construct a well-defined Fock space over a quasifree vacuum state, provided that the two-point functions satisfies the so-called Hadamard condition. Hadamard states are states, for which the two-point function has the following singularity structure

$$\omega\big(\phi(x)\phi(y)\big) \equiv \omega_2(x,y) = \frac{u}{\sigma} + v\log\sigma + w, \tag{6}$$

where $\sigma(x,y)$ is the square of the geodesic distance of x and y and u, v, w are smooth functions on $\mathcal{M}$. It has been shown that if ω_2 has the Hadamard singularity structure in a neighbourhood of a Cauchy surface, then it has this form everywhere [11]. To show that, one uses that ω_2 satisfies the wave equation. This result can then be used to show that on a globally hyperbolic spacetime there is a wide class of states whose two-point functions have the Hadamard singularity structure.

The two-point function ω_2 must be positive,

$$\omega\big(\phi(f)^\dagger\phi(f)\big) = \int d\mu(x)d\mu(y)\ \bar{f}(x)\omega_2(x,y)f(y) \geq 0,$$

and must obey the Klein-Gordon equation. These requirements determine u and v uniquely and put stringent conditions on the form of w. In a globally hyperbolic spacetime there are unique retarded and advanced Green functions

$$\Delta_{ret}(x,y) \quad , \quad \Delta_{adv}(x,y) \quad \text{with} \quad \text{supp}(\Delta_{ret}) = \{(x,y); x \in J_+(y)\},$$

where $J_+(y)$ is the causal future of y. The *Feynman Green function* is related to ω_2 and the advanced Green function as

$$i\Delta_F(x,y) = \omega_2(x,y) + \Delta_{adv}(x,y).$$

Since Δ_{adv} is unique, the ambiguities of Δ_F are the same as those of ω_2. The *propagator function*

$$i\Delta(x,y) = [\phi(x),\phi(y)] = \Delta_{ret}(x,y) - \Delta_{adv}(x,y)$$

determines the antisymmetric part of ω_2,

$$\omega_2(x,y) - \omega_2(y,x) = i\Delta(x,y),$$

so that this part is without ambiguities. For a scalar field without self-interaction we expect that

$$\omega\big(\phi(x_1)\ldots\phi(x_{2n-1})\big) = 0, \quad \omega\big(\phi(x_1)\ldots\phi(x_{2n})\big) = \sum_{\substack{i_1<i_2\cdots<i_n \\ j_1<j_2\cdots<j_n}} \prod_{k=1}^{n} \omega\big(\phi(x_{i_k})\phi(x_{j_k})\big).$$

A state ω fulfilling these conditions is called *quasifree*. Now one can show that any choice of $\omega_2(x,y)$ fulfilling the properties listed above gives rise to a well-defined Fock space $\mathcal{F} = \oplus\mathcal{F}_n$ over a quasifree vacuum state. The scalar product on the 'n-particle subspace' $\mathcal{F}_n$ in

$$\mathcal{F}_n = \{\psi \in \mathcal{D}(\mathcal{M}^n)_{symm}/\mathcal{N}\}^{completion}, \quad n = 0,1,2,\ldots, \tag{7}$$

where $\mathcal{D}(\mathcal{M}^n)$ denotes the smooth symmetric functions on $\mathcal{M} \times \cdots \times \mathcal{M}$ (n factors) with compact support, is

$$(\psi_1,\psi_2) = \int d\mu(x_1,..,x_n,y_1,..,y_n) \prod_{i=1}^{n} \omega_2(x_i,y_i)\bar\psi_1(x_1,..,x_n)\psi_2(y_1,..,y_n),$$

where $d\mu(x_1,x_2,..) = d\mu(x_1)d\mu(x_2)\ldots$. Since ω_2 satisfies the wave equation, the functions in the image of $\Box + m^2$ have zero norm. The set of zero-norm states $\mathcal{N}$ has been divided out in order to end up with a positive definite Hilbert space. The smeared field operator is now defined in the usual way: $\phi(f) = a(f)^\dagger + a(\bar f)$, where

$$\big(a(\bar f)\psi\big)_n(x_1,..,x_n) = \sqrt{n+1}\int d\mu(x,y)\omega_2(x,y)f(x)\psi_{n+1}(y,x_1,..,x_n)$$

$$\big(a(f)^\dagger\psi\big)_n(x_1,..,x_n) = \frac{1}{\sqrt{n}}\sum_{k=1}^{n} f(x_k)\psi_{n-1}(x_1,..,x_{k-1},x_{k+1}..,x_n), \quad n>0$$

and $(a(f)^\dagger\psi)_0 = 0$. It is now easy to see that ω_2 is just the Wightman function of ϕ in the vacuum state ψ_0: $\omega_2(x,y) = (\psi_0,\phi(x)\phi(y)\psi_0)$.

3   The Unruh Effect

We may ask the question how quantum fluctuations appear to an accelerating observer? In particular, if the observer was carrying with him a robust detector, what would this detector register? If the motion of the observer undergoing constant (proper) acceleration is confined to the x^3 axis, then the world line is a hyperbola in the x^0, x^3 plane with asymptotics $x^3 = \pm x^0$. These asymptotics are *event horizons* for the accelerated observer. To find a natural comoving frame we consider a family of accelerating observers, one for each hyperbola with asymptotics $x^3 = \pm x^0$. The coordinate system is then the comoving one in which along each hyperbola the space coordinate is constant while the time coordinate τ is proportional to the proper time as measured from an initial instant $x^0 = 0$ in some inertial frame. The world lines of the uniformly accelerated particles are the orbits of one-parameter group of Lorentz boost isometries in the 3-direction:

$$\begin{pmatrix} x^0 \\ x^3 \end{pmatrix} = \rho \begin{pmatrix} \sinh \kappa t \\ \cosh \kappa t \end{pmatrix} = e^{\kappa \omega t} \begin{pmatrix} 0 \\ \rho \end{pmatrix}, \qquad (\omega^\mu_\nu) = \begin{pmatrix} 0 & 1 \\ 1 & 0 \end{pmatrix}.$$

In the comoving coordinates (t, x^1, x^2, ρ)

$$ds^2 = \kappa^2 \rho^2 dt^2 - d\rho^2 - (dx^1)^2 - (dx^2)^2.$$

so that the proper time along a hyperbola ρ =const is $\kappa \rho t$. The orbits are tangential to the *Killing field*

$$K = \partial_t = \kappa(x^3 \partial_0 + x^0 \partial_3) \quad \text{with} \quad (K, K) = (\kappa \rho)^2 = g_{00}. \tag{8}$$

Some typical orbits are depicted in figure (1). Since the proper acceleration on the orbit with $(K, K) = 1$ or $\rho = 1/\kappa$ is κ, it is conventional to view the orbits of K as corresponding to a family of observers associated with an observer who accelerates uniformly with acceleration $a = \kappa$.

The coordinate system t, ρ covers the Rindler wedge R on which K is time-like future directed. The boundary H^+ and H^- of the wedge is given by $\rho = 0$ and appears as a *Killing horizon*, on which K becomes null. Beyond this event horizon the Killing vector field becomes space-like in the regions F, P and time-like past directed in L. The parameter κ plays the role of the *surface gravity*. To see that, we set $r - 2M = \rho^2/8M$ in the Schwarzschild solution and linearise the metric near the horizon $r \sim 2M$. One finds that

$$ds^2 \sim \underbrace{(\kappa \rho)^2 dt^2 - d\rho^2}_{\substack{\text{2-dim Rindler} \\ \text{spacetime}}} - \underbrace{\frac{1}{4\kappa^2} d\Omega^2}_{\substack{\text{2-sphere of} \\ \text{radius } 1/2\kappa}}$$

contains the line element of two-dimensional Rindler spacetime, where $\kappa = 1/4M$ is indeed the surface gravity of the Schwarzschild black hole.

Killing horizons and surface gravity. The notion of Killing horizons is relevant for the Hawking radiation and the thermodynamics of black holes and can

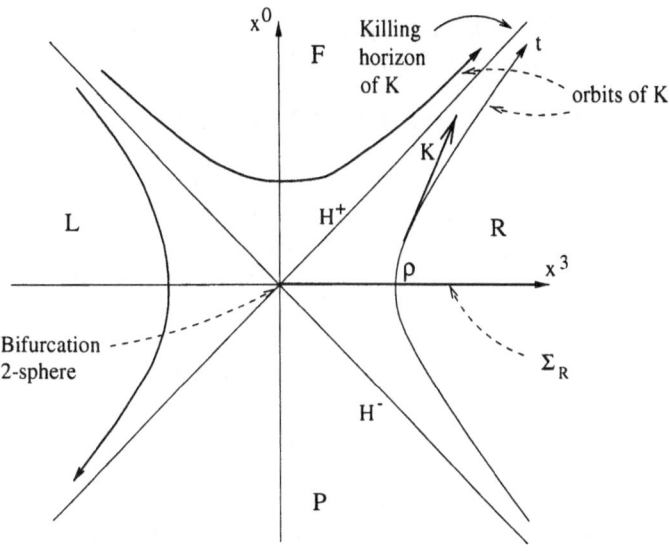

Fig. 1 *A Rindler-observer sees only a quarter of Minkowski space*

already be illustrated in Rindler spacetime. Let $S(x)$ be a smooth function and consider a family of hyper-surfaces $S(x) = $ const. The vector fields normal to the hyper-surfaces are

$$l = g(x)(\partial^\mu S)\partial_\mu,$$

with arbitrary non-zero function g. If l is null, $l^2 = 0$, for a particular hyper-surface $\mathcal{N}$ in the family, $\mathcal{N}$ is said to be a *null hypersurface*. For example, the normal vectors to the surfaces $S = r - 2M = $ const in Schwarzschild spacetime have norm

$$l^2 = g^2 g^{\mu\nu}\partial_\mu S \partial_\nu S = g^2\left(1 - \frac{2M}{r}\right),$$

and the horizon at $r = 2M$ is a null hypersurface.
Let $\mathcal{N}$ be a null hypersurface with normal l. A vector t tangent to $\mathcal{N}$ is characterised by $(t, l) = 0$. But since $l^2 = 0$, the vector l is itself a tangent vector, i.e.

$$l^\mu = \frac{dx^\mu}{d\lambda}, \quad \text{where} \quad x^\mu(\lambda) \quad \text{is a null curve on} \quad \mathcal{N}.$$

Now one can show, that $\nabla_l l^\mu|_{\mathcal{N}} \sim l^\mu$, which means that $x^\mu(\lambda)$ is a geodesic with tangent l. The function g can be chosen such that $\nabla_l l = 0$, i.e. so that λ is an affine parameter. A null hypersurface $\mathcal{N}$ is a *Killing horizon* of a Killing field K if K is normal to $\mathcal{N}$.

Let l be normal to $\mathcal{N}$ such that $\nabla_l l = 0$. Then, since on the Killing horizon $K = fl$ for some function f, it follows that

$$\nabla_K K^\mu = fl^\nu \nabla_\nu(fl^\mu) = fl^\mu l^\nu \partial_\nu f = (\nabla_K \log|f|)K^\mu \equiv \kappa K^\mu \quad \text{on} \quad \mathcal{N}. \quad (9)$$

One can show, that the *surface gravity* $\kappa = \frac{1}{2}\nabla_K \log f^2$ is constant on orbits of K. If $\kappa \neq 0$, then $\mathcal{N}$ is a bifurcate Killing horizon of K with bifurcation 2-sphere B. In this non-degenerate case κ^2 is constant on $\mathcal{N}$. For example, for the Killing field in Rindler spacetime (8) $\nabla_K K = \pm\kappa K$ on the Killing horizon and the bifurcation 'sphere' is at $\rho = 0$. If $\mathcal{N}$ is a Killing horizon of K with surface gravity κ, then it is also a Killing horizon of cK with surface gravity $c^2\kappa$. Thus the surface gravity depends on the normalisation of K. For asymptotically flat spacetimes there is the natural normalisation $K^2 \to 1$ and K future directed as $r \to \infty$. With this normalisation the surface gravity is the acceleration of a static particle near the horizon as measured at spatial infinity.

A Killing field is uniquely determined by its value and the value of its derivative $F_{\mu\nu} = \nabla_{[\mu}K_{\nu]}$ at any point $p \in M$. At the bifurcation point p of a bifurcate Killing horizon K vanishes and hence is determined by $F_{\mu\nu}(p)$. In two dimensions $F_{\mu\nu}(p)$ is unique up to scaling. The infinitesimal action of the isometries α_t generated by K takes a vector v^μ at p into

$$L_K v^\mu = F^\mu_\nu v^\nu. \quad (10)$$

The nature of this map on T_p depends upon the signature of the metric. For Riemannian signature it is an infinitesimal rotation and the orbits of α_t are closed with a certain period. For Lorentz signature (10) is an infinitesimal Lorentz boost and the orbits of α_t have the same structure as in the Rindler case. A similar analysis applies to higher dimensions.

The Rindler wedge R is globally hyperbolic with Cauchy hypersurface Σ_R (see fig. (1)). Thus it may be viewed as a spacetime in its own right, and we may construct a quantum field theory on it. When we do that, we obtain a remarkable conclusion, namely that the standard Minkowski vacuum Ω_M corresponds to a thermal state in the new construction. This means, that an accelerated observer will feel himself to be immersed in a thermal bath of particles with temperature proportional to his acceleration a [10],

$$kT = \hbar a/2\pi c.$$

The noise along a hyperbola is greater than that along a geodesic, and this excess noise excites the Rindler detector: A uniformly accelerated detector in its ground state may jump spontaneously to an excited state. Note that the temperature tends to zero when $\hbar$ tends to zero. Such a radiation has non-zero entropy. Since the use of an accelerated frame seems to be unrelated to any statistical average, the appearance of a non-vanishing entropy is rather puzzling. The Unruh effect shows, that at the quantum level there is a deep relation between the theory of relativity and the theory of fluctuations associated with states of thermal equilibrium, two major aspects of Einstein's work: The distinction between quantum zero-point and thermal fluctuations is not an invariant one, but depends on the motion of the observer. Note that the temperature is proportional to the acceleration a of the observer. Since $a = 1/\rho$ this means that $T\rho = \text{const} \Longleftrightarrow T\sqrt{g_{00}} = \text{const}$. This is just the *Tolman-Ehrenfest relation* [12] for the temperature in a fluid in hydrostatic equilibrium in a gravitational field. The factor $\sqrt{g_{00}}$ guarantees that no work can be gained by transferring radiation between two regions at different gravitational potentials.

Let us calculate the number of 'Rindler-particles' in Minkowski vacuum. To simplify the analysis, we consider a zero-mass scalar field in two-dimensional Minkowski space. In the Heisenberg picture, the expansions in terms of annihilation and creation operators are

$$\phi = \int dk \left(a_k u_k + h.c. \right), \quad \text{where} \quad u_k = \frac{1}{\sqrt{4\pi\omega}} e^{-i\omega x^0 + ikx^3}, \quad \omega = |k|$$

and

$$\phi = \int dp \left(b_p v_p + h.c. \right), \quad \text{where} \quad v_p = \frac{1}{\sqrt{4\pi\epsilon}} \rho^{ip/\kappa} e^{-i\epsilon t}, \quad \epsilon = |p|.$$

The β-coefficients are found to be

$$\beta(p,k) = -(\bar{u}_k, v_p) = \frac{1}{4\pi} \int_0^\infty \left(\sqrt{\frac{\omega}{\epsilon}} - \sqrt{\frac{\epsilon}{\omega}} \frac{1}{\kappa\rho} \right) e^{ik\rho} \rho^{ip} d\rho,$$

where we have evaluated the time-independent 'scalar-product' at $t = 0$ for which $x^0 = 0$. Using the formula

$$\int_0^\infty dx\, x^{\nu-1} e^{-(\alpha+i\beta)x} = \Gamma(\nu)(\alpha^2 + \beta^2)^{-\nu/2} e^{-i\nu \arctan(\beta/\alpha)} \tag{11}$$

we arrive at

$$\beta(p,k) = -\frac{\Gamma(ip/\kappa)}{4\pi\kappa} \omega^{-ip/\kappa} \left(\sqrt{\frac{\epsilon}{\omega}} \pm \frac{p}{\sqrt{\epsilon\omega}} \right) e^{\mp\pi p/2\kappa} \quad \text{for} \quad \frac{k}{\omega} = \pm 1,$$

or at

$$|\beta(p,k)|^2 = \frac{1}{2\pi\kappa\omega} \frac{1}{e^{2\pi\epsilon/\kappa} - 1}.$$

The Minkowski spacetime vacuum is characterised by $a_k \Omega_M = 0$ for all k. Assuming that this is the state of the system, the expectation value of the occupation number as defined by the Rindler observer, $n_p \equiv b_p^\dagger b_p$, is found to be

$$(\Omega_M, n_p \Omega_M) = \int dk |\beta(p,k)|^2 = \text{volume} \times \frac{1}{e^{2\pi\epsilon/\kappa} - 1}. \tag{12}$$

Thus for an accelerated observer the quantum field seems to be in an equilibrium state with temperature proportional to $T = \kappa/2\pi = a/2\pi$. An observer with $a = 10^{21} \text{cm/sec}^2$ feels a temperature $T \sim 1^0 K$. Since T tends to zero as $\rho \to \infty$ the Hawking temperature (i.e. temperature as measured at spatial ∞) is actually zero. This is expected, since there is nothing inside which could radiate. But for a black hole $T_{local} \to T_H$ at infinity and the black hole must radiate at this temperature.

Let us finally see, how the (massless) Feynman-Green function in Minkowski spacetime,

$$i\Delta_F(x, x') = \langle 0|T\left(\phi(x)\phi(x')\right)|0\rangle = \frac{i}{4\pi^2}\frac{1}{(x - x')^2 - i\epsilon},$$

appears to an accelerated observer. Let $x = (t, \rho)$ and $x' = (t', \rho)$ be two events on the world line of an accelerated observer. Since the invariant distance of these two events is $2\rho \sinh \frac{\kappa}{2}(t-t')$, one arrives at the following spectral representation of the Feynman-propagator as seen by this observer:

$$\Delta_F(x, x') = \frac{1}{(2\pi)^4}\left(\frac{\kappa}{\rho}\right)^2 \int d^4p\, e^{-iE(t-t')} \left(\frac{1}{p^2 + i\epsilon} - 2\pi i \frac{\delta(p^2)}{e^{\beta|E|} - 1}\right). \quad (13)$$

This is the finite temperature propagator. It follows, that atoms dragged along the world line find their excited levels populated as predicted by a temperature $\beta^{-1} = a/2\pi$.

4   The Stress-Energy Tensor

Semiclassically one would expect that back-reaction is described by the 'semi-classical Einstein equation'

$$G_{\mu\nu} = 8\pi G \langle T_{\mu\nu}\rangle,$$

where the right-hand side contains the expectation value of the energy-momentum tensor of the relevant quantised field in the chosen state. If the characteristic curvature radius L in a region of spacetime is much greater then the Planck length l_{pl}, then in the calculation of $\langle T_{\mu\nu}\rangle$ one can expand in the small parameter $\epsilon = (l_{pl}/L)^2$ and retain only the terms up to first order in ϵ (one-loop approximation). The term of order ϵ, containing a factor $\hbar$, represents the main quantum correction to the classical result. In the one-loop approximation or free fields the contributions of all fields to $\langle T_{\mu\nu}\rangle$ are additive and thus can be studied independently.

The difficulties with defining $\langle T_{\mu\nu}\rangle = \omega(T_{\mu\nu})$ are present already in Minkowski spacetime. The divergences are due to the vacuum zero-fluctuations. The methods of extracting a finite, physically meaningful part, known as renormalisation procedures, were extensively discussed in the literature [14]. A simple cure for this difficulty is (for free fields) the *normal ordering* prescription. We first consider the ill-defined object $\phi^2(x)$, which is part of the stress-energy tensor. We may split the points and consider first the object $\omega(\phi(x)\phi(y))$ which solves the Klein-Gordon equation. This bi-distribution makes perfectly good sense. For physically reasonable states ω in the Fock space (e.g. states with a finite number of particles) the singular behaviour of this bi-distribution is the same as that belonging to the vacuum state, $\omega_0(\phi(x)\phi(y))$. For such states the difference

$$F(x,y) = \omega(\phi(x)\phi(y)) - \omega_0(\phi(x)\phi(y))$$

is a smooth function of its arguments. Hence, after performing this 'vacuum subtraction' the coincidence limit may be taken. We then define

$$\omega(\phi^2(x)) = \lim_{x \to y} F(x,y).$$

The same prescription can be used for the stress-energy tensor. We define

$$\omega(T_{\mu\nu}(x)) = \lim_{x \to x'} D_{\mu\nu'} F(x,x'), \quad D_{\mu\nu'} = \partial_\mu \partial_{\nu'} - \frac{1}{2} g_{\mu\nu} [\partial_\alpha \partial^{\alpha'} - m^2]. \quad (14)$$

In curved spacetime some restrictions should be expected on the class of states on which $\langle T_{\mu\nu} \rangle$ can be defined this way. The *Hadamard condition* provides a restriction of exactly this sort of states.

Although (14) is not a physical definition of expectation values of the stress-energy tensor itself (no preferred vacuum state, vacuum polarisation), it sensibly defines the *differences* of the expected stress energy between two states. In the absence of an obvious prescription it is useful to take an axiomatic approach. Wald showed that a renormalised stress tensor satisfying certain reasonable physical requirements is essentially unique [13]. Its ambiguity can be absorbed into redefinitions of the coupling constants in the (generalised) gravitational field equation. Wald's requirements are:

Consistency: Whenever $\omega_1(\phi(x)\phi(y)) - \omega_2(\phi(x)\phi(y))$ is a smooth function, then $\omega_1(T_{\mu\nu}) - \omega_2(T_{\mu\nu})$ is well-defined and should be given by the above 'point-splitting' prescription.

Conservation: There is a regularisation which respects the diffeomorphism invariance, so that $\nabla_\nu T^{\mu\nu} = 0$ holds. This property is needed for consistency of Einstein's gravitational field equation.

Normalisation: In Minkowski spacetime, we have $(\Omega_M, T_{\mu\nu} \Omega_M) = 0$.

Causality: For a fixed in-state in an asymptotically static spacetime $\omega_{in}(T_{\mu\nu}(x))$ is independent of variations of $g_{\mu\nu}$ outside the past light cone of x. For a fixed out-state, $\omega_{out}(T_{\mu\nu})$ is independent of metric variations outside the future light cone of x.

The Causality axiom can be replaced by a locality property, which does not assume an asymptotically static spacetime. The first and last properties are the key ones, since they uniquely determine the expected stress-energy tensor up to the addition of local curvature terms:

Uniqueness theorem: Let $T_{\mu\nu}$ and $\tilde{T}_{\mu\nu}$ be operators on a globally hyperbolic spacetime satisfying the axioms of Wald. Then the difference $U_{\mu\nu} = T_{\mu\nu} - \tilde{T}_{\mu\nu}$ is a multiple of the identity operator, is conserved, $\nabla_\nu U^{\mu\nu} = 0$ and is a local tensor of the metric. That is, it depends only on the metric and its derivatives, via the curvature tensor, at the same point x. As a consequence $\omega(T_{\mu\nu}) - \omega(\tilde{T}_{\mu\nu})$ is independent of the state ω and depends only locally on curvature invariants. The proofs of these properties are rather simple and can be found in the standard textbooks.

Calculating the stress-energy tensor. A 'point-splitting' prescription where one subtracts from $\omega(\phi(x)\phi(y))$ the expectation value $\omega_0(\phi(x)\phi(y))$ in some fixed state ω_0 fulfils the consistency requirement, but cannot fulfil the first and third axiom at the same time. However, if one subtracts a locally constructed bi-distribution $H(x,y)$ which satisfies the wave equation, has a suitable singularity structure and is equal to $(\Omega_M, \phi(x)\phi(y)\Omega_M)$ in Minkowski spacetime, then all four properties will be satisfied.

To find a suitable bi-distribution one recalls the singularity structure (6) of $\omega_2(x,y)$. In Minkowski spacetime and for massless fields $w = 0$ and this suggests that we take the bi-distribution

$$H(x,y) = \frac{u(x,y)}{\sigma} + v(x,y)\log\sigma$$

For massless fields the resulting stress-energy obeys all properties listed above (for massive fields a slight modification is needed).

Effective action. The classical metric energy momentum tensor

$$^{cl}T_{\mu\nu}(x) = \frac{2}{\sqrt{|g|}}\frac{\delta S}{\delta g^{\mu\nu}(x)}$$

is symmetric and conserved (for solutions of the field equation) for a diffeomorphism-invariant classical action S. If we could construct a diffeomorphism-invariant *effective action* Γ, whose variation with respect to the metric yields an expectation value of the energy momentum tensor,

$$\langle T_{\mu\nu}(x)\rangle = \frac{2}{\sqrt{|g|}}\frac{\delta\Gamma}{\delta g^{\mu\nu}(x)},$$

then $\langle T_{\mu\nu}\rangle$ would be conserved by construction. There exists a number of procedures for regularising $\langle T_{\mu\nu}\rangle$, i.e. dimensional, point-splitting or zeta-function regularisation, to mention the most popular ones. Unfortunately the 'divergent' part' of $T_{\mu\nu}$ cannot be completely absorbed into the parameters already present in the theory, i.e. gravitational and cosmological constant and parameters of the field theory under investigation. One finds that one must introduce new, dimensionless parameters.

The regularisation and renormalisation of the effective action is more transparent. The divergent geometric parts of the effective action, $\Gamma = \int \eta\gamma_{div} + \Gamma_{finite}$ have in the one-loop approximation the form

$$\gamma_{div} = A + BR + C(\text{Weyl})^2 + D\big[(\text{Ricci})^2 - R^2\big] + E\nabla^2 R + FR^2.$$

Only the part containing A and B can be absorbed into the classical action of gravity. The remaining terms with dimensionless parameters $C - F$ lead, upon variation with respect to the metric, to a 2-parameter ambiguity in the expression for $T_{\mu\nu}$.

Effective actions and $\langle T_{\mu\nu} \rangle$ in two dimensions. In two dimensions there are less divergent terms in the effective action. They have the form $\gamma_{div} = A + BR$. The last topological term does not contribute to $T_{\mu\nu}$ and the first one leads to an ambiguous term $\sim A g_{\mu\nu}$ in the energy momentum tensor.

The symmetric stress-energy tensor has 3 components, two of which are (almost) determined by $T^{\mu\nu}_{;\nu} = 0$. As independent component we choose the trace $T = T^{\mu}_{\mu}$ which is a scalar of dimension L^{-2}. The ambiguities in the reconstruction of $T^{\mu\nu}$ from its trace is most transparent if we choose isothermal coordinates for which

$$ds^2 = e^{2\sigma} \left((dx^0)^2 - (dx^1)^2 \right).$$

This is possible in two dimensions. Introducing null-coordinates

$$u = \frac{1}{2}(x^0 - x^1) \quad \text{and} \quad v = \frac{1}{2}(x^0 + x^1) \Rightarrow ds^2 = 4e^{2\sigma} dudv,$$

the non-vanishing Christoffel symbols are $\Gamma^u_{uu} = 2\partial_u \sigma$, $\Gamma^v_{vv} = 2\partial_v \sigma$ and the Ricci scalar reads $R = -2e^{-2\sigma} \partial_u \partial_v \sigma$. Rewriting the conservation in null-coordinates we obtain

$$\partial_u \langle T_{vv} \rangle + e^{2\sigma} \partial_v \langle T \rangle = 0 \quad , \quad \partial_v \langle T_{uu} \rangle + e^{2\sigma} \partial_u \langle T \rangle = 0, \tag{15}$$

where $T = T^{\mu}_{\mu} = e^{-2\sigma} T_{uv}$. The trace $\langle T \rangle$ determines $\langle T_{vv} \rangle$ up to a function $t_v(v)$ and $\langle T_{uu} \rangle$ up to a function $t_u(u)$. These free functions contain information about the state of the quantum system.

In the case of a classical conformally invariant field, $^{cl}T^{\mu}_{\mu} = 0$. An important feature of $\langle T_{\mu\nu} \rangle$ is that its trace does not vanish any more. This trace-anomaly is a state-independent local scalar of dimension L^{-2} and hence must be proportional to the Ricci scalar,

$$\langle T \rangle = \frac{c}{24\pi} R = -\frac{c}{12\pi} e^{-2\sigma} \partial_u \partial_v \sigma,$$

where c is the *central charge*. Inserting this trace anomaly into (15) yields

$$\langle T_{uu,vv} \rangle = -\frac{c}{12\pi} e^{\sigma} \partial^2_{u,v} e^{-\sigma} + t_{u,v} \quad \text{and} \quad \langle T_{uv} \rangle = -\frac{c}{12\pi} \Box_0 \sigma. \tag{16}$$

Formally, the expectation value of the stress-energy tensor is given by the path integral

$$\langle T_{\mu\nu}(x) \rangle = -\frac{1}{Z[g]} \int D\phi \, \frac{2}{\sqrt{g}} \frac{\delta}{\delta g^{\mu\nu}} e^{-S[\phi]} = \frac{2}{\sqrt{g}} \frac{\delta}{\delta g^{\mu\nu}} \Gamma[\phi],$$

where the effective action is given by

$$\Gamma[g] = -\log Z[g] = -\log \int \mathcal{D}\phi \, e^{-S[\phi]} = \frac{1}{2} \log \det(-\triangle_c)$$

and we made the transition to Euclidean spacetime (which is allowed for the $2d$ models under investigation). For arbitrary spacetimes the spectrum of $\triangle_c$ is not known. However, the variation of Γ with respect to σ in $g_{\mu\nu} = e^{2\sigma}\hat{g}_{\mu\nu}$ is proportional to the expectation value of the trace of the stress-energy tensor,

$$\frac{\delta\Gamma}{\delta\sigma(x)} = -2g^{\mu\nu}(x)\frac{\delta\Gamma}{\delta g^{\mu\nu}(x)} = -\sqrt{g}\langle T^{\mu}_{\mu}(x)\rangle$$

and can be calculated for conformally coupled particles in conformally flat spacetimes. From the conformal anomaly one can (almost) reconstruct the effective action. In particular, in two dimensions the result is the *Polyakov effective action*

$$\Gamma[g] - \Gamma[\delta] = \frac{c}{96\pi} \int \sqrt{g} R \frac{1}{\triangle} R,$$

where the central charge c is 1 for uncharged scalars and Dirac fermions[2]. The $\langle T_{\mu\nu}\rangle$ is found by differentiation with respect to the metric. The covariant expression is

$$\langle T_{\mu\nu}\rangle = \frac{c}{48\pi}\left(2g_{\mu\nu}R - 2\nabla_{\mu}\nabla_{\nu}S + \nabla_{\mu}S\cdot\nabla_{\nu}S - \frac{1}{2}g_{\mu\nu}\nabla^{\alpha}S\cdot\nabla_{\alpha}S\right), \quad (17)$$

with $S = \frac{R}{\triangle}$ and in isothermal coordinates this simplifies to (16), as it must be. This energy-momentum tensor is consistent, conserved, and causality restricts the choice of the Green function $1/\triangle$. The ambiguities in inverting the wave operator in (17) shows up in the free functions $t_{u,v}$. A choice of these functions is equivalent to the choice of a state.

Let us now apply these results to the (t,r) part of the Schwarzschild black hole

$$ds^2 = \alpha(r)dt^2 - \frac{1}{\alpha(r)}dr^2, \qquad \alpha(r) = 1 - \frac{2M}{r}, \qquad (G = 1)$$

which we treat as two-dimensional black hole[3]. We use the 'Regge-Wheeler tortoise coordinate' $r_* = r + 2M\log(r/M - 2)$, such that the metric becomes conformally flat, $ds^2 = \alpha(dt^2 - dr_*^2)$ and introduce null-coordinates $2u = t - r_*$ and $2v = t + r_*$. Using $\partial_{r_*} = \alpha\partial_r$ we obtain for the light-cone components (16) of the energy momentum tensor

$$\langle T_{uu,vv}\rangle = -\frac{c}{12\pi}\left(\frac{2M\alpha}{r^3} + \frac{M^2}{r^4}\right) + t_{u,v}, \qquad \langle T_{uv}\rangle = -\frac{c}{12\pi}\frac{2M\alpha}{r^3}$$

---

[2] see [15] for modifications of this result, for a spacetime with nontrivial topology.

[3] The resulting energy-momentum tensor is not identical to the tensor that one gets when one quantises only the s-modes in the four-dimensional Schwarzschild metric [16].

or for $\langle T_{\mu\nu}\rangle$ in the $x^\mu = (t, r_*)$ coordinate system

$$\langle T_\mu{}^\nu\rangle = -\frac{cM}{24\pi r^4}\begin{pmatrix} 4r + \frac{M}{\alpha} & 0 \\ 0 & -\frac{M}{\alpha}\end{pmatrix} + \frac{1}{4\alpha}\begin{pmatrix} t_u + t_v & t_u - t_v \\ t_v - t_u & -t_u - t_v\end{pmatrix}. \tag{18}$$

The *Boulware state* is the state appropriate to a vacuum around a static star and contains no radiation at spatial infinity $\mathcal{J}^\pm$. Hence t_u and t_v must vanish. This state is singular at the horizon. To see that, we use regular Kruskal coordinates:

$$U = -e^{-u/2M} \quad\text{and}\quad V = e^{v/2M} \quad\text{so that}\quad ds^2 = \frac{16M^3}{r}e^{-r/2M}dU\,dV. \tag{19}$$

With respect to these coordinates the energy-momentum tensor takes the form

$$\langle T_{UU}\rangle = 4\Big(\frac{M}{U}\Big)^2\langle T_{uu}\rangle, \qquad \langle T_{VV}\rangle = 4\Big(\frac{M}{V}\Big)^2\langle T_{vv}\rangle \quad\text{and}\quad \langle T_{UV}\rangle = -4\frac{M^2}{UV}\langle T_{uv}\rangle.$$

For the Boulware vacuum $t_u = t_v = 0$ and $\langle\ldots\rangle$ is singular at the past horizon at $V = 0$ and future horizon at $U = 0$. The component $\langle T_{UU}\rangle$ is regular at the future horizon if $M^2 t_u = c/192\pi$ and $\langle T_{VV}\rangle$ is regular at the past horizon if $M^2 t_v = c/192\pi$. The state regular at both horizons is the *Israel-Hartle-Hawking* state. In this state the asymptotic form of the energy-momentum tensor is

$$\langle 0_{HH}|T_\mu{}^\nu|0_{HH}\rangle \sim \frac{c}{384\pi M^2}\begin{pmatrix} 1 & 0 \\ 0 & -1\end{pmatrix} = \frac{c\pi}{6}(kT)^2\begin{pmatrix} 1 & 0 \\ 0 & -1\end{pmatrix} \tag{20}$$

with $T = 1/8\pi kM = \kappa/2\pi k$. This is the stress-tensor of a *bath* of thermal radiation at temperature T. Finally, demanding that energy-momentum is regular at the future horizon and that there is no incoming radiation, i.e. $M^2 t_u = c/192\pi$ and $t_v = 0$, results in

$$\langle 0_U|T_\mu{}^\nu|0_U\rangle \sim \frac{c}{768\pi M^2}\begin{pmatrix} 1 & 1 \\ -1 & -1\end{pmatrix} = \frac{c\pi}{12}(kT)^2\begin{pmatrix} 1 & 1 \\ -1 & -1\end{pmatrix}. \tag{21}$$

The *Unruh state* is regular on the future horizon and singular at the past horizon. It describes the Hawking evaporation process with only outward flux of thermal radiation.

Euclidean Black Holes. The most elegant and powerful derivation of the Hawking radiation involves an adaption of the techniques due to Kubo to show that the Feynman propagator for a spacetime with stationary black hole satisfies the KMS condition. Consider a system with time-independent Hamiltonian H. The time evolution of an observable A in the Heisenberg picture is $A(z) = e^{izH}Ae^{-izH}$, where $z = t + i\tau$ is complex time. For $\tau = 0$ ($t = 0$) it is the time-evolution in a static spacetime with Lorentzian (Euclidean) signature. If $\exp(-\beta H), \beta > 0$ is trace class, one can define the equilibrium state of temperature $T = 1/\beta$:

$$\langle A\rangle_\beta = \frac{1}{Z}\mathrm{tr}\,e^{-\beta H}A, \qquad Z = \mathrm{tr}\,e^{-\beta H}. \tag{22}$$

Let us introduce the finite temperature correlation functions

$$G_+^\beta(z, \vec{x}, \vec{y}) = \langle \phi(z, \vec{x})\phi(0, \vec{y})\rangle_\beta = \frac{1}{Z}\mathrm{tr}\left(e^{i(z+i\beta)H}\phi(0, \vec{x})e^{-izH}\phi(0, \vec{y})\right)$$

$$G_-^\beta(z, \vec{x}, \vec{y}) = \langle \phi(0, \vec{y})\phi(z, \vec{x})\rangle_\beta = \frac{1}{Z}\mathrm{tr}\left(\phi(0, \vec{y})e^{izH}\phi(0, \vec{x})e^{-i(z-i\beta)H}\right).$$

We have used the cyclicity under the trace. Both exponents in G_+ have negative real parts if $-\beta < \tau < 0$; for G_- the condition reads $0 < \tau < \beta$. Therefore, these formulae define holomorphic functions in those respective strips with boundary values $G_\pm^\beta(t, \vec{x}, \vec{y})$. It follows immediately, that

$$G_-^\beta(z, \vec{x}, \vec{y}) = G_+^\beta(z - i\beta, \vec{x}, \vec{y}) \tag{23}$$

which is the KMS condition. This condition is now accepted as a definition of 'thermal equilibrium at temperature $1/\beta$'.

So far the analytic functions $G_\pm$ have been defined in disjoint, adjacent strips in the complex time plane. The KMS-condition states that one of these is the translate of the other and this allows us to define a periodic function throughout the complex plane, with the possible exception of the lines $\tau = \Im(z) = n\beta$. Because of locality $\phi(x)$ and $\phi(y)$ commute for space-like separated events and

$$[\phi(t, \vec{x}), \phi(0, \vec{y})] = 0 \quad \text{for} \quad t \in I \subset R.$$

Then the boundary values of $G_\pm^\beta$ coincide on I and we conclude (by the edge-of-the-wedge theorem) that they are restrictions of a single holomorphic, periodic function, $\mathcal{G}^\beta(z, \vec{x}, \vec{y})$, defined in a connected region in the complex time plane except parts of the lines $\tau = n\beta$.

With these preparations we are now ready to show that the Green function in Schwarzschild spacetime satisfies the KMS-condition. Starting with the analytically continued Schwarzschild metric

$$ds^2 = \alpha dz^2 - \frac{1}{\alpha}dr^2 - r^2 d\Omega^2, \qquad \alpha = 1 - 2M/r, \quad z = t + i\tau,$$

we perform the same coordinate transformation to (complex) Kruskal coordinates as we did for the Lorentzian solution:

$$Z = V + U = 2e^{r_*/4M}\sinh\frac{z}{4M} \quad \text{and} \quad X = V - U = 2e^{r_*/4M}\cosh\frac{z}{4M}.$$

The line element reads

$$ds^2 = \frac{16M^3}{r}e^{-r/2M}\left(dZ^2 - dX^2\right) - r^2 d\Omega^2$$

and the Killing field takes the form

$$K = \partial_z = \frac{1}{4M}\left(Z\partial_X + X\partial_Z\right) = \frac{1}{4M}\left(V\partial_V - U\partial_U\right).$$

Setting $Z = T + i\mathcal{T}$ the orbits of K are

$$\begin{pmatrix} T \\ X \end{pmatrix} = 2e^{r_*/4M} \begin{pmatrix} \sinh t/4M \\ \cosh t/4M \end{pmatrix} \quad \text{and} \quad \begin{pmatrix} \mathcal{T} \\ X \end{pmatrix} = 2e^{r_*/4M} \begin{pmatrix} \sin \tau/4M \\ \cos \tau/4M \end{pmatrix},$$

in the Lorentzian and Euclidean slices, respectively. As expected from the general properties of bifurcation spheres, these are Lorentz-boosts and rotations, respectively. Since the Euclidean slice is periodic in τ, the analytic Green function $\mathcal{G}(z = t + i\tau, \vec{x}, \vec{y})$ is periodic in imaginary time τ with period $8\pi M$. This corresponds to a temperature $T = 1/8\pi M$, the Hawking temperature.

The vector field (with affine parametrisation) normal to the Killing horizon $\mathcal{N}$ (the past and future horizons) is $l = \partial_V$ on the future horizon and $l = \partial_U$ on the past horizon. It follows that the surface gravity κ (see (9)) is $1/4M$ on the future horizon and $-1/2M$ on the past horizon.

Energy-momentum tensor near a black hole. In any vacuum spacetime $R_{\mu\nu}$ vanishes and so do the two local curvature terms which enter the formula for $T_{\mu\nu}$ with undetermined coefficients. Hence $T_{\mu\nu}$ is well-defined in the Schwarzschild spacetime. The symmetry of $\langle T_\nu^\mu \rangle$ due to the $SO(3)$ symmetry of the spacetime of a non-rotating black hole and the conservation $\nabla_\nu \langle T^{\mu\nu} \rangle$ reduce the number of independent components of $\langle T_\nu^\mu \rangle$. Christensen and Fulling [18] showed that in the coordinates (t, r_*, θ, ϕ) the tensor is block diagonal. The (t, r_*) part admits the representation

$$\langle T_\nu^\mu \rangle = \begin{pmatrix} \frac{T}{2} - \frac{H+G}{\alpha r^2} - 2\Theta & 0 \\ 0 & \frac{H+G}{\alpha r^2} \end{pmatrix} + \frac{W}{4\pi\alpha r^2} \begin{pmatrix} 1 & -1 \\ 1 & -1 \end{pmatrix} + \frac{N}{\alpha r^2} \begin{pmatrix} -1 & 0 \\ 0 & 1 \end{pmatrix} \quad (24)$$

and the (θ, ϕ)-part has the form

$$\langle T_\nu^\mu \rangle = \left(\frac{T}{4} + \Theta \right) \begin{pmatrix} 1 & 0 \\ 0 & 1 \end{pmatrix}. \quad (25)$$

Here N and W are two constants and

$$\alpha(r) = \left(1 - \frac{2M}{r} \right), \quad T(r) = \langle T_\mu^\mu \rangle, \quad \Theta(r) = \langle T_\theta^\theta \rangle - \frac{1}{4} T(r)$$

$$H(r) = \frac{1}{2} \int_{2M}^{r} (r' - M) T(r') dr', \quad G(r) = 2 \int_{2M}^{r} (r' - 3M) \Theta(r') dr'.$$

The energy-momentum tensor is characterised unambiguously by fixing two functions $T(r), \Theta(r)$ and two constants N, W. The constant W gives the intensity of radiation of the black hole at infinity and N vanishes if the state is regular on the future horizon.

The radiation intensity W is non-vanishing only in the *Unruh vacuum*. It has been calculated for the massless scalar field ($s = 0$), two-components neutrino field ($s = 1/2$), electromagnetic field ($s = 1$) and gravitational field ($s = 2$) by Page and Elster [19]:

M^2W_0	$M^2W_{1/2}$	M^2W_1	M^2W_2
$7.4 \cdot 10^{-5}$	$8.2 \cdot 10^{-5}$	$3.3 \cdot 10^{-5}$	$0.4 \cdot 10^{-5}$

The coefficient N vanishes for the Unruh and Israel-Hartle-Hawking states.

The calculation of the functions in (24,25) meets technical difficulties connected with the fact that solutions of the radial mode equation (see below) are not expressed through known transcendental functions and, consequently, one needs to carry out renormalisation in divergent integrals within the framework of numerical methods. The results for $\langle T_t^t \rangle$ and $\langle T_r^r \rangle$ for the Israel-Hartle-Hawking and the Unruh states have been calculated by Howard/Candelas and Elster [20].

In the Hartle-Hawking state the Kruskal coordinate components of $\langle T_{\mu\nu} \rangle$ near the horizon are found to be of order $1/M^4$. The energy flux into the black hole is negative, as it must be since the 'Hartle-Hawking vacuum' is time independent and the energy flux at future infinity is positive. This is possible since $\langle T_{\mu\nu} \rangle$ need not satisfy the energy conditions.

s-wave contribution to $\langle T_{\mu\nu} \rangle$. The covariant perturbation theory for the $4d$ effective action Γ as developed in [23] is very involved for concrete calculations. Here we shall simplify the problem by considering s-modes of a minimally coupled massless scalar field propagating in an arbitrary (possibly time-dependent) spherically symmetric four-dimensional spacetime. The easiest way to perform this task is to compute the contribution of these modes to the effective action. We choose adapted coordinates for which the Euclidean metric takes the form

$$ds^2 = \gamma_{ab}(x^a) \, dx^a dx^b + \Omega^2(x^a) \omega_{ij} dx^i dx^j,$$

where the last term is the metric on S^2. Now one can expand the (scalar) matter field into spherical harmonics. For s-waves, $\phi = \phi(x^a)$, the action for the coupled gravitational and scalar field is

$$S = -\frac{1}{4} \int \left[\Omega^2 \, {}^\gamma\mathcal{R} + {}^\omega\mathcal{R} + 2(\nabla\Omega)^2 \right] \sqrt{\gamma} d^2 x + 2\pi \int \Omega^2 (\nabla\phi)^2 \sqrt{\gamma} d^2 x,$$

where ${}^\gamma\mathcal{R}$ is the scalar curvature of the $2d$ space metric γ_{ab}, ${}^\omega\mathcal{R} = 2$ is the scalar curvature of S^2 and $(\nabla\Omega)^2 = \gamma^{ab}\partial_a\Omega\partial_b\Omega$. The purely gravitational part of the action is almost the action belonging to $2d$ dilatonic gravity with two exceptions: first, the numerical coefficient in front of $(\nabla\Omega)^2$ is different and second, the action is not invariant under Weyl transformation due to the ${}^\omega\mathcal{R}$ term. The action is quite different from the actions usually considered in $2d$ (string-inspired) field theories, because of the unusual coupling of ϕ to the dilaton field Ω. Choosing isothermal coordinates, $\gamma_{ab} = e^{2\sigma}\gamma_{ab}^f$, where γ_{ab}^f is the metric of the flat $2d$ space, one arrives with ζ-function methods at the following exact result for the effective action for the s-modes [16]

$$\Gamma_s = {}^{(n)}\Gamma + {}^{(i)}\Gamma$$

$$^{(n)}\Gamma[\sigma, \Omega] = \frac{1}{8\pi} \int \left(\frac{1}{12} {}^{\gamma}\mathcal{R} \frac{1}{\triangle_\gamma} {}^{\gamma}\mathcal{R} - \frac{\triangle_\gamma \Omega}{\Omega} \frac{1}{\triangle_\gamma} {}^{\gamma}\mathcal{R} \right) \sqrt{\gamma} d^2 x$$

$$^{(i)}\Gamma[\Omega] = \Gamma_s[\sigma = 0, \Omega] = \frac{1}{2} \log \det \left(-\triangle_f + \frac{\triangle_f \Omega}{\Omega} \right).$$

The second contribution $^{(i)}\Gamma$ is invariant under $2d$ Weyl transformation, whereas the first one is not. Unfortunately, the determinant cannot be calculated exactly and one must resort to some perturbation expansion. For details I refer to [16]. Ignoring backscattering one finds

$$^{(1)}\Gamma = \frac{1}{8\pi} \int \left(\frac{1}{12} {}^{\gamma}\mathcal{R} \frac{1}{\triangle_\gamma} {}^{\gamma}\mathcal{R} - \frac{\triangle_\gamma \Omega}{\Omega} \times \left[1 + \log \frac{\triangle_\gamma \Omega}{\mu^2 \Omega} \right] \right) \sqrt{\gamma} d^2 x.$$

Due to backscattering one needs to add the following term:

$$^{(2)}\Gamma = -\frac{\xi}{12 \cdot 8\pi} \int \left({}^{\gamma}\mathcal{R} \frac{1}{\triangle_\gamma} {}^{\gamma}\mathcal{R} + \text{local terms} \right) \sqrt{\gamma} d^2 x,$$

where $\xi \sim 0.9$. From the action $\Gamma_2 = {}^{(1)}\Gamma + {}^{(2)}\Gamma$ one obtains $\langle T_{\mu\nu} \rangle$ by variation with respect to the metric. To get the flux of the Hawking radiation we need to continue back to Lorentzian spacetime by changing the signs in the appropriate places. According to [23] we arrive at the in-vacuum energy-momentum tensor by replacing $-1/\triangle$ by the retarded Green function. Neglecting backscattering, the luminosity of the black hole is found to be

$$L = -\frac{\pi}{12} \frac{1}{(8\pi M)^2}.$$

This coincides with the total s-wave flux of the Hawking radiation obtained with other methods [5] without taking backscattering effects into account. With backscattering, the Hawking radiation is modified and compares well with that obtained by other means [24].

5   Wave equation in Schwarzschild spacetime

We study the classical wave propagation of a Klein-Gordon scalar field in fig.2. At late times, one expects that every solution will propagate into the black hole region II and/or propagate to $\mathcal{J}^+$.
In the spherically symmetric spacetime we may set

$$\phi = \frac{f(t, r)}{r} Y_{lm} e^{-i\omega t}$$

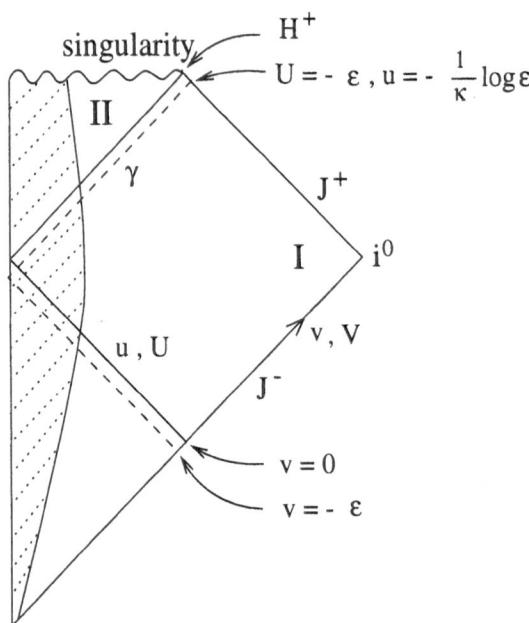

Fig. 2 *The propagation of particles in the geometric optics approximation.*

and the wave equation $(\Box + m^2)\phi = 0$ reduces to the radial equation

$$\frac{\partial^2 f}{\partial t^2} - \frac{\partial^2 f}{\partial r_*^2} - V(r_*)f = 0, \quad V(r_*) = \left(1 - \frac{2M}{r}\right)\left(\frac{2M}{r^3} + \frac{l(l+1)}{r^2} + m^2\right), \quad (26)$$

where M is the mass of the black hole and m that of the Klein-Gordon field. As $r_* \to -\infty$ (i.e. $r \to 2M$) the potential falls off exponentially, $V \sim \exp(r_*/2M)$, and as $r_* \to \infty$ the potential behaves as $\sim m^2 - 2Mm^2/r_*$ in the massive case and $\sim l(l+1)/r^2$ in the massless case. In the asymptotic region $r \to \infty$ this equation possesses outgoing solution $\sim e^{i\omega r_*}$ and ingoing solutions $\sim e^{-i\omega r_*}$. In terms of the null-coordinates the asymptotic solutions look like

$$f_\omega^{out} \sim e^{-2i\omega u} \quad \text{and} \quad f_\omega^{in} \sim e^{-2i\omega v}. \quad (27)$$

Consider a geometric optics approximation in which a particle's world line is a null ray, γ, of constant phase u and trace this ray backwards in time from $\mathcal{J}^+$. The later it reaches $\mathcal{J}^+$ the closer it must approach H^+. As $t \to \infty$ the ray γ becomes a null geodesic generator γ_H of H^+. We specify γ by its affine distance from γ_H along an ingoing null geodesic through H^+ (see fig.3a). The affine parameter on the ingoing null geodesic is U, so that according to (19)

$$U = -\epsilon \Rightarrow u = -\frac{1}{2\kappa}\log\epsilon, \quad f_\omega^{out} \sim \exp\left(\frac{i\omega}{\kappa}\log\epsilon\right).$$

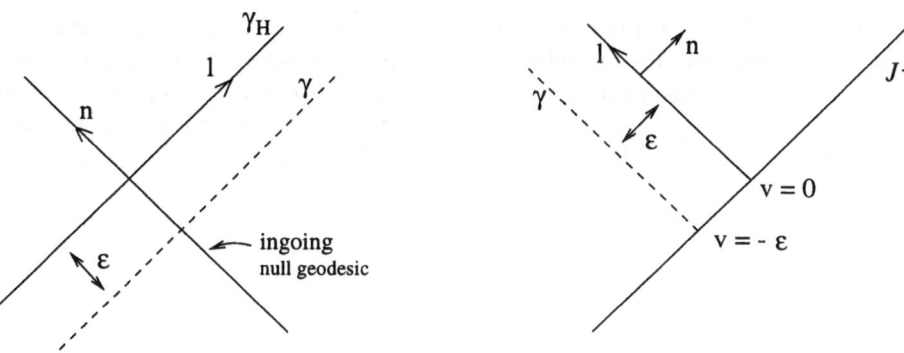

Fig. 3 *The particle's world line γ in relation to γ_H and the parallel-transport of n and l along the continuation of γ_H back to $\mathcal{J}^-$.*

This oscillates rapidly at later times t and this justifies the geometric optics approximation. Now we must match f_ω^{out} onto a solution near $\mathcal{J}^-$. In our approximation we just need to parallel-transport n and l along the continuation of γ_H back to $\mathcal{J}^-$. We choose v such that this continuation meets $\mathcal{J}^-$ at $v = 0$. The continuation of γ will meet $\mathcal{J}^-$ at an affine distance ϵ along an outgoing null geodesic on $\mathcal{J}^-$. Since $ds^2 = 4dudv + \ldots$ on $\mathcal{J}^-$ the coordinate $2v$ is the affine parameter measuring this distance, so $2v = -\epsilon$ on γ and

$$f_\omega \sim \exp\left(\frac{i\omega}{\kappa}\log(-2v)\right)\theta(-v),$$

where we took into account, that null rays with $v > 0$ do not reach $\mathcal{J}^+$. Now we take the Fourier transform

$$\tilde{f}_\omega(\omega') = \int\limits_{-\infty}^{0} e^{2i\omega' v} f_\omega(v)\, dv = \frac{1}{2}\int\limits_{0}^{\infty} \tilde{v}^{i\omega/\kappa} e^{-i\omega' \tilde{v}}\, d\tilde{v}, \quad \omega' > 0.$$

Using (11) one sees, that

$$\tilde{f}_\omega(\omega') = -e^{\pi\omega/\kappa}\tilde{f}_\omega(-\omega') \quad \text{for} \quad \omega' > 0.$$

It follows, that a mode of positive frequency ω on $\mathcal{J}^+$ matches onto mixed positive and negative frequency modes on $\mathcal{J}^-$. We see, that the Bogolubov coefficients are related by $\beta_{ij} = -\exp(-\pi\omega_i/\kappa)\alpha_{ij}$. From the Bogolubov relations (4) one then gets

$$\left(\beta\beta^\dagger\right)_{ii} = \frac{1}{e^{2\pi\omega_i/\kappa} - 1}. \tag{28}$$

For calculating the late time particle flux through $\mathcal{J}^+$ we need the inverse β-coefficients, $\beta' = -\beta^t$. One easily finds, that $\langle N_i\rangle_{\mathcal{J}^+} = (\beta'^\dagger\beta')_{ii} = (\beta\beta^\dagger)_{ii}$. This is the Planck-distribution at the Hawking temperature $T_H = \hbar\kappa/2\pi$.

The detailed form of the potential in (26) is irrelevant in the geometric optics
approximation. But the incoming waves will partially scatter off the gravitational
field (on the l-dependent potential V in (26)) to become a superposition of
incoming and outgoing waves. The backscattering is a function of ω and the
spectrum is not precisely Planckian. The total luminosity of the hole is given by

$$L = \frac{1}{2\pi} \sum_{l=0}^{\infty} (2l+1) \int_0^{\infty} d\omega \; \omega \frac{\Gamma_{\omega l}}{e^{8\pi M\omega} - 1}. \qquad (29)$$

A black hole is actually grey, not black. The dependency on the angular momentum (and spin) of the particles resides in the grey-body factor $\Gamma_{\omega l}$.

6   Back-reaction

The main effect of the quantum field will be a decrease of M at the rate at
which energy is radiated to infinity by particle creation. Since the spacetime is
static outside the collapsing matter, the expected energy current $J_\mu = \langle T_{\mu\nu} \rangle K^\nu$
is conserved in that region. The calculation showed, that there will be a steady
nonzero flux F. In [21] the contribution of the different particle species to this
flux has been determined. The contribution of massive particles of rest mass m
is exponentially small if $m > \kappa$. Black holes of mass $M > 10^{17}$g can only emit
neutrinos, photons and gravitons. Black holes of mass $5 \cdot 10^{14}$g$\leq M \leq 10^{17}$g
can also emit electrons and positrons. Black holes of smaller mass can emit
heavier particles. A non-rotating black hole emits almost as a body heated to
the temperature

$$T[^0\text{K}] = \frac{\hbar\kappa}{2\pi c} = \frac{\hbar c^3}{8\pi GkM} \sim 10^{26} \frac{1}{M[\text{g}]}.$$

The deviation from thermal radiation is due to the frequency dependence of
the penetration coefficient $\Gamma_{s\omega l}$. This coefficient is also strongly spin-dependent,
$\Gamma_{s\omega l} \sim \omega^{2s+1}$. As spin increases, the contribution of particles to the radiation of
a non-rotating black hole diminishes. The distribution of the radiated particles
in different mass-intervals is shown in the following table:

M [g]	$L \left[\frac{\text{erg}}{\text{sec}}\right]$	particles radiated
$M > 10^{17}$	$3.5 \times 10^{12} \left(\frac{10^{17} g}{M}\right)^2$	81.4% $\nu_e, \bar{\nu}_e, \nu_\mu, \bar{\nu}_\mu$ 16.7% γ 1.9% g
$10^{17} > M > 5 \times 10^{14}$	$6.3 \times 10^{16} \left(\frac{10^{15} g}{M}\right)^2$	45% $\nu_e, \bar{\nu}_e, \nu_\mu, \bar{\nu}_\mu$ 9% γ 1% g 45% e^-, e^+
$10^{14} > M > 10^{13.5}$	$10^{19} \left(\frac{10^{14} g}{M}\right)^2$	48% $\nu_e, \bar{\nu}_e, \nu_\mu, \bar{\nu}_\mu$ 28% e^-, e^+ 11% γ 1% g 12% $N, \bar{N}$

The following formula describes the rate of mass loss

$$-\frac{dM}{dt} \sim 4 \cdot 10^{-5} f \cdot \left(\frac{m_{pl}}{M}\right)^2 \frac{m_{pl}}{t_{pl}} = 7.7 \cdot 10^{24} f \cdot \left(\frac{1}{M[\text{g}]}\right)^2 \frac{\text{g}}{\text{sec}} = \frac{\alpha}{M^2}. \quad (30)$$

The contributions of the (massless) particle species are encoded in $f(M)$. From Page we take

$$f = 1.02 h(\frac{1}{2}) + 0.42 h(1) + 0.05 h(2),$$

where $h(s)$ is the number or distinct polarisations of spin-s particles. The rate equation (30) is easily integrated to yield

$$M(t) = \left(M_0^3 - 3\alpha t\right)^{1/3},$$

We see that a black hole radiates all of its mass in a finite time $\tau \sim M_0^3/3\alpha$. Inserting for α yields

$$\tau \sim 10^{71} \left(\frac{M}{M_\odot}\right)^3 \text{sec}.$$

If primordial black holes of mass $\sim 5 \cdot 10^{14}$g were produced in the early universe, they would be in the final stages of evaporation now. Primordial black hole of smaller mass would have already evaporated and contributed to the γ-ray background. See the review of Carr [22] for the possibility of observing quantum explosions of small black holes.

The magnitude of the Kruskal coordinate components of $\langle T_{\mu\nu}\rangle_H$ near the black hole are found to be of order $1/M^4$ in Planck units, as expected on dimensional grounds. Since the background curvature is of order $1/M^2$ the quantum field should only make a small correction to the structure of the black hole for $M \gg 1$, or $M \gg 10^{-5}$g.

7   Generalisations and Discussion

In the previous section we have studied the Hawking effect in the case of the Schwarzschild black hole. Lets us consider now different generalisations of this effect and its possible consequences.

Hawking radiation of rotating and charged holes. The *Kerr solution* has null-hypersurfaces at

$$r = r_\pm = M \pm \sqrt{M^2 - a^2},$$

where $a = J/M$, which are Killing horizons of the Killing fields

$$K_\pm = k + \Omega m = k + \left(\frac{a}{r_\pm^2 + a^2}\right)m \qquad k = \partial_t, \quad m = \partial_\phi,$$

with surface gravities

$$\kappa_\pm = \frac{r_\pm - r_\mp}{2(r_\pm^2 + a^2)}.$$

For the extreme Kerr solution with $a^2 = M^2$ the surface gravity vanishes.

For a Schwarzschild hole the number of particles per unit time in the frequency range ω to $\omega + d\omega$ passing out through a surface of the sphere is

$$\frac{1}{e^{8\pi M\omega} - 1}\frac{d\omega}{2\pi}.$$

For a Kerr Black hole ω is replaced by $\omega - m\Omega$ in this formula, where m is the azimuthal quantum number of the spheroidal harmonics, and Ω is the angular speed of the event horizon. Hence, the Planck factor at J^+ becomes

$$\frac{1}{e^{2\pi(\omega - m\Omega)/\kappa} \pm 1}, \qquad + \text{fermions}, \quad - \text{bosons}.$$

The emission is stronger for positive m than for negative m. In the boson case the Planck factor becomes negative when $\omega < m\Omega$ and super-radiance occurs: the effect of radiation amplifies the incoming classical wave with positive m. The result admits the following interpretation: Consider a rotating black hole enclosed in a mirror-walled cavity. A scattering of a 'particle' in a super-radiant mode by the black hole increases the number of quanta. After reflection by the mirror, these quanta are again scattered on the black hole and their number increases again, and so on. No stationary equilibrium distribution is possible for such modes. However, if the size of the cavity is not too large, $r < 1/\Omega$, then the super-radiative modes are absent and equilibrium is possible. A related effect is that the rotation of the hole enhances the emission of particles with higher spins.

For a charged hole with *Reissner-Nordström metric*

$$ds^2 = \alpha(r)dt^2 - \frac{1}{\alpha(r)}dr^2 + r^2 d\Omega^2, \qquad \alpha(r) = 1 - \frac{2M}{r} + \frac{q^2}{r^2}$$

the event horizon is at $r = r_+ = M + \left(M^2 - q^2\right)^{1/2}$ and the surface gravity is found to be

$$\kappa = \frac{1 - 16\pi^2 q^4 / A^2}{4M},$$

where $A = 4\pi r_+^2$ is the area of the horizon. If follows that the presence of the charge depresses the temperature $kT_H = \kappa/2\pi$ of the hole. For an extremal hole with charge $q = M$ or with $a^2 = M^2$ the Hawking temperature is zero, whereas the area is not ($A = 4\pi M^2$ for the extreme Reissner-Nordström hole). In the laws of black hole thermodynamics the entropy of a black hole is $S = A/4$ and hence non-vanishing for extreme black holes. The formulation of the third law, namely that $S \to 0$ as $T \to 0$, is not true for extremal holes[4]. The failure of the formulation of the third law may not be too disturbing. There other quantum systems with a degenerate ground state for which it fails as well.

Loss of Quantum Coherence. Consider the behaviour of the quantum field in the spacetime of a collapse, fig.4 in which back-reaction effects are not taken into account. The state of the field at late times in region I, and in particular the flux of thermal particles reaching infinity, must be described by a density matrix. The particles which entered the black hole at early times are correlated with the particles in region I. There is always a loss of information whenever one performs an inclusive[5] measurement outside the horizon. Such entropy increase is common to all inclusive measurements in physics. Perhaps we can understand this situation better if we recall the resolution of the well-known question raised by Einstein, Podolsky and Rosen. A pure quantum state is defined globally; its coherence may extend over field variables located at well-separated points on a space-like surface.

Let us distinguish between the set of out-states corresponding to particles moving away from the black hole (the visible ones) and those falling into the hole (the invisible ones). When one calculates expectation values $\langle A \rangle = (\psi, A\psi)$ of operators A depending only on the creation and annihilation operators belonging to the visible modes, this expectation value can be written as $\langle A \rangle = \operatorname{tr} \rho A$. In a Fock space construction one can derive an explicit formula for the density matrix ρ in terms of the pure state ψ. Here it suffices to sketch the emergence of a mixed state from a pure one. Let $\psi = \psi_i^I \otimes \psi_i^{II}$ be orthonormal pure states in the big Hilbert space $\mathcal{H} = \mathcal{H}_I \otimes \mathcal{H}_{II}$. Let us further assume that the observable A is the identity in $\mathcal{H}_{II}$. Then the expectation value

$$(\psi, A\psi) \quad \text{in the pure state} \quad \psi = \sum \alpha_i \psi_i^I \otimes \psi_i^{II}, \quad \sum |\alpha_i|^2 = 1$$

becomes

$$(\psi, A\psi) = \sum_{ij} \bar{\alpha}_i \alpha_j \left(\psi_i^I \otimes \psi_i^{II}, A\psi_j^I \otimes \psi_j^{II}\right) = \sum p_i(\psi_i^I, A\psi_i^I) = \operatorname{tr}(\rho A),$$

---

[4] see the contribution of Claus Kiefer: the canonical theory of gravity predicts $S(T \to 0) = 0$, whereas superstring-theory predicts $S(T \to 0) = A/4$.

[5] not all commuting observables are measured

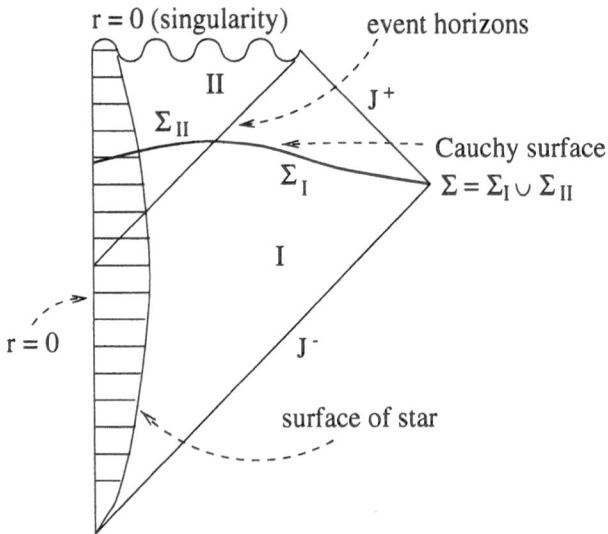

Fig. 4 *A conformal diagram of the spacetime resulting from a complete collapse of a spherical body. The region II lies outside of the chronological past of J^+.*

where $p_i = |\alpha_i|^2$ and $\rho = \sum p_i P_i$. The P_i are the projectors on the states ψ_i^I. We have used, that the ψ_i^{II} are orthonormal. Thus, if we are only measuring observables in the region I outside of the black hole and ignore the information about the inside, then pure states become indeed mixed states. For a black hole $\alpha_i \sim \exp(-\pi\omega_i/\kappa)$ (see (28)) and ρ is the thermal state. As is also clear, for operators A which are not the identity in $\mathcal{H}_{II}$ the expectation values $(\psi, A\psi)$ cannot be written as $\operatorname{tr} \rho A$.

Consider now the spacetime fig.5 in which back-reaction causes the black hole to 'evaporate'. The visible particles propagating to infinity can be described by a (thermal) density matrix. The particle creation and scattering will be described by a unitary S-matrix, provided that the invisible particles are represented in the 'out'-Hilbert space. What happens now when the black hole disappears from the spacetime? Apparently at late times, if one takes the 'out'-Hilbert space to be the Fock space associated with visible particles, the entire state of the field is mixed. Then one cannot describe particle creation and scattering by a unitary S-matrix, since an initial pure state evolved into a density matrix. This is the phenomenom of *loss of quantum coherence*. What are the possible ways out of this problem? A complete calculation including all back-reaction effects might resolve the issue, but even this is controversial, since the resolution very probably requires an understanding of the Planck scale physics. For example, QFT predicts that $T_{loc} \to \infty$ on the horizon of a black hole. This should not be believed when T reaches the Planck energy. The quantum aspects of gravity cannot be any longer ignored and this temperature is then of the order of the

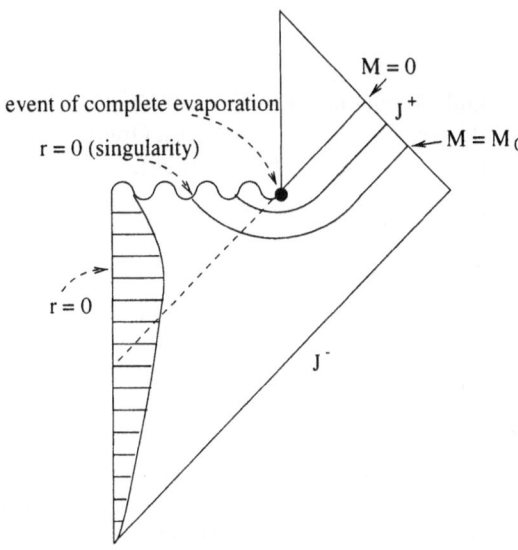

Fig. 5 *A conformal diagram of a spacetime in which black hole formation and evaporation occurs. The contour labelled $M = 0$ lies at the (retarded) time corresponding to the final instant of evaporation.*

maximum (Hagedorn) temperature of string theory[6].

A natural approach to dealing with this situation is to consider 'toy models', for example in two spacetime dimensions, in which the semiclassical analysis could be done. In lower dimensions one adds a 'dilaton' field to render gravity non-trivial (this field naturally arises in low energy string theory). The resulting two-dimensional theories are dynamically nontrivial and mimic many features of four-dimensional general relativity: they possess black-hole solutions, Hawking radiation and there exist laws of black hole thermodynamics which are completely analogous to the laws in four dimensions. Callen et.al [25] studied the model

$$S = \frac{1}{2\pi} \int d^2x \sqrt{-g} \left(e^{-2\sigma} \left[R + 4(\nabla\sigma)^2 + 4\lambda^2 \right] + \frac{1}{2}(\nabla f)^2 \right), \qquad (31)$$

containing a metric field $g_{\mu\nu}$, a dilaton field σ and a matter field f. The Hawking radiation of the f-'particles' can be calculated the way we explained in our two-dimensional model calculations above. So far these model calculations have not resolved the problems with the final stage of the black hole evaporations (the problems are the same as those with the Liouville theory at strong-coupling). A further simplification of (31) has been discovered by Russo, Susskind and Thorlacius [26]. Rather recent calculations seem to indicate[7] that information is not destroyed, but slowly released as the black hole decays back to vacuum [27].

---

[6] See the contribution of G. 't Hooft.

[7] See the contribution of C. Kiefer.

Bibliography

[1] Brunetti, R. and Fredenhagen, K. (1997), Los-Alamos e-print gr-qc/9701048, (Talk given at the Conference on Operator Algebras and Quantum Field Theory, Rome Italy, 1–6 July 1996).

[2] Bunch, T.S., (1981), Annals of Physics 131, 118.

[3] Parker, L. (1969), Phys. Rev. D 183, 1057.

[4] Hawking, S. (1975), Commun. Math. Phys. 43, 199.

[5] Birrell, N.D. and Davies, P.C.W. (1982), 'Quantum Fields in Curved Space', CUP; Fulling, S.A. (1989), 'Aspects of Quantum Field Theory in Curves Spacetime', CUP; Wald, R.M. (1994), 'Quantum Theory in Curved Spacetime and Black Hole Thermodynamics', The University of Chicago Press; Novikov, I.D. and Frolov, V.P. (1989), 'Physics of Black Holes', Kluver Acad. Publ., London.; Audretsch, J. (1990), in 'Quantum Mechanics in Curved Spacetime', eds. Audretsch, and de Sabbata, Plenum, New York

[6] Geroch, R. (1979), J. Math. Phys. 11, 437.

[7] Hawking, S.W. and Ellis, G.F.R. (1973), ' The Large Scale Structure of Spacetime', CUP.; Townsend, P.K. (1997), Los Alamos e-print gr-qc/9707012, (Lecture Notes given as part of the Cambridge University Mathematical Tripos).

[8] Ashtekar, A. and Magnon, A. (1975), Proc. Roy. Soc. London A346, 375.

[9] Kay, B. (1978), Commun. Math. Phys. 62, 55.

[10] Unruh, W.G. (1976), Phys. Rev. D14, 870.

[11] Fulling, S.A., Sweeny, M., and Wald, R.M (1978), Commun. Math. Phys. 63, 257.

[12] Tolman, R.C. and Ehrenfest, P. (1930), Phys. Rev. 36, 1791.

[13] Wald, R.M. (1984), Phys. Rev. D17, 1477.

[14] De Witt, B.S. (1975), Phys. Rep. C19, 297; Grib, A.A., Mamayev, S.G. and Mostepanenko, V.M. (1994), 'Vacuum Quantum Effects in Strong Fields', Friedman Laboratory Publishing, St. Petersburg; Christensen, S.M. (1978), Phys. Rev. D17, 946.

[15] Wipf, A. and Sachs, I., (1995), Annals of Physics 249, 380.

[16] Mukhanov, V., Wipf, A., and Zelnikov, A. (1994), Phys. Lett. B 332, 283.

[17] Kubo, R. (1957), Phys. Soc. Japan 12, 570; Martin, P.C. and Schwinger, J. (1959), Phys. Rev. 115, 1342.

[18] Christensen, S.M. and Fulling S.A., (1977), Phys. Rev. D15, 2088.

[19] Page, D.N. (1982), Phys. Rev. D25, 1499; Elster, T. (1983), Phys. Lett. A94, 205.

[20] Howard, K.W. and Candelas, P. (1984), Phys. Rev. Lett. 53, 403; Elster, T. (1984), Class. Quant. Grav. 1, 43.

[21] Page, D.N. (1977), Phys. Rev. D14, 1360.

[22] Carr, B.J. (1983), in M.A. Markov and P.C. West (eds.), Quantum Gravity, Plenum Press, N.Y., p. 337.

[23] Barvinsky, A.O. and Vilkovisky, G.A. (1990), Nucl. Phys. B333, 471; Barvinsky, A.O. and Vilkovisky, G.A. (1987), Nucl. Phys. B282, 163.

[24] Simkins, R.D. (1986), Massive Particle Emission from Schwarzschild Black Hole, Thesis, Pennsylvania State University.

[25] Callen, C.G., Giddings, S.B., Harvey, J.A., and Strominger, A. (1992), Phys. Rev. D45, 1005.

[26] Russo, J.G., Susskind, L., and Thorlacius, R. (1993), Phys. Rev. D47, 533.

[27] Polchinsky, J. and Strominger, A. (1994), Phy. Rev. D50, 7403.

Towards a Full Quantum Theory of Black Holes

Claus Kiefer

Fakultät für Physik, Universität Freiburg, Hermann-Herder-Straße 3, D-79104
Freiburg, Germany

Abstract. This review gives an introduction to various attempts to understand the
quantum nature of black holes. The first part focuses on thermodynamics of black holes,
Hawking radiation, and the interpretation of entropy. The second part is devoted to
the detailed treatment of black holes within canonical quantum gravity. The last part
adds a brief discussion of black holes in string theory and quantum cosmology.

1 Introduction and Summary

It is of fundamental importance to obtain a full quantum description of black
holes. The reasons are of a technical, conceptual, and observational nature. Technical, because it provides a highly nontrivial application of quantum gravitational equations in the full, non-perturbative, regime. One of the main open
issues thereby is what substitutes the classical singularities in quantum theory.
Conceptual, because the present status of semiclassical approaches leads to problems such as the information loss problem, which can be satisfactorily dealt with
only in the full theory. Observational, because apart from potential cosmological data this is probably the only window to directly test a quantum theory of
gravity.

This goal has not yet been reached, since a consistent theory of quantum
gravity has not yet been constructed. Many quantum aspects of black holes,
however, have been understood in the last 25 years, which could lead the way
to a full understanding. This review article is intended to give a pedagogical
introduction to results which have been obtained in the framework of present
approaches towards quantum gravity.

In Sect. 2, I shall review the key issues which lead to the conclusion that
black holes are quantum objects. The issues are thermodynamics of black holes,
Hawking radiation, and the interpretation of black hole entropy. Since many of
these topics are discussed at great length by other lecturers, in particular by
't Hooft, Israel, Neugebauer, and Wipf, I shall present only those issues which I
consider to be of particular relevance.

Sect. 3 presents one approach towards a theory of quantum gravity in some
detail – the canonical quantisation of general relativity. This approach by itself
most likely leads to an effective theory only, but it is the most straightforward
approach available and offers by itself interesting insights into possible quantum
aspects of black holes. The issues addressed cover both applications of the "full"
theory (such as a wave function for the eternal Reissner-Nordström hole) and

the semiclassical expansion (such as the description of Hawking radiation and black hole entropy in the context of the Wheeler-DeWitt equation).

Sect. 4, finally, gives a brief introduction to superstrings and the issue of black hole entropy as being obtained from counting the states of D-branes. I shall also offer some speculations about the role of black holes in quantum cosmology.

2   Why Black Holes Are Quantum Objects

2.1   Thermodynamics of Black Holes

In the beginning of the seventies, a surprising analogy was discovered between black holes and thermodynamical systems in the framework of general relativity, see the lectures by Israel and Neugebauer in this volume. (Other reviews are, e.g., Bekenstein (1980), Wald (1994, 1998), and Kiefer (1997a)). This analogy is summarised in Table 1 (with an obvious notation):

Table 1 The laws of black hole mechanics

Law	Thermodynamics	Stationary Black Hole
Zeroth	T constant on a body in thermal equilibrium	surface gravity κ constant on the horizon of a black hole
First	$dE = TdS - pdV + \mu dN$	$d(mc^2) = \frac{\kappa c^2}{8\pi G}dA + \Omega dJ - \phi dq$
Second	$dS \geq 0$	$dA \geq 0$
Third	$T = 0$ cannot be reached	$\kappa = 0$ cannot be reached

In the following I shall mostly deal with nonrotating holes ($J = 0$), but often keep a nonvanishing charge q. This is not realistic from an astrophysical point of view, but provides an interesting nontrivial example which mimics in many examples the relevant case of rotating holes.

Some comments are appropriate for the Third Law, because this will also be relevant for Sect. 3. In ordinary thermodynamics, there exist various inequivalent formulations of this law. One version frequently used was introduced by Planck in 1911: The entropy S goes to zero (or a material-dependent constant) as the temperature T goes to zero. From this (and some mild assumptions) follows a weaker version: $T = 0$ cannot be reached in a finite number of steps, see e.g. Wilks (1961) for details. It is *this* version of the Third Law that was proven by Israel (1986) for black holes and that is stated in Table 1. (In the proof the validity of the weak energy condition for matter in a neighbourhood of the apparent horizon was used.)

$S \to 0$ as $T \to 0$ is very helpful in thermodynamics, since it allows one to determine the entropy from measurements of specific heats, C. It follows from Planck's version of the Third Law that $C \to 0$ as $T \to 0$, but not vice versa (as is sometimes erroneously stated). Planck's version is not always fulfilled; it is violated, for example, for glasses (which have a higher disorder than the corresponding cristalline state). Other examples include the molecule CH_3D (Straumann 1986) or a gas confined to a circular string at zero temperature (Wald 1997). From the point of view of quantum statistics it is clear that Planck's version holds if there is a unique non-degenerate ground state at $T = 0$. This is violated in these examples.

The above analogy between black hole mechanics and ordinary thermodynamics holds in a much more general framework than general relativity, see Iyer and Wald (1994, 1995), and Wald (1998). If one only assumes that the field equations follow from a diffeomorphism covariant Lagrangian, L, the First Law holds (whether a generalisation of the area theorem holds is not clear).

The term[1] $\kappa dA/8\pi G$ occurring in the First Law is replaced by

$$d \int_{\mathcal{C}} Q = \frac{\kappa}{2\pi} d\mathcal{S}, \quad \text{with } \mathcal{S} \equiv -2\pi \int_{\mathcal{C}} \frac{\delta L}{\delta R_{abcd}} n_{ab} n_{cd} \ , \tag{1}$$

where Q is the 2-form associated with the Killing field ξ normal to the horizon, where the presence of a bifurcate Killing horizon is assumed ($\mathcal{C}$ is the bifurcation surface); n_{ab} denotes the binormal to $\mathcal{C}$ ($\nabla_a \xi_b = \kappa n_{ab}$). For the special case of general relativity, $L = R\sqrt{-g}/16\pi G$, the corresponding expression in Table 1 is recovered. If one, on the other hand, assumes beforehand that $S \propto A$, the Einstein field equations must hold (Jacobson 1995).

For generalisations of the laws of black hole mechanics to cases where non-abelian matter fields are present I refer to Heusler (1996), and the references therein.

For completeness I want to mention another, different, analogy between black holes and statistical mechanics: Choptuik (1993) discovered through numerical studies that if a spherical wave packet of a massless scalar field collapses, there exists a critical parameter (characterising the strength of the ensuing gravitational self-interaction of the field) above which no black hole forms. In the vicinity of this critical parameter there is a universal relation for the black hole mass like in the vicinity of a critical point in statistical mechanics.

2.2   Hawking Radiation

The analogies between ordinary thermodynamics and black hole mechanics, summarised in Table 1, were first regarded as purely formal, since classically a black hole cannot radiate (it behaves like an ideal absorber). Can quantum theory change this conclusion? One could imagine that $T_{BH} \propto \hbar$ and $S_{BH} \propto \hbar^{-1}$; in

---

[1] From now on we set $c = 1$.

fact, from dimensional arguments one recognises that to achieve $T_{BH} \neq 0$ one would have $T_{BH} \propto \hbar\kappa/k_B$ and $S_{BH} \propto k_B A/G\hbar$, since no other fundamental constants are at one's disposal (at least within standard physics).

Using quantum field theory on a curved background spacetime, Hawking (1975) was able to show that black holes *do* in fact radiate and have a finite entropy. The temperature is

$$T_{BH} = \frac{\hbar\kappa}{2\pi k_B} \; , \tag{2}$$

and the entropy therefore from the First Law

$$S_{BH} = \frac{k_B A}{4G\hbar} \; . \tag{3}$$

This is a very general result, since no use of particular gravitational field equations was made.

This is the reason why black hole thermodynamics seems to hold in a much wider framework, see (1). One there has the formal expression $S_{BH} = k_B S/\hbar$, which would thus give a general local geometric notion of black hole entropy. However, no quantum field theoretical calculation has been made to justify this interpretation.

For later convenience I give the explicit expressions for a Reissner-Nordström black hole (a charged spherically symmetric black hole),

$$T_{BH} = \frac{\hbar}{8\pi k_B G m} \left(1 - \frac{q^4}{R_0^4} \right) \; , \tag{4}$$

where

$$R_0 = Gm + \sqrt{(Gm)^2 - q^2} \tag{5}$$

is the radius of the event horizon. The entropy is

$$S_{BH} = \frac{k_B}{G\hbar} \pi R_0^2 \; . \tag{6}$$

An extremal hole is defined by $|q| = Gm$; its temperature thus vanishes, while its entropy appears to be nonvanishing (and not a constant). It thus seems as if Planck's version of the Third Law were violated, but the situation for extremal holes is more subtle, as will be discussed in Sect. 3. Holes with $|q| > Gm$ exhibit a naked singularity and are therefore generally excluded from consideration, although their role within quantum gravity is unclear.

How can one interpret Hawking radiation? The central point is that the notion of vacuum (and therefore also the notion of particles) loses its invariant meaning in the presence of a dynamical background. Incoming modes of the quantum field are redshifted while propagating through the collapsing geometry, which is why the quantum state of the outgoing modes is different. If the initial state is a vacuum state, the outgoing state contains "particles". The redshift is especially high near the horizon, where the modes spend a long time before escaping to infinity. This is the reason why Hawking radiation is present very long after the collapse is finished for a comoving observer, contrary to what one would naively expect. The presence of the horizon is also responsible for the thermal nature of the radiation, since no particular information about the details of the collapse can enter. It turns out that the vacuum expectation value of the energy-momentum tensor of the quantum field is negative near the horizon, corresponding to a flux of negative energy *into* the hole (this is the basis for the pictorial interpretation of the Hawking effect, where one partner of a pair of virtual particles can fall into the hole, thus enabling the other partner to become real and escape to infinity, where it can be observed as Hawking radiation). For details of this scenario, I refer to e.g. Wipf (this volume), 't Hooft (1996, and this volume), Birrell and Davies (1982), Wald (1994), and the references therein. The negativity of this expectation value is, like the Casimir effect, a genuine quantum feature.

This negative energy flux leads to a decrease of the black hole mass and is equal to the positive flux of the Hawking radiation at infinity. From a simple application of Stefan-Boltzmann's law, one can heuristically estimate that the time $t(m_P)$ for the hole to lower its mass to roughly the Planck mass $m_P \equiv \sqrt{\hbar/G}$ is $t(m_P) \propto m_0^3$, where m_0 is the initial mass of the hole. After this stage is reached, the semiclassical calculations used by Hawking (1975) are expected to break down. It is one of the most interesting open features of a full quantum gravity to provide a detailed understanding of this final phase.

How can one observe the Hawking effect? It is easy to estimate that for an initial mass of about one solar mass, $m_0 \approx m_\odot$, $t(m_P) \approx 10^{65}$yrs, which is much longer than the age of the Universe. Before this time the radiation is much too weak to be noticeable. The effect can thus *not* be observed for black holes originating from stellar collapse. Only if primordial black holes were left over from the Big Bang, would there be a hope of observation (if the initial mass of the hole is $m_0 \approx 10^{15}$g,[2] the final stages of the primordial hole would occur "today"). The amount of primordial holes is strongly constrained by the smoothness of the Big Bang, see Sect. 4. It is thus not clear whether this effect is observable at all. Bousso and Hawking (1997) have investigated pair creation of black holes during an inflationary phase in the early Universe. By applying the no-boundary proposal of Hartle and Hawking (1983), they estimated that no significant number of neutral holes having sufficient initial mass survive inflation.

If "hot" black holes were around, they would contribute to the observed γ-ray background. Before the final evaporation (about which nothing is known), the spectrum should according to (2) be thermal. Since this is not true for the γ-ray background, one finds from observations that the number of primordial holes must be less than about 10^4 per pc^3 (Page and Hawking 1976). Wright (1996) estimated from the anisotropy component of the γ-ray background in the halo of the Milky Way an upper limit of 0.4 explosions of primordial holes per pc^3 and year.

It may also be possible that the existence of primordial black holes can be inferred

---

[2] The size of such a hole would be only about 10^{-13}cm!

from the variation of quasar luminosities (Hawkins 1993), although this is at present a contentious issue.

It must be mentioned that there exists an effect analogous to the Hawking effect in Minkowski space, discovered by Unruh (1976). An observer with uniform acceleration a observes thermal radiation in the Minkowski vacuum state with a temperature

$$T_U = \frac{\hbar a}{2\pi k_B} . \tag{7}$$

The common feature with the black hole case is the presence of a *horizon* which in particular is responsible for the thermal nature of the radiation. In fact, (7) directly follows from (2) upon replacing the surface gravity κ by a. Israel (1976) showed that observers whose observations are limited by a horizon see a "thermal vacuum state". This follows after summing over the unobservable states behind the horizon. It must be emphasised that near the horizon the black hole geometry resembles the geometry of Rindler spacetime ('t Hooft 1996), which is the spacetime appropriate for an accelerated observer.

For a quasistationary observer near a black hole (i.e., at a fixed radial distance r from the hole), Hawking effect and Unruh effect are intertwined through the formula

$$T_{BH}(r) = \frac{\hbar \kappa}{2\pi k_B \chi(r)} , \tag{8}$$

where $\chi(r)$ is the redshift factor of the black hole geometry, and the spherically-symmetric case was assumed. (A position-dependent temperature is a typical feature of gravitational systems.) In the limit $r \to \infty$ the Hawking effect (2) is recovered (thermal radiation at infinity), while for $r \to R_0$, the effect is purely one of acceleration and (7) is recovered. This "thermal atmosphere" near the horizon plays an important role in many discussions of black hole entropy, see below.

An interesting connection between the Unruh effect and the Schwinger effect (pair creation of charged particles in an external electric field) was discussed by Parentani and Massar (1997). This analogy enabled them to associate a formal entropy with the Unruh effect, $S_U = k_B \pi M^2 / e\mathcal{E}\hbar$, where $\mathcal{E}$ is the constant accelerating electric field, and M is the mass of the charged particle. With $a = e\mathcal{E}/M$ one has $S_U \propto M^2$ and $T_U \propto M^{-1}$, i.e. a formal analogy to the Hawking effect (although with a different interpretation, since here M refers to the quantum field, while in the black hole case, m refers to the classical black hole mass).

Can the Unruh effect (7) be observed? Bell and Leinaas (1987) discussed the motion of electrons in storage rings. For such circular motion, the effect is not purely thermal, since there is no horizon. Still, this effect leads to a change in the spin polarisation of the electron, which may be obervable. However, present measurements of this polarisation are not precise enough to unambiguously uncover such an effect from the data.

A related effect (quantum radiation by moving interfaces between different dielectrics) could be responsible for sonoluminescence (light emission by sound-driven air bubbles in water), which until now remains unexplained, see Eberlein (1996). This is undecided at the moment.

2.3   Interpretation of Entropy

If black holes can be attributed a genuine entropy, see (3), the question arises whether a generalised Second Law of the kind

$$\frac{\mathrm{d}}{\mathrm{d}t}(S_{BH} + S_M) \geq 0 \tag{9}$$

holds, where S_M denotes the entropy of ordinary matter. This was investigated in many special situations, and numerous gedankenexperimente have shown that (9) in fact holds, i.e. that there exists no perpetuum mobile of the second kind in black hole physics. A typical situation is one where a box containing thermal radiation (this maximises the matter entropy) is lowered in a quasistationary manner towards a black hole, into which the radiation is then thrown, see Bekenstein (1980), and Israel (this volume). Unruh and Wald (1982) have shown that there is a minimal change of entropy if the box is opened at the floating point given by the Archimedean principle (weight of box is equal to the buoyancy from the Unruh radiation), which is just enough to save the Second Law (9). In this discussion the relation (8) plays an important role.

Frolov and Page (1993) have given a proof for the generalised Second Law (9) under the assumptions that one remains within the semiclassical approximation and that a special initial state (no correlation between modes coming out of the past horizon and modes coming in from past null infinity) is chosen. The choice of a special initial state is of course a necessary prerequisite for any derivation of a Second Law, see Zeh (1992) and Sect. 4.

The above discussion remains fully within the context of phenomenological thermodynamics (similar to discussions in the last century before the advent of the molecular hypothesis). A most interesting question is then whether S_{BH} can be derived from quantum statistical considerations,

$$S_{BH} \stackrel{?}{=} -k_B \mathrm{Tr}(\rho \ln \rho) \equiv S_{SM} \tag{10}$$

with an appropriate density matrix ρ. This is a key issue in the process of understanding black holes in quantum gravity. Does black hole entropy, for example, correspond to the large number of states which may be hidden behind the horizon? Or does it correspond to the large number of possible initial states? *Where* is the entropy located (if at all)? These question may indicate the kind of questions that arise.

Using a flat space example (with a surface that separates two regions and that mimics a horizon), Bombelli et al. (1986), and Srednicki (1993) have argued that the entropy is located near the horizon. This may also be suggested by the presence of the thermal atmosphere there, see the discussion after (8). In the black hole context, this was investigated by Frolov and Novikov (1993).

They showed that by counting internal degrees of freedom one gets $S_{SM} \propto A$. All these authors found, however, a divergent prefactor. Although lying inside, these degrees of freedom are located mainly in the vicinity of the horizon. An attempt to show that (10) can be derived from the number of possible initial configurations of the hole was made by Zurek and Thorne (1985).

A concrete realisation of the ideas of Frolov and Novikov (1993) was done by Barvinsky, Frolov, and Zelnikov (1995). They consider a quantum state for the black hole and make the ansatz that this state is constructed from the no-boundary proposal of Hartle and Hawking (1983). The wave function is defined on three-dimensional geometries and matter fields thereon, see Sect. 3. The three-geometry is taken to be the Einstein-Rosen bridge $\Sigma \equiv \mathbb{R} \times S^2$.[3] The density matrix ρ_{in} of the black hole is then obtained from this pure state by tracing out all degrees of freedom outside the horizon. For the statistical mechanical entropy this leads to

$$S_{SM} = -k_B \mathrm{Tr} \left(\rho_{in} \ln \rho_{in} \right) = k_B \frac{A}{360 \pi l^2} \ , \tag{11}$$

where l is a cutoff parameter (proper distance to the horizon). One recognises that one gets a divergent result for $l \to 0$. (Taking for l the Planck length $l_P \equiv \sqrt{G\hbar}$ would yield a result proportional to (3).) It is speculated that a finite result is obtained after the quantum gravitational "uncertainty" of the horizon is taken into account, see also Sect. 4.

Since

$$\mathrm{Tr} \left(\rho_{in} \ln \rho_{in} \right) = \mathrm{Tr} \left(\rho_{out} \ln \rho_{out} \right)$$

(see e.g. p. 297 in Giulini et al. (1996)), the result $S \propto A$ also follows in approaches where the degrees of freedom lie *outside* the horizon. An example is the "brick wall model" of 't Hooft (1996), see also his contribution to this volume.

The above result by Barvinsky, Frolov, and Zelnikov (1995) arises entirely from the "one-loop level" of the wave function (that is the level of the WKB prefactor). Usually, however, S_{BH}, Eq. (3), is recovered solely from the classical action, which corresponds to the "tree level" of approximation. Since this latter type of derivation plays a crucial role in many discussions, and will in particular be of some relevance in Sect. 3, a brief overview will now be given.

The origin of these discussions goes back to Gibbons and Hawking (1977) who extended the analogy between path integrals and partition sums to gravitational systems. This analogy, on the other hand, was introduced within ordinary statistical mechanics by Feynman and Hibbs (1965).

Consider the partition sum of the canonical ensemble,

$$e^{-\beta F} \equiv Z = \mathrm{Tr} e^{-\beta \hat{H}} \ , \tag{12}$$

---

[3] It is shown that this state is equal to the so-called Hartle-Hawking vacuum state which is relevant for eternal holes, see Hartle and Hawking (1976). This thus provides an example where both types of "Hartle-Hawking" agree.

where $\beta = (k_B T)^{-1}$, and F is the free energy. On the other hand, the quantum mechanical kernel of the evolution operator reads

$$G(x,t;x',0) = \langle x|e^{-it\hat{H}/\hbar}|x'\rangle = \int \mathcal{D}x(\tau)\ e^{iS[x(\tau)]/\hbar}\ ,\qquad(13)$$

where also its expression in terms of path integrals is given (the paths going through x' at time 0 and through x at time t). For simplicity, I have suppressed all indices which may be attached to x.

The partition sum Z can be evaluated in this way, if one transforms $t \to -i\beta\hbar$ and performs a trace:

$$Z = \int dx\ G(x,-i\beta\hbar;x,0) = \int \mathcal{D}x(\tau)\ e^{-I[x(\tau)]/\hbar}\ .\qquad(14)$$

The paths go now from x at "time" 0 back to x at "time" $\beta\hbar$. (I denotes the euclidean action.) To express Z in this way is especially suited for perturbation theory, see Feynman and Hibbs (1965). If the Hamiltonian has the standard form

$$\hat{H} = \frac{\hat{p}^2}{2m} + V(\hat{x})\ ,\qquad(15)$$

one finds in perturbation theory (the "small" parameter being $\beta\hbar$) for the free energy the expression (see standard books on statistical mechanics)

$$F = F_0 + \frac{\hbar^2\beta^2}{24m}\langle(V'(x))^2\rangle\ ,\qquad(16)$$

where the expectation value is performed with respect to the canonical ensemble. The first term, F_0, gives the classical value for the free energy ("tree level"). It follows from evaluating the classical action upon classical trajectories. Because the action contains an integration from 0 to $\beta\hbar$, for small $\beta\hbar$ (corresponding to $\hbar \to 0$ or $T \to \infty$) the result for F_0 is independent of $\hbar$. The second term in (16) describes the "one-loop level" of the perturbation. It follows from an evaluation of the quadratic fluctuations around the classical action. (There is no term linear in $\hbar$.)

If Z (or F) is known, all other thermodynamic quantities (in particular the entropy) can be calculated. The mean value of the Hamiltonian is

$$\langle\hat{H}\rangle \equiv E = -\frac{\partial \ln Z}{\partial \beta}\ ,\qquad(17)$$

the entropy is given by

$$S = k_B(\ln Z + \beta E) = \frac{E-F}{T} = -\frac{\partial F}{\partial T}\ .\qquad(18)$$

One also has $S \approx k_B \ln g(E)$, where $g(E)$ is the number of states in the energy interval given by the mean square deviation of the energy. The specific heats can be inferred from second derivates of the partition sum,

$$C = k_B \beta^2 \frac{\partial^2 \ln Z}{\partial \beta^2} = k_B (\Delta \hat{H})^2 \beta^2 = -\beta \frac{\partial S}{\partial \beta} . \tag{19}$$

Gibbons and Hawking (1977) now used a (formal) quantum gravitational path integral to evaluate the partition sum in the gravitational context, see also Hawking (1979) and Hawking and Penrose (1996). In contrast to the above standard context, the euclidean viewpoint is there assumed to be fundamental and not just a convenient rewriting of the original lorentzian theory.

The path integral cannot, of course, be evaluated exactly (and it is unclear, whether it can be rigorously defined in quantum gravity). One can, however, resort to a steepest descent (saddle point) approximation, where only the first (and sometimes the second) contribution is taken into account. The first contribution is just the classical action evaluated for a classical solution of Einstein's equations. The next order takes into account the standard WKB-prefactor.

The euclidean action of vacuum general relativity without cosmological constant reads

$$I = -\frac{1}{16\pi G} \int d^4x \; R\sqrt{g} + \frac{1}{8\pi G} \int d^3x \; (K - K^0)\sqrt{h} . \tag{20}$$

In the volume term, R denotes the four-dimensional Ricci scalar, and g the determinant of the four-dimensional metric. In the boundary term, K denotes the trace of the extrinsic curvature, and h the determinant of the three-dimensional metric. For purposes of regularisation in the asymptotically flat case, the trace of the extrinsic curvature K^0 of the same boundary embedded in *flat* space has to be subtracted.

If one considers spherically symmetric uncharged black holes, one has to evaluate (20) for the euclidean Schwarzschild solution (the generalisation to $q \neq 0$ is straightforward). For this solution $R = 0$, and there is thus no contribution from the volume term. The whole contribution (which I shall call I^*) thus arises from the boundary which here is the t-axis times a sphere of large radius. This is a typical feature of black hole physics, which we shall encounter again in the course of this lecture.

To evaluate the partition sum one has to start from the expression (14), where one has to sum over all four-dimensional metrics instead of just paths $x(\tau)$. In the saddle point approximation one has (denoting with g symbolically the four-dimensional metric),

$$Z = \int \mathcal{D}g(x)e^{-I[g(x)]/\hbar} \approx \exp(-I^*/\hbar) = \exp\left(-\frac{(\beta\hbar)^2}{16\pi G\hbar}\right) . \tag{21}$$

It is due to the fact that only the *boundary term* of the euclidean action contributes to (21), that the lowest order approximation of the path integral (the "tree level") depends already quadratically on β. As one recognises from (16) and the discussion following it, in the standard situation β occurs quadratically only at the next order.

From (17) one immediately finds

$$\langle \hat{H} \rangle = E = \frac{\hbar\beta}{8\pi G} = m \tag{22}$$

which leads to the expression (4) for the temperature (with $q = 0$). From (18) one finds for the entropy

$$S = k_B (\ln Z + \beta m) = \frac{\hbar\beta^2}{16\pi G} = \frac{k_B A}{4G\hbar} = S_{BH} \ . \tag{23}$$

If $\ln Z$ had only a linear dependence on β, the entropy would turn out to be zero.

From (19) one finds $C = -\beta^2 \hbar / 8\pi G$ and thus a *negative specific heat!* This is in particular in conflict with the positivity of $(\Delta \hat{H})^2$ und means, of course, that the black hole is unstable in asymptotically flat space, as can immediately be inferred from the inverse mass dependence of the Hawking temperature (4). As such, this is not very surprising, since instability is typical for gravitational phenomena (Zeh 1992). This negativity is therefore not an artifact of the tree-level approximation.

Davies (1977) showed that for rotating or charged holes, the specific heat can become positive for $J/m \gtrsim 0.68Gm$ (rotating holes, where J is the angular momentum) and $q \gtrsim 0.86Gm$ (charged holes).

In the attempt to find a thermodynamically stable situation, Gibbons and Perry (1978) considered a microcanonical ensemble of a black hole immersed in a bath of radiation with fixed volume: They found that at a sufficiently high energy density a black hole will nucleate from a box containing radiation, in the same way as a liquid drop can condense out of saturated vapour. However, to obtain stability the black hole mass m must be about 98% of the total energy, which means that the radiation cannot serve as a heat bath for the hole.

In a canonical ensemble description, the specific heat can be made positive if the black hole is put into a box (York 1986, 1991). At the boundary of the box, boundary conditions must be specified, i.e. in the Schwarzschild case one can fix the temperature of the box and its radius r_B. It follows then that stability can be achieved for $2Gm < r_B < 3Gm$, i.e. only for a very small box.

Alternatively, one can use a microcanonical description, where the energy (and other extensive variables) are fixed at the boundary (Brown and York 1993). This is very natural for gravitating systems where energy can be expressed as a surface integral. Instead of the euclidean path integral (14) for the canonical partition sum, one can express the density of states $\nu(E)$ directly as a *lorentzian* path integral,

$$\nu(E) = \int \mathcal{D}x(t)\ e^{iS_E[x(t)]/\hbar} \ ,$$

where S_E is Jacobi's action in which the energy is fixed. The sum goes over all paths that are periodic in real time. This path integral may be defined even in cases where the canonical partition function (which follows via an integral transform) is divergent. Brown and York (1993) showed that $\ln \nu \approx A/4G\hbar$, as long as the black hole can be described semiclassically by any real stationary axisymmetric black hole.

If the hole is charged, one must in addition fix the charge at the boundary or, alternatively, the electric potential, see Braden et al. (1990).

Iyer and Wald (1995) gave a comparison between the Noether charge approach, see (1), and various euclidean approaches. They showed that the results agree in their respective domains of applicability, see also Brown (1995). It is interesting that $\exp(S_{BH})$ also gives the enhancement factor for the rate of black hole pair creation relative to ordinary pair creation, in accordance with the heuristic interpretation of this factor as the number of internal states of the hole.

Can these derivations of black hole entropy at the *tree* level be reconciled with the above-mentioned derivations at the *one-loop* level, see (11)? Problems arise due to the UV-divergencies connected with one-loop calculations: For renormalisation one needs to subtract the infinite quantity $S_{BH}(G_{bare})$ evaluated at the "bare" gravitational constant G_{bare}, a quantity that has no clear statistical mechanical meaning. As Frolov, Fursaev, and Zelnikov (1997) have shown, this difficulty can be avoided in theories where $G_{bare}^{-1} = 0$, such as Sakharov's induced gravity, see also Frolov and Fursaev (1998) for a review: If one includes there non-minimally coupled scalar fields or additional vector fields, one obtains a finite entropy that is equal to S_{BH}. In induced gravity, the dynamical degrees of freedom of the gravitational field arise from collective quantum excitations of heavy matter fields. The same fields produce S_{BH}, since the gravitational action is already itself a "one-loop effect". This result may also indicate why superstring theory, another "effective theory of gravity", allows one to reproduce S_{BH} from the counting of quantum states, see Sect. 4.

It was the intention of this section to give convincing arguments that black holes must be quantum objects and that they can be fundamentally understood only in the framework of quantum gravity. Before I shall discuss some approaches to quantum gravity in more detail, I want to remark that one can already speculate from the above results about some possible features of the full theory. One result of such a speculation is the intriguing feature of a possible area (and thus mass) quantisation for a black hole, see e.g. Bekenstein (1997), and the references therein. It was suggested from heuristic considerations that

$$A = 16\pi(Gm)^2 = 4G(\ln 2)\hbar n, \quad n \in \mathbb{N} \ . \tag{24}$$

This would already in the semiclassical theory change drastically the spectrum of black hole radiation. For example, no quanta would be emitted with frequencies lower than some fundamental frequency $(\ln 2)/8\pi Gm$, in contrast to the thermal nature of Hawking radiation. One could thus test this effect of quantum gravity already for $m \gg m_P$ (provided that primordial holes exist).

The result (2) of a *thermal* spectrum of black hole radiation was obtained in the semiclassical limit, where gravity is treated classically. If it were true even in the full theory of quantum gravity, it would mean that "information" were lost in the following sense: Since one can in principle start from any initial quantum state (even a pure one), its exact evolution into a thermal state would contradict the unitary evolution law of standard quantum theory. In this case, a theory of quantum gravity would possess some radical new features. Since, however, the full theory is not yet known, the answer to this *problem of information loss* is also not yet known (see, for example, the review in Giddings 1994). This "problem" may, however, serve as a useful leitmotif in the search for a full theory. How even the semiclassical limit might be altered has been mentioned in the context of (24). The effect of quantum gravitational corrections on this information loss will be briefly discussed in Sect. 3.2.

3   Black Holes in Canonical Quantum Gravity

3.1   A Brief Introduction into Canonical Gravity

Canonical quantum gravity is obtained via the application of standard canonical quantisation rules to the theory of general relativity (or some other classical theory, but I shall restrict myself to general relativity). Since this does not provide a unified description of all fields, it is expected that the resulting framework is only an effective theory. There is, however, the hope that canonical quantum gravity may reflect many of the features of a genuine quantum theory of gravity. Its formulation must be intrinsically non-perturbative, since general relativity is known to lead to a non-renormalisable quantum theory at the perturbative level. A perhaps more serious candidate for a genuine quantum theory of gravity unifying all interactions, superstring theory, is briefly described in the next section.

The canonical framework assumes that the classical spacetime $\mathcal{M}$ is globally hyperbolic, $\mathcal{M} = \Sigma \times \mathbb{R}$, such that a $3 + 1$ decomposition (a foliation into spacelike hypersurfaces) can be performed. This is already of relevance for the classical theory because it allows one to pose a well-defined Cauchy problem (e.g. in numerical relativity, see the contribution of Seidel to this volume). A $3 + 1$ formulation is required because the canonical approach is a Hamiltonian formulation of the theory. Due to the four-dimensional diffeomorphism invariance ("coordinate invariance" in spacetime), the classical theory contains four constraints at each space point, one Hamiltonian!constraint,

$$\mathcal{H} \approx 0, \tag{25}$$

and three spatial diffeomorphism constraints ("coordinate invariance" on the three-dimensional spatial hypersurface Σ),

$$\mathcal{D}_a \approx 0. \tag{26}$$

Here, as usual, $\approx$ denotes the weak equality in the sense of Dirac.

The canonical configuration variable can be chosen to be the three-dimensional metric $h_{ab}(\vec{x})$ on Σ, and the canonical momentum is then a linear function of the extrinsic curvature of Σ. To this one can add any matter fields in the standard manner. This constitutes the traditional, geometrodynamic, approach. Alternatively, one may choose a complex connection or so-called loop variables on Σ for the configuration variables. This brings in many formal similarities to Yang–Mills theories. I want to emphasise that the constraint structure (25, 26) is typical for all versions of canonical theories that possess a diffeomorphism invariance on the classical level, even if the specific form is different. This is the basis for the hope that these versions have important common features. Also superstring theory has a constraint structure, although its interpretation is somewhat different from here.

In the following I want to restrict myself to the quantisation method proposed by Dirac. This means to *formally* transform the above constraint equations into operator equations acting on physical states Ψ,

$$\hat{\mathcal{H}}\Psi = 0 \ , \tag{27}$$

and

$$\hat{\mathcal{D}}_a\Psi = 0 \ . \tag{28}$$

The wave functional Ψ depends, in the geometrodynamic approach, on the three-metric (as well as on non-gravitational fields), in the other approaches mentioned above on the complex connection or on loop variables.[4] Due to the constraints (28), the wave functional is invariant under three-dimensional coordinate transformations. This is often indicated by writing $\Psi[^3\mathcal{G}]$, where $^3\mathcal{G}$ means "three-geometry", although this is a loose notation, since Ψ cannot explicitly be given in this form.

If space is compact, there are no further constraints. If not, additional constraints arise from variables living at boundaries. This will be of particular relevance for our treatment of black holes, see Sect. 3.2.

It cannot be the purpose of this article to give a detailed introduction into this approach and its problems. A comprehensive reference is Ehlers and Friedrich (1994). A recent report on the connection and loops approaches can be found, for example, in Ashtekar (1997); a recent report on conceptual problems in Isham (1997). A comprehensive review of canonical quantum gravity as applied to cosmology is Halliwell (1991). The black hole examples discussed below may also be thought to give illustrative examples for the full framework.

A helpful analogy between ordinary (quantum) mechanics and (quantum) general relativity is given in Table 2.

The most important conceptual lesson from the above comparison is that spacetime has no fundamental meaning in canonical quantum gravity, in the

---

[4] In the latter cases there are also additional constraints coming from triad rotations.

Table 2 Comparison of mechanics and general relativity

Mechanics of one particle	General relativity
position q	geometry ${}^3\mathcal{G}$ of a three-dimensional space
trajectory $q(t)$	spacetime $\{{}^3\mathcal{G}(t)\} \equiv {}^4\mathcal{G}$
uncertainty between position and momentum	uncertainty between "space and time" (three-geometry and extrinsic curvature)
$\psi(q,t)$	$\Psi[{}^3\mathcal{G}, t] \equiv \Psi[{}^3\mathcal{G}]$

same way as a particle trajectory has no fundamental meaning in quantum mechanics. This fact lies behind the so-called "problem of time" in quantum gravity – the absence of any external time parameter in the constraint equations (27, 28), and the related problem of which Hilbert space (if any) to choose in quantum gravity. (This is way the quantum gravitational wave function in Table 2 is t-independent.) To a large extent, these issues are open, see e.g. Kiefer (1997b). Fortunately, in the black hole case, the "rest of the Universe" can be assumed to be in a semiclassical regime where a concept of time exists, so that some of the above conceptual problems don't have to be dealt with in the first place. These problems are, however, relevant if the whole Universe including the black hole is described in quantum terms, see Sect. 4.

A frequently employed approximation scheme is to perform a semiclassical expansion of the equations (27, 28), see Kiefer (1994). One writes the full wave functional as $\Psi \equiv \exp(\mathrm{i}S/\hbar)$ with some arbitrary complex function S which is expanded into powers of the gravitational constant: $S = G^{-1}S_0 + S_1 + GS_2 + \ldots$. This is then inserted into (27, 28), leading to equations at consecutive orders of G. It must be emphasised that this can be done only in a formal way, since it is unclear how to rigorously define the equations (27, 28). For finite-dimensional models it was shown by Barvinsky and Krykhtin (1993) and Barvinsky (1993) how up to "one loop" a consistent factor ordering and a consistent Hilbert space structure can be obtained. The important open issue is to find a consistent, anomaly-free, regularisation for their equations in the field theoretic case.

The highest order (G^1) yields the gravitational Hamilton-Jacobi equation for S_0. This is equivalent to the classical Einstein equations and corresponds to the "tree level" of the theory. A special solution S_0 thus corresponds to a family of classical spacetimes. The next order (G^0) leads to a functional Schrödinger equation for non-gravitational fields in a given background. It corresponds to the "one-loop" limit of quantum field theory in an external background, the limit in which the Hawking radiation is derived. Higher orders in G then lead to genuine quantum gravitational correction terms as well as back reaction terms from the non-gravitational fields onto the semiclassical background.

The approximation scheme sketched above is not unique. Alternative schemes can be found, e.g., in Bertoni, Finelli, and Venturi (1996), and Kim (1997). They differ from the above in the treatment of the back reaction of the non-gravitational fields.

The next section is devoted to the application of canonical methods to a particular situation: spherically symmetric black holes.

3.2 Quantisation of Spherically Symmetric Black Holes

The first model which I shall briefly describe is the case of spherically symmetric black holes. I shall begin with the so-called "eternal hole", where only the gravitational degrees of freedom (and, in the Reissner-Nordström case, the electromagnetic field) are taken into account. The more realistic case where additional dynamical fields (such as a scalar field) are present is discussed thereafter.

The eternal Schwarzschild hole was first discussed by Thiemann and Kastrup (1993), Kastrup and Thiemann (1994) within the connection dynamical approach and then by Kuchař (1994) in the geometrodynamical approach, see also Cavaglià, de Alfaro, and Filippov (1996). I shall follow the geometrodynamical approach and generalise it to include the Reissner-Nordström case, see also Louko and Winters-Hilt (1996). "Eternal" refers to the time-symmetric case where both a past and a future horizon are present ("complete Kruskal diagramme"). Such holes cannot result from a collapse. Although thus being unrealistic from an astrophysical point of view, eternal holes provide a useful (and relatively simple) framework for questions of principle.

Starting point is the ADM form for general spherical symmetric metrics on $\mathbb{R} \times \mathbb{R} \times S^2$:

$$ds^2 = -N^2(r,t)dt^2 + \Lambda^2(r,t)(dr + N^r(r,t)dt)^2 + R^2(r,t)d\Omega^2 . \quad (29)$$

The lapse function N encodes the possibility to perform arbitrary reparametrisations of the time parameter, while the shift function N^r is responsible for reparametrisations of the radial coordinate (this is the only freedom in performing spatial coordinate transformations that is left after spherical symmetry has been imposed). The parameter r is only a label for the spatial hypersurfaces; if the hypersurface extends from the left to the right wedge in the Kruskal diagramme, one takes $r \in (-\infty, \infty)$. If the hypersurface originates at the bifurcation point where path and future horizon meet, $r \in (0, \infty)$. If one has in addition a spherically symmetric electromagnetic field, one makes the following ansatz for the one-form potential:

$$A = \phi(r,t)dt + \Gamma(r,t)dr . \quad (30)$$

In the Hamiltonian formulation, ϕ as well as N and N^r are Lagrange multipliers whose variations yield the constraints of the theory. Variation of the Einstein-Hilbert action with respect to N yields the Hamiltonian constraint (25) which for the spherically symmetric model reads

$$\mathcal{H} = \frac{G}{2}\frac{\Lambda P_\Lambda^2}{R^2} - G\frac{P_\Lambda P_R}{R} + \frac{\Lambda P_\Gamma^2}{2R^2} + G^{-1}V_G \approx 0 \ , \tag{31}$$

where the gravitational potential term reads, explicitly,

$$V_G = \frac{RR''}{\Lambda} - \frac{RR'\Lambda'}{\Lambda^2} + \frac{R'^2}{2\Lambda} - \frac{\Lambda}{2} \ . \tag{32}$$

(A prime denotes differentiation with respect to r.) Variation with respect to N^r yields one (radial) diffeomorphism constraint (26),

$$\mathcal{D}_r = P_R R' - \Lambda P_\Lambda' \approx 0 \ . \tag{33}$$

One recognises from this constraint that R transforms as a scalar, while Λ transforms as a scalar density.

Variation of the action with respect to ϕ yields as usual the Gauß constraint

$$\mathcal{G} = P_\Gamma' \approx 0 \ . \tag{34}$$

The constraint (33) generates radial diffeomorphisms for the fields R, Λ and their canonical momenta. It does not generate diffeomorphisms for the electromagnetic variables. This can be taken into account if one uses the multiplier $\tilde{\phi} = \phi - N^r \Gamma$ instead of ϕ and varies with respect to $\tilde{\phi}$ (Louko and Winters-Hilt 1996), but for our purposes it is sufficient to stick to the above form (33).

The model of spherical symmetric gravity can be embedded into a whole class of models usually referred to as "two-dimensional dilaton gravity theories". This terminology comes from effective two-dimensional theories (usually motivated by string theory) which contain in the gravitational sector a scalar field (the "dilaton") in addition to the two-dimensional metric (of which only the conformal factor is relevant). Interest in such models arose after Callan et al. (1992) studied one model in detail (now called the CGHS model), in which they addressed the issues of Hawking radiation and back reaction[5]. This was facilitated by the fact that this model is classically soluble even if another, conformally coupled, scalar field is included. The canonical formulation of this model can be found, e.g., in Louis-Martinez, Gegenberg, and Kunstatter (1994) and Demers and Kiefer (1996). The dilaton field is analogous to the field R from above, while the conformal factor of the two-dimensional metric is analogous to Λ.

The dilaton model contains one non-trivial parameter, the constant λ which has the dimension of an inverse length. The corresponding Hawking temperature and entropy are given by, respectively,

$$T_{BH} = \frac{\hbar\lambda}{2\pi k_B}, \quad S_{BH} = \frac{2\pi k_B m}{\hbar\lambda} \ .$$

---

[5] A detailed review of two-dimensional black holes is Strominger (1995).

Note that the temperature is here independent of the black hole mass m, and that therefore the entropy is linear in m. This is also the reason why some aspects of this models are unrealistic from the four-dimensional point of view.

Coming back to the spherically symmetric model, consider first the boundary conditions for $r \to \infty$. (If $r \in (-\infty, \infty)$, there are analogous conditions for $r \to -\infty$ which will be ignored here, see Kuchař (1994).) For $r \to \infty$ one has in particular

$$\Lambda(r,t) \to 1 + \frac{Gm(t)}{r}, \ R(r,t) \to r, \ N \to N(t) , \tag{35}$$

as well as

$$P_\Gamma(r,t) \to q(t), \quad \phi(r,t) \to \phi(t) . \tag{36}$$

From the variation with respect to Λ one then finds the boundary term $G \int dt \ N\delta m$. In order to avoid the unwanted conclusion $N = 0$ (no evolution at infinity), one has to compensate this term in advance by adding the boundary term

$$-G \int dt \ Nm$$

to the classical action. Note that m is just the ADM!mass. The need to include such a boundary term was recognised by Regge and Teitelboim (1974). Similarly, for charged holes, one has to add the term

$$-\int dt \ \phi q$$

to compensate for $\int dt \ \phi\delta q$ which arises from varying P_Γ. If one wished instead to consider q as a given, external parameter, this boundary term would be obsolete.

As long as restriction is made to the eternal hole, appropriate canonical transformations allow to simplify the classical constraint equations considerably (Kuchař 1994, Louko and Winters-Hilt 1996). One gets

$$(\Lambda, P_\Lambda; R, P_R; \Gamma, P_\Gamma) \longrightarrow (\mathcal{M}, P_\mathcal{M}; \mathcal{R}, P_\mathcal{R}; Q, P_Q) .$$

In particular,

$$\mathcal{M}(r,t) = \frac{P_\Gamma^2 + P_\Lambda^2}{2R} + \frac{R}{2} \left(1 - \frac{R'^2}{\Lambda^2} \right) \overset{r \to \infty}{\Longrightarrow} m(t) \tag{37}$$

$$Q(r,t) = P_\Gamma \overset{r \to \infty}{\Longrightarrow} q(t) . \tag{38}$$

(I note that $\mathcal{R} = R$ and that the expression for $P_\mathcal{R}$ is somewhat lengthy and will not be given here.)

The new constraints, which are equivalent to the old ones, read

$$\mathcal{M}' = 0 \quad \Rightarrow \quad \mathcal{M}(r,t) = m(t), \tag{39}$$

$$Q' = 0 \quad \Rightarrow \quad Q(r,t) = q(t), \tag{40}$$

$$P_{\mathcal{R}} = 0 \quad . \tag{41}$$

Note that $N(t)$ and $\phi(t)$ are prescribed functions that must not be varied; otherwise one would be led to the unwanted restriction that $m = 0 = q$. This can be remedied if the action is parametrised, bringing in new dynamical variables,

$$N(t) =: \dot{\tau}(t),$$
$$\phi(t) =: \dot{\lambda}(t) \ . \tag{42}$$

Here, τ is the proper time that is measured with standard clocks at infinity, and λ is the variable conjugate to charge; λ is therefore connected with the elctromagnetic gauge parameter at the boundaries. In the canonical formalism one has to introduce momenta conjugate to these variables, which will be denoted π_τ and π_λ, respectively. This, in turn, requires the introduction of additional constraints linear in momenta,

$$\mathcal{C}_\tau = \pi_\tau + Gm \approx 0, \tag{43}$$

$$\mathcal{C}_\lambda = \pi_\lambda + q \approx 0 \tag{44}$$

which have to be added to the action:

$$-G \int dt \ m\dot{\tau} \quad \rightarrow \quad \int dt \ (\pi_\tau \dot{\tau} - N\mathcal{C}_\tau), \tag{45}$$

$$- \int dt \ q\dot{\lambda} \quad \rightarrow \quad \int dt \ (\pi_\lambda \dot{\lambda} - \phi\mathcal{C}_\lambda) \ . \tag{46}$$

The remaining constraints in this model are thus (41) and (43,44).

Quantisation proceeds then in the way sketched in Sect. 3.1 by acting with an operator version of the constraints on wave functionals $\Psi[\mathcal{R}(r); \tau, \lambda]$. Since (41) leads to $\delta\Psi/\delta\mathcal{R} = 0$, one is left with a purely quantum *mechanical* wave function $\psi(\tau, \lambda)$. The implementation of the constraints (43,44) then yields

$$\frac{\hbar}{i}\frac{\partial\psi}{\partial\tau} + m\psi = 0, \tag{47}$$

$$\frac{\hbar}{i}\frac{\partial\psi}{\partial\lambda} + q\psi = 0 \tag{48}$$

which can readily be solved to give

$$\psi(\tau,\lambda) = \chi(m,q)e^{-i(m\tau+q\lambda)/\hbar} \tag{49}$$

with an arbitrary function $\chi(m,q)$. Note that m and q are here considered as being fixed. The reason for this is that up to now we have restricted attention to one semiclassical component of the wave function only (eigenstates of mass and charge). Superpositions of states with different m and q can be made, and I shall make some remarks on this below.

If the hypersurface goes through the whole Kruskal diagramme of the eternal hole, only the boundary term at $r \to \infty$ (and an analogous one for $r \to -\infty$) contributes. Of particular interest in the black hole case, however, is the case where the surface originates at the *bifurcation surface* ($r \to 0$) of past and future horizons. This makes sense since data on such a surface suffice to construct the whole right Kruskal wedge, which is all that is accessible to an observer in this region. Moreover, this mimics the situation where a black hole is formed by collapse, in which the regions III and IV of the Kruskal diagramme are absent.

What are the boundary conditions that are adopted at $r \to 0$? They are chosen in such a way that the classical solutions have a nondegenerate horizon and that the hypersurfaces begin at $r = 0$ asymptotic to hypersurfaces of constant Killing time (Louko and Whiting 1995). In particular,

$$N(r,t) = N_1(t)r + \mathcal{O}(r^3), \tag{50}$$
$$\Lambda(r,t) = \Lambda_0(t) + \mathcal{O}(r^2), \tag{51}$$
$$R(r,t) = R_0(t) + R_2(t)r^2 + \mathcal{O}(r^4) . \tag{52}$$

Variation leads, similarly to the situation at $r \to \infty$, to a boundary term at $r = 0$:

$$-N_1 R_0 (G\Lambda_0)^{-1} \delta R_0 .$$

If $N_1 \neq 0$, this term must be subtracted ($N_1 = 0$ corresponds to the case of extremal holes, $|q| = Gm$, which is characterised by $\partial N/\partial r(r = 0) = 0$.) Introducing the notation $N_0 \equiv N_1/\Lambda_0$, the boundary term to be added to the classical action reads

$$(2G)^{-1} \int \mathrm{d}t \; N_0 R_0^2 .$$

The quantity

$$\alpha \equiv \int_{t_1}^{t} \mathrm{d}t \; N_0(t) \tag{53}$$

can be interpreted as a "rapidity" because it boosts the normal vector to the hypersurfaces $t = constant$, n^a, in the way described by

$$n^a(t_1) n_a(t) = -\cosh \alpha , \tag{54}$$

see Hayward (1993). To avoid fixing N_0, one introduces an additional parametrisation (Brotz and Kiefer 1997)

$$N_0(t) = \dot{\alpha}(t) . \tag{55}$$

Similarly to (45,46) above, one must replace in the action

$$(2G)^{-1} \int dt \, R_0^2 \dot{\alpha} \quad \rightarrow \quad \int dt \, (\pi_\alpha \dot{\alpha} - N_0 \mathcal{C}_\alpha) \ , \tag{56}$$

with the new constraint

$$\mathcal{C} = \pi_\alpha - \frac{A}{8\pi G} \approx 0 \ , \tag{57}$$

where $A = 4\pi R_0^2$ is the surface of the bifurcation sphere. One notes that α and A are canonically conjugate variables, see Carlip and Teitelboim (1995).

Quantisation then leads to (taking all constraints into account)

$$\psi(\alpha, \tau, \lambda) = \chi(A, m, q) \exp\left(\frac{i}{\hbar} \left[\frac{A}{8\pi G}\alpha - m\tau - q\lambda\right]\right) \ . \tag{58}$$

Since A occurs in the state (58), one may suspect that also the entropy comes into play here, see (3). However, (58) is a pure quantum state, which possesses vanishing entropy, and A is only part of its phase. The relation to entropy can only be achieved after an appropriate euclideanisation is performed, compare Sect. 2.3. This will be done below. (The wave function for a Reissner-Nordström hole, if an additional complex scalar field is coupled, can be found in Moniz (1997). In contrast to our model, his situation describes a dynamical evolution.)

The classical equations are found from (58) in the standard way by finding the extremum of the phase with respect to the parameters. For this to work, only two of the three parameters A, m, q can be considered as independent. (I shall choose m and q.) Differentiating the phase with respect to m and setting the result to zero yields

$$\alpha = 8\pi G \left(\frac{\partial A}{\partial m}\right)^{-1} \tau \ . \tag{59}$$

From Table 1 one recognises the occurrence of the surface gravity κ on the right-hand side of (59):

$$\alpha = \kappa\tau \ , \tag{60}$$

which is just the classical relation for the rapidity, see Brotz (1997). This is not surprising since it is known that boundary terms in the classical action are important in the derivation of the First Law of black hole mechanics (Wald 1998). Generally, conjugate quantities in thermodynamics (extensive – intensive) correspond to conjugate variables in the Hamiltonian formalism.

Differentiating the phase of (58) with respect to q and setting the result to zero yields

$$\phi = \frac{\kappa}{8\pi G}\frac{\partial A}{\partial q} = -\frac{\partial m}{\partial q}\Big|_A = -\frac{q}{R_0} \ , \tag{61}$$

another "thermodynamical" relation.[6] This completes the solution of the eternal Reissner-Nordström hole.

I shall now turn to the more realistic case where an additional dynamical field is present. This can be used to "form" the black hole in the first place, and leads to the emergence of interesting features such as Hawking radiation. It also provides an interesting application of the semiclassical expansion presented in Sect. 3.1. I denote the scalar field by f, see e.g. Romano (1995), Demers and Kiefer (1996) and Kuchař et al. (1997) for details of the formalism.

At order G^0, the total wave functional is of the form

$$\Psi \approx C^g e^{iS_0^g/\hbar} \bar{\chi} ,\tag{62}$$

where C^g and S_0^g depend only on the gravitational (and electromagnetic) variables. These variables comprise the functions $\Gamma(r), R(r), \Lambda(r)$ as well as the boundary variables α, τ, λ. The functional $\bar{\chi}$ depends, in addition, on the scalar field f. The important point is that $\bar{\chi}$ obeys a functional Schrödinger equation with respect to the background found from S_0^g.

As in the general case, S_0^g obeys the Hamilton-Jacobi equation for gravity (plus electromagnetic field). An explicit solution reads (Brotz and Kiefer 1997)

$$S_0^g = \int_0^\infty dr \left(q\Gamma + G^{-1}\Lambda F - G^{-1}\frac{RR'}{2} \ln \frac{R'/\Lambda + F/R}{R'/\Lambda - F/R} \right)$$
$$+ \frac{A\alpha}{8\pi G} - m\tau - q\lambda ,\tag{63}$$

where

$$F = R\sqrt{\frac{R'^2}{\Lambda^2} + \frac{2M(r)}{R} - 1}\tag{64}$$

and

$$M(r) = m - \frac{q^2}{2R(r)} .\tag{65}$$

Note that S_0^g depends parametrically on m and q which are just the mass and the charge of the hole, respectively. Expression (65) is nothing but the total energy of the hole. Inspection of (63) exhibits that the electromagnetic part in (62) from S_0^g is given by

$$\exp i \left(\int_0^\infty dr \Gamma - \lambda \right) q .$$

This expression can be understood as follows. The electromagnetic potential (30) changes under a gauge transformations according to

$$A \to \phi dt + \Gamma dr + d\xi = (\phi + \dot{\xi})dt + (\Gamma + \xi')dr .\tag{66}$$

---

[6] It corresponds to $\partial S/\partial N = -\mu/T$ with $\mu = -\phi$ and $N = q$ (N is the particle number and μ the chemical potential.)

Therefore,

$$\int_0^\infty dr\, \Gamma(r) \to \int_0^\infty dr\, \Gamma(r) + \xi(\infty) - \xi(0) \ .$$

Now, $\xi(\infty) - \xi(0)$ may be absorbed into λ, since λ itself was interpreted as the boundary gauge parameter.

Since the full theory is linear, one can perform arbitrary superpositions of states (62) with *different* values for m and q. These describe situations where the hole has neither a definite charge nor a definite mass. However, such superpositions can only be distinguished from a corresponding mixture if one could "measure" the variables conjugate to m and q, i.e. τ and λ. Otherwise, effective "superselection rules" would result, see Giulini, Kiefer, and Zeh (1995), and Chap. 6 of Giulini et al. (1996).

Another interesting situation is described by a superposition of the state (62) with its complex conjugate (this is possible since the full Wheeler-DeWitt equation is real). Such superpositions may follow in a natural way from appropriate boundary conditions (Hajicek 1992). It was shown in Demers and Kiefer (1996) that these superpositions (which can be heuristically interpreted as representing a superposition of a black hole with a white hole) become indistinguishable locally from a mixture after the irreversible interaction with the Hawking radiation is taken into account – a process known as decoherence (Giulini et al. 1996).

How can the Hawking radiation be found from a state such as (62)? This was clarified in Demers and Kiefer (1996) in the context of dilaton gravity (the extension to spherically symmetric gravity should be straightforward). One solves the functional Schrödinger equation obeyed by $\bar\chi$ in a background describing the collapse to a black hole. The initial state is taken to be a Gaussian (a "vacuum state"). During the evolution, this state remains a Gaussian, but with a different "width". This just expresses the fact, as mentioned in Sect. 2.2, that the notion of a vacuum becomes ambiguous in such a situation. Using the initial state as the reference vacuum state also at late times, the evolved state contains "particles" with respect to that vacuum. One has

$$\langle \bar\chi | \hat n(k) | \bar\chi \rangle = \frac{1}{\exp\left(\frac{\hbar|k|}{k_B T_{BH}}\right) - 1} \ , \tag{67}$$

where $\hat n$ denotes the "particle number operator" for the mode of wave number k with respect to the original vacuum. Note that, although $\bar\chi$ is a *pure* quantum state, the expectation value (67) is a Planckian distribution with respect to the Hawking temperature T_{BH}. The difference of $\bar\chi$ to a genuine mixture will be noticed if other expectation values (of "higher order operators") are performed.

For the important case where the surfaces are fixed at the bifurcation sphere, it turns out that the field f must vanish at this point for the state $\bar\chi$ to be normalisable. Thus, the bifurcation sphere acts like a "mirror" for this field. This is why the quantum state turns out to be a pure one. Other surfaces which penetrate the interior of the hole lead to a mixed state outside after the interior degrees of freedom are "traced out" (as in Israel 1976).

Can one go beyond the order of approximation (62)? This is in fact possible, but so far only in a formal way, without addressing in detail the issue of regularisation (Kiefer 1994). Still, however, qualitative features can be studied. At oder G^1, correction terms to the functional Schrödinger equation obeyed by $\bar{\chi}$ are obtained. Among these terms, there is an imaginary term, $i \mathrm{Im} H_m$, contributing to the effective matter Hamiltonian. In the case of collapse to a black hole, $\mathrm{Im} H_m < 0$ (Kiefer, Müller, and Singh 1994). Since the following equation holds for the density matrix ρ,

$$\frac{\mathrm{d}}{\mathrm{d}t} \left([\mathrm{Tr}\rho]^2 - \mathrm{Tr}\rho^2 \right) = 4 \mathrm{Tr} \left([\rho \mathrm{Tr}\rho - \rho^2] \mathrm{Im} H_m \right) \ , \tag{68}$$

one finds from $\mathrm{Im} H_m < 0$ that the difference between $(\mathrm{Tr}\rho)^2$ and $\mathrm{Tr}\rho^2$ decreases, corresponding to an increase in "purity" for the quantum state. Whether this may indicate a quantum gravitational "recovery of information" from the hole can of course only be judged from the full, as yet elusive, theory. This result at least demonstrates what kind of effects one might expect to see in higher orders of the semiclassical approximation.

At order G^1, also back reaction terms from the matter fields (here from the f-field) onto the gravitational background are found (Kiefer 1994). These can be evaluated only in special cases, for example in the toy model of a 2+1-dimensional black hole coupled to a conformal scalar field (Brotz 1998).

An interesting point is of course whether there are situations where the semiclassical approximation breaks down in the first place. This would mean that quantum gravity effects can become important below the Planck scale. Keski-Vakkuri et al. (1995), for example, arrived at the conclusion that the semiclassical approximation breaks down at the black hole horizon, in the sense that tiny fluctuations of the black hole mass may produce an immense change in the matter state. The physical implications of this result are not yet fully clear. It can also not be excluded that anomalies in quantum gravity spoil the above semiclassical limit and demand for an explicit modification of the constraints, see e.g. Cangemi, Jackiw, and Zwiebach (1996).

I emphasised above that there is not yet any connection with a notion of entropy for the pure quantum state (58). This can be established after some "euclideanisation" is performed, see the discussion in Sect. 2.3. How does this work? From (55) it is clear that the rapidity α is connected with the lapse function. Therefore, going to the euclidean regime means both $\tau \to -i\beta\hbar$, see (14), and $\alpha \to -i\alpha_E$. Regularity of the line element then demands that $\alpha_E = 2\pi$ (Brotz and Kiefer 1997). Consequently, the euclidean version of the quantum state (62) contains the term

$$\exp \left(-\beta m + \frac{A}{4\hbar G} \right) \ . \tag{69}$$

There is in addition the euclideanised version of the integral in (63) and the term containing $\lambda \to \lambda_E = -i\hbar\beta\phi$.

This does of course not yet yield a partition sum. However, after the whole semiclassical part is evaluated at the classical value for the Hamilton-Jacobi functional and a trace is performed, one finds by applying (17) that the second term in (69) is just the Bekenstein-Hawking entropy (3). Alternatively, one can interpret (69) as directly giving the enhancement factor for the rate of black hole pair creation relative to ordinary pair creation. Here my focus was just to show how the expression for S_{BH} emerges in the canonical formalism.[7]

Consider now the case of an extremal hole, where $|q| = Gm$. As can be immediately inferred from the discussion after (52), there is no surface term to consider, since $N_0 = 0$. Thus, $\alpha = 0$, and there is no A-term in (58). This would also mean that the entropy is zero. Recalling our discussion in Sect. 2.1, this shows that Planck's version of the Third Law is fulfilled. This result was also found in a variety of other approaches, see the references in Brotz and Kiefer (1997). It is *not* fulfilled in string theory, where $S_{extreme} = A/4\hbar G$, see Sect. 4. It is also not fulfilled for the extreme (Kerr) black hole which occurs in the transition from the disk of dust solution to the rotating black hole solution, see Neugebauer's contribution to this volume.

The above derivation of entropy via boundary terms suggests the following natural interpretation in terms of "missing information". For surfaces which in the classical spacetime correspond to slices through the *full* Kruskal diagramme, this "information" is maximal in the sense that one can recover the full spacetime from data on this surface. Since no boundary (except at infinity) is present, the entropy is zero. For slices that start at the bifurcation sphere, this information is less than maximal for Schwarzschild black holes and for non-extreme Reissner-Nordström black holes. They are therefore attributed the entropy $A/4\hbar G$. In contrast, the maximum information (for the full spacetime up to the Cauchy horizon) is already available for such slices in the extreme case, as can be easily recognised from the corresponding Penrose diagramme. Extreme holes are therefore attributed a vanishing entropy. A somewhat related interpretation was given in the path integral framework by Martinez (1995). An interesting point was raised by Ghosh and Mitra (1997) who argued that $S_{extreme} \neq 0$ follows from extremisation after quantisation, while $S_{extreme} = 0$ holds for extremisation before quantisation.

Can the quantisation of mass (or area), as described by (24), be found within the canonical formalism? This is, unfortunately, an open issue. One can, for example, postulate Bohr-Sommerfeld type of quantisation rules in the euclidean theory (Kastrup 1996). This would lead to

$$nh = \oint \pi_\alpha \, \mathrm{d}\alpha = \int_0^{2\pi} \frac{A}{8\pi G} \, \mathrm{d}\alpha = \frac{A}{4G} \; . \tag{70}$$

This is similar to (24), albeit with a different numerical factor. Whether a similar

---

[7] Due to Smarr's formula, (69) is consistent with (21).

result can be found in the physically relevant lorentzian theory remains open.

Other interesting developments can only be mentioned here. Carlip (1997) was able to give a statistical mechanical origin for the black hole entropy in the case of a 2+1-dimensional black hole. There it results from "would-be-gauge" degrees of freedom becoming dynamical at the horizon. Using the loop approach to canonical quantum gravity, Rovelli (1996) found that $S_{BH} \propto A$, although with a numerical coefficient different from (3).

To summarise, canonical quantum gravity can offer the tool to understand quantum features of black holes such as entropy and Hawking radiation. Still, however, the main problems are not yet solved: Can the Bekenstein-Hawking entropy for four-dimensional black holes be derived by counting appropriate degrees of freedom? What is the final evolution of a black hole, after the semiclassical approximation breaks down?

4   Further Developments

In Sect. 3 I discussed canonical quantum gravity as a possible framework to understand black holes. A different approach to quantum gravity is superstring theory. It necessarily contains gravity and gauge theories, and must implement supersymmetry for reasons of consistency.

Like canonical quantum gravity, string theory follows through the quantisation of a classical theory (a propagating string in some background spacetime), but is itself interpreted in a drastically different way: It is supposed to give a fundamental theory where all interactions including gravity are unified in a quantum framework. The background spacetime used in the construction of the theory plays only an auxiliary role. Like canonical quantum gravity, string theory suffers from the "problem of time", although this is not always stated clearly. The notion of spacetime again emerges only in an appropriate semiclassical limit. (The role of the semiclassical expansion parameter is here played by the string length, see below). An important fact in string theory is that consistency conditions (the absence of a Weyl anomaly) severely *restricts* the number of dimensions of this semiclassical spacetime, e.g. to $D = 10$ for the superstring. This, then, enforces the implementation of an appropriate mechanism to encurl the superfluous dimensions in a Kaluza-Klein type manner to avoid contradiction with observation. Whether the level of canonical quantum gravity, as discussed in Sect. 3, follows from string theory in an appropriate limit is not yet clear. It must, however, lead to *some* quantum gravitational corrections to the ordinary functional Schrödinger equation, and may thus lead to the possibility both to test the theory and to discriminate it from competitors like the approach presented in Sect. 3.

A detailed introduction into string theory can be found, for example, in Polchinski (1994, 1996), and the references therein. Here I only want to briefly sketch some intriguing recent developments aiming at a derivation of the black hole entropy (3) by counting quantum states, see Horowitz (1998) for a review.

String theory contains two important parameters: The string length l_S and the string coupling $g_S^2 \equiv \exp(2\varphi)$. Here, φ denotes the dilaton field which appears in the two-dimensional string action. It gives rise to the string coupling, since g_S^2 appears as a "gravitational constant" in the effective action (arising in the semiclassical approximation to lowest order in l_S) for the background spacetime and background fields. The Planck length, l_P, then appears as a *derived* quantity,

$$l_P \propto g_S l_S , \qquad (71)$$

and similar relations follow for other "coupling constants". It is important to note that the semiclassical approximation, and with it the notion of a spacetime metric, breaks down for curvatures bigger than l_S^{-2}.

How does the entropy of a black hole come into play? First, assigning an entropy to an excited string state by counting its degeneracy, it turns out that this entropy is (for high excitations) proportional to the energy (mass) of that state and not to the mass squared. It would thus seem as if a string had not enough states to yield the entropy of a black hole. The crucial point, however, is that the Planck length, and therefore the gravitational constant, *depends* on the string coupling, see (71). Thus, if g_S is increased, Gm is increased, too, and a black hole is formed at some stage (Horowitz 1998). Comparing, then, the black hole mass with the string mass at $l_S = R_0$ (R_0 is the Schwarzschild radius), it turns out that the black hole entropy becomes proportional to the string entropy. A string may thus possess enough states to give the Bekenstein-Hawking entropy.

For a quantitative comparison, one must give a precise calculation. It is most straightforward in this respect to first consider states which obey a relation similar to $q = Gm$ in the Reissner-Nordström case (although with generalised charges). Such states are called BPS states. At weak coupling ($g_S \ll 1$), one has bound states of so-called D-branes (Polchinski 1996) in flat space, and the number of these states can be counted. D-branes are dynamical objects of various dimensions, which are a necessary ingredient of string theory. As the coupling increases, the BPS-relation between mass and charges is preserved, and the number of states remains unchanged. For high coupling ($g_S \gg 1$), one thus obtains an extremal black hole with the same number of states. Surprisingly, its Bekenstein-Hawking entropy exactly coincides with the entropy of the D-branes in the flat space description (Strominger and Vafa 1996). The original calculation was for five-dimensional black holes and then generalised to four-dimensional holes. One may thus interpret the D-branes as giving the desired microscopic description for the black hole entropy. Since it turns out in this approach that $S_{extreme} = A/4\hbar G \neq 0$, string theory leads to a different result than the canonical treatment presented in Sect. 3.2.

The calculations have been extended to the case of near-extremal black holes which, in contrast to the extremal ones, exhibit Hawking radiation (here interpreted as the emission of closed strings from D-branes). It could be shown that even the *rate* of Hawking radiation agrees with the decay amplitude for the corresponding D-brane configuration (see e.g. Das 1997). Since all string calculations preserve unitarity, it seems that there is no violation of unitarity also in the black hole radiation. Consequently, there would be no "loss of information". Of course, to get a non-vanishing entropy in the first place, some coarse-graining must be involved, and the process of decoherence will again play a crucial role (Myers 1997). There will thus only be the apparent non-unitarity connected with the neglect of degrees of freedom be present – the total system evolves unitarily (Giulini et al. 1996).

Whether the above string result also holds for general black holes, i.e. far away from extremality (such as for the Schwarzschild black hole), is not yet clear. It must also be emphasised that all results are obtained in lowest order of l_S, i.e. in the lowest order of the semiclassical approximation where a background structure is available. The full, non-perturbative, evolution of a black hole therefore still remains mysterious.

In the semiclassical approximation to canonical quantum gravity, as presented in Sect. 3, a crucial role for the interpretation of entropy is played by the presence of boundary conditions at the bifurcation sphere (where the two horizons in the Kruskal diagramme meet). This, however, cannot be extended to the full theory in a straightforward manner. The main reason is that the horizon of a black hole is a *classical concept*. As I emphasised in Sect. 3.1, the canonical theory does not possess any notion of spacetime at the fundamental level, in the same way as ordinary quantum theory does not possess any notion of particle trajectories in the full theory. A horizon, however, is a genuine spacetime concept. Therefore, the results presented in Sect. 3.2 only hold as far as a notion of spacetime can be applied at least in some approximation.

That the concept of an event horizon is a classical artifact, becomes especially obvious in quantum cosmology (Zeh 1992). Consider, for example, the case of a Friedmann universe that classically recollapses. Since the entropy content of the present universe is far from maximal, it must have been very tiny at the big bang – the big bang was extremely smooth (which is why one would not expect to find many primordial black holes). This led Penrose (1979) to the formulation of his Weyl tensor hypothesis that the universe is homogeneous at the big bang, but not at the big crunch. In quantum gravity, however, there is no external time parameter which could possibly distinguish between big bang and big crunch. If entropy is small near the big bang, it must also be small near the big crunch, since both regions correspond to the *same* region of the quantum gravitational configuration space. The consequences of this fact for the arrow of time and for black holes were investigated in Kiefer and Zeh (1995). Entropy is always growing with increasing size of the Universe, leading to a (formal) reversal of the arrow of time near the classical turning point. The same boundary condition of low entropy at small size necessarily leads to the fact that neither an event horizon nor a singularity (naked or hidden) forms for a black hole. Cosmic censorship would thus be automatically implemented. Although still speculative, this scenario at least demonstrates what qualitatively new features emerge from quantum gravity if one leaves the semiclassical sector.

Acknowledgements. I am grateful to Thorsten Brotz, Valeri Frolov, Domenico Giulini, and H.-Dieter Zeh for a critical reading of this manuscript.

Bibliography

[1] Ashtekar, A. (1997): Polymer geometry at Planck scale and quantum Einstein equations. In *General Relativity and Gravitation*, edited by M. Francaviglia, G. Longhi, L. Lusanna, and E. Sorace (World Scientific, Singapore), p. 3–29

[2] Barvinsky, A.O. (1993): Operator ordering in theories subject to constraints of the gravitational type. Class. Quantum Grav. **10**, 1985–1999

[3] Barvinsky, A.O., Frolov, V.P., Zelnikov, A.I. (1995): The wave function of a black hole and the dynamical origin of entropy. Phys. Rev. D **51**, 1741–1763

[4] Barvinsky, A.O., Krykhtin, V. (1993): Dirac and BFV quantization methods in the 1-loop approximation: closure of the quantum constraint algebra and the conserved inner product. Class. Quantum Grav. **10**, 1957–1984

[5] Bekenstein, J.D. (1980): Black-hole thermodynamics. Physics Today (January), 24–31

[6] Bekenstein, J.D. (1997): Quantum black holes as atoms. Report gr-qc/9710076, to appear in the Proccedings of the VIII Marcel Grossmann Meeting, ed. by R. Ruffini and T. Piran (World Scientific, Singapore)

[7] Bell, J.S., Leinaas, J.M. (1987): The Unruh effect and quantum fluctuations of electrons in storage rings. Nucl. Phys. B **284**, 488–508

[8] Bertoni, C., Finelli, F., Venturi, G. (1996): The Born-Oppenheimer approach to the matter-gravity system and unitarity. Class. Quantum Grav. **13**, 2375–2383

[9] Birrell, N.D., Davies, P.C.W. (1982): *Quantum fields in curved space* (Cambridge University Press, Cambridge)

[10] Bombelli, L., Koul, R.K., Lee, J., Sorkin, R.D. (1986): Quantum source of entropy for black holes. Phys. Rev. D **34**, 373–383

[11] Bousso, R., Hawking, S.W. (1997): Black holes in inflation. Nucl. Phys. B (Proc. Suppl.) **57**, 201–205

[12] Braden, H.W., Brown, J.D., Whiting, B.F., York, J.W., Jr. (1990): Charged black hole in a grand canonical ensemble. Phys. Rev. D **42**, 3376–3385

[13] Brotz, T. (1998): Quantization of Black Holes in the Wheeler-DeWitt Approach. Phys. Rev. D **57**, 2349–2362

[14] Brotz, T., Kiefer, C. (1997): Semiclassical black hole states and entropy. Phys. Rev. D **55**, 2186–2191

[15] Brown, J.D. (1995): Black hole entropy and the Hamiltonian formulation of diffeomorphism invariant theories. Phys. Rev. D **52**, 7011–7026

[16] Brown, J.D., York, J.W., Jr. (1993): Microcanonical functional integral for the gravitational field. Phys. Rev. D **47**, 1420–1431

[17] Callan, C.G., Jr., Giddings, S.B., Harvey, J.A., Strominger, A. (1992): Evanescent black holes. Phys. Rev. D **45**, R1005–R1009

[18] Cangemi, D., Jackiw, R., Zwiebach, B. (1996): Physical States in Matter-Coupled Dilaton Gravity. Ann. Phys. (N.Y.) **245**, 408–444

[19] Carlip, S. (1997): Statistical mechanics and black hole thermodynamics. Nucl. Phys. B (Proc. Suppl.) 57, 8–12

[20] Carlip, S., Teitelboim, C. (1995): The off-shell black hole. Class. Quantum Grav. **12**, 1699–1704

[21] Cavaglià, M., de Alfaro, V., Filippov, A.T. (1996): Quantization of the Schwarzschild Black Hole. Int. J. Mod. Phys. D **5**, 227–250

[22] Choptuik, M.W. (1993): Universality and Scaling in Gravitational Collapse of a Massless Scalar Field. Phys. Rev. Lett. **70**, 9–12

[23] Das, S.R. (1997): D-brane decay and Hawking Radiation. Report hep-th/9709206

[24] Davies, P.C.W. (1977): The thermodynamic theory of black holes. Proc. R. Soc. Lond. A **353**, 499–521

[25] Demers, J.-G., Kiefer, C. (1996): Decoherence of black holes by Hawking radiation. Phys. Rev. D **53**, 7050–7061

[26] Eberlein, C. (1990): Theory of quantum radiation observed as sonoluminescence. Phys. Rev. A **53**, 2772–2787

[27] Ehlers, J., Friedrich, H. (1994), eds.: *Canonical Gravity: From Classical to Quantum* (Springer, Berlin)

[28] Feynman, R.P., Hibbs, A.R. (1965): *Quantum Mechanics and Path Integrals* (McGraw-Hill, New York)

[29] Frolov, V.P., Fursaev, D.V. (1998): Thermal Fields, Entropy, and Black Holes. Report hep-th/9802010

[30] Frolov, V.P., Fursaev, D.V., Zelnikov, A.I. (1997): Statistical origin of black hole entropy in induced gravity. Nucl. Phys. B **486**, 339–352

[31] Frolov, V., Novikov, I. (1993): Dynamical origin of the entropy of a black hole. Phys. Rev. D **48**, 4545–4551

[32] Frolov, V., Page, D.N. (1993): Proof of the Generalized Second Law for Quasistationary Semiclassical Black Holes. Phys. Rev. Lett. **71**, 3902–3905

[33] Ghosh, A., Mitra, P. (1997): Understanding the Area Proposal for Extremal Black Hole Entropy. Phys. Rev. Lett. **78**, 1858–1860

[34] Gibbons, G.W., Hawking, S.W. (1977): Action integrals and partition functions in quantum gravity. Phys. Rev. D **15**, 2752–2756

[35] Gibbons, G.W., Perry, M.J. (1978): Black holes and thermal Green functions. Proc. R. Soc. Lond. A **358**, 467–494

[36] Giddings, S.B. (1994): Quantum Mechanics of Black Holes. Report hep-th/9412138

[37] Giulini, D., Joos, E., Kiefer, C., Kupsch, J., Stamatescu, I.-O., Zeh, H.D. (1996): *Decoherence and the Appearance of a Classical World in Quantum Theory* (Springer, Berlin)

[38] Giulini, D., Kiefer, C., Zeh, H.D. (1995): Symmetries, superselection rules, and decoherence. Phys. Lett. **199A**, 291–298

[39] Hajicek, P. (1992): Quantum Mechanics of Gravitational Collapse. Commun. Math. Phys. **150**, 545–559

[40] Halliwell, J.J. (1991): Introductory lectures on quantum cosmology. In *Quantum cosmology and baby universes*, ed. by S. Coleman, J.B. Hartle, T. Piran, and S. Weinberg (World Scientific, Singapore), p. 159–243

[41] Hartle, J.B., Hawking, S.W. (1976): Path-integral derivation of black-hole radiance. Phys. Rev. D **13**, 2188–2203

[42] Hartle, J.B., Hawking, S.W. (1983): Wave function of the Universe. Phys. Rev. D **28**, 2960–2975

[43] Hawking, S.W. (1975): Particle Creation by Black Holes. Commun. Math. Phys. **43**, 199–220

[44] Hawking, S.W. (1979): The path-integral approach to quantum gravity. In *General Relativity*, ed. by S.W. Hawking and W. Israel (Cambridge University Press, Cambridge), p. 746–789

[45] Hawking, S.W., Penrose, R. (1996): *The Nature of Space and Time* (Princeton University Press, Princeton)

[46] Hawkins, M.R.S. (1993): Gravitational microlensing, quasar variability and missing matter. Nature (London) **366**, 242–245

[47] Hayward, G. (1993): Gravitational action for spacetimes with nonsmooth boundaries. Phys. Rev. D **47**, 3275–3280

[48] Heusler, M. (1996): *Black Hole Uniqueness Theorems* (Cambridge University Press, Cambridge)

[49] Horowitz, G.T. (1998): Quantum States of Black Holes. In: *Black Holes and Relativistic Stars*, ed. by R.M. Wald (The University of Chicago Press, Chicago), p. 241–266

[50] Isham, C.J. (1997): Structural issues in quantum gravity. In *General Relativity and Gravitation*, ed. by M. Francaviglia, G. Longhi, L. Lusanna, and E. Sorace (World Scientific, Singapore), p. 167–209

[51] Israel, W. (1976): Thermo-field dynamics of black holes. Phys. Lett. **57A**, 107–110

[52] Israel, W. (1986): Third Law of Black-Hole Dynamics: A Formulation and Proof. Phys. Rev. Lett. **57**, 397–399

[53] Iyer, V., Wald, R.M. (1994): Some properties of the Noether charge and a proposal for dynamical black hole entropy. Phys. Rev. D **50**, 846–864

[54] Iyer, V., Wald, R.M. (1995): Comparison of the Noether charge and Euclidean methods for computing the entropy of stationary black holes. Phys. Rev. D **52**, 4430–4439

[55] Jacobson, T. (1995): Thermodynamics of Spacetime: The Einstein Equation of State. Phys. Rev. Lett **75**, 1260–1263

[56] Kastrup, H.A. (1996): The quantum levels of isolated spherically symmetric gravitational systems. Phys. Lett. B **385**, 75–80

[57] Kastrup, H.A., Thiemann, T. (1994): Spherically symmetric gravity as a completely integrable system. Nucl. Phys. B **425**, 665–686

[58] Keski-Vakkuri, E., Lifschytz, G., Mathur, S.D., Ortiz, M.E. (1995): Breakdown of the semiclassical approximation at the black hole horizon. Phys. Rev. D **51**, 1764–1780

[59] Kiefer, C. (1994): The semiclassical approximation to quantum gravity. In: Ehlers and Friedrich (1994), p. 170–212

[60] Kiefer, C. (1997a): Quanteneigenschaften Schwarzer Löcher. Physik in unserer Zeit **28**, 22–30

[61] Kiefer, C. (1997b): Does time exist at the most fundamental level? In: *Time, Temporality,. Now*, ed. by H. Atmanspacher and E. Ruhnau (Springer, Berlin), p. 227-240

[62] Kiefer, C., Müller, R., Singh, T.P. (1994): Quantum Gravity and Non-Unitarity in Black Hole Evaporation. Mod. Phys. Lett. A **9**, 2661–2669

[63] Kiefer, C., Zeh, H.D. (1995): Arrow of time in a recollapsing quantum universe. Phys. Rev. D **51**, 4145–4153

[64] Kim, S.P. (1997): Problem of unitarity and quantum corrections in semi-classical quantum gravity. Phys. Rev. D **55**, 7511–7517

[65] Kuchař, K.V. (1994): Geometrodynamics of Schwarzschild black holes. Phys. Rev. D **50**, 3961–3981

[66] Kuchař, K.V., Romano, J.D., Varadarajan, M. (1997): Dirac constraint quantization of a dilatonic model of gravitational collapse. Phys. Rev. D **55**, 795–808

[67] Louis-Martinez, D., Gegenberg, J., Kunstatter, G. (1994): Exact Dirac quantization of all 2D dilaton gravity theories. Phys. Lett. B **321**, 193–198

[68] Louko, J., Whiting, B.F. (1995): Hamiltonian thermodynamics of the Schwarzschild black hole. Phys. Rev. D **51**, 5583–5599

[69] Louko, J., Winters-Hilt, S.N. (1996): Hamiltonian thermodynamics of the Reissner-Nordström-anti-de Sitter black hole. Phys. Rev. D **54**, 2647–2663

[70] Martinez, E.A. (1995): Microcanonical functional integral and entropy for eternal black holes. Phys. Rev. D **51**, 5732–5741

[71] Moniz, P.V. (1997): Wave function for the Reissner-Nordström black hole. Mod. Phys. Lett. A **12**, 1491–1505

[72] Myers, R. (1997): Pure states don't wear black. Gen. Rel. Grav. **29**, 1217–1222

[73] Page, D.N., Hawking, S.W. (1976): Gamma rays from primordial black holes. Astrophys. Journ. **206**, 1–7

[74] Parentani, R., Massar, S. (1997): Schwinger mechanism, Unruh effect, and production of accelerated black holes. Phys. Rev. D **55**, 3603–3613

[75] Penrose, R. (1979): Singularities and Time-Asymmetry. In *General Relativity*, ed. by S.W. Hawking and W. Israel (Cambridge University Press, Cambridge), p. 581–638

[76] Polchinski, J. (1994): What is string theory? Report hep-th/9411028

[77] Polchinski, J. (1996): TASI Lectures on D-branes. Report hep-th/9611050

[78] Regge, T., Teitelboim, C. (1974): Role of Surface Integrals in the Hamiltonian Formulation of General Relativity. Ann. Phys. (N.Y.) **88**, 286–318

[79] Romano, J.D. (1995): Spherical Symmetric Scalar Field Collapse: An example of the Spacetime Problem of Time. Report gr-qc/9501015

[80] Rovelli, C. (1996): Black Hole Entropy from Loop Quantum Gravity. Phys. Rev. Lett. **77**, 3288–3291

[81] Srednicki, M. (1993): Entropy and Area. Phys. Rev. Lett. **71**, 666–669

[82] Straumann, N. (1986): *Thermodynamik* (Springer, Berlin), p. 100

[83] Strominger, A. (1995): Les Houches Lectures on Black Holes. In: *Proceedings of the Les Houches Summer School, Session LXII, 2 August-9 September 1994: Fluctuating Geometries in Statistical Mechanics and Field Theory*, ed. by F. David, P. Ginsparg, and J. Zinn-Justin (Elsevier, Amsterdam)

[84] Strominger, A., Vafa, C. (1996): Microscopic origin of the Bekenstein-Hawking entropy. Phys. Lett. B **379**, 99–104

[85] 't Hooft, G. (1996): The scattering matrix approach to the quantum black hole: an overview. Int. J. Mod. Phys. A **11**, 4623–4688

[86] Thiemann, T., Kastrup, H.A. (1993): Canonical quantization of spherically symmetric gravity in Ashtekar's self-dual representation. Nucl. Phys. B **399**, 211–258

[87] Unruh, W.G. (1976): Notes on black-hole evaporation. Phys. Rev. D **14**, 870–892

[88] Unruh, W.G., Wald, R.M. (1982): Acceleration radiation and the generalized second law of thermodynamics. Phys. Rev. D **25**, 942–958

[89] Wald, R.M. (1994): *Quantum Field Theory in Curved Spacetime and Black Hole Thermodynamics* (University of Chicago Press, Chicago)

[90] Wald, R.M. (1998): Black Holes and Thermodynamics. In: *Black Holes and Relativistic Stars*, ed. by R.M. Wald (The University of Chicago Press, Chicago), 155–176 gr-qc/9702022

[91] Wald, R.M. (1997): The "Nernst Theorem" and Black Hole Thermodynamics. Phys. Rev. D **56**, 6467–6474

[92] Wilks, J. (1961): *The third law of thermodynamics* (Oxford University Press, Oxford)

[93] Wright, E.L. (1996): On the density of primordial black holes in the galactic halo. Astrophys. Journ. **459**, 487–490

[94] York, J.W., Jr. (1986): Black-hole thermodynamics and the Euclidean Einstein action. Phys. Rev. D **33**, 2092–2099

[95] York, J.W., Jr. (1991): Black holes and Partition Functions. In *Conceptual Problems of Quantum Gravity*, ed. by A. Ashtekar and J. Stachel (Birkhäuser, Boston), p. 573–596

[96] Zeh, H.D. (1992): *The Physical Basis of the Direction of Time* (Springer, Berlin)

[97] Zurek, W.H., Thorne, K.S. (1985): Statistical Mechanical Origin of the Entropy of a Rotating, Charged Black Hole. Phys. Rev. Lett. **54**, 2171–2175

Quantum Information on the Black Hole Horizon

Gerard 't Hooft

Institute for Theoretical Physics, University of Utrecht, Princetonplein 5, 3584 CC Utrecht, The Netherlands

Abstract. The scattering matrix approach to the black hole quantization problem is introduced and further elaborated. There appears to be a general concensus about the quantum degeneracy of black holes increasing exponentially with the horizon area. Attempts are described to reproduce the individual quantum states, in particular by exploiting an operator algebra that results from considerations of the gravitational back reaction.

1   Introduction

One would expect that the quantum mechanical properties of a black hole should follow naturally by applying large scale physics. Only the space-time region at the side of the observer, the "physical side" of the horizon, should be relevant. Indeed one can calculate accurately the quantum mechanical effects near a large black hole, as seen by an outside observer, by first studying what an infalling observer would experience, and then performing the appropriate general coordinate transformation. As is to be expected from quantum mechanical calculations, one finds "probabilities": chances that particles of certain types, with certain momenta, energies or other quantum numbers, emerge at certain places. It is when one wants to interpret these outcomes in terms of some Schrödinger equation for the black holes as a whole, that the first genuine problems emerge[1, 2]. In these lectures, the problems we encounter are exposed, and roads towards their resolution are explored.

2   The Space-time Metric of a Black Hole Under Formation

The space-time metric of a stationary, non-rotating and electrically neutral black hole is the *Schwarzschild metric*:

$$ds^2 = -\left(1 - \frac{2M}{r}\right)dt^2 + \frac{dr^2}{1 - 2M/r} + r^2 d\Omega^2 , \qquad (2.1)$$

where

$$d\Omega^2 \equiv d\theta^2 + \sin^2\theta\, d\phi^2 . \qquad (2.2)$$

Here, M stands for Gm_{BH}, where G is Newton's constant and m_{BH} is the black hole mass. Often we will employ the *Kruskal coordinates*[3], which we will write as (x, y, θ, ϕ), with

$$\left(\frac{r}{2M} - 1\right)e^{r/2M} = xy; \tag{2.3}$$

$$e^{t/2M} = x/y.$$

In terms of these coordinates we have (see Fig. 1)

$$\frac{dx}{x} + \frac{dy}{y} = \frac{dr}{2M(1 - 2M/r)}; \qquad \frac{dx}{x} - \frac{dy}{y} = \frac{dt}{2M}; \tag{2.4}$$

$$ds^2 = \frac{32M^3}{r}e^{-r/2M}\,dx dy + r^2 d\Omega^2.$$

The apparent singularity at the horizon, $r = 2M$, has disappeared. The only true singularities are at the curves $xy = -1$, where $r = 0$. The region $\{x > 0,\ y > 0\}$ is the "outside region", the only region from which distant observers can obtain any information. The line $y = 0$, where $r = 2M$, is the "future event horizon"; the line $x = 0$, where also $r = 2M$, is the "past event horizon".

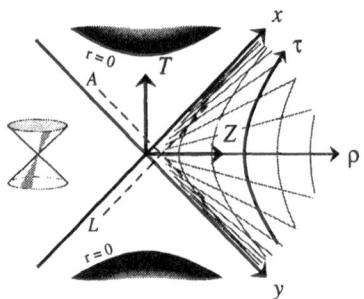

Fig. 1 Various coordinates used to describe the Schwarzschild metric. The local light cones are oriented everywhere as indicated at the left.

In the region $r \approx 2M$ one can write the metric as

$$ds^2 \approx \frac{16M^2}{e}dx dy + 4M^2 d\Omega^2 \tag{2.5}$$

and with the coordinate substitution

$$\frac{4M}{\sqrt{e}}x = Z + T, \qquad \frac{4M}{\sqrt{e}}y = Z - T, \tag{2.6}$$

$$2M(\theta - \tfrac{1}{2}\pi) = X, \qquad 2M\phi = Y,$$

at small X, Y, one finds that in terms of these coordinates space-time is approximately flat:

$$ds^2 \approx -dT^2 + dZ^2 + dX^2 + dY^2 .$$ (2.7)

The transformation

$$Z = \varrho \cosh \tau , \qquad T = \varrho \sinh \tau ,$$ (2.8)

brings us back to the Schwarzschild coordinates (close to the horizon), apart from normalization factors:

$$t/2M = 2\tau , \qquad 8M(r - 2M) = \varrho^2 .$$ (2.9)

The description of a flat space-time (2.7) in terms of the coordinates (2.8) is called "Rindler space"[4]. We see that close to the horizon, the Schwarzschild coordinates r and t behave as Rindler space coordinates.

To see that black holes can actually be formed by ordinary matter we have to study time-dependent solutions. For details concerning construction of such solutions we refer to[5]. The Penrose diagram for a configuration with both ingoing and outgoing matter is shown in Fig. 2a. In a classical black hole, there is only ingoing matter. In a coordinate frame that is flat at some distance from the black hole, the configuration looks as in Fig. 2b.

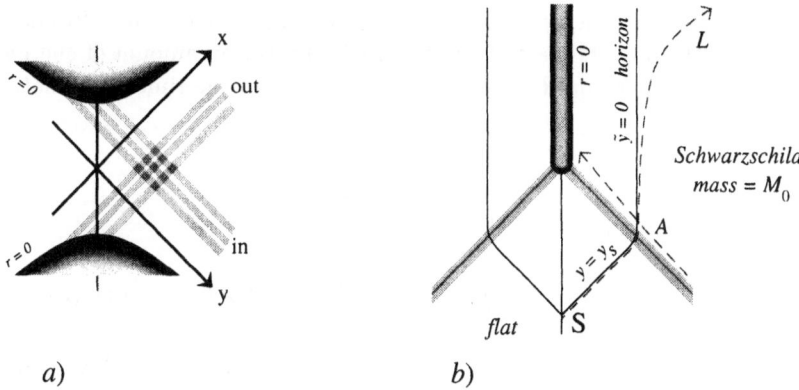

Fig. 2 a) Spherically symmetric configuration of matter radially moving inward and outward with the speed of light. b) Spherically symmetric black hole formed by radially inmoving lightlike matter.

3 Hawking Radiation[6, 7]

Consider the Minkowski coordinate frame $\{T, X, Y, Z\}$, or $\{T, \mathbf{X}\}$ for short, and a scalar field $\Phi(T, \mathbf{X})$. Let this field simply obey a Klein-Gordon equation,

$$(\partial^2 - m^2)\Phi = 0. \tag{3.1}$$

The quantum theory is written in the Heisenberg representation, which means that the states $|\psi\rangle$ are space-time independent, but the fields are operators depending both on space and on time. Usually, a complete set of solutions of (3.1) is written in terms of the Fourier modes with respect to the Minkowski space coordinates, and one gets

$$\Phi(\mathbf{X}, T) = \int \frac{d^3\mathbf{k}}{\sqrt{2k^0(\mathbf{k})(2\pi)^3}} \left(a(\mathbf{k})e^{i\mathbf{k}\cdot\mathbf{X}-ik^0T} + a^\dagger(\mathbf{k})e^{-i\mathbf{k}\cdot\mathbf{X}+ik^0T} \right), \tag{3.2}$$

$$\dot{\Phi}(\mathbf{X}, T) = \int \frac{-ik^0 d^3\mathbf{k}}{\sqrt{2k^0(\mathbf{k})(2\pi)^3}} \left(a(\mathbf{k})e^{i\mathbf{k}\cdot\mathbf{X}-ik^0T} - a^\dagger(\mathbf{k})e^{-i\mathbf{k}\cdot\mathbf{X}+ik^0T} \right). \tag{3.3}$$

Here $k^0(\mathbf{k}) = \sqrt{\mathbf{k}^2 + m^2}$, and the transformation from a and $a^\dagger$ to Φ and $\dot{\Phi}$ has been designed such that the following commutation rules are maintained:

$$[\Phi(\mathbf{X}, T), \Phi(\mathbf{X}', T)] = 0, \qquad [\Phi(\mathbf{X}, T), \dot{\Phi}(\mathbf{X}', T)] = i\delta^3(\mathbf{X} - \mathbf{X}'), \tag{3.4}$$

$$[a(\mathbf{k}), a(\mathbf{k}')] = 0, \qquad [a(\mathbf{k}), a^\dagger(\mathbf{k}')] = \delta^3(\mathbf{k} - \mathbf{k}'). \tag{3.5}$$

Not only do these commutation rules ensure that $a^\dagger$ and a act as creation and annihilation operators, but also the time dependence in (3.2) and (3.3) implies that the objects created and annihilated carry an amount of energy equal to k^0.

The operator H_M that generates boosts in the time coordinate T,

$$\frac{\partial \Phi}{\partial T} = -i[\Phi, H_M], \tag{3.6}$$

is the Minkowski-Hamiltonian

$$H_M = \int \mathcal{H}_M(\mathbf{X})d^3\mathbf{X} = \int d^3\mathbf{k}\, k^0(\mathbf{k})a^\dagger(\mathbf{k})a(\mathbf{k}). \tag{3.7}$$

We need first the transition to light-cone coordinates, and we define

$$a(\mathbf{k})\sqrt{k^0} = a_1(\tilde{k}, k^+)\sqrt{k^+}. \tag{3.8}$$

where $k^+ = \frac{1}{\sqrt{2}}(k^0 + k^3)$, and $\tilde{k}$ is the transverse component of $\mathbf{k}$. Since

$$\frac{\partial k^+}{\partial k^3}\bigg|_{\tilde{k}} = \frac{1}{\sqrt{2}}\left(1 + \frac{k^3}{\mu}\right) = \frac{k^+}{k^0}, \qquad \mu \equiv \sqrt{\tilde{k}^2 + k_3{}^2 + m^2} \tag{3.9}$$

the new operators a_1, $a_1^\dagger$ are normalized by

$$[a_1(\tilde{k}, k^+), a_1^\dagger(\tilde{k}', k^{+'})] = \delta^2(\tilde{k} - \tilde{k}')\delta(k^+ - k^{+'}). \tag{3.10}$$

To obtain operators a_2 that transform neatly under time boosts in Rindler space (i.e., Lorentz boosts in flat space-time), we define them as Fourier transforms with respect to $\ln(k^+)$:

$$a_2(\tilde{k}, \omega) = (2\pi)^{-1/2} \int_0^\infty \frac{dk^+}{\sqrt{k^+}} a_1(\tilde{k}, k^+) e^{i\omega \ln \left(\frac{k^+\sqrt{2}}{\mu}\right)}, \tag{3.11}$$

of which the inverse is:

$$a_1(\tilde{k}, k^+)\sqrt{k^+} = (2\pi)^{-1/2} \int_{-\infty}^\infty d\omega a_2(\tilde{k}, \omega) e^{-i\omega \ln \left(\frac{k^+\sqrt{2}}{\mu}\right)}. \tag{3.12}$$

With the normalization factors chosen in (3.11, 12), the operators a_2 and $a_2^\dagger$ again obey

$$[a_2(\tilde{k}, \omega), a_2^\dagger(\tilde{k}', \omega')] = \delta^2(\tilde{k} - \tilde{k}')\delta(\omega - \omega'), \tag{3.13}$$

Let the Minkowski-Hamiltonian H_M obey Eq. (3.7). The Rindler-Hamiltonian, H_R is then defined to be

$$H_R = H_R^I - H_R^{II}, \quad H_R^I = \int_{\varrho>0} d^3\mathbf{X}\, \varrho\mathcal{H}_M \,;\quad H_R^{II} = \int_{\varrho<0} d^3\mathbf{X}\, |\varrho|\mathcal{H}_M \tag{3.14}$$

(note that the Rindler Hamiltonian H_R is dimensionless). If the region $\varrho > 0$ (see Fig. 1) is called quadrant I and $\varrho < 0$ quadrant II, we see that all observables made of fields in quadrant II commute with H_R^I and *vice versa*. One finds that

$$H_R = \int_{-\infty}^\infty d\omega\, \omega\, a_2^\dagger(\tilde{k}, \omega)a_2(\tilde{k}, \omega)\,; \tag{3.15}$$

$$H_R^I = \int_0^\infty d\omega\, \omega\, a_I^\dagger(\tilde{k}, \omega)a_I(\tilde{k}, \omega)\,;\quad H_R^{II} = \int_0^\infty a_{II}^\dagger(\tilde{k}, \omega)a_{II}(\tilde{k}, \omega),$$

where a_I and a_{II} are related to a_2 and $a_2^\dagger$ as follows:

$$\begin{pmatrix} a_I(\tilde{k}, \omega) \\ a_{II}(\tilde{k}, \omega) \\ a_I^\dagger(-\tilde{k}, \omega) \\ a_{II}^\dagger(-\tilde{k}, \omega) \end{pmatrix} = \frac{1}{\sqrt{1 - e^{-2\pi\omega}}} \begin{pmatrix} 1 & 0 & 0 & e^{-\pi\omega} \\ 0 & 1 & e^{-\pi\omega} & 0 \\ 0 & e^{-\pi\omega} & 1 & 0 \\ e^{-\pi\omega} & 0 & 0 & 1 \end{pmatrix} \begin{pmatrix} a_2(\tilde{k}, \omega) \\ a_2(\tilde{k}, -\omega) \\ a_2^\dagger(-\tilde{k}, \omega) \\ a_2^\dagger(-\tilde{k}, -\omega) \end{pmatrix}. \tag{3.16}$$

They obey

$$[a_I(\tilde{k},\omega), a_I^\dagger(\tilde{k}',\omega')] = [a_{II}(\tilde{k},\omega), a_{II}^\dagger(\tilde{k}',\omega')] = \delta(\omega - \omega')\delta^2(\tilde{k},\tilde{k}');$$
$$[a_I, a_{II}] = [a_I, a_{II}^\dagger] = 0. \tag{3.17}$$

Thus we observe that the corresponding Hilbert space is separable into two factor spaces: $\mathcal{H} = \mathcal{H}_I \otimes \mathcal{H}_{II}$. The space $\mathcal{H}_I$ is described by the Hamiltonian H_R^I and $\mathcal{H}_{II}$ is described by the Hamiltonian $-H_R^{II}$.

The Rindler- or Boulware vacuum state $|0, 0\rangle$ is defined by

$$a_I|0,0\rangle = a_{II}|0,0\rangle = 0. \tag{3.18}$$

This is not the same as the vacuum experienced by a freely falling observer, who is said to experience the Minkowski- or Hawking vacuum $|\Omega\rangle$, which obeys

$$a_2(\tilde{k},\omega)|\Omega\rangle = 0. \tag{3.19}$$

It is not difficult to express this state in terms of the basis generated by a_I and a_{II}:

$$a_I(\tilde{k},\omega)|\Omega\rangle = e^{-\pi\omega} a_{II}^\dagger(-\tilde{k},\omega)|\Omega\rangle, \tag{3.20}$$
$$a_{II}(\tilde{k},\omega)|\Omega\rangle = e^{-\pi\omega} a_I^\dagger(-\tilde{k},\omega)|\Omega\rangle,$$

so that

$$|\Omega\rangle = \prod_{\tilde{k},\omega} \sqrt{1 - e^{-2\pi\omega}} \sum_{n=0}^{\infty} |n\rangle_I |n\rangle_{II} e^{-\pi n\omega}, \tag{3.21}$$

where the square root is added for normalization.

Notice that

$$H_R|\Omega\rangle = (H_R^I - H_R^{II})|\Omega\rangle = 0, \tag{3.22}$$

which confirms that $|\Omega\rangle$ is Lorentz invariant; remember that H_R is the generator of Lorentz boosts.

If one does not have the means to observe any features at $\varrho < 0$ this implies that one only has at one's disposal operators $\mathcal{O}$ composed of the fields in region I, that is, the operators a_I and $a_I^\dagger$. These act only in the factor space $\mathcal{H}_I$ but are proportional to the identity operator in $\mathcal{H}_{II}$:

$$\mathcal{O}\left(|\psi\rangle_I|\psi'\rangle_{II}\right) = |\psi'\rangle_{II}\left(\mathcal{O}|\psi\rangle_I\right). \tag{3.23}$$

Let us limit ourselves momentarily to a single point $(\tilde{k},\omega)$. There the expectation value for such an operator in the state $|\Omega\rangle$ is

$$\langle\Omega|\mathcal{O}|\Omega\rangle = (1 - e^{-2\pi\omega}) \sum_{n_1,n_2} {}_{II}\langle n_1|\,{}_I\langle n_1|\mathcal{O}|n_2\rangle_I\,|n_2\rangle_{II} e^{-\pi\omega(n_1 + n_2)}$$

$$= \sum_{n\geq 0} {}_I\langle n|\mathcal{O}|n\rangle_I e^{-2\pi n\omega}(1 - e^{-2\pi\omega}) = \mathrm{Tr}(\mathcal{O}\,\varrho_\Omega), \tag{3.24}$$

where ϱ_Ω is the *density matrix* $Ce^{-\beta H_I}$ corresponding to a thermal state at the temperature[6] $T = 1/2\pi$. Note that in Rindler space time, energy and temperature are dimensionless. If we scale with the appropriate factor $4M$ as in Eq. (2.9) we find the *Hawking temperature*,

$$T_H = 1/8\pi M = 1/8\pi G m_{\mathrm{BH}}. \tag{3.25}$$

This result is highly independent of the way the black hole was formed. In case the collapse took place in a less symmetric way, or at various steps and intervals, one still finds that an observer falling in the black hole should observe the Hawking vacuum state there, and this necessarily leads to the density matrix ϱ_Ω. In particular, one could assume that the collapsing matter was in a *pure* quantum state, and even in that case, the outgoing radiation appears to be mixed according to the matrix ϱ_Ω. The question to be asked is how literally this result is to be taken. One could conclude

i) that black holes must be fundamentally different from other objects in nature. They do not obey a single Schrödinger equation (which after all would allow pure states to evolve only into pure states), but instead obey probabilistic equations of motion that are not purely quantum mechanical[2].

According to this view, a more basic theory at the Planck scale would show no quantum mechanical features of the familiar kind. Alternatively, perhaps,

ii) black holes do obey a Schrödinger equation, but this equation requires knowledge of all inaccessible observables behind the horizon, so that a black hole forms an infinitely degenerate state. In this case the black hole can never decay completely[7], but it decays into stable, infinitely degenerate, final states with masses of the order of the Planck mass, called *remnants*. Thirdly, however, one may suspect that

iii) the density matrix derivation depended on certain hidden assumptions of a statistical nature, such that the answer may be correct in a statistical sense, but more precise treatments may yield a purely quantum mechanical description of a black hole that nevertheless has only a finite degeneracy. This is the scattering matrix assumption, which we will further investigate from section 7 onwards.

Thus, one expects the system as a whole to react just as any other physical system does: when it absorbs infalling material, or even just infalling radiation, it should react some way or other, and enter into a state that is orthogonal to what it would have evolved into if the infalling material had been in a different mode or totally absent. This is just the experimentally observed fact that all known evolution laws in small-scale physics can be written in terms of a *unitary* evolution matrix. It is hard to understand how the world at the scale of ordinary atomic and elementary particle physics could behave quantum mechanically and evolve in a unitary way, if quantum mechanics were not at the basis of the laws of dynamics at the smallest distance scales.

It appears that the derivation of the density matrix ϱ_Ω in Eq. (3.24) cannot be exactly right, since it implies that infalling material of whatever variety should not affect the outgoing radiation at all (linearized quantum field theory was used). This would violate unitarity.

The density matrix ϱ_Ω has to be replaced by a pure state.

4   Black Hole Entropy, and its Interpretation in Terms of Quantum States

The fact that the radiation emitted, as described by Eq. (3.25), is *thermal*, opens up the possibility to approach this phenomenon from a thermodynamical point of view. Taking m_{BH} to be the energy and $T = T_H$ the temperature, one readily derives the *entropy* S:

$$T dS = dm_{BH}; \qquad dS = 8\pi G m_{BH} dm_{BH}; \qquad S = 4\pi G m^2_{BH} + C, \quad (4.1)$$

where C is an unknown integration constant, to be referred to as the "entropy normalization constant".

It is important to note that the expression obtained for the entropy S, apart from the integration constant, is always equal to $\frac{1}{4} A/G$, where A is the *area* of the horizon, a finding that will be very much at the center of our discussions.

Connecting the entropy to the *density of quantum mechanical states*[8], must be done with considerable care, since there will be two kinds of divergences: at the horizon and at spacelike infinity. In fact, one may very well question the mere *existence* of such quantum levels. This, however, is the key assumption of this paper: not only is the quantum mechanics of black holes meaningful, it can also be *derived*, and the constant C in Eq. (4.1) is finite and of order one (apart from subdominant terms). In order to enable us to judge the relation between the entropy just derived, and the density of quantum states, we now present a direct argument concerning the density of states, an argument that will also show that any infinities at the *horizon* must be absorbed in C, but the "infrared" infinities arising from spacelike infinity should be excluded; the latter represent the radiation field far from the black hole.

The spectral density of a black hole can be derived from its Hawking temperature by applying time reversal invariance[9]. We have to our disposal both the *emission rate* (the Hawking radiation intensity), and the *capture probability*, or the effective cross section of the black hole for infalling matter.

The cross section σ is approximately determined by the radius r^+ of the horizon:

$$\sigma = 2\pi r_+^2 = 8\pi M^2, \quad (4.2)$$

and slightly more for objects moving in slowly. The emission probability $W dt$ for a given particle type, in a given quantum state, in a large volume $V = L^3$ is:

$$W dt = \frac{\sigma(\mathbf{k})v}{V} e^{-E/T} dt, \quad (4.3)$$

where **k** is the wave number characterizing the quantum state, v is the particle velocity, and E is its momentum.

Now we *assume* that the process is also governed by a Schrödinger equation. This means that there exist quantum mechanical transition amplitudes,

$$\mathcal{T}_{\text{in}} = {}_{\text{BH}}\langle M + GE| \; |M\rangle_{\text{BH}}|E\rangle_{\text{in}},$$

$$\text{and} \quad \mathcal{T}_{\text{out}} = {}_{\text{BH}}\langle M|\,{}_{\text{out}}\langle E| \; |M + GE\rangle_{\text{BH}}, \tag{4.4}$$

where the states $|M\rangle_{\text{BH}}$ represent black hole states with mass M/G, and the other states are energy eigenstates of particles in the volume V. In terms of these amplitudes, using the so-called Fermi Golden Rule, the cross section and the emission probabilities can be written as

$$\sigma = |\mathcal{T}_{\text{in}}|^2 \varrho(M + GE)/v, \tag{4.5}$$

$$W = |\mathcal{T}_{\text{out}}|^2 \varrho(M)\frac{1}{V}, \tag{4.6}$$

where $\varrho(M)$ stands for the level density of a black hole with mass M. The factor v^{-1} in Eq. (4.5) is a kinematical factor, and the factor V^{-1} in W arises from the normalization of the wave function.

Now, time reversal invariance relates $\mathcal{T}_{\text{in}}$ to $\mathcal{T}_{\text{out}}$ (through complex conjugation). To be precise, all one needs is $PC\hat{T}$ invariance, since the parity transformation P and charge conjugation C have no effect on our calculation of σ. Dividing the expressions (4.5) and (4.6), and using (4.3), one finds:

$$\frac{\varrho(M + GE)}{\varrho(M)} = e^{E/T} = e^{8\pi ME} \tag{4.7}$$

(naturally, we assume the energy E to be small compared to the black hole mass M, so that the E^2 terms are relatively insignificant). This is easy to integrate:

$$\varrho(M) = e^{4\pi M^2/G + C} = e^S. \tag{4.8}$$

For large enough black holes, Eq. (4.8) may be rewritten as

$$\varrho(M) = 2^{A/A_0}, \tag{4.9}$$

where A is the horizon area and A_0 is a fundamental unit of area,

$$A_0 = 4G \ln 2. \tag{4.10}$$

This suggests a spin-like degree of freedom on all surface elements of size A_0.

As stated earlier, the importance of this derivation is the fact that the expressions used as starting points are the *actual* Hawking emission rate and the *actual* black hole absorption cross section. This implies that, if in more detailed considerations divergences are found near the horizon, these divergences should not be used as arguments to adjust the relation between entropy and level density

by large renormalization factors. Furthermore, the Golden Rule argument can
be used only to deal with one emitted particle at the time. Hence, we should not
take the outside volume V so large that the dominant emission mode contains
very many particles. Therefore, any divergences found when the outside volume
is taken to infinity should be subtracted.

Extension to the more general Kerr-Newman solutions is straightforward.

5   The Brick Wall Model[9]

In this Section, we now present a model in which only *low energy* quantum
fluctuations of the fields are taken into account. We apply quantum field theory
up to some point r_1 close to the horizon: $r_1 = r_+ + h$, $h > 0$. For simplicity we
only consider scalar fields $\Phi_i(r, \theta, \phi, t)$, whose only interaction is the gravitational
one with the metric; generalization towards spinor, vector or even perturbative
gravitational field excitations will be straightforward. To simplify things, we just
represent all those by giving the fields Φ_i a multiplicity N, so $i = 1 \ldots, N$. At
$r = r_1$ we impose a boundary condition:

$$\Phi_i(r, \theta, \phi, t) = 0 \quad \text{if} \quad r \leq r_1. \tag{5.1}$$

The quanta of the fields will be given a temperature $T = T_H$. The question one
may ask is: which value should one assign to the cutoff parameter h, such that
the entropy of this system precisely takes the value (4.1), so that the density
of quantum states corresponds to (4.8)? We will need an infrared cutoff in the
form of a box with radius L:

$$\Phi_i(r, \theta, \phi, t) = 0 \quad \text{if} \quad r \geq L. \tag{5.2}$$

To determine the thermodynamic properties of this system, one must com-
pute the energy levels $E(n, \ell, \ell_3)$ of the bosons Φ_i. The Lagrange density $\mathcal{L}$ in
the metric (2.1) is given by

$$2\mathcal{L}(\mathbf{x}, t) = \left(1 - \frac{2M}{r}\right)^{-1}(\partial_t \Phi_i)^2 - \left(1 - \frac{2M}{r}\right)(\partial_r \Phi_i)^2 - r^{-2}(\partial_\Omega \Phi_i)^2 - m_i^2 \Phi_i^2. \tag{5.3}$$

In the approximation

$$m_i^2 \ll 2M/\beta^2 h, \quad L \gg 2M, \tag{5.4}$$

the main contributions to free energy at a temperature $T = \beta^{-1}$ is found to
be[9, 5]

$$F \approx -\frac{2\pi^3 N}{45h}\left(\frac{2M}{\beta}\right)^4 - \frac{2}{9\pi}L^3 N \int_m^\infty \frac{dE(E^2 - m^2)^{3/2}}{e^{\beta E} - 1}. \tag{5.5}$$

The second part is the usual contribution from the vacuum surrounding the black hole at great distances, and as argued before, should be discarded. The first part is an intrinsic contribution from the horizon, and it is seen to diverge linearly as $h \downarrow 0$.

The contribution of the horizon to the total energy U and the entropy S are

$$U = \frac{\partial}{\partial \beta}(\beta F) = \frac{2\pi^3}{15h}\left(\frac{2M}{\beta}\right)^4 N, \tag{5.6}$$

$$S = \beta(U - F) = \frac{8\pi^3}{45h}2M\left(\frac{2M}{\beta}\right)^3 N. \tag{5.7}$$

When this is adjusted to the Hawking value, Eq. (4.1), with $\beta = 1/T_H = 8\pi M$, we find that the cutoff parameter h must be chosen to be

$$h = \frac{NG}{720\pi M}. \tag{5.8}$$

The total energy U of the thermally excited particles is given by

$$GU = \tfrac{3}{8}M, \tag{5.9}$$

independently of N. Alternatively, one could have tuned the energy U to be equal to m_{BH}, which would yield the same order of magnitude for h, but adjusting the physical degrees of freedom, i.e., the entropy S, appears to us more sensible. Clearly, it makes little sense to allow $h\overset{?}{\to}0$, since then both the entropy and the energy would diverge.

We refer to the cutoff near the horizon as a "brick wall". The physical distance between the brick wall and the horizon is

$$\int_{r=2M}^{r=2M+h} \mathrm{d}s = \int_{2M}^{2M+h} \frac{\mathrm{d}r}{\sqrt{1-2M/r}} = 2\sqrt{2Mh} = \sqrt{\frac{NG}{90\pi}}, \tag{5.10}$$

which is independent of the mass m_{BH} of the black hole. The brick wall should be a property of any horizon of arbitrary size. If N is not too large, the brick wall thickness is of the order of the Planck length.

The brick wall model, with the values of β and h fixed according to Equations (3.25) and (4.1), actually reproduces the thermodynamic properties of a black hole quite nicely, and could have served as a realistic model for a black hole that fully obeys Schrödinger's equation and preserves quantum coherence, except for the fact that it also preserves all symmetries of the underlying quantum field theory; it could generate chemical potentials for the various globally conserved quantum numbers. Thus, not only the temperature must be constrained to keep the Hawking value, but also the chemical potentials are constrained to be zero. In principle, this is easy to realize, simply by introducing symmetry breaking effects in the brick wall boundary condition, but probably one would then be pushing this model too far; anyway, its most important deficiency is that we completely gave up invariance under general coordinate transformations near the horizon.

The most important lesson to be learned from the brick wall model is that Hawking radiation can indeed be seen to be compatible with quantum mechanical purity, if only one could introduce a cutoff at the Planck scale.

6   The Aichelburg-Sexl Metric near a Black Hole

The gravitational effect of an infalling particle in the Schwarzschild metric can be understood when we transform to a locally flat space-time, Eqs. (2.6). Consider the coordinate frames of Section 2. As Schwarzschild time t, or equivalently, Rindler time τ, evolves, the infalling particle is Lorentz boosted, as we see in Eq. (2.8). In terms of the flat coordinates, therefore, the energy of the particles increases exponentially, and thus it quickly reaches values where gravitational effects can no longer be ignored. These effects are easy to calculate in the approximation that the source particle moves with the speed of light[10, 11].

For simplicity, consider the case that the surrounding space-time is completely flat. In the rest frame we can approximate the metric as

$$ds^2 = dx^2 + \frac{2\mu}{r}dt^2 + \frac{2\mu}{r}dr^2 , \qquad (6.1)$$

$$r \equiv \sqrt{x_1^2 + x_2^2 + x_3^2} , \qquad dx^2 \equiv -dt^2 + dr^2 + r^2 d\Omega^2 .$$

where $\mu = Gm$, and m is the mass of the source particle. This we rewrite as

$$ds^2 = dx^2 + \frac{2\mu}{r}(u \cdot dx)^2 + \frac{2\mu}{r}dr^2 , \qquad r = \sqrt{x^2 + (u \cdot x)^2} , \qquad (6.2)$$

where

$$u = (0,0,0,i) ; \qquad u^2 = -1 . \qquad (6.3)$$

In these expressions, we have neglected all effects that are of higher order in the particle mass μ, since we choose μ to be small. The particle's Schwarzschild radius r_+ is very small, and the Lorentz boost to be considered next will only further reduce the particle's dimensions.

The advantage of the notation chosen in Eq. (6.2) is of course that now the Lorentz boost is straightforward. In the boosted frame we can take

$$m u^\mu \Rightarrow (0,0,p,ip) = p^\mu , \qquad Gp = \frac{\mu v}{\sqrt{1 - v^2/c^2}} \gg \mu . \qquad (6.4)$$

In the limit $\mu \Rightarrow 0$, p fixed, one has $r \Rightarrow |x \cdot u|$. It will turn out to be useful to compare the metric then obtained with the flat space-time metric in two coordinate frames $y_{(\pm)}^\mu$, defined as

$$y_{(\pm)}^\mu = x^\mu \pm 2\mu u^\mu \log r . \qquad (6.5)$$

We have:

$$dy^2_{(\pm)} = dx^2 \pm \frac{4\mu}{r}(u \cdot dx)\,dr - 4\mu^2\,\frac{dr^2}{r^2}\,; \qquad (6.6)$$

$$ds^2 - dy^2_{(\pm)} = \frac{2\mu}{r}\,d[r \mp (u \cdot x)]^2 + 4\mu^2(d\log r)^2\,. \qquad (6.7)$$

Now consider the limit (6.4). We keep p fixed but let μ tend to zero. We now claim that when $(p \cdot x) > 0$, the metric ds^2 approaches the flat metric $dy^2_{(+)}$, whereas when $(p \cdot x) < 0$, we have $ds^2 \Rightarrow dy^2_{(-)}$, and at the plane defined by $(p \cdot x) = 0$ these two flat space-times are glued together according to

$$y^\mu_{(+)} = y^\mu_{(-)} + 4\mu\,u^\mu \log|\tilde{x}|\,, \qquad (6.8)$$

where $\tilde{x} = (X, Y, 0, 0)$, the transverse part of the coordinates y^μ.

Verifying the flatness of space-time away from the plane $(p \cdot x) = 0$, is easy, but to ascertain the connection formula (6.8), is a bit more delicate. One can show[5] that

$$ds^2 \;\to\; dy^2_{(+)} \qquad \text{if } (p \cdot x) \gtrsim 0\,, \qquad (6.9A)$$

$$ds^2 \;\to\; dy^2_{(-)} \qquad \text{if } (p \cdot x) \lesssim 0\,, \qquad (6.9B)$$

$$y_{(+)} = y_{(-)} + 4\mu\,u^\mu \log r \quad \text{in the region } (p \cdot x) \approx 0\,, \qquad (6.9C)$$

which is equivalent to Eq. (6.8). This defines the Aichelburg-Sexl metric[10].

The effect a fast moving particle has on the surrounding space-time, is visualized in Fig. 3. In terms of light cone coordinates, we have the connection formula

$$x^+{}_{(+)} - x^+{}_{(-)} = 4Gp^+ \log|\tilde{x}| = 4\sqrt{2}\,Gp \log|\tilde{x}|\,;$$
$$x^-{}_{(+)} - x^-{}_{(-)} = 0\,. \qquad (6.10)$$

Here, $\tilde{x}$ is the transverse distance from the source particle, which is moving (highly relativistically) along the line $x^- = \tilde{x} = 0$. The r.h.s. of Eq. (6.10) is a Green function, $-\sqrt{2}f(\tilde{x})p$, satisfying the equation

$$\bar{\partial}^2 f(\tilde{x}) = -8\pi G\,\delta^2(\tilde{x})\,, \qquad (6.11)$$

where the sign is chosen such that f is large for small values of $\tilde{x}$.

This result can be generalized to the case of a particle moving into a finite size black hole. More details can be found in refs[11, 5]. For most purposes, however, it is sufficient to look at Rindler space, which corresponds to an infinite size black hole. There, the fast particle produces the Aichelburg-Sexl metric as described above.

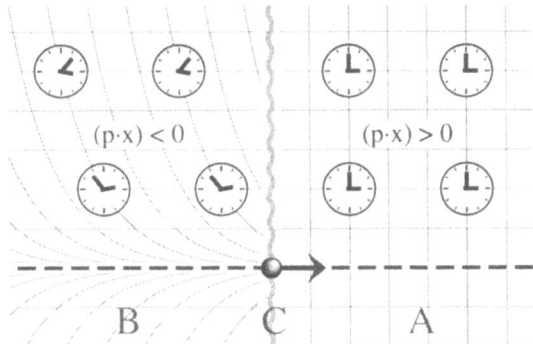

Fig. 3 Snapshot of the gravitational shock wave caused by a highly relativistic particle. If we have a rectangular grid and synchronized clocks before the particle passed by (region A), then, behind the particle (region B), the grid will be deformed, and the clocks desynchronized. The shift is proportional to the logarithm of the transverse distance.

7   Constructing the S-Matrix from the Gravitational Back Reaction

The postulate that scattering of particles against a black hole can be described by a quantum mechanical scattering matrix is an *assumption* that cannot be proved from the principles of quantum field theory and general relativity alone. Indeed, it may well be at variance with these theories, if the latter would be extrapolated to beyond the Planck scale. The S-matrix Ansatz applied here may be seen as a new physical principle, perhaps comparable to Max Planck's new postulate in his 1900 paper, that energies are quantized. The S-matrix Ansatz reads as follows[12]:

> *All physical interaction processes that begin and end with free, stable particles moving far apart in an asymptotically flat space-time, therefore also all those that involve the creation and subsequent evaporation of a black hole, can be described by one scattering matrix S relating the asymptotic outgoing states* |out⟩ *to the ingoing states* |in⟩.

In essence, the Ansatz will be used in the following way[13, 12]: consider one state $|\text{in}_0\rangle$ and one state $|\text{out}_0\rangle$, with a possible black hole in their connecting history. We assume some value for the transition amplitude $\langle\text{out}_0|\text{in}_0\rangle = \mathcal{N}$. This means that we replaced the out-state produced by the Hartle-Hawking!vacuum, which actually was a quantum mechanical mixture of states, by one arbitrary choice

$|\text{out}_0\rangle$. Then, using all the physical laws that we know and trust, we compute neighboring S-matrix elements, $\langle\text{out}_0 + \delta_{\text{out}}|\text{in}_0 + \delta_{\text{in}}\rangle$.

If there were no interactions, the effects from δ_{in} onto the out-states would not have been discernable. All amplitudes would have to be equal, and the scattering matrix thus obtained could never be unitary. Since in the calculations of Section 5, interactions between the Φ particles were ignored, those calculations were not good enough to give us our S-matrix. In this section, we will take only one type of interaction into account: the gravitational shift computed in the previous section. Thus, we only consider particles moving in and out in the longitudinal direction, with hyper-relativistic speeds when they are near the horizon. Far away from the horizon, as soon as $r - 2M = \mathcal{O}(2M)$, they will be allowed to go slower, indeed, out-moving particles may turn around to fall back in again. What has to be done in order to accommodate for such possibilities, is to define the S-matrix to consist of three ingredients:

$$S = S_{\text{out}}S_{\text{hor}}S_{\text{in}}, \tag{7.1}$$

where S_{in} relates the asymptotic in-states to wave packets moving inwards very near the horizon, S_{out} connects wave packets moving outwards very near the horizon to the asymptotic out-states, and S_{hor} is the really important part telling us how particles moving inwards very near the horizon affect the outgoing particles very near the horizon. S_{in} and S_{out} follow unambiguously from known laws of low-energy physics, and require little discussion.

In the limit $M \to \infty$, the region near the horizon can be described as a Rindler space. The angles θ and ϕ are replaced by flat transverse coordinates, and we rescale the momentum p accordingly. We recover the shift (6.10), determined by the Green function f of Eq. (6.11).

To begin our construction of the S-matrix, let us take δ_{in} to be one extra particle going in with momentum δp_{in}^-, at the transverse position $\tilde{x}'$. Since we use the conventions of Section 4, the value of δp_{in}^- is negative. The outgoing particles, at points $\tilde{x}$ near to the point $\tilde{x}'$, are shifted inwards, so that δx_{out}^- is negative, and

$$x_{\text{out}}^- \to x_{\text{out}}^- + \delta x_{\text{out}}^-(\tilde{x}), \qquad \delta x_{\text{out}}^-(\tilde{x}) = f(\tilde{x} - \tilde{x}')\delta p_{\text{in}}^-, \tag{7.2}$$

where f obeys Eq. (6.11), or, if from now on $8\pi G = 1$,

$$\tilde{\partial}^2 f(\tilde{x}) = -\delta^2(\tilde{x}). \tag{7.3}$$

We now temporarily suppress the superscripts $\{\pm\}$, since the subscripts in and out suffice, and later we want to reintroduce $\{\pm\}$ with different sign conventions. Any outgoing particle has a wave packet ψ, oscillating as $e^{ip_{\text{out}}x_{\text{out}}}$. With the shift δx_{out}, this wave turns into

$$e^{ip_{\text{out}}x_{\text{out}} - ip_{\text{out}}\delta x_{\text{out}}} = e^{-i\int \mathrm{d}^2\tilde{x}\left[\delta x_{\text{out}}(\tilde{x})\hat{P}_{\text{out}}(\tilde{x})\right]}\psi, \tag{7.4}$$

where $\hat{P}_{\text{out}}(\tilde{x})$ is the operator that generates a shift at transverse position $\tilde{x}$. It is also the total momentum density of the outgoing particles at transverse position $\tilde{x}$.

Now combining this with Eq. (7.2), we see that

$$\dot{\psi} \Rightarrow \psi' = e^{-i\int \mathrm{d}^2\tilde{x}\left[\delta p_{\text{in}} f(\tilde{x} - \tilde{x}')\hat{P}_{\text{out}}(\tilde{x})\right]} \psi. \tag{7.5}$$

Repeating this many times, adding (or removing) different ingoing particles in the in-state, with momenta adding all up to $P_{\text{in}}(\tilde{x}')$ at the transverse position $\tilde{x}'$, we see that the total effect is:

$$\psi' = e^{-i\int \mathrm{d}^2\tilde{x}\mathrm{d}^2\tilde{x}'\left[P_{\text{in}}(\tilde{x}')f(\tilde{x} - \tilde{x}')\hat{P}_{\text{out}}(\tilde{x})\right]} \psi. \tag{7.6}$$

Notice the complete symmetry between in- and outgoing particles. $P_{\text{in}}(\tilde{x}')$ refers to all momenta of particles going in during a certain epoch where we have control over the ingoing particles. $\hat{P}_{\text{out}}(\tilde{x})$ refers to all particles seen going out during a similar epoch of observations. Before or after these two epochs we do not have the opportunity to observe or control. The states there are kept fixed as much as is possible. Of course, both P_{in} and $\hat{P}_{\text{out}}$ are operators; from now on we omit the hat $\hat{(\,)}$.

Noting that, according to the result of Section 4, the total number of quantum states should be finite, we have reasons to believe that, by adding or subtracting a sufficient number of particles, we can generate *all* in-states from $|\text{in}_0\rangle$, and for the out-states it is even more natural to have $P_{\text{out}}(\tilde{x})$ refer to *all* outgoing particles. It is suggested to describe the in- and out states exclusively by giving the functions $P_{\text{in}}(\tilde{x})$ and $P_{\text{out}}(\tilde{x})$. One then obtains

$$\langle\{P_{\text{out}}(\tilde{x})\}|\{P_{\text{in}}(\tilde{x})\}\rangle = \mathcal{N} \exp\left[-i\int \mathrm{d}^2\tilde{x}\,\mathrm{d}^2\tilde{x}'\,P_{\text{in}}(\tilde{x}')f(\tilde{x} - \tilde{x}')P_{\text{out}}(\tilde{x})\right], \tag{7.7}$$

where $\mathcal{N}$ is a common normalization factor. The magnitude of this factor is fixed by requiring S to be unitary; its phase cannot be determined, but in most cases it will be a freely adjustable parameter anyway, since our amplitude tends to violate global conservation laws.

This scattering matrix is indeed unitary, if one imposes the inner product

$$\langle\{P_{\text{in}}(\tilde{x})\}|\{P_{\text{in}}'(\tilde{x})\}\rangle = \mathcal{N}' \prod_{\tilde{x}} \delta\left(P_{\text{in}}(\tilde{x}) - P_{\text{in}}'(\tilde{x})\right), \tag{7.8}$$

for the in-states, with again some normalization parameter $\mathcal{N}'$, and we impose a similar inner product rule for the out-states.

We should hasten to add, that the S-matrix (7.7) cannot be the ultimate result of our theory, since the states $|\{P_{\text{in}\atop\text{out}}(\tilde{x})\}\rangle$ with the inner product (7.8) form a *continuum* of states, and this is not the result we want. What this really means is that we still expect some cut-off mechanism when $|\tilde{x} - \tilde{x}'|$ approaches the Planck length. Indeed, if $|\tilde{x} - \tilde{x}'|$ approaches the Planck length, our present result is invalid, since then the *transverse* components of the momenta also produce shifts, and those have not been taken into account. If, however, we limit ourselves to a "coarse grained" description, specifying only features that are large compared to the Planck length, and if it could indeed be accepted that restricting oneself to the gravitational interaction forces only (and of those only the longitudinal ones), is reasonable, then (7.7) seems to be a reasonable approximation to the S-matrix that we are looking for. In view of this, let us first further analyze what this S-matrix implies.

8 Functional Operator Algebra on the Horizon

Consider the Hilbert space of in-states $|\{P_{\text{in}}(\tilde{x})\}\rangle$ with inner product (7.8), and define an operator $U_{\text{in}}(\tilde{x})$ that is canonically conjugated to $P_{\text{in}}(\tilde{x})$:

$$[P_{\text{in}}(\tilde{x}),\, U_{\text{in}}(\tilde{x}')] = -i\delta^2(\tilde{x} - \tilde{x}'), \tag{8.1}$$

$$[P_{\text{in}}(\tilde{x}),\, P_{\text{in}}(\tilde{x}')] = [U_{\text{in}}(\tilde{x}),\, U_{\text{in}}(\tilde{x}')] = 0 \tag{8.2}$$

(We regard all these operators as acting on in-states). The eigenstates of $U_{\text{in}}(\tilde{x})$ are the functional Fourier transforms of the eigenstates $|\{P_{\text{in}}(\tilde{x})\}\rangle$ of the operators P_{in}:

$$|\{U_{\text{in}}(\tilde{x})\}\rangle = \mathcal{N}'' \int \mathcal{D}P_{\text{in}} e^{-i \int d\tilde{x}\, P_{\text{in}}(\tilde{x}) U_{\text{in}}(\tilde{x})} |\{P_{\text{in}}(\tilde{x})\}\rangle, \tag{8.3}$$

where $\mathcal{N}''$ is again a normalization factor.
    Writing this as

$$\langle\{U_{\text{in}}(\tilde{x})\}|\{P_{\text{in}}(\tilde{x})\}\rangle = \mathcal{N}''' e^{i \int d\tilde{x}\, P_{\text{in}}(\tilde{x}) U_{\text{in}}(\tilde{x})}, \tag{8.4}$$

we find that the states $|\{U_{\text{in}}(\tilde{x})\}\rangle$ can be expressed in terms of the states $|\{P_{\text{out}}(\tilde{x})\}\rangle$, by using Equ. (7.7).
    We find:

$$U_{\text{in}}(\tilde{x}') = -\int d\tilde{x}\, f(\tilde{x} - \tilde{x}') P_{\text{out}}(\tilde{x}), \tag{8.5}$$

and similarly:

$$U_{\text{out}}(\tilde{x}') = \int d\tilde{x}\, f(\tilde{x} - \tilde{x}') P_{\text{in}}(\tilde{x}), \tag{8.6}$$

where U_{out} is the operator canonically conjugated to P_{out}, since in addition to Eqs. (8.1) and (8.2) we have for the out-states:

$$[P_{\text{out}}(\tilde{x}),\, U_{\text{out}}(\tilde{x}')] = -i\delta^2(\tilde{x} - \tilde{x}'), \tag{8.7}$$

$$[P_{\text{out}}(\tilde{x}),\, P_{\text{out}}(\tilde{x}')] = [U_{\text{out}}(\tilde{x}),\, U_{\text{out}}(\tilde{x}')] = 0 \tag{8.8}$$

By virtue of the fact that Eqs (8.5) and (8.6) relate operators on in-states to operators on out-states, we say that these generate the S-matrix. Rewriting the equations as

$$\tilde{\partial}^2 U_{\text{in}}(\tilde{x}) = P_{\text{out}}(\tilde{x}) , \qquad \tilde{\partial}^2 U_{\text{out}}(\tilde{x}) = -P_{\text{in}}(\tilde{x}), \qquad (8.9)$$

underlines the local nature of these equations with respect to the transverse coordinates $\tilde{x}$. Also:

$$\langle\{U_{\text{out}}(\tilde{x})\}|\{U_{\text{in}}(\tilde{x})\}\rangle = \mathcal{N}'''' \exp\left[-i\int \mathrm{d}^2\tilde{x}\, \tilde{\partial}U_{\text{out}}(\tilde{x}) \cdot \tilde{\partial}U_{\text{in}}(\tilde{x})\right]. \qquad (8.10)$$

Because of its local nature, this equation may be suspected to be more elementary than Eq. (7.7), which was derived earlier. Combining (8.10) with (8.4) and the analogous inner product between the U_{out} and P_{out} eigenstates, we rewrite Eq. (7.7) as

$$\langle\{P_{\text{out}}(\tilde{x})\}|\{P_{\text{in}}(\tilde{x})\}\rangle = \mathcal{N}\int \mathcal{D}U_{\text{in}}(\tilde{x})\int \mathcal{D}U_{\text{out}}(\tilde{x}) \qquad (8.11)$$

$$\exp\left[\,i\int \mathrm{d}^2\tilde{x}\{\,-\tilde{\partial}U_{\text{out}}(\tilde{x})\cdot\tilde{\partial}U_{\text{in}}(\tilde{x}) + P_{\text{in}}(\tilde{x})U_{\text{in}}(\tilde{x}) - P_{\text{out}}(\tilde{x})U_{\text{out}}(\tilde{x})\}\right],$$

where $\mathcal{N}$ is again a different but universal normalization factor (henceforth, we write such factors simply as $\mathcal{N}$.)

Imagine now that both the in- and the out-state can be completely composed of a finite number, $N = N_{\text{in}} + N_{\text{out}}$, of particles. Let us denote the momenta of the ingoing particles as $p^{-,i}_{\text{in}}$, $i = 1, \ldots, N_{\text{in}}$, entering at transverse coordinates $\tilde{x}^i$, and those of the outgoing particles, at transverse coordinates $\tilde{x}^j$, as $-p^{+,j}_{\text{out}}$, $j = N_{\text{in}} + 1, \ldots, N$. The reason for the minus sign here, is that now the total momentum going into the horizon can be seen as the sum of all 4-vectors p^μ of the in- and outgoing particles, as it is usually done in field theory. The operators $U_{\text{in} \atop \text{out}}$ are put in a Lorentz vector x^μ without sign changes:

$$x^+ = U_{\text{in}} , \qquad x^- = U_{\text{out}} . \qquad (8.12)$$

Substituting

$$P_{\text{in}}(\tilde{x}) = \sum_{i=1}^{N_{\text{in}}} p^{-,i}_{\text{in}} \delta^2(\tilde{x} - \tilde{x}^i) , \qquad P_{\text{out}}(\tilde{x}) = -\sum_{j=N_{\text{in}}+1}^{N} p^{+,j}_{\text{out}} \delta^2(\tilde{x} - \tilde{x}^j), \qquad (8.13)$$

one obtains

$$\langle\text{out}|\text{in}\rangle = \mathcal{N}\int \mathcal{D}x^+(\tilde{x})\int \mathcal{D}x^-(\tilde{x}) \qquad (8.14)$$

$$\exp\left[i\int \mathrm{d}^2\tilde{x}\{\,-\tfrac{1}{2}\tilde{\partial}x^\mu(\tilde{x})\tilde{\partial}x^\mu(\tilde{x})\} + i\sum_{i=1}^{N} p^{\mu,i}x^\mu(\tilde{x}^i)\right].$$

Here, the transverse components of x^μ are not functionally integrated over; they are the transverse coordinates. The factor $\frac{1}{2}$ compensates for double counting. The contribution of the transverse components of x^μ to the integrand must be subtracted, which corresponds to a renormalization of $\mathcal{N}$.

It is, however, more realistic to put the external particles in wave functions that are eigenstates of momenta only. Therefore, we must convolute this expression by transverse wave functions $e^{i\tilde{p}^i \cdot \tilde{x}^i}$, where the transverse components of the momenta, $\tilde{p}^i$, must be kept small compared to the Planck energy (otherwise, it would have been illegal to ignore the transverse gravitational shifts.) We then obtain

$$\langle \text{out}|\text{in}\rangle = \mathcal{N}\Big(\prod_i \int \mathrm{d}^2\tilde{x}^i\Big) \int \mathcal{D}x^+(\tilde{x}) \int \mathcal{D}x^-(\tilde{x})$$

$$\exp\Big[\ i\int \mathrm{d}^2\tilde{x}\{-\tfrac{1}{2}\tilde{\partial}x^\mu(\tilde{x})\cdot\tilde{\partial}x^\mu(\tilde{x})\} + i\sum_{i=1}^{N} p^{\mu,i}x^\mu(\tilde{x}^i)\Big]\,, \qquad (8.15)$$

where now the effects of the wave functions are included in the contributions of the external momenta $p^{\mu,i}$ to the 'vertex insertions'. Thus, in contrast to Eq. (8.14), $p^{\mu,i}$ here have transverse components.

It is here that the striking resemblence to string amplitudes should be pointed out. We have the string integrand (for closed strings), as well as the integration over moduli space, which here is formed by the points $\tilde{x}^i$ where the particles cross the horizon. The fact that the action is linearized is understandable, since all transverse dimensions have been kept large compared to the longitudinal ones. What is more surprising is the value of the string constant: it is equal to i, in units where $8\pi G = 1$.

The way in which here the black hole horizon is identified with a string worldsheet is sketched in Fig. 4. At $t \to -\infty$ we have ingoing closed "strings". Arriving at the horizon these strings exchange a string, whose world sheet wraps around the horizon exactly once. The edges of the holes left behind are the outgoing closed strings.

At this point, let us once again focus on the nature of the Hilbert space of in- and outgoing particles. Suppose that, for simplicity, we discretize the transverse coordinates $\tilde{x}$. The functional integrals then become finite-dimensional. What distinguishes this space from the usual Fock space is now, that at every point $\tilde{x}$ exactly one "particle" is allowed. The only way to mimic the usual Fock space is to assume that every elementary point particle must be given a *different* value for its transverse coordinate $\tilde{x}$. This constraint may be considered to be negligible, if the $\tilde{x}$ are sufficiently fine-grained, but it is somewhat puzzling how to maintain this constraint in an infinite-volume limit. Apparently, unlike ordinary Fock space, a state with two or more particles at the *same* transverse position $\tilde{x}$, with momenta $p^{\mu,1}, \ldots, p^{\mu,k}$, is indistinguishable from the state with just a single particle there, whose momentum is $\sum_{i=1}^{k} p^{\mu,i}$. This may seem to be odd, but it should be noted that this situation is identical to what one has in string theory, where the integrand for a many-particle amplitude is identical to the

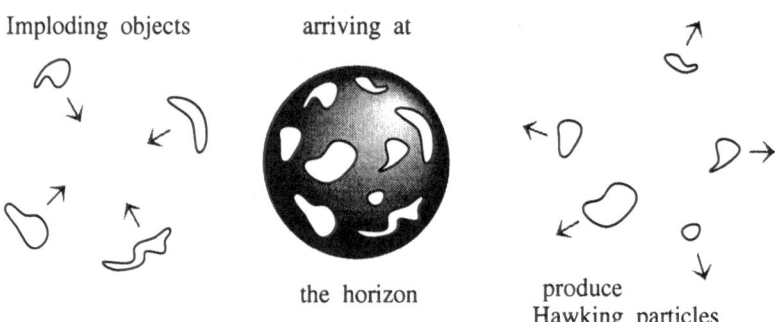

Imploding objects   arriving at

the horizon   produce
Hawking particles

Fig. 4 The horizon as a string world sheet. Three snapshots of a collision event with a black hole intermediate state.

amplitude for fewer particles, when two or more of the vertex insertions happen to coincide in the string world sheet.

The operators $x^\mu(\tilde{x})$ may be regarded as an "average position" operator for *all* particles ever entering or leaving the horizon at that point. This may (partly) explain why this information never "disappears" behind the horizon: there are always sufficient numbers of particles to be seen outside. This is in fact guaranteed by our brick wall model: the number of particles at distance greater than h from the horizon are sufficient to represent all "information" concerning the state of the black hole.

There is a special interpretation for the commutation rule

$$[U_{\text{out}}(\tilde{x}), U_{\text{in}}(\tilde{x}')] = \int \mathrm{d}^2\tilde{x}'' f(\tilde{x} - \tilde{x}'')[P_{\text{in}}(\tilde{x}''), U_{\text{in}}(\tilde{x}')] = -if(\tilde{x} - \tilde{x}').$$

(8.16)

We could decide to interpret $-U_{\text{out}}(\tilde{x})$ as indicating the position of the horizon with respect to the particles seen to emerge from the black hole, and similarly, $-U_{\text{in}}(\tilde{x})$ as the *time reverse* of this: the position of the past horizon with respect to the ingoing particles. Eq. (8.16) implies an uncertainty relation for these two quantities. For ordinary black holes, $U_{\text{out}}(\tilde{x})$ is usually precisely defined, as it is determined by the momentum distribution of the ingoing particles that actually formed the black hole. $U_{\text{in}}(\tilde{x})$ is the horizon of the time-reversed, or "white hole". In our picture, the white hole is the object formed by the Hawking particles if we follow these backwards in time. This is usually spread quantum mechanically over a large range of values. In our view, white holes are nothing but quantum superpositions of black holes. They relate to black holes just like the momentum and the position of a quantum particle are related to each other.

9 The Transverse Gravitational Force; A Discrete Spectrum of States[12]

So-far, the transversal component of the gravitational force has not been taken into account. One may suspect that this is the reason why our algebra is still represented by a continuum of states.

Unfortunately, as we will show, including the transverse gravitational force is difficult. We here only give an indication as to how one could proceed along these lines, so as to further improve our theory.

Let us recapitulate our algebra. From Section 8:

$$[P_{\mathrm{in}}(\tilde{x}), P_{\mathrm{in}}(\tilde{y})] = 0 = [P_{\mathrm{out}}(\tilde{x}), P_{\mathrm{out}}(\tilde{y})] ; \tag{9.1}$$

$$[U_{\mathrm{in}}(\tilde{x}), U_{\mathrm{in}}(\tilde{y})] = 0 = [U_{\mathrm{out}}(\tilde{x}), U_{\mathrm{out}}(\tilde{y})] ; \tag{9.2}$$

$$[P_{\mathrm{in}}(\tilde{x}), U_{\mathrm{in}}(\tilde{y})] = -i\delta^2(\tilde{x} - \tilde{y}\ ) = [P_{\mathrm{out}}(\tilde{x}), U_{\mathrm{out}}(\tilde{y})] ; \tag{9.3}$$

$$P_{\mathrm{out}}(\tilde{x}) = \tilde{\partial}^2 U_{\mathrm{in}}(\tilde{x}) ; \quad P_{\mathrm{in}}(\tilde{x}) = -\tilde{\partial}^2 U_{\mathrm{out}}(\tilde{x}) ; \tag{9.4}$$

$$[U_{\mathrm{in}}(\tilde{x}), U_{\mathrm{out}}(\tilde{y})] = if(\tilde{x} - \tilde{y}) ; \quad [P_{\mathrm{in}}(\tilde{x}), P_{\mathrm{out}}(\tilde{y})] = -i\tilde{\partial}^2 \delta^2(\tilde{x} - \tilde{y}). \tag{9.5}$$

As our starting point we again use Eqs. (9.1) – (9.5), but assume these to be valid only when the functions $U_{\mathrm{in}}(\tilde{x})$ and $U_{\mathrm{out}}(\tilde{x})$ are slowly varying. For later convenience, we rename the transverse coordinates on the horizon as (σ^1, σ^2), and now define a 2-surface $x^\mu(\tilde{\sigma})$ embedded in 4-space:

$$x^+ = U_{\mathrm{in}} , \qquad x^- = U_{\mathrm{out}} , \qquad \tilde{x} = \tilde{\sigma} . \tag{9.6}$$

The orientation of the surface is given by the tensor

$$W^{\mu\nu}(\tilde{\sigma}) = -W^{\nu\mu}(\tilde{\sigma}) = \varepsilon^{ab} \frac{\partial x^\mu}{\partial \sigma^a} \frac{\partial x^\nu}{\partial \sigma^b} . \tag{9.7}$$

We have

$$\frac{\partial \tilde{x}^a}{\partial \sigma^b} = \delta^a_b . \tag{9.8}$$

Now first consider the case that $x^\pm$ are slowly varying. This implies

$$W^{12} = 1 ; \qquad W^{1\pm} = \frac{\partial x^\pm}{\partial \sigma^2} ;$$

$$W^{2\pm} = -\frac{\partial x^\pm}{\partial \sigma^1} ; \qquad W^{+-} = \mathcal{O}(\partial_\sigma x^\pm)^2 . \tag{9.9}$$

Commutation rules follow from Eqs. (9.2) and (9.5):

$$[W^{1+}(\tilde{\sigma}), W^{2-}(\tilde{\sigma}')] = [W^{2+}(\tilde{\sigma}), W^{1-}(\tilde{\sigma}')] = i\partial_1\partial_2 f(\tilde{\sigma} - \tilde{\sigma}');$$

$$[W^{1+}(\tilde{\sigma}), W^{1-}(\tilde{\sigma}')] = -i\frac{\partial^2}{\partial \sigma^{2^2}} f(\tilde{\sigma} - \tilde{\sigma}') ; \tag{9.10}$$

$$[W^{2+}(\tilde{\sigma}), W^{2-}(\tilde{\sigma}')] = -i\frac{\partial^2}{\partial \sigma^{1^2}} f(\tilde{\sigma} - \tilde{\sigma}') .$$

As a special case, we have

$$[W^{\mu+}(\tilde{\sigma}),\, W^{\mu-}(\tilde{\sigma}')] = -i\tilde{\partial}^2 f(\tilde{\sigma}-\tilde{\sigma}') = i\delta^2(\tilde{\sigma}-\tilde{\sigma}'), \qquad (9.11)$$

where the index μ is summed over. It is this equation that we can reformulate in a manifestly Lorentz covariant form. One then may hope that not only the longitudinal, but also the shifts in all other directions will have been accommodated for. Since according to Eq. (9.9), W^{12} is the dominating component of the tensor $W^{\mu\nu}$, one may rewrite the right hand side of Eq. (9.11) as

$$\varepsilon^{+-12}W_{12}(\tilde{\sigma})\delta^2(\tilde{\sigma}-\tilde{\sigma}') \approx \tfrac{1}{2}\varepsilon^{+-\mu\nu}W_{\mu\nu}(\tilde{\sigma})\delta^2(\tilde{\sigma}-\tilde{\sigma}'), \qquad (9.12)$$

with $\varepsilon^{+-12} = i\varepsilon^{3412} = i$. The covariant generalization is then:

$$[W^{\mu\alpha}(\tilde{\sigma}),\, W^{\mu\beta}(\tilde{\sigma}')] = \tfrac{1}{2}\delta^2(\tilde{\sigma}-\tilde{\sigma}')\varepsilon^{\alpha\beta\mu\nu}W^{\mu\nu}(\tilde{\sigma}). \qquad (9.13)$$

This equation, as well as (9.11), is invariant under all continuous reparametrizations of the $\tilde{\sigma}$ coordinates (note that $W^{\mu\nu}$, as defined by Eq. (9.7), transforms as a density.)

It is tempting to assume Eq. (9.13) to have a wider range of validity than the non-covariant Eqs. (9.1) – (9.10). After all, Lorentz invariance guarantees that Eq. (9.13) continues to hold when the derivatives of $x^{\pm}(\tilde{x})$ are arbitrarily large. Unfortunately, the equations (9.11) do not form a closed algebra, since at the left hand side the index μ is still summed over. One can, however, limit oneself to the self-dual part:

$$K^{\mu\nu} = i(W^{\mu\nu} + \tfrac{1}{2}\varepsilon^{\mu\nu\kappa\lambda}W^{\kappa\lambda}), \qquad (9.14)$$

which has only three independent components:

$$K_1 = i(W^{23} + W^{14}) ; \quad K_2 = i(W^{31} + W^{24}) ; \quad K_3 = i(W^{12} + W^{34}) , \qquad (9.15)$$

and, indeed, their algebra closes. From Eq. (9.14) we derive:

$$[K_a(\tilde{\sigma}),\, K_b(\tilde{\sigma}')] = i\varepsilon_{abc}K_c(\tilde{\sigma})\delta^2(\tilde{\sigma}-\tilde{\sigma}'). \qquad (9.16)$$

The operators $K_a(\tilde{\sigma})$ are distributions. In order to construct representations of the algebra (9.16), we introduce test functions $f(\tilde{\sigma})$, $g(\tilde{\sigma})$, and write

$$L_a^{(f)} \stackrel{\text{def}}{=} \int K_a(\tilde{\sigma})f(\tilde{\sigma})\mathrm{d}^2\tilde{\sigma}, \qquad (9.17)$$

$$\left[L_a^{(f)},\, L_b^{(g)}\right] = i\varepsilon_{abc}L_c^{(fg)}. \qquad (9.18)$$

Restricting oneself to test functions f with $f^2 = f$, which only take the values 0 or 1, we find that the operators $L_a^{(f)}$ obey the commutation rules of the angular momenta:

$$[L_a^{(f)}, L_b^{(f)}] = i\varepsilon_{abc} L_c^{(f)}. \tag{9.19}$$

Notice that, since these operators L_a are obtained by integrating K_a over the region(s) where $f = 1$, and because the definition of K_a can be traced back to Eq. (9.7), one can rewrite $L_a^{(f)}$ as a contour integral:

$$L_1^{(f)} = i \oint_{\delta f} (x^2 dx^3 + x^1 dx^4), \qquad \text{etc.,} \tag{9.20}$$

where δf stands for the boundary of the support of f.

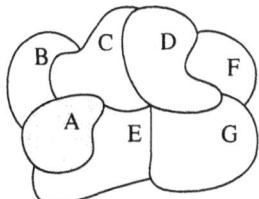

Fig. 5 Domains on the horizon corresponding to a representation of the algebra (9.16).

Suppose now that we have a set of test functions f which are equal to 1 on domains A or B, etc., and zero elsewhere. The domains form a lattice (of our choice) on the horizon, see Fig. 5. In each domain we have a set of three operators L_a that commute as angular momentum operators. The states could be formed out of the $|\ell, m\rangle$ eigenstates of $\mathbf{L}^2$ and L_3. If we combine domains to form some larger domain, the corresponding angular momentum operators must be added to form the new $\mathbf{L}$ operators, by the use of Clebsch-Gordan coefficients. Actually, if any of the ℓ values is larger than the minimal value $\frac{1}{2}$ (or perhaps, in some cases, 1), one can imagine splitting the corresponding domain into smaller ones with each the ℓ value $\frac{1}{2}$. Thus, one may end up with a lattice where on each site one has $m = \pm\frac{1}{2}$. It would not make much sense to maintain domains which have $\mathbf{L} = 0$, because the vanishing of the integrals (9.20) would imply that these regions have no spatial extent.

At first sight, this looks like a complete resolution of our problems. If each domain could be attributed an area equal to $4G \ln 2$ (see Eq. (4.10)), we exactly reproduce Eq. (4.9) for the level density. Unfortunately, life is not so simple. In Eq. (9.15), $W^{i4} = iW^{i0}$ are antihermitean operators, but W^{12}, W^{23} and W^{31} are hermitean. Therefore, the hermitean conjugates of K_a, and those of L_a, are the *anti*self-dual parts of $W^{\mu\nu}$. The L_a operators are not hermitean, and therefore the ℓ and m quantum numbers need not be subjected to the usual constraints of being half-integer, nor to obey the usual inequalities $|m| \leq \ell$.

10   Black Hole Complementarity

Let us return to the argument at the end of Section 9 concerning the notion of causality[14]. It has often been raised as a point of criticism against our scattering matrix Ansatz, see Fig. 6. An observer A passes through an horizon, while also an onserver B detects Hawking radiation. If we were allowed to treat them as living in a space-time that is fixed by external conditions, these two observers could be considered to be spacelike separated, and therefore one could conclude that their measurement operators commute. Hilbert space can be factored into a space of states whose properties can be detected by A, and another space of states whose properties can be detected by B, and possible further factors that can be seen neither by A nor by B. If, however, this space were considered to be the horizon of a black hole, one would expect the states seen by A to be related to the states seen by B through an S-matrix, and hence no longer independent. For the black hole physicist, there is no contradiction. Any measurement made by B, implies the introduction of states obtained from the Hartle-Hawking!vacuum by acting on it with operators that create or remove particles seen by B, which for A would by outrageously energetic. These particles would cause gravitational shifts that seriously affect the ingoing objects, including the fragile detectors used by A. Thus, these observations cannot be independent. What is new here, even for any possible flat space-time observer, is that trans-Planckian particles are involved (with this term we mean particles whose energies are far beyond the Planck value). In short: the metric is *not* determined solely by external circumstances, but also by the particles under consideration.

Complementarity here stands for the idea that states in Hilbert space near a black hole will appear to be profoundly different when the ingoing observer compares them with the outside one. General coordinate transformations fail to relate the experiences of the Hawking observer to the ones of the ongoing observer. Nevertheless, we talk about the same Hilbert space of states. In the limit of the infinite size black hole, this implies that a mapping should exist between the states living in a flat background metric, and black hole states that have all their information mapped onto the horizon. Unfortunately, a completely coherent physical picture clarifying this situation is still lacking.

Apparently, new phenomena strongly affect the conventional form of quantum mechanical Hilbert space when trans-Planckian particles enter the scene. With trans-Planckian particles around, spacelike separated operators may no longer commute with each other.

11   Outlook

Even though the philosophy, adhered to in this paper, is completely straightforward, and should not present fundamental conceptual problems, it nevertheless turned out to be extremely difficult to implement it completely. The effects of

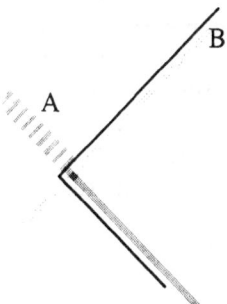

Fig. 6 The ingoing observer (A), and the Hawking observer (B).

transverse gravitational shifts were hard to implement, since these shifts do not commute with the longitudinal ones (because of their $\tilde{x}$-dependence). We have not mentioned another difficulty: the mass shell conditions for the in- and outgoing particles. We took these to be essentially massless, but most particles have a lower bound for their masses. Transverse momenta and masses, however, cause outgoing particles to fall back in again. The difficulty connected to this is the fact that, close to the horizon, ingoing and outgoing states will be difficult to distinguish. Presumably, the splitting of S according to $S = S_{out}S_{hor}S_{in}$ (Eq. 7.1), must be further refined.

The resemblance to string theory in our final results may suggest that one should readdress the black hole using string theory. Some caution however is called for. It is well-known that string theory requires a 10 or 26 dimensional target space, if tachyons and other unphysical features are to be avoided, but such arguments do not directly apply to our present aproach: unitarity and causality look very different, as is manifest from the observation that our string constant is purely imaginary. Secondly, by considering the "information content" of the states in our Hilbert space, we infer that a cut-off at the Planck scale is required that turns our world into a discrete one at that scale. This is quite unlike the starting points of string theory. Convergence of the various approaches may well be envisioned, but it is conceivable that two-dimensional conformal quantum field theory is no more (or less) relevant here than it is in certain statistical models such as the Ising model. We should keep in mind that QCD is also a theory that shows stringlike behavior, but clearly lives in four space-time dimensions, so that apparently the formal unitarity arguments are not applicable here.

The observation of black hole–white hole complementarity (Section 8) suggests an interesting relationship between the *black hole horizon* and the *white hole singularity*, and vice versa. After all, a white hole singularity would develop as soon as one allows Hawking particles to produce a gravitational field, as one would be tempted to do when contemplating time reversal invariance. Indeed, the point S in Fig. 2b, is not truly a point, but gets the extended shape of a caustic when ingoing matter is deprived of its spherical symmetry. The operator $U_{out}(\tilde{x})$ could be regarded as the one describing this caustic. When ingoing matter is allowed to enter during sufficiently large time intervals, this caustic becomes a true fractal. At the same time, this fractal may be relevant for the description of the singularity in the time-reversed black hole. A duality relationship between the black hole singularity and the horizon has been proposed in the framework of string theory.

Bibliography

[1] S.W. Hawking, Commun. Math. Phys. **43** (1975) 199; J.B. Hartle and S.W. Hawking, Phys. Rev. **D13** (1976) 2188.

[2] S.W. Hawking, Phys. Rev. **D14** (1976) 2460; *id.* Commun. Math. Phys. **87** (1982) 395; S.W. Hawking and R. Laflamme, Phys. Lett. **B209** (1988) 39; D.N. Page, Phys. Rev. Lett. **44** (1980) 301; R. Haag, H. Narnhofer and U. Stein, Commun. Math. Phys. **94** (1984) 219; R. Sorkin, Phys. Rev. Lett. 56 (1986) 1885, Phys. Rev. D **34** (1986) 373; P. Mitra, *Black hole Entropy*, Invited talk delivered at XVIII IAGRG Conference, Madras, February 1996, hep-th/9603184.

[3] S.W. Hawking and G.F.R. Ellis, *The Large Scale Structure of Space-time*, Cambridge: Cambridge Univ. Press, 1973.

[4] W. Rindler, Am.J. Phys. **34** (1966) 1174.

[5] G. 't Hooft, Int. J. Mod. Phys. **A11** (1996) 4623.

[6] R.M. Wald, Commun. Math. Phys. **45** (1975); *id.*, Phys. Rev. D **20** (1979) 1271; W.G. Unruh and R.M. Wald, Phys. Rev. D **29** (1984) 1047; W.G. Unruh, Phys. Rev. D **14** (1976) 870; S.W. Hawking and G. Gibbons, Phys. Rev. D **15** (1977) 2738.

[7] G. 't Hooft, Acta Physica Polonica **B19** (1988) 187-202.

[8] J.D. Bekenstein, Nuovo Cim. Lett. **4** (1972) 737; *id.*, Phys. Rev. D **7** (1973) 2333, D **9** (1974) 3292.

[9] G. 't Hooft, Nucl. Phys. **B256** (1985) 727-745.

[10] P.C. Aichelburg and R.U. Sexl, Gen. Rel. and Gravitation **2** (1971) 303; W.B. Bonner, Commun. Math. Phys. **13** (1969) 163.

[11] T. Dray and G. 't Hooft, Nucl. Phys. **B253** (1985) 173.

[12] G. 't Hooft, Proc. of the 4th seminar on Quantum Gravity, May 25-29, 1987, Moscow, USSR. Eds. M.A. Markov, V.A. Berezin and V.P. Frolov, World Scientific 1988, p. 551; *id.*, "Black holes as clues to the problem of quantizing gravity", Lecture notes for the CCAST/WL Meeting on *fields, strings and quantum gravity*, Beijing, June 1989. (Gordon and Breach, London); *id.* Nucl. Phys. **B335** (1990) 138; Physica Scripta T **36** (1991) 247.
id. "Scattering matrix for a quantized black hole", In book: *Black Hole Physics*, V. De Sabbata and Z. Zhang (eds.). 1992 Kluwer Academic Publishers, The Netherlands, p. 381; *id.* "S-Matrix theory for black holes", In Proceedings of a NATO Advanced Study Institute on *New Symmetry Principles in Quantum Field Theory*, held July 16-27, 1991 in Cargèse, France. Eds. J. Fröhlich, G. 't Hooft, A. Jaffe, G. Mack, P.K. Mitter and R. Stora NATO ASI Series, 1992 Plenum Press, New York, p. 275; *id.* "Black Holes, Hawking Radiation and the Information Paradox", Proceedings of *Trends in Astroparticle Physics*, Stockholm, Sweden, Sept. 22-25, 1994, THU-94/20.

[13] G. 't Hooft, Physica Scripta, **T15** (1987) 143-150.

[14] C.R. Stephens, G. 't Hooft and B.F. Whiting, Class. Quantum Grav. **11** (1994) 621.

Part VII

Panel Discussion

Panel Discussion: The Definitive Proofs of the Existence of Black Holes

Werner Collmar (Garching), Norbert Straumann (Zürich), Sandip K. Chakrabarti (Calcutta), Gerard 't Hooft (Utrecht), Edward Seidel (Potsdam), Werner Israel (Victoria)

It is our intention to give the non-expert reader a book at hand which enables him or her to recognize whether there are black holes around or not. In the various lectures of our school, the lecturers tried to address this question from theoretical as well as from observational points of view. In the panel discussion some of our lecturers were asked to sum up the present state of knowledge in the form of relatively short statements. They were explicitly requested to answer the fundamental question "Do black holes exist or not?". We hope that you will enjoy the output of our panel as much as we did.

The Editors

Werner Collmar:

As an astrophysical observer I want to summarize briefly (my knowledge on) the observational hints/evidence which we have for the existence of Black Holes (BHs) in outer space.

The most conclusive observations would be if we could directly image the astrophysical sites suspected to host a BH, like the center of active galaxies or of galactic binaries, for example. Active galactic nuclei (AGNs) are believed to contain massive BHs at their center. The Schwarzschild radius R_S of a $10^8 \, M_\odot$ BH is 3×10^8 km or ~ 2 AU (astronomical units). If a surrounding accretion disk extends out to $\sim 500 \, R_S$, its angular size at a distance of 10 Mpc is $\sim 2 \times 10^{-4}$ arcsec, which is well below the angular resolution of the Hubble Space Telescope. Since no resolved direct imaging is possible, the observational evidence has to come from 'indirect' means.

In AGNs large luminosities are generated in small volumes, like the total luminosity of our galaxy would be generated within our solar system, and would be switched on and off on timescales of days. This extraordinary fact is not explainable with nuclear fusion as the source of energy. However, it is most naturally explained by the release of gravitational energy close to a massive BH, which is the most efficient process we know.

The effects of the gravitational force of BHs on the surrounding stars or gas can be studied. From velocity and velocity-dispersion measurements near the center of galaxies, the enclosed mass can be estimated. The most promising cases for BHs are found in our 'Milky Way' with $\sim 2.45 \times 10^6 \, M_\odot$ within 0.015 pc [2] and in NGC 4258 from VLBI (Very Long Baseline Interferometry maser observations with a central mass of $3.6 \times 10^7 \, M_{\odot}$ within a volume of 0.13 pc [8]. Massive star clusters in such small volumes would not be stable for long times and therefore have to collapse to BHs. X-ray observations have revealed an asymmetric and redshifted iron K_{α} line in the Seyfert galaxy MCG-6-30-15 [10]. The shape and the redshift are most easily explained if the line is generated in a plasma which is located within the strong gravitational field close to a massive BH.

If a BH is a member of a binary system, its mass can be estimated (via the mass function) by the dynamics of the system. Several compact sources have lower limits on their masses which are significantly larger than $3 \, M_{\odot}$, the upper limit on the stability of a neutron star. The most promising stellar candidate to date is the X-ray nova V404 Cyg with an estimated mass for the compact object in the range between 10 and 15 $M_{\odot}$ [9].

Different spectral signatures in compact binaries have been proposed to be evidence for a BH as compact object instead of a neutron star, like spectral bumps at γ-ray energies or power-law spectra up to $\sim$1 MeV instead of spectra cutting off exponentially at hard X-rays. Because these interpretations are model dependent, their support for the evidence of BHs is less compelling than the mass estimates, for example.

In the future new instruments from the radio to the γ-ray band will come into operations with improved sensitivity and improved spectral and angular resolution. With these instruments we shall have better diagnostic tools to further investigate such promising topics like this X-ray line in MCG-60-30-15, for example. To my mind however, even these improved instruments will not provide a 'single key' observation definitely proving the existence of BHs. I rather believe that — as time progresses — the overall evidence for BHs will rise asymptotically to one.

Norbert Straumann:

The evidence of black holes in some X-ray binary systems and for supermassive black holes in galactic centers is still indirect, but has become overwhelming during the past few years. There is so far, however, very little evidence that these collapsed objects are described by the Kerr metric.

In my brief remarks I would like to tell those of you, who had no closer look at the observational situation, what I consider in both categories to be at present the very best candidates. I will indicate only one prospect of observing specific signatures associated with the Kerr metric. Others will have to say more on this.

Independent of the remarkable recent developments, I had never any serious doubts that black holes exist in great numbers in the astronomical universe. This is simply because 'cold' self-gravitating matter can only exist in a small mass range below a few solar masses. There is, on the other hand, absolutely no reason for all the many massive stars (or associations of stars) to get rid of their mass in the course of their evolution in some violent processes (supernovae), in order to be able to settle down in a cold final state (white dwarf, neutron star, quark star (?), ...).

The value of the largest possible mass, M_{max}, of a neutron star plays a decisive role in the observational identification of stellar-mass black holes. In view of the large uncertainties of the equation of state, it is important of having reliable limits of M_{max}. This is possible on the basis of general assumptions. For instance, if one accepts a fluid dynamic description of matter in a neutron star, causality implies that the velocity of sound should not exceed the velocity of light. (It is true that the sound velocity is only a phase velocity, but it provides the characteristics of the hydrodynamic equations for acceptable relativistic formulations.) For non-rotating stars, an upper bound of about $3.2M_\odot$ is obtained. On the basis of even weaker assumptions (microscopic stability, $dp/d\rho \geq 0$) one finds $M_{max} < 5M_\odot$ instead. Rotating neutron stars can have somewhat larger masses, see [6].

In binary X-ray systems the mass functions f can be measured with good accuracy. Since this provides a rigorous lower limit for the mass of the compact companion, the case for a black hole is extremely strong if f is bigger than $6M_\odot$, say. Now, this is the case for the *X-ray nova V404 Cygni*, which has the mass function

$$f = (6.08 \pm 0.06) \ M_\odot \,.$$

Other good cases have been mentioned in some of the lectures, but this is, as far as I can see, the most secure stellar-mass black hole candidate we have at the moment.

As far as supermassive black holes are concerned, I would like to emphasize that gas-dynamical evidence is not very strong in general, because gas can easily be pushed around. Therefore, the many examples of central gas and dust disks perpendicular to jets, like M87, do not establish the existence of black holes.

Thanks to the recent work of Genzel and coworkers [2, 3, 7], we now know that a dark mass of about $2.6 \times 10^6 M_\odot$ must reside within about a light week in Sgr A*. Its density is thus greater than $2 \times 10^{12} M_\odot/\mathrm{pc}^3$. There exist no stable configurations of normal stars, stellar remnants of sub-stellar entities at that density. This concentration of dark matter *in the center of the Milky Way* is now the best case for a supermassive black hole.

Another, almost equally good case is provided by NGC 4258. The following observations show that the central mass must be a black hole (or something even more exotic). The peculiar spiral galaxy NGC 4258 (distance $\sim 6.5\,\mathrm{Mpc}$) has in its core a disk extending from about 0.2 to 0.13 pc in radius. This was discovered with a very precise mapping of gas motions via the 1.3 cm maser-emission of H_2O with VLBI (Very Long Baseline Interferometry). The angular resolution of the array was better than 0.5 milliarc seconds. (This is 100 times better than the resolution of the HST (Hubble Space Telescope).) The rotational velocity distribution in the disk follows an exact Keplerian law around a compact dark mass, and the velocity of the inner edge is $1080\,\mathrm{km/s}$. From this one infers a dark mass concentration of $3.6 \times 10^7 \ M_\odot$. As in the case of the Milky Way, there are no long-lived star clusters with these extremal properties.

Finally, I would like to point out the possibility to study the relativistic region of a black hole with X-ray astronomy. Recent ASCA (Advanced Satellite for Cosmology and Astrophysics) observations of MCG-6-30-15 and other Seyfert 1 galaxies have revealed that the 6.4 keV fluorescent iron line is very broad and provides some evidence that the emitting region is orbiting close to a black hole. Gravitational effects are apparently skewing the line shape [10, 4]. With the greatly improved spectral resolution of XMM (X-ray Multi-mirror Mission), this might become an important tool to obtain specific information on the gravitational field in the relativistic X-ray emitting region. J. Wilms et al. have made detailed studies, see their contribution in Chap. 5, see also B.C. Bromley et al. [1].

Sandip K. Chakrabarti:

The problems at hand are (a) whether black holes exist, (b) whether the compact objects we call 'black holes' are actually those predicted by solutions of Einstein's equations and (c) what is the best way to identify black holes.

Since black holes are necessarily black (save Hawking radiations which are unobservable with present technology), their detections must be indirect. However, some of the evidences are more 'circumstantial' than others. For instance, the origin, acceleration and collimation of jets, fast time variabilities, stellar velocity dispersions etc. are far from convincing proofs. Measurements of the mass of the central objects such as in M87 ($4 \times 10^9 M_\odot$) by HST spectroscopy or in NGC4258 ($4 \times 10^7 M_\odot$) by water maser emission or in our Galactic Center ($2.5 \times 10^6 M_\odot$) by proper motion study of stars produce mass concentrations of $2.0 \times 10^7 M_\odot/pc^3$, $2.5 \times 10^9 M_\odot/pc^3$, and $6.5 \times 10^9 M_\odot/pc^3$ respectively. Similarly, the measurements of mass function of binary systems indicated a mass function of V404 Cyg to be $f(M) = (6.08 \pm 0.06)\ M_\odot$, a few other candidates have $f(M) \gtrsim 3.0\ M_\odot$. These objects may be strong candidates for black holes but the nagging issue still remains: Are they black holes as predicted by Einstein's equations or just some compact object with hitherto unknown equation of state? I.e., do these objects have horizons with all the associated properties?

In order to prove the existence of the horizons one must look for detailed spectral signatures, since radiations forming the spectra come out of infalling matter which respects the inner boundary condition (IBC) on the horizon. As is known, the IBC selects completely different hydrodynamic and radiative hydrodynamic branches of the global solutions of the governing equations. Unfortunately, till today not all the equations could be written down with certainty in black hole environment, what to talk about solving them. However, some of the predictions of the existing advective models are sufficiently robust and do not depend on the detailed models. For instance, the hard/soft transitions can be seen even in neutron stars, but almost *constancy of spectral slopes* with change of luminosity by factors of hundreds and particularly the photon index of -2.5 in weak power-law tail of soft states are *not seen* in neutron stars. Difference of neutron stars and black holes cannot be made in terms of total luminosity (even if through black hole horizons the entire energy can be advected and in principle makes the flow non-luminous) since any number of physical processes (such as outflows) in neutron stars may invalidate the argument completely.

With the recognition that we must rely on spectral signatures for a complete stock taking of the black holes in the universe, it has become easier to identify black holes. For instance, Cyg X-1, which is the first suspected candidate for a black hole does not satisfy the mass function criteria at all, its $f(M)$ is only (0.24 ± 0.01) $M_\odot$. But it shows the desired spectral slopes in both hard and soft candidates. Similarly, V404 Cygni and A0620-00 have large mass functions and also behave understandably in quiescence states as predicted by advective disk models. Supermassive black holes are difficult to be identified using spectral signatures, because their change of state would take thousands of years.

Gerard 't Hooft:

For an elementary particle physicist, the question of the existence of black holes has several aspects.

• From a philosophical point of view, we will never be able to prove the existence of anything with ultimate rigour. We cannot even prove that we exist ourselves. Clearly, what we are trying to do is to provide proofs "beyond all reasonable doubts".

• In quantum particle physics, it is of importance to know whether some object can exist in principle, even if its actual presence somewhere cannot be shown. A particle that can exist in principle, will represent an element in Hilbert space, and as such it will give important contributions to the Schrödinger equation. We call such particles *virtual* particles, and we want to know about them.

• This is why we want to know whether black holes with masses between 1 mg and 1 $M_\odot$ can exist, even if no astrophysical mechanism for their production was known. Furthermore, the importance of the existence, in the above sense, of these solutions of Einstein's equations is that they appear to be correct, acceptable solutions. If they did *not* exist, even in principle, this would imply a significant deficiency in our understanding of Nature. Not only the physics of the fundamental interactions would have to be revised, but even the laws at distance scales larger than centimeters, whereas these have been checked by many precision experiments.

• It is conceivable, however, that tiny black holes will turn out to be indistinguishable from more conventional forms of matter: the spectrum of "black hole states" might blend naturally into the spectrum of "elementary particles" (loosely speaking, black holes *are* elementary particles) and *vice versa*. In this case, large black holes (heavier than, say, a few milligrams) will still be so significantly different from other forms of matter that in practice confusion will be unlikely.

• For me, *astronomical* black holes represent the extreme other end of the scale. I am impressed by the evidence produced by the astrophysicists, and I believe that they have come extremely close to proving the existence of black holes.

• There is the question of the existence of *primordial* black holes, in particular the ones with masses in the range of planetoids. These should decay, and quantum theory predicts that once their masses have decreased to become that of a small asteroid, they turn themselves into radiation energy through a violent explosion. Whether these really exist in our universe (not just virtually) and whether these explosions can be or have already been observed, remains extremely dubious, at best.

• I would like to know from the astrophysicists whether three-body interactions in globular star clusters can produce black holes, which should come shooting out of the center, singlets as well as doublets (P. Hut, private communication). I would also like to know what the *smallest mass* is for a black hole that can be produced via conventional astrophysical processes.

Edward Seidel:

Evidence for the existence of black holes, once considered fantasy by many, is mounting these days on a monthly basis. Most astronomers now accept black holes as a standard part of their observed universe, while only a decade ago many were rather skeptical. As standard astronomical observations improve, more and more black hole candidates are found, while existing candidates are even more firmly thought to be actual black holes.

However, while such evidence is now very strong indeed, it is generally circumstantial. We still await direct and incontrovertible proof of the existence of black holes, i.e., a detectable signal emitted by a black hole that unambiguously identifies it. *Gravitational waves emitted by black hole interactions* should provide that "smoking gun" signal that not only proves that black holes exist, but that will also reveal essential properties of the black holes. In particular, the so-called ringing or quasinormal modes of the black hole are damped wavetrains that excited rather generically from a perturbed black hole, and the wavelength and damping time of these modes depending only on its mass and spin (and charge, which is probably not relevant). In all numerical studies of black hole formation, perturbation, and even collision, these modes have been strongly excited, and if detected they should uniquely identify the source as a black hole.

As gravitational wave observatories such as LIGO, VIRGO, and GEO600, are under construction around the world, the gravitational wave signals expected from black hole interactions is gaining increasing attention. Black hole mergers from binary coalescence is now considered one of the most promising sources of gravitational waves, and according to a recent study by Flanagan and Hughes [5], such systems may well provide the first gravitational waves to be seen after the detectors go online in a few years. In what follows I summarize the main points of their very detailed analysis.

There are three phases of black hole coalescence: (a) inspiral, (b) merger, and (c) ringdown. The inspiral is the adiabatic orbit phase where the holes slowly spiral together, before they are around $r = 6M$ apart, where M is the total mass of the system. At this time, the merger phase begins, where dynamic instabilities drive the holes towards each other in near free fall. This is a most violent stage of evolution for which numerical relativity will be essential to compute waveforms. Finally, a single, distorted black hole results, which will quickly settle in to the ringdown phase, where normal modes are emitted. According to Flanagan and Hughes, taking account of the sensitivity and bandwidth of the first generation detectors coming online by the year 2001, low mass coalescence (< 30 solar masses) should be seen via their inspiral waves, while high mass systems (between 100 and 700 solar masses) should be seen via their ringdown waves. Both types should be visible out to about 200 Mpc, and the event rate for the low mass systems should be of order several per year. By using numerical relativity calculations as templates for the merger phase, one can enhance the detection rate, and very significantly, one can analyze the signals to better understand the sources.

This is a very exciting time for black hole physics: gravitational wave astronomy is about to be born, and fortunately numerical relativity is on the track of simulations that will be needed to enhance and understand the upcoming observations. The next decade should provide many proofs of the existence of black holes, and a plethora of information about them as well.

Werner Israel:

Twenty years ago I spent a very pleasant sabbatical year at the ... Institute in ... . Although this story is true, I have blanked out the names because its significance is generic – it could have happened at any of dozens of institutions at that time. Shortly after my arrival there was a coffee party, and after some warm words of welcome, the Director of the Institute remarked, "Werner is going to be with us for a year. We should all talk to him and try to cure him of these silly notions he has about the possibility of black holes."

I was very well treated and enjoyed the most cordial personal relations with my hosts throughout that year, but in this one respect at least I'm afraid I proved something of a disappointment to them. The cure failed. As my wife will, I'm sure, readily confirm, I am the most unreasonably mulish of people, also fairly deaf and impervious to foreign languages.

I recall going to lunch one day with about a dozen members of the Institute – staff and graduate assistants – and I decided to take a poll, asking each in turn whether he believed that black holes are possible. There was a string of firm denials, except for one assistant, from whom I got the interesting response: "Not usually. But when I'm applying for a U.S. research grant, *then* I believe – temporarily".

I am still in touch with a few of the colleagues who gathered around the table that day. Not one has changed his views.

With the foreshortening that memory lends to past events, it now seems that in December 1967, John Wheeler declared, "Let there be black holes". and lo! there were black holes. But this is not how it really happened. Comprehension of this idea took years, acceptance much longer. I remember Felix Pirani's words in 1967: "The Schwarzschild singularity at $r = 2m$ is almost certainly a fraud – but we don't yet know what kind of fraud".

There was an incident in Canada in 1979 when the Winnipeg vice squad, raiding a video store, seized copies of (among others) the Walt Disney movie, "The Black Hole", on suspicion of obscenity – probably the only time a Disney film has ever suffered this indignity.

Of course, the disbelief of my Institute colleagues had nothing to do with observations. They were put off by the inherent absurdity of the *concept* – the inevitability of singularities, space turning into time, and so on. At that time, observational progress was lethargic..– It took ten years before the first serious black hole candidate Cygnus X-1, optically identified in 1972 by Tom Bolton of Toronto and by Webster and Murdin in England was joined by a second, LMC X-3, identified by a team at the Dominion Astrophysical Observatory, Victoria, B.C.

But, as we have heard at this School, in the last few years the observational arena has been transformed by the Hubble space telescope and sophisticated ground-based techniques. We now have compelling evidence for the presence of compact dark objects in galactic nuclei with masses ranging from millions to billions times that of the sun.

The most convincing and extraordinary case is NGC 4258. Here one is observing (by interferometric techniques in the microwave range) maser radiation from water molecules in a dusty torus orbiting just 0.3 light-year from the centre, indicating the presence of an invisible mass of 36 million solar masses within this radius.

In our own galaxy, the recent work of Eckart and Genzel, who measured proper motions of three dozen stars within 1/30 light-year from the centre, point convincingly to a central dark mass of 2 million suns.

At the 1996 "Texas" symposium in Chicago, there were even reports of the first direct observational evidence for general-relativistic strong-field effects and for event horizons. The Japanese X-ray satellite ASCA has found peculiar frequency shifts in the X-rays from galactic nuclei, which (it is claimed) can only be caused by strong gravity. Richard Mushotzky of the Goddard Space Institute reported on observations of X-ray novae during intervals when accretion is slow. Those thought (on the basis of mass) to contain black holes are dimmer than ones with neutron stars, presumably reflecting the difference between the hard surface of a neutron star and an event horizon.

Despite all of these developments, many who rejected black holes in the 1970s as conceptually absurd remain hard-core skeptics today. The moral I should like to draw is this. Nothing in science is 100 % certain. But a time arrives when the balance of probabilities has become so overwhelming that the burden of proof must pass to the opposition, and further attempts at rational argument are pointless. We *still* have flat-earth societies; 87 % of U.S. parents want creationism taught in schools; and, as for me, I still check my horoscope in the newspaper from time to time. Today, if one encounters a skeptic for whom non-existence of black holes is a matter of religious faith, perhaps it is kinder (and certainly less trouble) not to get involved in an argument.

Bibliography

[1] B.C. Bromley, K. Chen, W.A. Miller, ApJ **475** (1997) 57

[2] A. Eckart and R. Genzel, Nature **383** (1996) 415

[3] A. Eckart and R. Genzel, MNRAS **284** (1997) 576

[4] A.C. Fabian, K. Nandra, C.S. Reynolds, W.N. Brandt, C. Otani, Y. Tanaka, H. Inoue, K. Iwasawa, Los Alamos eprint archive astro-ph/9507061

[5] E.E. Flanagan and S.A. Hughes, Phys.Rev. **D57** (1998) 4535

[6] J.L. Friedman in: *Black Holes and Relativistic Stars. A Symposium in Honor of S. Chandrasekhar*, Ed. R. Wald (University of Chicago Press, Chicago 1998)

[7] R. Genzel, A. Eckart, T. Ott and F. Eisenhauer, MNRAS **291** (1997) 219

[8] L.J. Greenhill et al., ApJ **440** (1995) 619

[9] T. Shahbaz et al., MNRAS **271** (1994) 10

[10] Y. Tanaka et al., *Nature* **375** (1995) 659

Exercises

1   Exercises to the lectures of W. Collmar/V. Schönfelder

1.1 The Eddington limit plays a fundamental role in astrophysical sources which are powered by accretion of matter. It follows from the requirement that the radiation pressure does not overcome the force of gravity.

Show that for spherical symmetric geometry

$$L_{\mathrm{Edd}} = 4\pi G M m_{\mathrm{p}} c\sigma_{\mathrm{T}} \approx 1.3 \times 10^{38} M_{\odot} \mathrm{erg/sec}$$

where M is the mass of the accreting object, m_{p} the proton mass, c the velocity of light and σ_{T} the Thomson crossection. (Assume that the material consists of hydrogen and is fully ionised and that the radiation scatters only on ambient electrons.)

1.2 In the field of AGNs the broad band energy spectra from radio to γ-ray energies are very often plotted as νF_ν versus ν-plots, where ν is the frequency and $F_\nu = \mathrm{d}F(\nu)/\mathrm{d}\nu$ represents the differential energy flux at frequency ν per unit interval of phton frequency (F_ν has the dimensions $\mathrm{erg\,cm^{-2}\,sec^{-1}\,Hz^{-1}}$). Very often, instead of the differential energy flux, the differential photon number flux $N_\nu = \mathrm{d}N(\nu)/\mathrm{d}\nu$ is given, where N_ν is the measured photon number flux per unit frequency interval.

Show that

$$\nu F_\nu = \nu^2 N_\nu = \frac{\mathrm{d}F(\nu)}{\mathrm{d}(\ln \nu)},$$

hence that νF_ν is the energy flux per natural logarithmic frequency interval (not per decade of photon frequency).

2   Exercises to the lectures of N. Straumann

2.1 Repeat the $3+1$ reduction of Maxwell's equations for a *static* spacetime (e.g. the Schwarzschild solution). Make little use of the notes. (This was already achieved by Einstein during his Prag time.)

2.2 Specialize Ex. 2.1 to electrostatics:

$$d(\alpha\mathcal{E}) = 0, \quad d(*\mathcal{E}) = 4\pi\rho.$$

With the first equation one can introduce a scalar potential Φ with $\alpha\mathcal{E} = -d\Phi$.

a) Consider the Schwarzschild background and write this differential equation for Φ in terms of the standard coordinates r, ϑ, φ. Insert an ℓ-pole ansatz

$$\Phi = f(r) Y_{lm}(\vartheta, \varphi)$$

and derive the differential equation for $f(r)$ outside the charges. Can you solve this?

b) Perhaps, somebody is able to find Φ for a point charge e located outside the horizon. The solution of this problem was found early by E. Copson, Proc. R. Soc. **A118**, 184 (1928); see also R. Hanni and R. Ruffini, Phys. Rev. **D8**, 3258 (73).

It is given by:

$$\Phi(r, \vartheta) = \frac{em}{ar} +$$

$$\frac{e}{ar} \frac{(r-m)(a-m) - m^2 \cos \vartheta}{[(r-m)^2 + (a-m)^2 - m^2 - 2(r-m)(a-m)\cos\vartheta + m^2 \cos^2 \vartheta]^{1/2}}.$$

2.3 Specialize to magnetostatics and consider an axially symmetric purely poloidal field

$$\mathcal{B} = \frac{1}{2\pi} d\Psi \wedge d\varphi \qquad (\Psi\text{: flux function}).$$

Derive the differential equation for Ψ on a Schwarzschild background. Try to find Ψ for an asymptotically homogeneous $\vec{B}$-field.

2.4 For comparison with the solution in section 3 of the lectures, consider in ordinary electrodynamics a rotating, ideally conducting, uncharged sphere in a homogeneous magnetic field (unipolar induction). Determine the $\vec{E}$-field. Compute the voltage between the north pole and the equator.

2.5 Show that $\mathrm{div}\vec{\beta} = 0$ for the Kerr solution.

2.6 Fill in the details which lead to the $\vec{E}$-field in section 3 of the lectures.

2.7 Consider this $\vec{E}$-field along the symmetry axis and show that $\vec{E} = E_r \vec{e}_r$,

$$E_r = -\frac{\partial \Phi}{r}, \qquad \Phi = aB_0 \left[1 - \frac{2Mr}{r^2 + a^2}\right].$$

A charged test particle is thus accelerated. Find a solution of Maxwell's equations with a total charge $Q \neq 0$ such that there is no force on the test particle. (Answer: $Q = -2B_0 J$, sign?). Put in numbers and show that this charging up (by an exterior plasma) is astrophysically of no interest.

2.8 Energy conservation for a charged test particle in a combined stationary field g, F: Let ξ be a Killing field such that $L_\xi F = 0$. Choose a gauge such that also $L_\xi A = 0$ ($F = dA$). Show that

$$(\pi, \xi) = \text{const.}, \qquad \pi = mu + A,$$

where u is the 4-velocity. Let $A = -\Phi dt + \mathcal{A}$ and $\xi = \partial_t$.

Then this observation implies

$$E \equiv -m(u, \xi) + e\Phi = \text{const.}$$

2.9 Analyse the idealized electric motor with a black hole as rotator in Fig. 8 of the lectures.

2.10 Prove the identity $\delta(m \wedge \mathcal{X}) = -m \wedge \delta\mathcal{X}$ in section 9 of the lectures.

2.11 Supplement to sections 8 and 9 of the lectures:

Let u be the 4-velocity of the plasma and $u = \gamma(e_0 + \vec{v})$ its decomposition relative to the FIDOs.

(a) Write the particle number conservation $\nabla \cdot (nu) = 0$ (n =number density) in the form

$$(\partial_t - L_{\vec{\beta}})(\gamma n) + \vec{\nabla} \cdot (\gamma \alpha n \vec{v}) = 0, \tag{1}$$

which reduces in the stationary and axisymmetric case to

$$\vec{\nabla} \cdot (\alpha n \vec{u}) = 0, \quad \vec{u} = \gamma \vec{v}. \tag{2}$$

(b) Set

$$\vec{u}^{\text{pol}} = \frac{\eta}{\alpha n} \vec{B}^{\text{pol}} \tag{3}$$

and show that $\mathcal{E} = i_{\vec{v}}B$ (ideal MHD condition) leads to

$$\vec{u} = \frac{\eta}{\alpha n} \vec{B} + \gamma \frac{\Omega_F - \omega}{\alpha} \vec{\partial}_\varphi. \tag{4}$$

Write the toroidal part of $\vec{u}$ as in section 8.11 (with angular velocity Ω) and conclude

$$\gamma(\Omega - \Omega_F)\tilde{\omega} = \frac{\eta}{n} B_\varphi. \tag{5}$$

(c) Insert (4) into (2) and show that one finds

$$d\eta \wedge d\Psi = 0 \quad \Longrightarrow \quad \eta = \eta(\Psi). \tag{6}$$

(d) Assume a toroidal velocity field $\vec{v}$ and deduce from $\mathcal{E} = i_{\vec{v}}B$ and $\alpha\mathcal{E} = -d\Phi + i_{\vec{\beta}}A$ that

$$d\Phi = \frac{\Omega}{2\pi} d\Psi,$$

which implies $d\Omega \wedge d\Psi = 0$, and hence $\Omega = \Omega(\Psi)$.

2.12 ADM formalism for electrodynammics

Insert the $3+1$ reductions for F and $*F$ into the Maxwell Lagrangian $-\frac{1}{2}F\wedge *F$. Use the representation of $\mathcal{E}$ and $\mathcal{B}$ in terms of the potentials Φ and $\mathcal{A}$, and show that

$$-\frac{1}{2}F\wedge *F = dt \wedge \mathcal{L},$$

where $\mathcal{L}$ is the 3-form

$$\mathcal{L} = \Phi d*\mathcal{E} - \partial_t \mathcal{A}\wedge *\mathcal{E} - \frac{1}{2}\alpha\left(\mathcal{E}\wedge *\mathcal{E} + \mathcal{B}\wedge *\mathcal{B}\right)+(i_\beta \mathcal{B})\wedge *\mathcal{E} + \text{exact differential.}$$

Regard in the action principle

$$S(\Phi,\mathcal{E},\mathcal{A}) = \int \mathcal{L}\wedge dt, \quad \text{with} \quad \mathcal{B} = d\mathcal{A},$$

the fields Φ, $\mathcal{E}$, and $\mathcal{A}$ as independent, and verify that one obtains the correct field equations:

$$d*\mathcal{E} = 0, \qquad\qquad \text{(constraint eq.)} \qquad\qquad (7)$$

$$\left.\begin{array}{l} \partial_t \mathcal{A} + d\Phi + \alpha\mathcal{E} - i_\beta d\mathcal{A} = 0, \\ (\partial_t - L_\beta)*\mathcal{E} = d(\alpha * d\mathcal{A}), \end{array}\right\} \quad \text{(dynamical eq.)} . \qquad (8)$$

Φ can be described arbitrarily. The dynamical evolution of $\mathcal{E}$ and $\mathcal{B}$ determined by (8), is independent of Φ. Prove this fact (which reflects gauge invariance).

2.13 Invent other exercises.

3 Exercises to the Lectures by M. Heusler

3.1 Ricci scalar of a stationary metric

Derive the formula

$$R = R^{(3)} - \frac{2}{S}\Delta^{(3)}S + \frac{S^2}{2}\left(da\,,\,da\right)$$

for the Ricci scalar with respect to the stationary metric

$$g = -S^2(dt + a)\otimes(dt + a) + g^{(3)} .$$

Hint: Solve the Cartan structure equations for the metric $g_{i0} = 0$, $g_{00} = -1$, $g_{ij} = g_{ij}^{(3)}$ and the tetrad $\theta^0 = S(dt + a)$, $\theta^i = dx^i$.

3.2 Conformal transformations

Consider the conformal transformation

$$\bar{g} = \Omega^2 g^{(3)}$$

of the three-dimensional metric $g^{(3)}$. Show that the Ricci scalar and the Laplacian of a function S with respect to $\bar{g}$ become

$$\Omega^2 \bar{R} = R^{(3)} - \frac{4}{\Omega} \Delta^{(3)} \Omega + \frac{2}{\Omega^2} (d\Omega, d\Omega),$$

$$\Omega^2 \bar{\Delta} S = \Delta^{(3)} S + \frac{1}{\Omega} (d\Omega, dS),$$

where $(\,,\,)$ is the inner product with respect to $g^{(3)}$. Use these formulas with $\Omega = S$, and the result of the previous exercise, to obtain the expression (3) for the Ricci scalar with respect to the metric (1).

3.3 Target space for static vacuum gravity

Use the transformation (11) to establish (12). Then use the stereo-graphic projection, $\mathrm{Re}(\varepsilon) = x(z+1)^{-1}$, $\mathrm{Im}(\varepsilon) = y(z+1)^{-1}$, to show that this is the metric of the pseudo-sphere (a space of constant negative curvature): $PS^2 = \{(x, y, z) \in \mathbb{R}^3 \mid -z^2 + x^2 + y^2 = -1\}$.

3.4 Stationary Maxwell equations

Compute the Maxwell equations, $d*F = 0$, for the stationary gauge potential (13) with respect to the metric (1). *Result:*

$$dB = 0, \quad d\left(\sigma^{-1} \bar{*} E + a \wedge B\right) = 0,$$

where

$$E \equiv d\phi, \quad B \equiv \sigma \bar{*}(\bar{F} + \phi f).$$

Show that E and B coincide with the usual definitions of the electric and the magnetic one-forms, $E = -F(\partial_t, \cdot)$, $B = (*F)(\partial_t, \cdot)$. Also use the definition (2) and the general identity $d * (\omega/\sigma^2) = 0$ to establish the four-dimensional form of the stationary Maxwell equations,

$$d * \left(\frac{d\phi}{\sigma} - 2\psi \frac{\omega}{\sigma^2}\right) = 0, \quad d * \left(\frac{d\psi}{\sigma} + 2\phi \frac{\omega}{\sigma^2}\right) = 0,$$

where ψ is the magnetic potential, $B \equiv d\psi$.

3.5 Effective action for the stationary EM system

Use the decomposition (13) and the metric (1) to establish the effective action (16). *Hint:* It remains to show that

$$F \wedge *F = dt \wedge \left[\sigma(\bar{F} + \phi f) \wedge \bar{*}(\bar{F} + \phi f) - \sigma^{-1} d\phi \wedge \bar{*} d\phi\right].$$

3.6 Effective action in terms of scalar potentials

Use the Lagrange multiplier method (with the constraints $d\bar{F} = 0$ and $df = 0$) and the definitions (19) and (20) to pass from the effective action (16) to the effective action (21), involving only scalar potentials. *Hint:* First substitute $\bar{A}$ by ψ, and then a by Y.

3.7 Ernst potentials

Derive the equations (27) for the Ernst potentials from the effective action (26).

3.8 Sigma-model form of the EM action

Establish the expression (30) for the effective action of the EM system in the presence of a Killing symmetry. *Hint:* Use the definition (28) of Φ to compute the current matrix J^A_B. *Result:*

$$J^A_B = 2\,\text{sig}(N) \left(v_B d\bar{v}^A - \bar{v}^A dv_B \right) + 4\,\bar{v}^A v_B\, v^C d\bar{v}_C\,.$$

(Also note that $\bar{v}^C v_C \equiv \eta^{AB}\bar{v}_A v_B = -\text{sig}(N)$; and hence, $\bar{v}^C dv_C = -v^C d\bar{v}_C$.) Show that

$$\frac{1}{4} \langle J^A_B, J^B_A \rangle = 2\,\text{sig}(N) \langle d\bar{v}^A, dv_A \rangle + 2\,v_A \bar{v}_B \langle d\bar{v}^A, dv^B \rangle,$$

and use the definition (29) of the Kinnersley vector in terms of the Ernst potentials to complete the derivation.

3.9 Sigma-model equations

Show that the variation of the effective action (30) with respect to the matrix Φ yields the matrix current conservation law

$$d\bar{*}J = 0\,, \quad \text{i.e.,} \quad \frac{1}{\sqrt{\bar{g}}} \left(\sqrt{\bar{g}}\bar{g}^{ij}\, (J_i)^A_B \right)_{,j} = 0\,.$$

Hint: First establish $\delta J = J\Phi^{-1}\delta\Phi - \Phi^{-1}\delta\Phi J + d(\Phi^{-1}\delta\Phi)$.

3.10 Surface gravity of a static horizon

Show that the surface gravity of the Killing horizon in a static spacetime is constant. *Hint:* First establish the identity

$$2\,di_\tau i_\xi *\omega = -i_\tau(dN \wedge d\xi) + d\,(\tau, \xi) \wedge dN + \xi \wedge di_\tau dN\,,$$

for an arbitrary vector field τ and a Killing field ξ with norm $N = (\xi, \xi)$ and twist $\omega = \frac{1}{2}*(\xi \wedge d\xi)$. (Here i_τ denotes the interior product, e.g. $[i_\tau(\alpha \wedge \beta)]_\nu = \tau^\mu(\alpha_\mu \beta_\nu - \beta_\mu \alpha_\nu)$ for one-forms α and β.) Now use the fact that the LHS vanishes identically if the spacetime is static. Finally, use the definition (35) of κ in order to evaluate the RHS on the horizon. (*Result:* $0 = (\tau, \xi)\,[\xi \wedge d\kappa]$.)

3.11 Ricci identity for Killing fields

Prove the Killing field identities

$$\mathrm{d} * \xi = 0, \quad L_\xi * = * L_\xi,$$

$$\Delta \xi = 2 R(\xi),$$

where ξ is an arbitrary Killing one-form, L_ξ is the Lie derivative with respect to ξ, $\Delta \equiv -(*\mathrm{d}*\mathrm{d}+\mathrm{d}*\mathrm{d}*)$ is the Laplacian, and $R(\xi)$ is the Ricci one-form with components $R(\xi)_\mu \equiv R_{\mu\nu}\xi^\nu$. *Hint:* Use the Killing identity, $\nabla_\mu k_\nu + \nabla_\nu k_\mu = 0$, and the general identity (for arbitrary vector fields) $(\nabla_\nu \nabla_\mu - \nabla_\mu \nabla_\nu)\xi^\beta = R_{\mu\nu\alpha}{}^\beta \xi^\alpha$.

3.12 A twist identity

Use the results of Exercise 11 to prove the identity

$$\mathrm{d}\omega = *[\xi \wedge R(\xi)],$$

where ω is the twist of the Killing field ξ, $\omega \equiv \frac{1}{2} * (\xi \wedge \mathrm{d}\xi)$. *Hint:* For arbitrary p-forms α and vector fields (one-forms) τ one has the identity

$$*(\alpha \wedge \tau) = i_\tau * \alpha.$$

Solution:

$$\mathrm{d}(*\xi \wedge \mathrm{d}\xi) = \mathrm{d}i_\xi * \mathrm{d}\xi = (L_\xi - i_\xi\mathrm{d}) * \mathrm{d}\xi = *L_\xi\mathrm{d}\xi - i_\xi\mathrm{d} * \mathrm{d}\xi$$
$$= i_\xi * \Delta\xi = *(\Delta\xi \wedge \xi) = 2 * [\xi \wedge R(\xi)].$$

3.13 The vacuum staticity theorem

Complete the proof of the vacuum staticity theorem. In particular, establish the differential identity (44) for a Killing field (one-form) k with norm $\sigma = -(k, k)$ and twist $\omega = \frac{1}{2} * (k \wedge \mathrm{d}k)$. *Hint:* First show that

$$\mathrm{d}\left(\frac{k}{\sigma}\right) = \frac{2}{\sigma^2} * (k \wedge \omega) = -\frac{2}{\sigma^2} i_k * \omega.$$

3.14 The Israel theorem

Complete the proof of the Israel theorem:
(i) Compute G_{00}, G_{11} and R_{00} for the Israel metric (50).
(ii) Show that the vacuum equations and the expressions (51)-(53) imply the inequalities (54), (55). (iii) Establish the asymptotic behavior (57). *Hint:* Asymptotic flatness implies the existence of asymptotic coordinates $\{x^\mu\}$ such that $S^2 = 1 - 2M/r + \mathcal{O}(r^{-2})$, where $r^2 = \delta_{\mu\nu}x^\mu x^\nu$. Also use $\rho^{-2} =$

(dS, dS), $K_{ab} = (2\rho)^{-1}\partial g_{ab}/\partial S$, and $g_{ab} \to r^2 d\Omega^2$, as $r \to \infty$.
(iv) Establish the horizon behavior (58) of K/S. *Hint:* Derive the formula

$$\frac{1}{8}R_{\alpha\beta\gamma\delta}R^{\alpha\beta\gamma\delta} = \frac{1}{S^2\rho^2}\left(K^2 + K_{ab}K^{ab} + 2\frac{\left(\tilde\nabla\rho, \tilde\nabla\rho\right)}{\rho^2}\right),$$

for the curvature invariant, to conclude that K and K_{ab} vanish on the horizon. Use this in the G_{11} vacuum equation.

3.15 Komar mass of a static spacetime

Derive the mass formula (60) for a static vacuum black hole spacetime. *Hint:* The Komar mass is defined by the asymptotic flux integral

$$M = -\frac{1}{8\pi G}\int_{S_\infty^2} *dk.$$

Use Stokes' theorem, the Ricci identity $[d * dk = 2 * R(k)]$, and the Einstein vacuum equations to convert this in an integral over the horizon. Evaluate the resulting expression in the Israel parametrization (50) of the metric.

3.16 Surface gravity in Israel parametrization

The surface gravity of a Killing horizon, generated by the Killing field (one-form) k, is obtained from the formula (see, e.g. Heusler 1996b)

$$\kappa^2 = -\frac{1}{4}(dk, dk) \quad \text{on } H[k].$$

Show that $\kappa = \rho_H^{-1}$ in the Israel parametrization (50) of the metric.

3.17 Ricci tensor of a static metric

Derive the components of the Ricci tensor with respect to the three-dimensional static metric $\bar g = -\rho^2 dt^2 + \tilde g$. *Result:*

$$\bar R_{tt} = \rho\,\tilde\Delta\rho, \quad \bar R_{ta} = 0,$$

$$\bar R_{ab} = \tilde R_{ab} - \rho^{-1}\tilde\nabla_b\tilde\nabla_a\rho.$$

3.18 Consistency of the reduced EM equations

Use (66) and (67) to show that

$$\tilde\nabla^b\tilde G_{ab} = \frac{1}{4}\text{Tr}\left\{\rho J^b\left(\tilde\nabla_b J_a - \tilde\nabla_a J_b\right) + J_a\,\tilde\nabla^b(\rho J_b)\right\}.$$

Conclude from this that the Bianchi identity for $\tilde G^{ab}$ and the definition $J^a = \Phi^{-1}\tilde\nabla^a\Phi$ imply the conservation laws (68), $\tilde\nabla_b(\rho J^b) = 0$.

3.19 EM equations in Weyl coordinates

Derive (70) and (71) from (68) and (67), respectively. Use Weyl coordinates and the metric

$$\tilde{g} = e^{2h(\rho,z)} \left(d\rho^2 + dz^2 \right).$$

3.20 Ernst equations

Use $\bar{g} = -\rho^2 dt^2 + \tilde{g}$ and the transformations (74) to derive the Ernst equations (72), (73) from the three-dimensional equations (27). Also establish the form (77) of the Ernst equations with respect to the metric $\tilde{g} = \text{factor} \cdot [(x^2 - 1)^{-1} dx^2 + (1 - y^2)^{-1} dy^2]$. Show that ε and λ according to (78) are solutions to the Ernst equations.

3.21 Explicit representation of the electrovac coset

Derive the formula (88) for the trace of the relative difference $\Psi \equiv \Phi_2 \Phi_1^{-1} - \mathbb{1}$. *Hint:* Use the definitions (28) and (29) to express Φ in terms of the Ernst potentials $\mathcal{E}$ and Λ. *Result:*

$$\Phi = \frac{1}{2X} \begin{pmatrix} |\mathcal{E}|^2 + 2|\Lambda|^2 + 1 & |\mathcal{E}|^2 + (\bar{\mathcal{E}} - \mathcal{E}) - 1 & 2\Lambda(\bar{\mathcal{E}} - 1) \\ |\mathcal{E}|^2 + (\mathcal{E} - \bar{\mathcal{E}}) - 1 & |\mathcal{E}|^2 - 2|\Lambda|^2 + 1 & 2\Lambda(\bar{\mathcal{E}} + 1) \\ 2\bar{\Lambda}(\mathcal{E} - 1) & 2\bar{\Lambda}(\mathcal{E} + 1) & 2|\Lambda|^2 - (\mathcal{E} + \bar{\mathcal{E}}) \end{pmatrix}.$$

4 Exercises to the lectures by W. Israel

4.1

For the two following examples (and for other purposes) it is useful to note (and even check) that, for the 2-metric

$$ds^2 = A^2 dx_1^2 + B^2 dx_2^2,$$

the curvature scalar is

$$R = -\frac{2}{AB} \left\{ \left(\frac{A_2}{B} \right)_{,2} + \left(\frac{B_1}{A} \right)_{,1} \right\}.$$

The subscripts denote partial derivatives.

4.2 Vacuum polarization and Hawking radiation in 2-dimensional black holes

Take the 2-metric in the form

$$ds^2 = \frac{dr^2}{f(r)} - f(r)dt^2,$$

with an event horizon at $r = r_0$, so that $f(r_0) = 0$, $f(r_0) = 2\kappa$. Introduce retarded and advanced times u, v by

$$dt \pm dr/f(r) = \begin{cases} dv = -n_\mu dx^\mu \\ du = -\ell_\mu dx^\mu \end{cases}$$

and the corresponding Kruskal coordinates U, V by

$$dU = -\kappa U\, du = -L_\mu dx^\mu, \quad dV = \kappa V\, dv = -N_\mu dx^\mu.$$

Take the stress-energy tensor in the form

$$T^{\mu\nu} = \frac{1}{2}T_\alpha{}^\alpha g^{\mu\nu} + E(\ell^\mu \ell^\nu + n^\mu n^\nu) + F\ell^\mu \ell^\nu$$

so that $T_\alpha{}^\alpha$ corresponds to "vacuum polarization", E is an isotropic radiation field and F the net outward flux.

By re-expressing $T^{\mu\nu}$ in term of the Kruskalized lightlike vectors L^μ, N^μ, show that:

$$T^{\mu\nu} \text{ regular on } H^+ \quad \Longleftrightarrow \quad (T_\alpha{}^\alpha, E) = O(1), E + F = O(U^2),$$
$$T^{\mu\nu} \text{ regular on } H^- \quad \Longleftrightarrow \quad (T_\alpha{}^\alpha, F) = O(1), E = O(V^2).$$

Check that the conservation laws require

$$F'(r) = 0, \quad E'(r) = -\frac{1}{4}f(r)\frac{d}{dr}T_\alpha{}^\alpha(r).$$

For a massless scalar field, $T_\alpha{}^\alpha$ is known to be given by the "trace anomaly"

$$T_\alpha{}^\alpha = \frac{\hbar}{24\pi}R = -\frac{\hbar}{24\pi}f''(r).$$

The "Boulware vacuum" state of the field appears empty (modulo irremovable vacuum polarization) to stationary observers (i.e. the notion of positive frequency is defined with respect to the stationary time coordinate t). The condition on the stress-energy is that $T^{\mu\nu} \to 0$ as $r \to \infty$. Show that, for the Boulware vacuum,

$$E_B = \frac{1}{48\pi}\left(\frac{1}{2}ff'' - \frac{1}{4}f'^2\right), \quad F_B = 0 \quad (\hbar = 1).$$

Show that the Boulware stress-energy is that of a stationary fluid with energy density and pressure given by

$$\rho_B = \frac{1}{24\pi}\left(f'' - \frac{f'^2}{4f}\right), \quad P_B = -\frac{1}{24\pi}\frac{f'^2}{4f}.$$

What is the origin of the negative terms? (Note that stationary observers have acceleration $a = f'/2\sqrt{f}$.) Show that the Boulware stress-energy is singular on both horizons.

The Hartle-Hawking "vacuum" appears empty (modulo vacuum polarization) to free-falling observers on the horizon. The condition is that $T^{\mu\nu}_{HH}$ is regular on both horizon sheets. Show that $E_{\text{HH}} = \frac{1}{48\pi}\left(\frac{1}{2}ff'' + \kappa^2 - \frac{1}{4}f'^2\right)$ and that

$$\rho_{\text{HH}} = \frac{1}{24\pi}\left(f'' + \frac{\kappa^2 - \frac{1}{4}f'^2}{f}\right), \quad P_{\text{HH}} = \frac{1}{24\pi}\frac{\kappa^2 - \frac{1}{4}f'^2}{f}.$$

At infinity, this represents a uniform distribution of black-body radiation with

$$P_{\text{HH}} \approx \rho_{\text{HH}} \approx \frac{\kappa^2}{24\pi}.$$

Compare this with the result

$$P = \rho = \frac{\pi}{6}T_\infty^2$$

for zero-mass scalar Planck radiation in two dimensions, to obtain $T_\infty = \kappa/2\pi$.

The Hartle-Hawking state thus represents a black hole in equilibrium with its own radiation, confined within a large spherical box.

Verify that the Hartle-Hawking state truly represents "material" in thermal equilibrium by showing that the entropy density $S(r)$, defined by $S = (\rho + P)/T$ (appropriate for zero-mass "particles" with vanishing chemical potential), satisfies Gibbs law $T\mathrm{d}S = \mathrm{d}\rho$. ($T(r)$ is the locally-measured temperature.)

The Unruh vacuum is the state which appears empty to stationary observers on $\mathcal{I}^-$ (using time v) and to free-falling observers on the future horizon (using time V). It is the state appropriate for a black hole formed in a gravitational collapse (so that the past horizon is not a physical part of the spacetime). The boundary conditions on $T^{\mu\nu}$ are: no incoming radiation from $\mathcal{I}^-$ and $T^{\mu\nu}$ regular on H^+. Show that

$$E_{\text{U}} = \frac{1}{48\pi}\left(\frac{1}{2}ff'' - \frac{1}{4}f'^2\right), \quad F = \frac{\pi}{12}T_\infty^2.$$

Conclude that, in the Unruh state, the black hole is emitting radiation at the Hawking temperature.

Show that the stress-energy tensor for the three states are related by

$$(T^{\mu\nu})_{\text{U}} = (T^{\mu\nu})_{\text{B}} + \frac{\kappa^2}{48\pi}\ell^\mu\ell^\nu,$$

$$(T^{\mu\nu})_{\text{HH}} = (T^{\mu\nu})_{\text{U}} + \frac{\kappa^2}{48\pi}n^\mu n^\nu.$$

Comment on the effect of these added lightlike fluxes on (i) regularity of horizons and (ii) the conservation laws.

If the black hole has an inner horizon at $r = r_-$, show that all three stress tensors become singular on its ingoing sheet (the Cauchy horizon) and that in the Unruh state this is due to a blueshifted influx of negative energy $-\frac{\pi}{12}T_{\mathrm{CH}}^2 n^\mu n^\nu$ at the Cauchy horizon, where T_{CH} is the Hawking temperature of the inner horizon.

4.3 Every 2-metric can be writen in conformally flat form

$$ds^2 = F^{-2}\left(dx^2 - dt^2\right).$$

The stress tensor for a quantized massless scalar field in this geometry is

$$T_\mu^\nu = F^2 \left(T_\mu^\nu\right)_{\mathrm{flat}} - \frac{\hbar}{12\pi}\left\{F\left(F^\nu_{\ ,\mu} - \delta_\mu^\nu F^\alpha_{\ ,\alpha}\right) + \frac{1}{2}\delta_\mu^\nu F_{,\alpha}F^{,\alpha}\right\}.$$

(The comma denotes partial differentiation, and indices on the right-hand side are raised with the flat metric $\eta^{\mu\nu}$.) Check that this expression has the correct trace and is conserved.

4.4 Moving mirror in 2-dimensional spacetime

Let $v = f(u)$ represent the path of a mirror in the spacetime

$$ds^2 = -du\, dv = dx^2 - dt^2.$$

Defining $F(u) = [f(u)]^{-1/2}$, show that the mirror velocity and acceleration are $u^\alpha = \left(\frac{du}{d\tau}, \frac{dv}{d\tau}\right) = (F, F^{-1})$ and $a^\alpha = F'(u)(F, -F^{-1})$.
Hence the (signed) magnitude of a^α is $a = -F'(u)$ (counted positive for acceleration to the right), and $da/d\tau = -FF''$. Let T_μ^ν represent the expectation value of stress-tensor for a massless scalar field in the in-vacuum (no influx from $\mathcal{I}^-$ on left and right sides of the mirror).

Now focus on the right side $v > f(u)$. Make the conformal transformation

$$\overline{ds}^2 = F^{-2}(u)ds^2$$

holding the path $v = f(u)$ fixed. Calculate the corresponding stress-tensor $\bar{T}_\mu^\nu$.

But in this new geometry the mirror is no longer accelerated! To see this, make the coordinate transformation $\bar{u} = f(u)$, $\bar{v} = v$. Then

$$\overline{ds}^2 = -d\bar{u}\, d\bar{v}$$

and the path of the mirror is $\bar{v} = \bar{u}$, i.e. $\bar{x} = 0$. Hence $\bar{T}_\mu^\nu = 0$. (This follows from staticity, conservation, tracelessness, and the in-vacuum condition.)

Deduce that the original stress tensor has the form

$$T^{\mu\nu} = -\frac{1}{12\pi}\frac{da}{d\tau}\ell^\mu\ell^\nu$$

and represents radiation streaming from the mirror to the right. It has negative energy if $da/d\tau > 0$.

What happens on the left?

What is the precise meaning of ℓ^μ in the above equation (including the normalization factor)?

Derive the force of back-reaction on the mirror, and comment on the possibility of a runaway solution.

4.5 Two concentric spherical massive thin shells move at the speed of light (one inwards and one outwards) in a spacetime with metric

$$ds^2 = \frac{dr^2}{f(r)} + r^2 d\Omega^2 - g(r)dt^2 \, .$$

Their intersection splits the spacetime into four sectors A, B, C, D according to

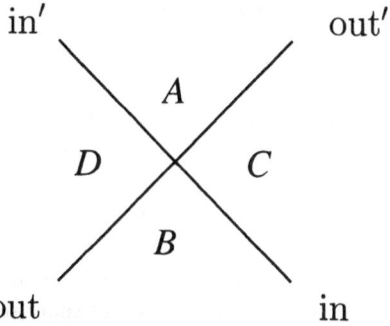

The functions f, g have different forms in these sectors. Prove that at the radius r_0 where the shells cross we have

$$f_A(r_0)f_B(r_0) = f_C(r_0)f_D(r_0)$$

(Dray-'t Hooft-Redmount relation).

(Hint: Let n^a, ℓ^a be ingoing and outgoing radial lightlike vectors at r_0, normalized by $\ell \cdot n = -1$. Writing the metric as

$$ds^2 = g_{ab}dx^a \, dx^b + r^2 d\Omega^2,$$

it follows that

$$f = g^{ab} \left(\partial_a r \right) \left(\partial_b r \right), \quad g^{ab} = -2\ell^{(a}n^{b)},$$

so that we have relations of the form

$$f = -2 \left(D_\ell r \right) \left(D_n r \right)$$

in each of the four sectors. Now $D_\ell r$ and $D_n r$ must be continuous across the outgoing and ingoing shells to keep their areas continuous.)

5 Exercises to the Lectures by G. 't Hooft

A. On the black hole metric without quantum mechanics

5.1 Given a spherically symmetric metric of the form

$$ds^2 = -P(r)dt^2 + \frac{dr^2}{P(r)} + r^2d\Omega^2,$$

where the function $P(r)$ has its first zero at $r = r^+$. Show that the general coordinate transformation

$$xy = \exp \int \frac{C}{P(r)}dr; \qquad x/y = \exp Ct,$$

leads to the generalised Kruskal metric

$$ds^2 = 2A(r)dx\,dy + r^2d\Omega^2,$$

and express $A(r)$ in terms of $P(r)$. Show that A is regular at $x = y = 0$.

5.2 Solve the Einstein equations for a spherically symmetric black hole that has one or several shells of massless moving particles moving in and out radially. In Eqs. (3.6) of [1] this means that the functions $g(x)$ (for ingoing matter) and $h(y)$ (for the outgoing mater) consist of one or several Dirac delta functions. Show that the resulting metric consists of several regions glued together, such that in each of these regions the original Schwarzschild metric holds with masses M_i. They are seperated by the dust shells. Where two dust shells cross ($r = r_0$), the four adjacent regions obey

$$(r_0 - 2M_1)(r_0 - 2M_3) = (r_0 - 2M_2)(r_0 - 2M_4).$$

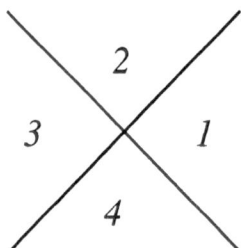

B. Quantum Field Theory in a curved metric

5.3 To the Rindler observer, the operators that annihilate particles at energies ω are a_I (in region I), and a_{II} (in region II). The relations between a_I, $a_I^\dagger$, a_{II}, $a_{II}^\dagger$ on the one hand, and the operators a and $a^\dagger$ relevant to the inertial frame on the other hand, are given by Eq. (5.27) in [1]. Consequently, the inertial vacuum state $|\Omega\rangle$, defined by

$$a\,|\Omega\rangle = 0,$$

obeys Eqs. (3.20) of my lecture:

$$a_I|\Omega\rangle = e^{-\pi\omega}a_{II}^\dagger|\Omega\rangle,$$
$$a_{II}|\Omega\rangle = e^{-\pi\omega}a_I^\dagger|\Omega\rangle.$$

Prove that, in terms of Rindler particles, $|\Omega\rangle$ is the entangled state

$$|\Omega\rangle = \sqrt{1 - e^{-2\pi\omega}}\sum_{n=0}^{\infty}|n\rangle_I\,|n\rangle_{II}\,e^{-\pi n\omega}.$$

5.4 Particle production at a cusp

Consider a spacetime with a cusp on the plane $x = t = 0$. This means that away from this plane, spacetime is everywhere flat, but on the plane we have (infinite) curvature: two particles, 1 and 2, passing the plane at opposite sides with initially the same velocity v, will have different velocities afterwards, such that one will be Lorentz-boosted with respect to the other (in the x-direction). The Lorentz boost parameter is γ. Show that, in a quantum field theory, if there is a vacuum at $t < 0$, there will be particles at $t > 0$. Compute the spectrum of these particles. This spectrum is ultra-violet divergent, a divergence that must be ascribed to the infinite sharpness of the cusp assumed.

C. Black Hole thermodynamics

5.5 The general Kerr-Newman black hole with mass M, angular momentum $J = Ma$, and charge Q has its horizon at

$$r = r_+ = M + \sqrt{M^2 - a^2 - Q^2}.$$

Its angular rotation velocity at the horizon is $\Omega = a/(r_+^2 + a^2)$, its electric potential at the horizon is $\phi = Qr_+/(r_+^2 + a^2)$ and its temperatur is

$$T = \frac{1}{\beta} = \frac{\sqrt{M^2 - a^2 - Q^2}}{2\pi(r_+^2 + a^2)}.$$

Show that its entropy S is 1/4 times the horizon surface area, or

$$S = \pi(r_+^2 + a^2),$$

and that (see [3])

$$M^2 = \frac{1}{2}Q^2 + \frac{\pi Q^4}{4S} + \frac{S}{4\pi} + \frac{\pi J^2}{S} = Q^2 + a^2 + 4T^2 S^2 ;$$
$$dM = TdS + \Omega dJ + \phi dQ .$$

Show that if we express the mass M as a function in terms of the temperature T, the charge Q, and the angular momentum J, a singularity develops at $4J^2 + Q^4 = \frac{4}{\pi}T^2 S^3$. This is sometimes interpreted as indicating a phase transition [3].

D. Gravitational shock wave

5.6 Consider two flat spacetimes glued together at a lightlike surface $x^- = 0$, such that there is a shift in x^+:

$$x_{(+)}^+ - x_{(-)}^+ = f(\tilde{x}),$$

where $\tilde{x}$ stands for the two transverse coordinates. Show that if $f(\tilde{x})$ is a linear function of $\tilde{x}$:

$$f(\tilde{x}) = a + \vec{b} \cdot \tilde{x},$$

then this spacetime is flat. In all other cases, there is a Dirac delta distributed curvature on the plane $x^- = 0$.

5.7 Prove that the Green function $F(\theta)$ obeying Eq. (10.24) of [1]:

$$-\tilde{\partial}_\Omega^2 F(\Omega, \Omega') + F(\Omega, \Omega') = \delta^2(\Omega - \Omega'),$$

can be written as Eq. (10.29),

$$F(\theta) = \frac{1}{2\pi\sqrt{2}\cosh\left(\frac{1}{2}\pi\sqrt{3}\right)} \int_0^{\pi-\theta} \frac{d\omega \cosh\left(\frac{1}{2}\omega\sqrt{3}\right)}{\sqrt{\cos\theta + \cos\omega}},$$

which proves that $F > 0$.

E. Horizon algebra

5.8 On our flat two-dimensional model for the horizon, we have defined the Green function $f(\tilde{x})$ obeying $\tilde{\partial}^2 f(\tilde{x}) = -\delta^2(\tilde{x})$. Derive all commutation rules (8.1 and 8.2 of my contribution) for the quantities $P_{\text{in,out}}(\tilde{x})$ and $U_{\text{in,out}}(\tilde{x})$ from the postulates

$$P_{\text{out}} = \tilde{\partial}^2 U_{\text{in}} ,$$
$$P_{\text{in}} = -\tilde{\partial}^2 U_{\text{out}} ,$$
$$[U_{\text{in}}(\tilde{x}), U_{\text{out}}(\tilde{y})] = if(\tilde{x} - \tilde{y}) . \tag{9}$$

5.9 Show that a local displacement operator $\tilde{P}_{\text{in}}(\tilde{x})$ can be defined,

$$\tilde{P}_{\text{in}} = P_{\text{in}}\tilde{\partial}U_{\text{in}}$$

such that it obeys all commutation rules required for the operator

$$D(\tilde{a}) \equiv \exp\left(i\int \tilde{a}(\tilde{x}) \cdot \tilde{P}_{\text{in}}(\tilde{x})\mathrm{d}^2\tilde{x}\right)$$

to generate an $\tilde{x}$ dependent translation $a(\tilde{x})$. Find the commutation rules between these operators $\tilde{P}_{\text{in}}$ and the other in-operators, including itself. The slight differences between the P and U commutation rules stem from the difference in their transformation rules: P is a density, U is a scalar.

Note that the out-operators commute non-locally with the $\tilde{P}_{\text{in}}(\tilde{x})$.

5.10 Define a new operator $\tilde{U}(\tilde{x})$ by

$$\tilde{P}_{\text{out}} - \tilde{P}_{\text{in}} \equiv \tilde{\partial}^2\tilde{U}\,,$$

and prove, with our definition of the operators $\tilde{P}$ of the previous exercise,

$$\partial_1 U_1 - \partial_2 U_2 = \partial_1 U_{\text{in}}\partial_1 U_{\text{out}} - \partial_2 U_{\text{in}}\partial_2 U_{\text{out}}\,, \qquad \text{and}$$
$$\partial_1 U_2 + \partial_2 U_1 = \partial_1 U_{\text{in}}\partial_2 U_{\text{out}} + \partial_2 U_{\text{in}}\partial_1 U_{\text{out}}\,.$$

From this, one can define *covariant* operators

$$X_\mu = X_\mu(U_{\text{in}}, U_{\text{out}}, \tilde{x} + \tilde{U})\,,$$

which tend to obey the covariant equation

$$\partial_i X^\mu \partial_j X_\mu = \lambda\delta_{ij}\,.$$

Bibliography

[1] G. 't Hooft, Int. J. Mod. Phys. **A11** (1996) 4623

[2] T. Dray and G. 't Hooft, Nucl. Phys. **B253** (1985) 173

[3] C. Lousto, Nucl. Phys. **B410** (1993) 155

List of Participants of the School

1. Marcus **Ansorg**　　　　　　　　　　`ansorg@tpi.uni-jena.de`
   Universität Jena, Germany

2. Werner **Benger**　　　　　　　　　　`werner@aei-potsdam.mpg.de`
   Max-Planck-Institut für Gravitationsphysik, Germany

3. Gerold **Betschart**　　　　　　　`gbetsch@physik-rzu.unizh.ch`
   Universität Zürich, Switzerland

4. Albrecht **Bischoff**
   WE-Heraeus-Stiftung, Germany

5. Thorsten **Brotz**　　　　　　　　　`brotz@physik.uni-freiburg.de`
   Universität Freiburg, Germany

6. Carsten van de **Bruck**　　　　　　　`cvdb@astro.uni-bonn.de`
   Universität Bonn, Germany

7. Sandip K. **Chakrabarti**　　　　　`chakraba@bose.ernet.in`
   S.N. Bose National Centre, Calcutta, India

8. Werner **Collmar**　　　　　　　`wec@mpe-garching.mpg.de`
   Max-Planck-Institut für Extraterrestrische Physik, Germany

9. Conrad **Cramphorn**　　　　　`conrad@mpa-garching.mpg.de`
   Max-Planck-Institut für Astrophysik, Germany

10. Tapas Kumar **Das**　　　　　　`tdas@boson.bose.res.in`
    S.N. Bose National Centre, Calcutta, India

11. Tammo **Diemer**　　　　　　`tammo@rhein.iam.uni-bonn.de`
    Universität Bonn, Germany

12. Martin **Dominik**　　　`dominik@hal1.physik.uni-dortmund.de`
    Universität Dortmund, Germany

13. Tim Oliver **Eynck**　　　`eynck@hal1.physik.uni-dortmund.de`
    Universität Dortmund, Germany

14. Helmut **Friedrich**　　　　　　`hef@aei-potsdam.mpg.de`
    Max-Planck-Institut für Gravitationsphysik, Germany

15. Alberto A. **Garcia**　　　　　`aagarcia@fis.cinvestav.mx`
    CINVESTAV, Mexico

16. Marcus **Gaul**　　　　　　　`mred@mppmu.mpg.de`
    Max-Planck-Institut für Physik, Germany

17. Valery **Gavrilov**   gavrilov@rz.uni-potsdam.de
    Universität Potsdam, Germany

18. Ralph **Gensheimer**   ralph.gensheimer@uni-konstanz.de
    Universität Konstanz, Germany

19. François **Gieres**   gieres@ipnl.in2p3.fr
    Université Claude Bernard, Lyon, France

20. Domenico **Giulini**   giulini@sonne.physik.unizh.ch
    Universität Zürich, Switzerland

21. Thomas **Görnitz**   goernitz@em.uni-frankfurt-de
    Universität Frankfurt, Germany

22. Andreas **Gross**   gross@ph-cip.uni-koeln.de
    Universität zu Köln, Germany

23. Friedrich W. **Hehl**   hehl@thp.uni-koeln.de
    Universität zu Köln, Germany

24. Christian **Heinicke**   heinicke@ph-cip.uni-koeln.de
    Universität zu Köln, Germany

25. Christian **Hettlage**   hettlage@uni-sw.gwdg.de
    Universität Göttingen, Germany

26. Markus **Heusler**   heusler@physik.unizh.ch
    Universität Zürich, Switzerland

27. Allen C. **Hirshfeld**   hirsh@hal1.physik.uni-dortmund.de
    Universität Dortmund, Germany

28. Gerard 't **Hooft**   g.thooft@fys.ruu.nl
    Universiteit Utrecht, The Netherlands

29. Sascha **Husa**   shusa@galileo.thp.univie.ac.at
    Universität Wien, Austria

30. Werner **Israel**   israel@uvphys.phys.uvic.ca
    University of Victoria, Canada

31. Michael O. **Katanaev**   katanaev@mi.ras.ru
    Steklov Mathematical Institute, Russia

32. Claus **Kiefer**   kiefer@phyq1.physik.uni-freiburg.de
    Universität Freiburg, Germany

33. Burkhard **Kleihaus**   kleihaus@darkstar.physik.uni-oldenburg.de
    Universität Oldenburg, Germany

34. Wilhelm **Kley**   wak@tpi.uni-jena.de
    Universität Jena, Germany

35. Gunar **Krenzer**   ogk@hpfs1.tpi.uni-jena.de
    Universität Jena, Germany

36. Christof J. **Kreuter**   christof.kreuter@cern.ch
    CERN, Switzerland

37. Matthias **Kunle**   kunle@tat.physik.uni-tuebingen.de
    Universität Tübingen, Germany

38. Kerstin Elena **Kunze**   `k.e.kunze@sussex.ac.uk`
University of Sussex, United Kingdom

39. Claus **Lämmerzahl**   `claus@spock.physik.uni-konstanz.de`
Universität Konstanz, Germany

40. Joachim **Lindig**   `lindig@itp.uni-leipzig.de`
Universität Leipzig, Germany

41. Jean-Pierre **Luminet**   `luminet@obspm.fr`
Observatoire de Paris-Meudon, France

42. Alfredo **Macias**   `amac@xanum.uam.mx`
Universidad Autonoma Metropolitana, Mexico

43. Bahram **Mashhoon**   `physgrav@cclabs.missouri.edu`
University of Missouri-Columbia, USA

44. Christian **Maulbetsch**   `maulbets@physik.fu-berlin.de`
Freie Universität Berlin, Germany

45. Ralph **Metzler**   `rjkm@thp.uni-koeln.de`
Universität zu Köln, Germany

46. Hinrich **Meyer**   `meyer@wpos7.physik.uni-wuppertal.de`
Universität Wuppertal, Germany

47. Declan **Moran**   `moran@tpi.uni-jena.de`
Universität Jena, Germany

48. Amir E. **Mosaffa**
Sharif University of Technology, Iran

49. Uwe **Münch**   `muench@ph-cip.uni-koeln.de`
Universität zu Köln, Germany

50. Alireza **Namazi**   `namazi@ph-cip.uni-koeln.de`
Universität zu Köln, Germany

51. Gernot **Neugebauer**   `neugebauer@tpi.uni-jena.de`
Universität Jena, Germany

52. Axel **Pelster**   `pelster@physik.fu-berlin.de`
Freie Universität Berlin, Germany

53. Herbert **Pfister**   `herbert.pfister@uni-tuebingen.de`
Universität Tübingen, Germany

54. Katja **Pottschmidt**   `katja@astro.uni-tuebingen.de`
Universität Tübingen, Germany

55. Hernando **Quevedo**   `quevedo@nuclecu.unam.mx`
Universidad Nacional Autonoma de México, Mexico

56. Voja **Radovanovic**   `rvoja@rudjer.ff.bg.ac.yu`
Belgrade University, Yugoslavia

57. Martin **Rainer**   `mrainer@aip.de`
Universität Potsdam, Germany

58. Josi-Luis **Rosales**   `rosales@phyq1.physik.uni-freiburg.de`
Universität Freiburg, Germany

59. Olivier **Sarbach**   sarbach@sonne.physik.unizh.ch
    Universität Zürich, Switzerland

60. Christoph **Schaab**   schaab@gsm.sue.physik.uni-muenchen.de
    Universität München, Germany

61. Udo **Schelb**   udo@lagrange.uni-paderborn.de
    Universität Paderborn, Germany

62. Sebastian **Schlicht**   schlicht@phyq1.uni-freiburg.de
    Universität Freiburg, Germany

63. Markus **Schneider**
    Germering, Germany

64. Rüdiger **Schopper**   schopper@usm.uni-muenchen.de
    Universitäts-Sternwarte München, Germany

65. Otto **Schwarz**   ottoschwarz.prien@t-online.de
    Prien am Chiemsee, Germany

66. Thomas **Schwarzweller**   thomas@hal1.physik.uni-dortmund.de
    Universität Dortmund, Germany

67. Edward **Seidel**   eseidel@aei-potsdam.mpg.de
    Max-Planck-Institut für Gravitationsphysik, Germany

68. Florian **Siebel**   siebel@cip.physik.uni-muenchen.de
    Universität München, Germany

69. Matthias **Soika**   msoika@astro.uni-bonn.de
    Universität Bonn, Germany

70. Abha **Sood**   sood@cygnus.physik.uni-oldenburg.de
    Universität Oldenburg, Germany

71. Roland **Speith**   speith@tat.physik.uni-tuebingen.de
    Universität Tübingen, Germany

72. Cristina **Stanciulescu**   stanciu@doppler.thp.univie.ac.at
    Universität Wien, Austria

73. Roland **Steinbauer**   stein@doppler.thp.univie.ac.at
    Universität Wien, Germany

74. Marcus **Strässle**   marcus@physik.unizh.ch
    Universität Zürich, Switzerland

75. Norbert **Straumann**   norbert@physik.unizh.ch
    Universität Zürich, Switzerland

76. Thomas **Strobl**   tstrobl@physik.rwth-aachen.de
    RWTH Aachen, Germany

77. Dietmar S. **Theiss**
    Lohmar, Germany

78. Jörg **Thorwart**   thorwart@ciphp01.physik.uni-freiburg.de
    Universität Freiburg, Germany

79. Joachim **Trümper**   jtruemper@mpe-garching.mpg.de
    Max-Planck-Institut für Extraterrestrische Physik, Germany

80. Dimitru N. **Vulcanov**   `vulcan@mitica.uvt.ro`
    West University of Timisoara, Romania
81. Jörn **Wilms**   `wilms@astro.uni-tuebingen.de`
    Universität Tübingen, Germany
82. Andreas **Wipf**   `aww@tpi.uni-jena.de`
    Universität Jena, Germany
83. Marion **Wirschins**   `wirschin@cygnus.physik.uni-oldenburg.de`
    Universität Oldenburg, Germany

Subject Index

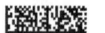